COMPUTING AND TECHNOLOGY ETHICS

COMPUTING AND TECHNOLOGY ETHICS

ENGAGING THROUGH SCIENCE FICTION

**EMANUELLE BURTON,
JUDY GOLDSMITH,
NICHOLAS MATTEI,
CORY SILER,**
AND
SARA-JO SWIATEK

The MIT Press

Cambridge, Massachusetts • London, England

The MIT Press would like to thank the anonymous peer reviewers who provided comments on drafts of this book. The generous work of academic experts is essential for establishing the authority and quality of our publications. We acknowledge with gratitude the contributions of these otherwise uncredited readers.

This book was set in Utopis by Westchester Publishing Services. Printed and bound in the United States of America.

Library of Congress Cataloging-in-Publication Data

Names: Burton, Emanuelle, author. | Goldsmith, Judith, author. | Mattei, Nicholas, author. | Siler, Cory, author. | Swiatek, Sara-Jo, author.
Title: Computing and technology ethics : engaging through science fiction / Emanuelle Burton, Judy Goldsmith, Nicholas Mattei, Cory Siler, and Sara-Jo Swiatek.
Description: Cambridge, Massachusetts : The MIT Press, [2023] | Includes bibliographical references and index.
Identifiers: LCCN 2022018577 (print) | LCCN 2022018578 (ebook) | ISBN 9780262048064 (hardcover) | ISBN 9780262374279 (epub) | ISBN 9780262374286 (pdf)
Subjects: LCSH: Technology—Moral and ethical aspects. | Computer scientists—Professional ethics. | Engineers—Professional ethics. | Ethics in literature. | Science fiction—History and criticism.
Classification: LCC BJ59 .B87 2023 (print) | LCC BJ59 (ebook) | DDC 170—dc23/eng/20220914
LC record available at https://lccn.loc.gov/2022018577
LC ebook record available at https://lccn.loc.gov/2022018578

10 9 8 7 6 5 4 3 2 1

Story Credits

"Dolly," by Elizabeth Bear. Elizabeth Bear.

"Message in a Bottle," by Nalo Hopkinson. Copyright 2005. First published in *Futureways*, edited by Rita McBride and Glen Rubsamen, Arsenal Pulp Press, Canada.

"The Gambler," by Paolo Bacigalupi. Paolo Bacigalupi.

"The Regression Test," by Wole Talabi. First published in *The Magazine of Fantasy and Science Fiction* (F&SF), January 2017.

"Apologia," by Vajra Chandrasekera. Vajra Chandrasekera.

"Asleep at the Wheel," by T. Coraghessan Boyle. Copyright © 2019 by T. Coraghessan Boyle. Originally published in the February 11, 2019, issue of *The New Yorker*. Reprinted by permission of Georges Borchardt, Inc., on behalf of the author.

"Codename: Delphi," by Linda Nagata. Copyright © 2014 by Linda Nagata. First published in *Lightspeed Magazine*, April 2014.

"Here-and-Now," by Ken Liu. Copyright © 2013 Ken Liu, first published in *Kasma SF*.

"Lacuna Heights," by Theodore McCombs. Theodore McCombs.

"Not Smart, Not Clever," by E. Saxey. E. Saxey.

"Today I Am Paul," by Martin L. Shoemaker. Copyright © 2015 by Martin L. Shoemaker.

"Welcome to Your Authentic Indian Experience™," by Rebecca Roanhorse. Copyright Rebecca Roanhorse as originally published in *Apex Magazine* on August 8, 2017.

Contents

PART TWO   ANTHOLOGY

Acknowledgments

We acknowledge the computer science departments of the University of Kentucky and the University of Illinois Chicago for encouraging us to teach the courses this book grew out of. And we celebrate all departments that support courses on technology ethics.

Thanks to the folks at the MIT Press for encouraging us and being patient when the pandemic slowed everything down. We would especially like to thank Marie Lufkin Lee, the editor who brought us on board; press assistant Alex Hoopes, for helping with story acquisitions; and editor Elizabeth Swayze, who brought us home.

We are grateful to all our authors whose short stories appear in the book. We would particularly like to thank Ken Liu for his enthusiasm, support, and conversations in the early days of this project.

Thanks to Kristel Clayville for being in conversation with the project, especially about responsibility ethics. Thanks to Tracy Weiner for conversation and excellent story recommendations. Thanks to Nedra McNeil and Zoe Brain for feedback, references, and sanity checks on the gender reassignment example in the introduction to chapter 4, and Nedra for one of the reflection questions for chapter 4. Thanks to Patrick Anderson for feedback and specialist assistance on chapter 2.

We gratefully acknowledge NSF grant IIS-1646887 for partial support of Burton, Siler, and Goldsmith in the early stages of writing this book. Any opinions, findings, and conclusions or recommendations expressed in this publication are those of the authors and do not necessarily reflect the views of the National Science Foundation.

Finally, we are grateful to our spouses and support systems that have carried us through book writing during a pandemic. Especially,

- Emanuelle Burton would like to thank her loving and supportive family for managing to remain interested in hearing about this project over many years of phone calls, and Kristel Clayville (additionally, again, always) for patience, faith, refreshment of the spirit, and a steady stream of paradigm shifts.

- Judy Goldsmith would like to thank Andy Klapper for his patience, love, and ability to offer support through many challenges, and her mother for her encouragement, and for not asking, "When is it going to be done?" And her friends, especially Wendy Frances, Beth Goldstein, Heather Laudan-Clark, Susan Pollack, and Susan Gardiner.

- Nicholas Mattei would like to thank his parents, Theresa and Mike, for their unwavering support over a lifetime, his brother Eric for keeping him focused, and his wife Liz for putting up with all his crap.

- Cory Siler would like to thank Sawyer Schmitt and Paige Campbell for many years of friendship and for the rage-inducing games of *Magic: The Gathering* that have kept the spark of humanity burning in her heart.

- Sara-Jo Swiatek would like to thank her husband Paul for always encouraging her to trust herself and her canine companions, Bisket and Rudy, who were by her side until their little hearts couldn't carry them any further.

TEXTBOOK

WHY ETHICS? WHY SCIENCE FICTION?

Learning Objectives

At the end of this chapter you will be able to:

1. *Explain the difference between the objectives of normative and descriptive ethics and what can be accomplished with each.*

2. *Summarize the key tensions present in ethical discussions including limited resources, competing kinds of good, and different ideas about what is good.*

3. *Contrast the differences between using case studies and science fiction for studying ethics and technology.*

4. *Discuss the difference between personal and professional ethics and list reasons why studying ethics is important for technologists.*

1.1   INTRODUCTION

Should autonomous weapons be legal? Will I be cared for by a robot in my old age? Does it matter if I submit buggy code for a beta release? When I ran the red light, was there a camera recording me, and was it able to read my license plate? If it couldn't get a clear image, will someone else get my ticket? Will other people be able to find out that I got a ticket?

Computing is a tremendously powerful field. Our daily lives, from travel to communication to medical care, have been shaped by computers and related technologies in ways that were barely imaginable twenty years ago. Computing technologies have already transformed the world, and they will continue to do so (Greenfield 2017).

These changes have not been wholly for the better. Social media has enabled people with common concerns and interests to find one another, which has benefited birdwatchers and indie music fans but also white supremacists. Putting financial records and transactions online has made banking more efficient and oftentimes more secure, but it has also left individuals and major institutions more vulnerable to hidden theft (Timberg et al. 2017). In addition, the growing technology gap between rich and poor countries could leave many developing economies at an even greater disadvantage than before (Baller 2016).

This mixture of good and bad is not unique to computing technology specifically, or even to technology in general. Anything that changes the world will have negative effects as well as positive ones. But when is a particular trade-off worth it? Which kinds of costs are acceptable, and how do we distinguish acceptable costs from unacceptable ones? Will this new product cross some sort of line established by religion, society, or the common sensibility of (some) people? Is it possible to design and execute a new project in a way that causes less harm?

These are hard questions, and they don't have "right" answers the way that many technical problems do. Technical problems, in ideal, stylized settings, offer clear benchmarks for success (e.g., efficiently produce this output from that input). But if you're trying to answer questions like the ones above (e.g., Should I build this system in this particular way? Should I build it at all?), the benchmarks are less clear. Those questions—ethical questions—demand that we evaluate the merits of the benchmarks themselves. They require us to consider why a project, or a particular way of building it, is helpful or beneficial; whom it helps; and what specific kinds of benefits it produces, or to articulate why it is the right thing to do. It also requires that we imagine how our projects affect the world, what kinds of harm they might cause, and to whom.

Technology has enormous potential to make human lives safer, freer, healthier, more interconnected, more full of beauty, and less plagued by poverty. It also has the power to dramatically increase inequality, to enable people with malicious intentions to harm others, and to create the conditions for people to make poor decisions. Which will your designs and implementations do? Asking the right questions, and answering them thoughtfully, is crucial to ensuring that your designs, systems, and work help make the world better.

The future in which we live, and the lives of those who come after us, will depend in part on your actions and choices as a technology development professional. That's why it's important that you understand what is potentially at stake in your work, whom it will impact, and how. You will be much better equipped to do these things if you take some time now—before these questions arise in your own working life—to clarify your own goals and values, to understand the questions you can ask, the kinds of choices you can make, and reflect on what it means to live up to those goals and values.

This book won't give you answers for how to be a good person or how to build good things. It will give you tools for *thinking* about problems as they arise, so that you can answer your own questions or talk productively to others in the field to figure out answers together.

This book introduces some of the persistent ethical problems and issues in the design, development, and deployment of technologies, especially those related to computing. However, our coverage of these issues is necessarily incomplete. Instead, we use these problems and issues as a jumping-off point to engage with the basic questions that one should ask in any situation where values are at stake. There are two reasons for this approach. The first is that technology changes all the time. Our daily lives are shaped by apps and platforms that did not exist ten or even two years ago or that existed in a very different form, with different capabilities and pitfalls. So rather than talking about specific issues related to specific technologies, which come and go and change rapidly, it is more useful to deal with the structural questions they raise.

The second reason for this focus on basic questions is that even though the ethics of technology and computing may present themselves in different forms, they are not necessarily different from other kinds of ethical questions. The problems examined in this book take their shape from recent technological innovations, but many of them are old problems, which resurface again and again in different areas of life. Accordingly, this book draws on philosophy, sociology, political science, literary studies, psychology, and the history of science. These cross-disciplinary insights only scratch the surface of relevant knowledge held by experts in these fields, but even these brief glances can give you a sense of how those experts could help you in your own future work as either consultants or collaborators. Our discussion in this chapter, as well as throughout much of the textbook, focuses primarily on computing technologies, both hardware and software. That said, many of the topics, problems, and questions apply more broadly to a range of specific issues concerning technology in the twenty-first century.

The point of this book is not to tell you what to think, nor is it to give you a formula for determining the correct ethical choice in situations you encounter. Instead, we hope to equip you with "*a means and a process for achieving [your] own moral judgments*" (Callahan 1980, 71; emphasis added). That is, we aim to train you to recognize ethical problems when you encounter them and to give you several sets of tools that you can use for evaluating and making decisions about what to do. As with any kind of problem-solving, the most important step is often a matter of figuring out what to ask.

Sidebar: This Is a Sidebar!
Throughout the book, you will encounter boxes like this one standing apart from the main text. We use these sidebars to highlight common misconceptions, elaborate on important but tangential points, offer alternative ways to understand issues, and recommend further reading. Many of the sidebars are meant to offer jumping off points for discussion or additional inquiry.

1.2   WHAT DOES IT MEAN TO SAY, "IS IT ETHICAL . . ."?

Conversations about ethics would be easy (or at least easier) if we all agreed either on what the world is like or what it should be like. But in practice, we don't agree—and in general, we aren't very good about talking about the reasons why we don't agree. Our ideas about good and bad are often rooted so deeply in our personal experience that we find it hard to imagine how a decent person could not share them. That's why, in practice, people often use the words "ethical" and "unethical" to mean "I agree with X" or "I don't like Y."

The language of ethics carries tremendous power, even when it is used clumsily or without reflection. If you've spent much time watching people argue about ethics on the internet, you have probably felt tempted at some point to give up on the language of ethics entirely. Although the authors of this textbook sympathize with that response, remember that there are also many good-faith attempts to discuss ethics (both on- and offline) and this discussion can be productive. Understanding the underlying logic of ethical assertions will help you better understand these discussions, even when they are sloppy or happening in bad faith. It will also equip you to better communicate with classmates or colleagues when working through ethically charged subjects.

Ethical statements typically come in two varieties: *normative* and *descriptive*. In a *normative* statement one provides an assessment of how things should be rather than how they are: for example, "he was wrong to do that" or "being kind is more important than being the most successful." Normative statements are rooted in value judgments about what is good and what is bad, or what is permissible and what is forbidden, or they are assessments of the relative value of different things. Ethics is often (mis)understood to be *only* normative: that is, to be aimed at establishing norms of thought, values, or conduct. This assumption is especially prevalent in many professional ethics courses, which are often used as a means to steer students' future behavior toward a set of professionally agreed-upon values, such as professionalism or honesty (Colby and Sullivan 2008; Martin and Weltz 1999). We discuss the professionally agreed-upon values for computer science, particularly some of the professional codes of ethics, in chapter 6.

But ethics can also be used as a tool for *description*, furnishing you, the decision maker, with a critical framework that enables you to understand what is happening in a given situation and what is at stake in any action you might take. Learning about the descriptive functions of ethics is as important as learning the professional norms of the field. Computer science is a field in which everyday practice and problem-solving takes place in a context that could barely be imagined a decade before. We cannot predict the ethical quandaries you will face. But with an education in ethical description, you will be equipped to engage in careful and substantive ethical reasoning when new and challenging problems confront you.

While description is an essential part of ethics, it is vital to recognize that making a judgment always necessarily involves norms. Although it is the normative element that often makes ethical judgments recognizable, the process that is used to arrive at ethical judgments is often described in a way that erases the norms themselves. When this erasure has happened, ethics appear to be simply a matter of knowing the relevant facts, and it can seem as if normative assessments follow automatically from an accounting of those facts. But a person who reasons in this way is not being objective, even though they may feel like they are. They are simply overlooking the foundations of their own observations and logic.

Even apart from the question of underlying norms, "simple factual descriptions" are not as simple as they first appear. There are many different ways to describe a single situation, and those differences often reflect normative judgments, intentional or otherwise, on the part of the person doing the describing. Some of those differences are a matter of which details you choose to include and which details you overlook or decide aren't important. That process of choosing details involves a normative judgment about what is relevant or useful for understanding a particular situation. But it's also important to realize that two different people describing exactly the same details might perceive those details in different ways. Those differences both reflect the normative judgments of the person who is describing and implicitly communicate normative judgments to others.

As an example of how normative judgments shape even "simple factual descriptions," imagine a newspaper article about a person who has been arrested near a crime scene. Imagine that this person could be accurately described either as a "suspect," a "loiterer," a "passerby," a "boy," or a "juvenile." The newspaper article could use any of those words and meet the criteria for simple factual description, and yet each word would communicate different normative judgments about what kind of person this is, what he might have done, and what he deserves. The newspaper's decision in this situation is unavoidable; any choice they make is normative and is possibly more persuasively so because the chosen word *appears* to be only descriptive. Every act of description involves similar choices, even if the describer is not aware that she is making them.

Ethics can be used as a tool for making normative judgements or for reflecting on the methods and criteria we use to reach those judgments. But, as stated above, ethics can also be a tool for description. The difference between descriptive and normative may seem clear in theory, but in practice, the distinction gets fuzzier the closer you look. Which facts are the most important in a given situation? How do we describe them? How do we fill in the gaps between the facts we have? These are all questions about how to describe a given situation, but any one of them could affect the ethical judgment you or others make in the end.

Every person has their own ideas about what is ethical and what is not, including you. Those ideas are shaped by the culture around us, but they are also specific to each of us. This book will help you to recognize the normative judgments that are part of the cultures around you and to understand their implications. It may also challenge you to reevaluate some of your beliefs about the world, in light of those values—or to reconsider your values, in light of your beliefs about the world.

1.3 WHY STUDY ETHICS?

It can be intimidating for those in the engineering and technology development disciplines to grapple seriously with ethical questions because, as these technologies ingrain themselves in almost every aspect of our lives, the questions seem too big and the stakes too high. As a reaction, you might find yourself tempted to focus solely on the technical dimensions of your work precisely because the social and ethical dimensions seem too big or too removed or even too grim for you to take on yourself. Many of us become scientists and engineers because we want to build things and to solve interesting problems, and the nontechnical questions raised by your work may feel insurmountable in a way that makes you want to focus on only the building part. However, you likely wouldn't release an untested program or drive across a bridge that had not been inspected, and building a system or a program without reflecting on its possible uses or impact is just as irresponsible. Many of the major avenues in computing, from autonomous robots to messaging apps, have an incredibly broad range of implications and applications, and many of those implications are still poorly understood. Even the developments that seem straightforwardly good, such as the digitization of medical records or easy access to live-updating navigation programs, emerge as more ethically ambiguous, or even harmful, when we look closer at who is using these technologies, how they are using them, and for what.

The good news is that in many cases you don't have to do this kind of thinking alone. In most settings, the best practices endorsed by professional societies and design professionals place emphasis on foregrounding multiple types of reflection in multidisciplinary teams, often referred to as *participatory design* (Lee et al. 2019; Vines et al. 2013). Development is a social enterprise, and we must confer with experts in other fields, best practice manuals, and focus groups of potential users to help think through design choices and to anticipate how a given design will be adopted and integrated when deployed in the world. Unfortunately, not all work environments are willing to dedicate time and resources to this kind of interdisciplinary consultation—and even when they are, you, the computing professional, still need to have a sense of what to pay attention to, and what to read about or ask about, and when.

Given the diversity of perspectives, values, and cultures in the world, dealing with ethical issues will never be easy. But there are ways to get better at identifying ethical problems and methods for determining more ethical courses of action. The ethical issues that arise in computing fields often look brand-new because they deal with phenomena that haven't existed before, such as intelligent robots or the unprecedented volume of available information. But, in fact, human beings have been wrestling with similar versions of these same questions for the whole of recorded history. When the ancient Greeks argued over the ideal values to instill in their children and which methods of education that would be most effective, they were grappling with many of the same issues that roboticists and programmers grapple with today when they contemplate the value and risks of artificial intelligence (AI). When educators and scholars in fifteenth-century Europe debated how the printing press might democratize (or destroy) established bodies of knowledge, they were struggling with many of the same issues that, in contemporary times, plague social media platforms like Facebook and Twitter (Greenfield 2017).

The absence of a foolproof formula for resolving ethical quandaries in all the situations you face doesn't mean that there are no codes or guidelines for ethical action. There are many different useful approaches to tackling ethical challenges. All of these are built on some assumptions about the world. What this means is that different ethical frameworks rely on different ways of conceptualizing the world and thus of conceptualizing the problem; part of the challenge is figuring out which approach best suits the problem you are confronting. This book will help you recognize ethical problems when you encounter them and give you concepts and tools that will help you analyze situations and make informed and responsible decisions about what to do.

Over the years, ethical theorists have come up with different systems of thought to help think through difficult ethical problems. Studying these ethical frameworks will give you different ways of asking questions about tough ethical challenges, as well as different ways of answering them. But as we discuss in chapter 2, strengthening our capacity to identify, analyze, and respond to ethical issues requires much more than applying concepts to specific situations. By learning about different kinds of ethical frameworks, you will be able to find the approaches that make the most sense to you and to use concrete language to explain your position. You will also be better able to understand the views of those who disagree with you and learn from their perspectives, even if you don't agree with them.

1.3.1 BASIC/PERENNIAL PROBLEMS IN ETHICS

There are some types of problems that come up repeatedly in ethics. Although every situation will be different, recognizing the *kind* of problem you are facing can be very helpful. It will allow you to focus on what is most important about the situation, and it will also enable you to think about how you or others have handled similar problems in the past.

It is often the case that there is no single best answer. In most cases, every possible choice has drawbacks of some kind. It is rare to be able to arrive at a clear-cut solution to a problem that is universally satisfying. But if you have a clear grasp of the nature of the problem you are considering, you will find that you are better able to conceive of and identify a way forward, with a better grasp of what you are sacrificing and why.

The first problem that we see often is the reality of *limited resources*. What should be done when the demand for something—such as jobs, oil, medicine, or teachers in the classroom—outstrips how much of it is available? When somebody donates a kidney, do you give it to the patient who is most seriously ill and will otherwise die in a few weeks? Or do you instead transplant the organ into someone who is still mostly healthy and who will likely recover more completely and live longer afterward? Note that this is not an abstract concept for those developing computing technologies—the largest paired organ exchanges in the world are being built and maintained by computer scientists (Purtill 2018).

Problems created by limited resources often seem as if they could be solved by balancing the scales, usually with technology: artificial hearts to replace the failing ones, more fuel-efficient cars so that oil is less in demand, or high-yield crops that can feed more people. In some cases, solutions like this can really help, but more often than not they do not solve the problem completely. Typically, there are other costs associated with the solution; for example, the crops deplete the soil quickly or additional costs of driving are made even worse because people drive more. But even more important for the study of ethics is that this kind of problem-solution structure assumes that everybody can agree on what the most important problem *is*, and that is rarely the case.

In making an ethical decision, there is often a need to consider *competing kinds of goods*: multiple goals, or good things (not necessarily objects), that come into conflict. At a basic level, the choice to attend one college instead of another is an example of competing goods. It might be a simple matter of which school you like better or which has a program for your major; on the other hand, it could be a question of what your family expects of you or what they can afford. Or maybe you have the chance to work in one of two research labs, each of which pursues a different kind of exciting projects, but you cannot work in both because each job would require you to commit all of your student work hours.

Even when people agree about how to think through a particular ethical question, they may still disagree because their aims are different: they have *different ideas about what is good*. Sometimes the clashes are obvious: abortion-rights advocates want it to be easy to terminate a pregnancy safely, and abortion opponents want to prevent this exact thing because these two groups have very different ideas about the morality of abortion and what is at stake. Other times, though, people who seem to share the same goals disagree significantly about how to achieve them: is nuclear power a good option for the environment or a terrible one? Two environmental advocates with the same set of facts and the same overall goal can come to different conclusions on nuclear power, because they disagree about how best to assess environmental well-being or have different ideas about what policies or practices could be changed.

In the case of the donated kidney given above, a doctor advocating for the sickest patient might say, "the whole point of transplant surgery is to save people's lives past the point when other kinds of medicine can help them; we should help the person most desperately in need." Another doctor might say, "the sicker patient will never be able to lead a normal life again, but the less-sick patient could live a long and healthy life after the surgery, so it is a better use of this scarce resource." These two doctors don't just disagree about which patient is more deserving; they disagree about *how to decide what counts* as most deserving. These kinds of fundamental disagreements are often invisible even to the people who are having the argument because they both describe their ethical vision in the same way: in this case, they both want to improve people's lives with transplant surgery. But they interpret that goal to mean different things.

What is good? What do we owe to others? How do we balance our obligations to others against our obligations to ourselves? And when we make judgments like these, where do our criteria come from? Studying ethics isn't about arriving at decisive answers to any of these questions. But studying ethics does make it easier to confront them, and it allows you to think more clearly about what is at stake in specific situations in which perennial issues arise and to explain your reasoning in an effective way.

1.3.2 SELF-INTEREST AND ETHICAL LIVING: CAN YOU DO BOTH AT ONCE?

There is a common misconception that good or ethical people ignore their own needs and sacrifice themselves for others. This is not true! Being ethical does not mean that you have to be selfless all the time or completely give up on your own success or pleasure. None of the ethical outlooks discussed in this textbook require the agent to forget about her own well-being. There are some ethical traditions, such as some forms of Christianity, that demand selflessness of this kind, but those ideas of good and right are specific to that value system rather than intrinsic to ethics (or even to any given ethical framework) in general.

Self-interest is not a single unified thing; it is a lot of kinds of things all tangled up together, and some of them are necessarily in conflict. There are also different ways that a thing can be good for you. Even when you're thinking only about your own interests and desires, you often have to choose between two good things (such as going on a fun vacation versus staying home to work on an important personal project). We often imagine ethical conflicts in binary terms: "Should I do the thing that's good for me and bad for you, or the thing that's good for you and bad for me?" But the options, and their stakes, are almost never so binary; there are often many possible choices. But more to the point, there are lots of ways to evaluate how something could be good or bad for you, or how it could be good or bad for the others who will be affected by your decision.

Some kinds of goods really are zero sum. For example, when you audition for a role in a play or a spot in a band, you are in direct competition with others who are auditioning for that same role. Their success in this particular area means your loss in that area, and vice versa.

But many, even most, other kinds of goods are not a matter of direct competition. It is entirely possible to take pleasure in the experience of helping, giving, or sharing scarce resources such as money, time, or cake and that pleasure can outweigh those resources that have been given away. For example, many people who volunteer their time talk about how much happier they are now that they are spending part of their free time improving the lives of others. It is also possible to create better situations so no one must sacrifice directly: for example, adding more members to your project team so that everyone can contribute and share in the success does not require any notable sacrifice.

Being ethical does not mean that you have to give up on your goals for your own life or give away the things you value. It does mean considering the well-being of others as well as your own. And in some cases, it means choosing to value mutually beneficial things as much as (or more than) benefits that come at the expense of others.

Sidebar: Ethics and Morality—What's the Difference?

It is not uncommon for the terms "ethics" and "morality" to be used interchangeably. However, the two terms may have different connotations for different people. In fact, moral philosophers sometimes disagree about whether ethics is a subset of morality or if morality is a subset of ethics (Piercey 2001). Before offering a definition of the two terms, it might be helpful to consider whether you think there is a difference between the questions "is it moral?" and "is it ethical"? If you decide the two terms are getting at something different, then consider: what is it, exactly, that marks the difference between these two terms?

Throughout this textbook we use the term "morals" to refer to a person's standards of behavior or beliefs concerning what is and is not acceptable for them to do. We use the term "ethics" in a similar way, but we additionally use the term to refer to the thoughtful reflection on and application of our standards of behavior and beliefs about what is acceptable to do and not do. The distinction between the two terms is intentionally subtle for two important reasons. First, as noted above, we draw from a range of different scholarly disciplines, and the two terms are not defined in the same way across all of these disciplines. Second, many of the references we use do not make a point to distinguish the two terms, and so forcing a distinction would potentially be misleading.

1.4 WHY THINK ABOUT TECHNOLOGY AND ETHICS TOGETHER?

You may be wondering, why do I need to make time for *ethics of computing technologies specifically*? Many ethics classes are located in arts or humanities departments and not in departments such as informatics, computing, or engineering (Davis 2009). There is increasing consensus among experts that focusing on specific problems related to those practitioners who design and develop technology and its use can give you the tools to succeed in your professional life. Moving these problems closer to the applied work you will do helps you learn how to recognize and deal with these issues as you become a professional (Narayanan and Vallor 2014; Rogaway 2015).

1.4.1 HOW HAVE RECENT ADVANCES IN TECHNOLOGY CHANGED THE CONDITIONS FOR ETHICS?

Many of the major issues involving the ethics of computing technologies can be understood as a subspecies of one of the major types of problems discussed in section 1.3.1: limited resources, competing kinds of good, and different ideas about what is good. In this sense, the ethical issues that arise from even the most cutting-edge technologies are rarely new problems. New technologies, from computers to irrigation to writing, frequently change the conditions in which ethical problems manifest, and these changes sometimes mean that one kind of problem is replaced by another, but it is often the case that the same basic patterns persist.

Yet it is also true that modern computing technology (including information processing and communications systems) has brought about some specific shifts in human social configurations on both a local and a global level. These changes in human society need to inform the way we ask and answer ethical questions.

In the 1980s, as information and communication technologies grew expansively, social critics, scientists, and engineers began to see the need for a specific focus on computer ethics to address the societal and individual ethical issues raised by computing technologies. James Moor (1985), one of the first of such critics, argued that there are features of computer technology that have changed, or soon would change, a broad range of human activities and transform long-standing social institutions. The problem is that we lack a conceptual framework for addressing these changes and in turn are faced with policy vacuums, in which regulations, laws, values, and norms have not been established.

One of the features of computing technologies that Moor argued makes them unique is what he called the "invisibility factor." Think for instance of the numerous computer chips and programs running silently in your car or on your smartphone. Moor targeted three issues that stem from the invisibility factor: (1) the malicious *invisible abuse* of the lower-level operations of a computer to invisibly exploit the user who is, for example, transferring money; (2) the *invisible programming values* present in any decision-making algorithm that may ensure that different people are invisibly advantaged or disadvantaged by the use of that algorithm; and (3) the largely unchallenged assumption that computers perform *invisible complex calculations* that no human is able to inspect and understand completely. We pick up many of these themes of invisibility throughout the book.

Technology and society scholar Deborah Johnson (1998) has argued that three specific changes have been brought about by the broader societal reconfigurations stemming from computer technologies:

- **Reproducibility**: Computing, information, and communication technology has radically changed the rate at which texts, images, video, and other kinds of information can be reproduced or transmitted to others. It has also made it much easier to collect, link, and preserve records. These developments have impacts in many areas of both privacy and society at large, and we explore them extensively in chapters 4 and 5.

- **Information flow**: Whereas earlier advances in communication technology have mostly improved the speed or effectiveness of one-to-one or one-to-many communication, the internet and other computing technologies have made it much easier to engage in many-to-many communication and for individuals to engage as the *one* in a one-to-many setup. These changes have had impacts in many areas of society on how we share knowledge and information; we explore these themes in chapters 3 and 5.

- **Identity conditions**: Our experiences as individuals in society are affected by advances in connectivity and computation because they enhance our ability to communicate with others; increase the likelihood that we are identified, recorded, and logged as we go about our daily lives; and, crucially, enable us to communicate via anonymous mechanisms. Developments in computing technology have made it easier to speak and act without any visible ties of these words or actions to their legal names or their public reputations, and they also have made it easier to track and identify individuals in many areas. We explore issues around identity and privacy in chapter 4.

These broad social changes are still very much in progress, and what happens next has not yet been decided. Much of it will depend on what you, your peers, and your coworkers build in the near future. It will also depend on how you choose to build it. If you are empathetic, creative, committed to your own values, and willing and able to think carefully about what and how you build, you can play a role in creating a better future.

1.4.2 WHY SHOULD COMPUTING PROFESSIONALS STUDY ETHICS?

There is no denying the fact that modern computing technologies have enhanced human life in various ways. In addition to helping human beings achieve certain goals, it can be argued that the mark of a great design is that it allows human beings to imagine new goals that would have been unthinkable without that particular piece of technology. But what kinds of goals are worth having? What makes a human life good or complete?

This question might seem more appropriate for a department of philosophy (or literature, or social science) than for a computer science or engineering department. It's not a question that a computer science student usually has to answer in order to complete a programming assignment or that an engineer must confront to fulfill a particular client order. But as we will discuss in this textbook, technology increasingly creates the conditions in which human beings live their lives; imagine how different our present world would be without social media! The choices and actions of technology creators can have a meaningful impact on the wider world—and those choices and actions are based to some degree in how they answer the basic philosophical question of "what is worth having or doing?" Because of the potential impact of their work, computing professionals have an obligation to think about that question and to learn about how others answer it. Human beings have a tendency to treat our own personal, individual answers to this question as ground truth, when in fact there are many ways to answer it. Different people may have widely varying ideas of what makes human life good and valuable.

Practitioners, professional societies, teachers, and leaders in the fields that deal with the design and development of computing technologies have a responsibility to drive the discussion about the impacts of their own work (Boyer 1997). The research and development community around big data, AI, and machine learning writ large has started to address this in a number of interesting ways, including the International Joint Conference on Artificial Intelligence (IJCAI) letter on autonomous weapons research (Future of Life Institute 2015) and the follow-on letter signed by CEOs of tech companies around the world (Conn 2017); technology companies releasing statements about their values in relationship to IT and AI (Pichai 2018); the Association for Computing Machinery (ACM) statement on Algorithmic Accountability (ACM US Public Policy Council 2017); the development of the Institute of Electrical and Electronics Engineers (IEEE) standard for algorithmic bias considerations (IEEE Standards Association 2021); and new conferences and research groups focused on fairness, accountability and transparency (ACM Conference on Fairness, Accountability, and Transparency 2021) as well as conferences focusing on the impact of AI on society (AAAI/ACM Conference on Artificial Intelligence, Ethics, and Society, n.d.). Many technical conferences encourage or require statements about the ethics of a project with paper submissions about the project. The utility of such debates is not that they result in standardized practices but rather that individual practitioners become more thoughtful and better informed about their work and its long-term effects. As in other areas of thought, this diversity of viewpoints is a strength when it can be harnessed to achieve a productive exchange of ideas and perspectives. The goal of this textbook is to foster your ability to fruitfully participate in such debates by informing you about the range of ethical, descriptive, and evaluative tools available to you.

As a professional who works with the design, implementation, and deployment of computing technologies, you will be able to do far more good in the world if you take the time to understand what other people value and why, instead of trying to make them conform to your own assumptions about what they should need or want. Although programs can specify value in a single variable, human values are complex. Although human desires and values can be influenced, they are deeply embedded, and humans often make decisions on the basis of deeply held values without analyzing those values or even being aware of their role in the decision.

There is also more than one way to talk about how human beings value things. Some argue that all human desires boil down to basic biological drives for food, sex, and physical safety. Others argue that more conceptual values—things like love, justice, or personal recognition—are distinct from the physical drives (Haidt 2012). But however they are measured or talked about, values and desires are an ineradicable part of human life. They play a fundamental role in how human beings think and act. Your own values undoubtedly play a role in your life and career goals, and in your "dream projects" as well.

Although modern computing, information, and communication technologies can definitely help to make human life better, they do not by themselves help us understand what makes human life better. People who work on the design, development, and deployment of computing technologies, much like everyone else in the world, typically default to the problem-solving approaches that they use most frequently and that have been helpful to them in the past. However, there are profound differences between humans and computers, hence treating questions about human good as an engineering problem is not always sufficient. That's why this book, and this course, will often demand that you think in ways that may feel unfamiliar or even uncomfortable. The critical skills required for responsible ethical thought are not the same as those required for good work in the design and development of computing technologies. But if computing is going to achieve the goal of making human lives easier and better, then computing professionals need to be good at both.

1.5 WHY USE SCIENCE FICTION TO STUDY ETHICS?

Why use science fiction to study ethics? It is often the case that introductory textbooks on professional ethics use case studies. Case studies are specific scenarios that capture common or representative challenges in ethics. Sometimes they are based on specific events that actually took place, and sometimes they are fictional. One reason that case studies are useful in the classroom is that they give everyone in the class a shared set of information to work with. Another reason is that a good case study captures some of the messiness and complexity of ethical quandaries when they occur in the midst of other experiences, as they do in real life. A good case study helps capture the reasons why a given problem is difficult to recognize or understand and why it is hard to figure out a good response. In this book we use stories to stand in for case studies, if you want to find additional case studies to supplement the stories, the National Academy of Engineering maintains a large list of ethics related material for professionals including numerous case studies (Online Ethics Center for Engineering and Science, n.d.).

A good science fiction story can function in the same way as a case study. They are also more fun! But that's not all. There are also reasons to think that science fiction can be more useful than case studies for honing your skills in ethical analysis. For one thing, stories are embedded in larger story-worlds, in which many other things are also happening. In a good story, you are not merely presented with a problem that requires solving. Instead, you get to see a particular individual (or set of individuals) with their own goals and interests and unrelated problems, who ultimately has to confront the ethical quandary. In this way, stories can better capture how challenging it can be to pay attention to ethical issues in the first place. They also can effectively capture the reality that ethical quandaries tend to crop up unexpectedly in the middle of a project, when you're thinking about something else and have other plans.

Another advantage of stories is their characters. Instead of the featureless figures of a case study, a story's characters have personalities of their own that shape their perceptions and choices. By watching a story's characters make decisions (and, in many cases, mistakes), we can gain perspective on how we might react in a similar situation, and why. Stories can also help us understand how a particular person might have interests, concerns, or vulnerabilities that we might not share and that are affected by the story's situation in ways that we otherwise might not have noticed. This combination of qualities also helps an engaged reader develop the *moral imagination* that is a key component of successful ethics education (Callahan 1980).

Finally, stories can be a useful way to get some distance from a charged issue. For example, many people are likely to have strongly held opinions about many technology-and-society issues such as warfare, elections, and appropriate forms of public speech. Because of that baggage, it can be hard to have a productive conversation about the underlying ethical issues or even to get to those issues in the first place. But if those same tensions and dynamics are relocated to an unfamiliar place and time, where nobody in that space has been affected, it can be easier to understand and appreciate a range of perspectives on the events that take place and the conditions that have created them. This productive estrangement is perhaps best described by a quote from philosopher Martha Nussbaum:

> [Fiction] reading frequently places us in a position that is both like and unlike the position we occupy in life; like, in that we are emotionally involved with the characters, active with them, and aware of our incompleteness; unlike, in that we are free of the sources of distortion that frequently impede our real-life deliberations. (Nussbaum 1990, 48)

Science fiction in particular is an easy fit for an ethics of computing technologies course. For at least the past 75 years, engineers and futurists have recognized that science fiction is, in key respects, better able than "realistic" fiction to reflect the near future (or possible futures) in which computing professionals work. Additionally, many science fiction writers are deeply immersed in the literature about cutting-edge technological progress. Some of the most famous and influential ones are, themselves, working scientists and engineers. As Alec Nevala-Lee writes of *Astounding*, one of several popular science fiction magazines that launched in the early twentieth century,

> [The] magazine counted Albert Einstein and the scientists of Bell Labs among its subscribers, and it made an indelible impression on such fans as the young Carl Sagan, who stumbled across it in a candy store. . . . [These] writers were creating nothing less than a shared vision of the future, which inevitably informs how we approach the present. Science fiction's track record for prediction is decidedly mixed, but at its finest, it as a proving ground for entire fields—such as artificial intelligence, which frequently invokes [science fiction author Isaac Asimov's] Three Laws of Robotics – that wouldn't exist for decades. (Nevala-Lee 2017, 11–12)

The science fiction stories collected in this book represent a range of possible futures. Some of the technologies in the stories are far beyond anything that exists right now; others could be developed and deployed in the next decade or the next month. The point of reading these stories is not to worry about the technical questions surrounding their design and implementation: some of them are not necessarily all that technically plausible, although they often resemble other present and future technologies that are plausible or might actually exist. Rather, the point is to immerse ourselves in different *sociotechnical* circumstances and to consider how society and the individuals in it adapt to that technology and are influenced by it—often in ways they don't expect.

Story Point: "Apologia," by Vajra Chandrasekera

> My work was to make him look larger than life, heroic, a proper vessel for the grand project of atoning and reconciliation that he represented for us all.

Throughout the text we have inserted *story points* such as this one. We elaborate more on how to integrate both the science fiction stories and the textbook in the "Introduction to the Story Bank" in part 2 of this book. Like a good set of case studies, the stories that we have selected for this book raise many questions and serve as examples, guides, and jumping-off points for discussion, description, and deeper investigation. These are some, but of course not all, places where the themes of the book intersect with the particular science fiction stories that we have included. Each story is prefaced with a *story frame* that highlights important topics, themes, or events of the story that one should read critically.

The future society of "Apologia" has perpetrated great wrongs in its past—racialized violence, colonization, and genocide—and the guilt weighs heavily. But because this future society possesses technology far more advanced than our own, they are able to conceive and execute an innovative solution to this guilt: they send a poet back through time to the moments of greatest horror, to apologize to the "natives" of those moments for the harm done. It's not hard to figure out that this solution is intended to benefit the denizens of the poet's own time rather than the people to whom he is apologizing: the story's narrator, the editor who splices together the public video feed of these apologies, says as much. By forefronting this reality and by taking us deeply into the social and cultural dynamics that lead to the Apology project and those that emerge from it, "Apologia" invites us to consider how and why such "solutions" come to pass—even though apologies after the fact cannot undo harm.

As is the case for several stories in this book, fanciful or improbable technologies offer us a fresh perspective on how humans use, inhabit, and/or are otherwise shaped by the technologies around them. On the basis of what we know of the world today and how technology is currently integrated into our lives, what would people be likely to do if new technological avenues became open to them? Stories like "Apologia" challenge us to move beyond shallow optimism and to imagine those possible technological futures with clarity, insight, and a more timely attention to potential harms.

1.6 PROFESSIONAL ETHICS AND GUIDELINES

Throughout most of this book, we discuss situations in which the answers to ethical questions are not clear-cut, and even the presence and nature of an ethical question is not always obvious. Often, it is not in the power of one person to make decisions about the appropriate design, implementation, and use of computer technology. Making progress on these questions is a long-term, society-wide effort, and deferring our judgment on them is often a better idea than deciding where we stand beforehand.

But as a practicing professional, you may face situations in which you do not have the luxury of deferring your judgment. In some cases, there *are* applicable principles that have been agreed upon by much of the profession or by a governing authority, even though you may personally disagree. In addition to the ethical norms of society at large, it is important to be familiar with the ethical norms of your field.

One of the roles of professional societies in any field is to articulate a code of ethics for practitioners in that field as well as to articulate the collective wisdom of the people who work in that field (Johnson 1998). Perhaps the most famous example of such a code is the Hippocratic Oath, which requires medical professionals to "use treatment to help the sick according to my ability and judgment, but never with a view to injury and wrong-doing" (American Medical Association, n.d.). These codes can provide guidance when the most accepted way to proceed needs to be stated unequivocally, or sometimes because the way forward is unclear.

Within computing-related fields, there are two major professional societies: the ACM (ACM Code of Ethics and Professional Conduct, n.d.) and the IEEE (Institute of Electrical and Electronics Engineers, n.d.). Each of these societies has a professional code of ethics and extensive online resources for professional ethics, as does the National Academy of Engineering (Online Ethics Center), which acts as an umbrella for most engineering disciplines.

Like all ethical codes, those of the IEEE and the ACM aim to strike a balance, offering consensus on some fundamental principles while still allowing for a range of views. Both sides of this equation are very important for any code of ethics. Without some basic principles, such as a commitment to honesty, a code of ethics isn't saying anything at all. But a code that has no flexibility and tries to define the correct point of view on every subject won't be useful or meaningful to the majority of people who supposedly use it. In chapter 6 we discuss each of these codes of ethics, how they came about, how one can employ them in one's professional life, and the limitations inherent in any code of ethics. However, this book strives to give you more ways of approaching problems than professional guidelines alone.

1.7   THINKING WITH ETHICAL FRAMEWORKS

Our goal in this textbook is to give you tools for thinking about dilemmas you may face as a computing professional. Ethical theory is included because it is one kind of tool. But rather than introduce you to only a handful of well-known ethical theories, we invite you to think with different ethical *frameworks*—frameworks that include important features of some ethical theories. These frameworks provide different resources for thinking about a variety of situations as they arise. By learning to describe problems using the principles and concepts provided by each framework, you will learn how to identify and articulate what is at stake and to which stakeholders in a broad range of ethical dilemmas. The various approaches to ethics presented in this book offer different ways to see the world, different vocabularies of significance, different axes along which to make distinctions between good and bad or right and wrong, and different conceptions of what it means to live a good life.

Moral philosophers sometimes invent new concepts, but some of their most important contributions have been rigorous accounts of the principles that people have already been using. In this book, we center discussions on several styles of ethical reasoning as philosophers have identified them: deontology, virtue ethics, communitarianism, utilitarianism, and some of the modern responses. By making these principles explicit and exploring them in depth, we can apply them to think about ethical dilemmas more methodically. Recognizing these different lines of thinking can shed light on disagreements—when two people talk past each other or reach different conclusions on what to do, it could result from ignorance or failure of reasoning, but perhaps the disagreement arises due to approaching the issue from fundamentally different premises.

1.8   LIFE AFTER ETHICS CLASS

This book introduces you to information and concepts that may be unfamiliar and gives you practice in using them, with the hope and expectation that you will continue to use them in your professional life. But continuing to make use of ethical concepts and insights requires a different kind of effort from the material for most other computing and technology courses. In fact, an ethics class is as practically useful as a programming class, and its usefulness is often complementary to the skills and tools you learn in other classes.

This course is also asking you to do something additional and more difficult in order to continue to apply what you will learn. If someone chooses to create or program a system to solve a problem and knows how to do it, there is little reason not to solve the problem in the most direct and efficient way possible. But ethical understanding requires an additional layer of commitment. It's one thing to reach a judgment about the best course of action while sitting in a classroom. It's another thing entirely to make that kind of choice in your own life *and stick to it*, instead of adopting an easier or more self-serving course. It's easy to pay attention to ethical questions in an ethics course because that's the whole point of being there. But when you're at your job and worried about pleasing your boss or your client or keeping costs down, it can be easy to forget even to think about ethics at all and to fail to recognize a developing ethical problem that's right in front of you.

In order to do the difficult thing of holding onto what you learn here, it may help to think of the course as a middle step, neither the beginning nor the ending of your development as an ethical thinker. It may be the first time that you sit down and think about ethics in a dedicated and systematic way, but it's almost certainly not the first time you've thought about ethics. If you have ever worried about whether you are really a good person or wondered whether a given choice is making the world better, then you have already begun.

This book is here to help you continue your development as an ethical thinker and to give you some tools to continue growing once your encounter with it is over. It is a concentrated window of time in which to learn about your own mind and about the views and values of your current and future colleagues in the field. Succeeding at ethics doesn't mean getting 100% on your ethics quiz; it means closing out your experience here with a clearer sense of what you value and an improved ability to articulate why you have these values and what it means to live up to them.

1.9 THE REST OF THIS BOOK

This book has two parts. Part 1 is a textbook in the traditional sense, and part 2 is an anthology of short science fiction stories that might be part of your ethics course, depending on how your instructor is teaching it. These stories have been selected because each of them raises vital issues in computer science or technology ethics in complex and illuminating ways. Each story is framed by a brief introduction that will help orient you to its core issues and a set of study questions you can use to provoke or guide your thinking about the story after you have read it. Throughout the textbook, you will find several *story points*, such as the one at the end of section 1.5, that direct you to particular stories in the anthology that explore the issues touched on in that part of the textbook.

- Chapter 2, "Ethical Frameworks," introduces several ethical frameworks: deontology, utilitarianism, virtue ethics, and communitarianism. Within each framework, we introduce two or more traditions, to give a brief sense of the breadth of approaches within each framework. We then look at three more recent developments, namely, responsibility ethics, feminist ethics, and the Capability Approach.

- Chapter 3, "Managing Knowledge," treats the topics of data, information, knowledge, and wisdom. It defines and addresses the relationships between these four concepts and raises some of the major challenges of working with them.

- Chapter 4, "Personhood and Privacy," looks at the intersection of two important concepts, personhood and privacy, and what impacts computing technology has on our understanding of these concepts.

- Chapter 5, "Technology and Society," investigates some of the common misconceptions about what people tend to think technology *is*. It then introduces science and technology studies in order to show how technology is inseparable from the context in which it is used and developed.

- Chapter 6, "Professional Ethics," looks at the history of professional ethics, what it means to be a professional, and what it means to be involved with the computing technology development process.

REFLECTION QUESTIONS

1. *What does it mean to say something is ethical? Give an example of a normative ethical assessment that you agree with and one that you do not. Give an example of a descriptive ethical statement.*

2. *Pick a piece of technology that you use every day. Make a list of the goals or values that you think the system is designed to achieve. For example: Does it connect people? Does it make something easier? Does it make something harder? Does it make something less expensive? Or does it make something impossible? What are the benefits? What are some of the potential drawbacks if the particular piece of technology achieves its goal?*

3. *Consider some ethical dilemmas that arise due to limited resources. Give an example of a resource that has become less limited due to advances in technology or computing, and give an example of a resource that has become more limited.*

4. *Think about some of the science fiction that you enjoy—pick one piece you are already familiar with and identify an ethical theme that appears in the work. Do all the characters in the story respond as you would? Pick one character that you feel embodies your "ideal" and one that does not. What can you learn from watching your ideal character? What can you learn from characters whose responses do not match the ones you would make?*

REFERENCES CITED IN THIS CHAPTER

AAAI/ACM Conference on Artificial Intelligence, Ethics, and Society. n.d. Accessed May 20, 2021. http://www.aies-conference.com/.

ACM Code of Ethics and Professional Conduct. n.d. Accessed June 1, 2022. https://www.acm.org/about-acm/acm-code-of-ethics-and-professional-conduct.

ACM Conference on Fairness, Accountability, and Transparency (ACM FAccT). 2021. New name as of ~2019. Accessed June 1, 2022. https://facctconference.org/.

ACM US Public Policy Council. 2017. Statement on algorithmic transparency and accountability. January 12. https://www.acm.org/binaries/content/assets/public-policy/2017_usacm_statement_algorithms.pdf.

American Medical Association. n.d. Delivering care: Ethics. Accessed May 5, 2021. https://www.ama-assn.org/delivering-care/ama-code-medical-ethics.

Baller, Silja, Soumitra Dutta, and Bruno Lanvin, eds. 2016. *The Global Information Technology Report 2016*. World Economic Forum. http://www3.weforum.org/docs/GITR2016/WEF_GITR_Full_Report.pdf.

Boyer, E. L. 1997. *Scholarship Reconsidered: Priorities of the Professoriate*. Jossey-Bass.

Callahan, Daniel. 1980. Goals in the teaching of ethics. In *Ethics Teaching in Higher Education*, edited by Daniel Callahan and Sissela Bok, 61–80. Springer.

Colby, A., and W. M. Sullivan. 2008. Ethics teaching in undergraduate engineering education. *Journal of Engineering Education* 97, no. 3 (July): 327–338.

Conn, Ariel. 2017. An open letter to the United Nations Convention on Certain Conventional Weapons. Future of Life Institute, August 20. https://futureoflife.org/autonomous-weapons-open-letter-2017.

Davis, B. G. 2009. *Tools for Teaching*. 2nd ed. Jossey-Bass.

Future of Life Institute. 2015. Autonomous weapons: An open letter from AI & robotics researchers. Open letter announced at the opening of the International Joint Conference on Artificial Intelligence, July 28. http://futureoflife.org/AI/open_letter_autonomous_weapons.

Greenfield, A. 2017. *Radical Technologies: The Design of Everyday Life*. Verso.

Haidt, Jonathan. 2012. *The Righteous Mind: Why Good People Are Divided by Politics and Religion*. Vintage.

IEEE Standards Association. 2021. IEEE P7003—Algorithmic Bias Working Group. Accessed May 20. http://sites.ieee.org/sagroups-7003/.

Institute of Electrical and Electronics Engineers (IEEE). n.d. IEEE Code of Ethics. Accessed June 1, 2022. https://www.ieee.org/about/corporate/governance/p7-8.html.

Johnson, Deborah G. 1998. *Ethics of Computing Technologies*. DIANE Publishing Company.

Lee, Min Kyung, Daniel Kusbit, Anson Kahng, Ji Tae Kim, Xinran Yuan, Allissa Chan, Daniel See, et al. 2019. WeBuildAI: Participatory framework for algorithmic governance. *Proceedings of the ACM on Human-Computer Interaction* 3, no. CSCW: 1–35.

Martin, C. D., and E. Y. Weltz. 1999. From awareness to action: Integrating ethics and social responsibility into the computer science curriculum. *ACM SIGCAS Computers and Society* 29, no. 2 (June): 6–14.

Moor, James H. 1985. What is computer ethics? *Metaphilosophy* 16, no. 4: 266–275.

Narayanan, A., and Vallor, S. 2014. Why software engineering courses should include ethics coverage. *Communications of the ACM* 57, no. 3 (March): 23–25.

Nevala-Lee, A. 2018. *Astounding: John W. Campbell, Isaac Asimov, Robert A. Heinlein, L. Ron Hubbard, and the Golden Age of Science Fiction*. Dey Street Books.

Nussbaum, Martha. 1990. *Love's Knowledge: Essays on Philosophy and Literature*. Oxford University Press.

Online Ethics Center for Engineering and Science. n.d. National Academy of Engineering. Accessed June 1, 2022. https://onlineethics.org/.

Pichai, Sundar. 2018. AI at Google: Our Principles. June 7. https://www.blog.google/technology/ai/ai-principles/.

Piercey, Robert. 2001. Not choosing between morality and ethics. *The Philosophical Forum* 32, no. 1: 53–72.

Purtill, Corrine. 2018. How AI changed organ donation in the US. Quartz, September 10. https://qz.com/1383083/how-ai-changed-organ-donation-in-the-us/.

Rogaway, P. 2015. *The Moral Character of Cryptographic Work*. Cryptology ePrint Archive, Report 2015/1162. https://www.cs.ucdavis.edu/~rogaway/papers/moral.html.

Timberg, Craig, Elizabeth Dwoskin, and Brian Fung. 2017. Data of 143 million Americans exposed in hack of credit reporting agency Equifax. *Washington Post*, September 27. https://www.washingtonpost.com/business/technology/equifax-hack-hits-credit-histories-of-up-to-143-million-americans/2017/09/07/a4ae6f82-941a-11e7-b9bc-b2f7903bab0d_story.html.

Vines, John, Rachel Clarke, Peter Wright, John McCarthy, and Patrick Olivier. 2013. Configuring participation: On how we involve people in design. In *Proceedings of the 2013 CHI Conference on Human Factors in Computing Systems*, 429–438. Association for Computing Machinery.

Williams, Bernard. [1986] 2011. *Ethics and the Limits of Philosophy*. Routledge.

CHAPTER 2

ETHICAL FRAMEWORKS

Learning Objectives

At the end of this chapter you will be able to:

1. *Explain why multiple frameworks have developed for understanding the world instead of one unified theory.*

2. *Describe the key points of ethical frameworks including deontology, utilitarianism, virtue ethics, and communitarianism.*

3. *Contrast the traditions described within each of the four frameworks and articulate how these different developments within the frameworks enable a wider breadth of application.*

4. *Use the contemporary frameworks of responsibility ethics, feminist ethics, and the capabilities approach to critique the classical frameworks.*

5. *Formulate the key ethical tensions within a story by drawing on the conceptual resources of the ethical frameworks.*

2.1 INTRODUCTION

What does it mean to "do ethics"?

On a surface level, it's easy: ethics is figuring out how to live well. But what does that mean? In trying to assess something another person has done or something that you might do, what is most important? Is it the intention behind an action or the consequences that come from that action? Is it necessary to know something about the character of a person before we can determine whether or not that person has acted morally? Is it important also to consider a person's religious commitments? What about societal laws and our responsibilities as citizens?

It is widely believed that ethics is only about decision making. But ethics also involves evaluation and description, and there are better and worse ways to do that. This chapter provides you with many different resources and processes for ethical reasoning. It is meant to broaden your ideas about what counts as ethics. By taking the time to learn about and think with various frameworks for "doing ethics," you will discover that the work of ethics begins long before you answer a question and even before you ask the question: it begins when you start to describe the world in which the question arises.

How do you identify an ethical problem? What language should be used to describe that problem? What, if anything, are we solving when we address an ethical conundrum? Are we trying to determine whether an action is forbidden or permitted? Or are we looking for principles so that we can determine how to act in the future? Perhaps all of these things; perhaps none of them.

In this chapter, you will become acquainted with several ethical frameworks. We use the language of "frameworks" rather than "theories" (which you are likely to encounter in other contexts) for a few different reasons. First, the word "theory" may seem to be detached from "real life" or from "practice." Whereas some ethical theories are intentionally abstract, they are intelligible only as far as they relate to actual experience.

Second, the theories that we do discuss in this chapter are the product of ongoing conversations about the good life, about what kinds of actions are permitted and which are not, and about how we justify our actions. "Framework" is a broader term. It captures the way a particular ethical lens is grounded by particular ideas about how to live, act, see, respond, and describe.

Each framework described in this chapter begins from different ideas about what is most important when it comes to ethics. Although there is some overlap, each framework has a different way of conceiving what it means to be ethical and uses different vocabulary to describe and evaluate ethical problems. Each will, therefore, yield different answers to questions like the ones above. By becoming familiar with these various frameworks, you will gain a broader perspective on the ethical challenges that you will encounter in your profession and a wider range of strategies for addressing them.

Within each framework we identify at least two different traditions. Some of these frameworks correspond to well-known ethical theories, as they are often packaged up for easy digestion. If you do your ethics research on the internet, you can find many resources that will tell you that there is one simple formula for being a deontologist and another one for being a utilitarian. But it is important to remember that these theories do not exist out of time; they have a history. They were categorized as theories merely as a way to organize ideas and to indicate certain patterns found within different intellectual traditions and communities. (We return to this point again below: see section 2.1.2.) They were labeled this way only after long conversations over time about what ethics are, how to evaluate values, and what it means to live a good life. Simplistic formulas aren't just a misrepresentation of a long and complex history; they also fail to capture why those conversations happened, and why they were so complex.

Ethics as a subject of study and mode of analysis goes beyond the academic field of philosophy. Especially in recent years, historians, psychologists, anthropologists, and even neurobiologists have made ethics a focus of research. Ethical theories can help us identify, describe, and analyze those perennial problems in ethics that we identified in the last chapter. Some ethical theories attempt to clarify and define principles in such a way that they can seem rather abstract, almost like formal logic. But most ethical theories draw upon a range of ideas and arguments, while also striving for coherency and consistency.

When it comes to "doing ethics," you as an individual do not have to start from scratch. We all are capable of analyzing a situation and discussing which aspects of that situation matter the most and using our imagination to think about how we might act or what we might think if we found ourselves in a similar situation. In this way we are all already "doing ethics," insofar as we make choices and live our lives (or try to) as we think we should. The various frameworks presented here will likely feel familiar or intuitive to you, at least in parts. In addition to supplying you with new approaches and perspectives, learning about ethical frameworks can help you become clearer or more consistent in the kinds of ethical reflection that you were already doing.

2.1.1 MULTIPLE FRAMEWORKS

The frameworks discussed at length in this chapter are *deontology, virtue ethics, communitarianism,* and *utilitarianism.* In addition, we offer a fifth section that briefly describes several contemporary developments in ethics. Each of these contemporary developments builds on some aspects of the prior four frameworks, while rejecting or revising other aspects.

Each of the four frameworks offers a different orientation toward the task of ethics. Stated briefly, deontology emphasizes moral obligation and prescribes or describes moral principles that govern action. Virtue ethics centers on human character as the locus of moral activity and pays special attention to how we develop and exercise good qualities. Communitarianism focuses on the interdependent nature of human life and examines how that interdependence shapes our possibilities for well-being and self-realization. Utilitarianism prioritizes the greatest happiness for the greatest number of people and therefore focuses on the outcomes of actions.

The thing that none of these frameworks offers is a single specific answer to any given problem. What unifies a given framework is not the answers it gives but rather are the terms on which disagreements take place and the kinds of methods and criteria that matter. Learning about a framework does not mean that you will be able to identify "the communitarian point of view" on a given situation or to determine that a certain kind of action is "good according to virtue ethics," simply because *these kinds of unified positions do not exist.* You will, however, be able to recognize a utilitarian (or deontological, or other framework-based) argument by paying attention to the methods and criteria that shape how that argument is being made.

To underscore the breadth and variety of each framework and acquaint you with how they operate in practice, we include a brief discussion of two or more *traditions* within each framework. The traditions within a given framework share many of the same basic assumptions and ideas with each other, but they interpret core principles differently or take different approaches to applying the methods of that framework. Learning about and comparing these traditions should help clarify that there is no set of authoritative teachings that must be maintained in order to claim a given framework or to understand its logic.

Under each framework you will also find a section called "Modalities for Judgment." These sections describe the patterns of thought and methods that characterize each framework. Ethical theories—and more broadly, ethical frameworks—operate with the presupposition that human life is complicated and that conflicts will be inevitable. In order to reconcile that complicated reality with their ideas about how to live well, people have reflected on and developed principles and conceptual tools for negotiating between conflicting interests and obligations. These are the modalities of judgment. We use the term "modality" to indicate that all ethical frameworks provide an orientation for determining which kinds of actions are ethical and which are not. A strong ethical framework is one that is flexible enough to address a wide range of ethical problems and can be mobilized in a variety of contexts.

It is important to see that the *process of describing and reasoning through* ethical problems can and should be distinguished from the *conclusions* that you reach. Just as a single ethical framework can admit many different and even conflicting judgments about the same problem, it is possible for two different frameworks to come to the same conclusion about what should be done. Similarly, two people using two different ethical frameworks might arrive at similar conclusions. These modalities for judgment are not explicit formulas or algorithms. We invite you to see them instead as lenses to evaluate the challenging problems that come with developments in technology.

Each ethical framework rests on certain metaphysical presuppositions, some of which we will identify and discuss below. Briefly speaking, metaphysics refers to our understanding of how the world works and the nature of reality, including what human beings are and are for. Even (or perhaps especially) when we are not aware of them, our metaphysical commitments shape the way we understand and approach ethics. Some of the frameworks and traditions in this chapter are grounded in metaphysical beliefs that may be unfamiliar to you. In order to make those frameworks and traditions easier to understand on their own terms, we have supplied some information about their metaphysical backgrounds.

Finally, we have chosen these particular frameworks because of their explanatory power, which is to say that they have the ability to capture a wide range of human actions and visions for the moral life. But in the end, no single moral framework can account for every kind of action or way of life, even if it seeks to do so. Each has been criticized over the years for missing some crucial element of ethical understanding and decision making. Each also takes certain things for granted and makes certain limiting assumptions about human life, even though it may purport to be universal (Oyěwùmí 1997). It is through comparing and contrasting these different perspectives that we begin to see the strengths and weaknesses of each approach and gain a clearer picture of when and how they can be helpful to us.

2.1.2 LIMITED FRAMES

The descriptions of ethical frameworks below are not meant to be exhaustive. While we both provide some context and history as well as identify main traditions within each framework, we cannot do full justice to the centuries of development, dialogue, and debate that has shaped these frameworks. The goal is to present some of the most important features of these frameworks so that you can think within them or at least with them.

Just as our treatment of these particular frameworks is limited, so too is our list of frameworks. We aim in this chapter to correct for the exclusively western focus of most ethics curricula today and to familiarize you with some recent developments in ethics that are particularly valuable for thinking about technology. Nonetheless, much of this textbook remains grounded in Anglo-American ethics, reflecting the biases of the academic discipline of ethics.

Additionally, we have largely bypassed talking about ethics from a specifically religious perspective. For many, the moral life is inseparable from religious beliefs and practices. Although we acknowledge throughout the religious grounding and teachings that inspire some of the traditions we discuss in this chapter, we have not made much space to unpack or analyze the religious concerns and ideas that underlie specific patterns and judgments.

It is also important to note that the frameworks and traditions that we discuss in this chapter still generate debate. This is in part why we introduce at least two different traditions within each of the frameworks. In reality, there are many traditions within each of these frameworks—more than we can include here—but having at least a few helps illuminate the different ways that the core concepts and mechanics of the framework can operate in practice. We include some pointers for additional reading in each of the sections for those who are interested.

2.1.3 HOW TO READ THIS CHAPTER

As you learn about these different approaches to ethics, you will likely discover similarities to your own existing beliefs, feelings, or habits of thought. You will also probably encounter approaches that are difficult to understand or maybe even offensive insofar as they conflict with your own values. Sometimes it is in learning what we disagree with that we get clearer insight into what we really value. And sometimes we find that what we value is not the same as what others value (Clarke 2010). That's OK, too. *It is entirely possible to learn about an ethical system without adopting it as your own philosophy!*

We encourage you to learn about these less intuitive approaches to ethics in the same way that you would learn a programming language or other specific skill that does not appeal to you: as an architecture of reasoning that is useful and appealing to others and that therefore helps to explain/interpret some things that you may encounter. Each of these frameworks can offer you resources for your own perception and reasoning even if you don't embrace them completely. It can also help you understand the reasoning of others, which is useful when you want to persuade them or understand why they have reached different conclusions from yours.

As mentioned in chapter 1, this book is not a guide or manual for how to do ethics. This point is worth repeating because the very notion that we can apply ethical theories to concrete situations is contestable (see MacIntyre 2013). The practical work of ethics is not about applying the rules of morality to social, corporate, or institutionalized subject matter in order to yield specific results. Nor is ethical thinking solely or even primarily about applying self-interpreting rules and laws to concrete situations in life. Even within deontology—an ethical framework that is known to emphasize laws and rules—things are never this easy. Human life is much more complex, and the task of ethics is for each of us to live the best life that we can. That task is complicated and challenging enough that most people decide, after some reflection, that they will take all the help they can get.

2.2 DEONTOLOGY

Deontology as an approach to ethics is best characterized by its focus on *duties*, *rights*, and *moral obligations*. Its two main presuppositions are that ethical evaluation primarily concerns the rightness or wrongness of actions and that ethical reasoning should help determine what we ought to do.

2.2.1 OVERVIEW OF DEONTOLOGY

The word "deontology" comes from the Greek word *deon*, meaning duty, obligation, or "that which is binding," and *-ology*, indicating a particular branch of knowledge. Deontology has existed in various forms. One of the most famous is associated with the eighteenth-century German philosopher Immanuel Kant. Kant's moral philosophy has been so influential that people sometimes refer to "Kantianism" as though it is its own ethical theory, similar to—although distinct from—deontology. Below we discuss Kantianism as an important development of an older tradition in which moral principles are said to be obtainable through human reason.

Despite its many variations, deontology does have some defining characteristics. Most notably, deontology emphasizes the rightness or wrongness of an action by reference to certain action-guiding principles. Depending on the context, these principles can be described as laws, rules, maxims, imperatives, or commands. Whatever the terms used, these principles are said to place certain constraints on human action. These constraints apply—or at the very least must be seriously considered—even in situations in which the consequences of an action are understood to be desirable or good.

Because deontology bears a minor resemblance to the simplistic, black-and-white thinking that many people associate with ethics, this complex and nuanced framework is frequently misunderstood in three specific ways. The first misunderstanding is that deontology is simple: you figure out what the law is, and then you do what it tells you to do or avoid doing what it does not permit you to do. For many people, this is what ethics is all about: adhering to basic rules and laws. But the demands of duty are complex, and balancing those many demands requires careful reflection, not just blind adherence.

The second misunderstanding is that because it is difficult to honor one's many duties at the same time, you can therefore pick and choose which laws to follow as a matter of individual choice or preference (or perhaps a matter of avoiding the least desirable punishments). But simply acting on preferences and regarding those preferences as if they are binding laws is not an accurate description of deontology. In fact, this kind of picking and choosing has more in common with moral relativism than it does with any of the ethical frameworks that we discuss in this chapter. A moral relativist is someone who believes that all moral judgments are based on individual viewpoints and that no one viewpoint ought to be privileged above any other—save that person's own, because most moral relativists are critical of anyone who disagrees with their position on the matter (Midgley 1991).

The third misunderstanding is that because deontology considers intention to be important, it therefore does not consider consequences to be important at all. The problem with this description of deontology is that it presents only a partial picture. It may help us understand the points of emphasis within a deontological framework, but it is a characterization that obscures deontology's many specificities and variations. Most forms of deontology acknowledge the ethical significance of the consequences and context of moral actions and choices (Rawls 1999, especially p. 26), even though they emphasize principles, laws, rules, and obligations that guide human action and decision making. Where these duties and moral obligations come from, and how they relate to each other, depend on the form or style of the deontology in question.

Unlike the simplistic and piecemeal approaches described above, deontology offers an approach to ethics that is morally practicable without abandoning the seriousness of moral laws. Deontology presumes that moral obligations are a real part of human life and cannot be dispensed with because they are inconvenient, even while it recognizes that honoring all of one's moral obligations is rarely a straightforward task. For a deontologist, the task of ethics is not to *choose* which obligations to follow but rather to consider all one's moral obligations in order to determine how to live and to act in light of all of them, especially in situations that put (or seem to put) these obligations in tension with one another.

Although deontology requires us to take account of all of the rules, laws, and duties that bind us, this does not mean that deontologists consider every single rule, law, or duty to be equally binding. For instance, both jaywalking and manslaughter are against the law in the United States, but very few people would be willing to argue that these two laws are equally meaningful or significant. However, most of the time the distinctions are more subtle, and even in situations in which people have acknowledged that a particular code of laws has legitimacy, it is not always easy to figure out what those laws and duties require of us at any given moment. Deontology does not demand that we follow the law regardless of circumstances—that wouldn't make any sense, because nothing in our lives ever happens apart from specific circumstances. Rather, the task of deontology is to understand how best to honor one's duties within those circumstances.

The laws proscribed and enforced by the government are one kind of rule, but of course they are not the only kind. Some rules or duties may be particular to a community, such as religious dietary laws, and others may apply only to a subgroup within a community, such as doctors' obligation to provide medical care in moments of need. (For more on profession-specific obligations—specifically those for programmers and technology developers—see chapter 6.) Different communities may have different ideas about which parts of human life must be guided by these rules and which parts are free of moral obligation.

It should come as no surprise that much of our lives is governed by rules, laws, and principles of actions. Many human actions and behaviors are explainable as following rules or laws, often without giving them much thought, such as walking on sidewalks instead of in the street or going to the end of the line at your favorite hot dog stand. There are also various rules of etiquette that we acknowledge, follow, and sometimes knowingly break, though for most deontologists these more day-to-day kinds of rules are not necessarily morally significant. Although it is easy to presuppose that rules and laws are burdens, especially in cultures in which freedom is highly valued, these less-significant rules make many aspects of our shared lives easier to negotiate and harmonize. Rules can even provide satisfactions of their own: after all, rules are what makes it possible to play many kinds of games and sports together.

A law, rule or duty will matter only if it is rooted in an authority that is recognized as legitimate. Not all deontologists recognize the same authorities, or even the same kinds of authority. Often, deontologists will specifically reject a law they take to be illegitimate—such as a state law that prohibits carrying weapons or criminalizes political protest—precisely because another system or authority they take to be more legitimate points them in a different direction. Also, deontologists can share a commitment to a particular law or duty, such as respecting the property of others or a prohibition against murder, but have different explanations for what makes that duty authoritative.

2.2.2 DEONTIC FORMS OF AUTHORITY AND TRADITIONS

In order to highlight the variation within deontology as a framework for thinking about ethics, below we introduce three different traditions that are deontological in their orientation. Each of them appeals to a particular kind of authority to justify and legitimate moral obligations and duties. The first tradition appeals to a "social contract," and therefore the authority of the law is grounded in a political claim about what it means to live together in a society under reasonable principles that can be applied to all. The next tradition grounds the authority of moral obligations and constraints on the existence of a god or gods, to whom duty is owed and who determine human beings' duties to others. Finally, we consider a tradition of deontology that insists that the basic principles of morality are to be derived from human reason. The feature that unifies all three of these traditions and warrants, describing them as deontological, is not a particular authority, duty, or even rule-governed action but is rather the belief that there is a difference between right and wrong and that this difference is supported by an authoritative claim about how we ought to act toward others and, in some cases, toward ourselves.

It is important to note that these traditions are not mutually exclusive. Thus, there is nothing preventing a deontologist from arguing that moral obligations can be grounded by all three forms of authority discussed in the following sections. These traditions were selected because in many ways they are defined by which authority they deem most important with regard to morally binding principles and norms.

Political Authority

Political authority comes from an organized human society. There are many forms of political authority, which vary with political systems. Authority is distinct from power, which in this context means the ability to materially enforce rules and punishment, irrespective of legitimacy. Most political systems and leaders claim to operate from a position of legitimate authority, serving the interests of the people, even if in practice they operate from a position of power.

An example of a deontological tradition grounded in political authority is contractarianism. Contractarianism begins from the presupposition that human beings are primarily, if not solely, driven by self-interest, and therefore the best strategy for deciding which institutions, principles, and social rules can legitimately place constraints on our otherwise selfish actions is first to find the ones on which all would agree. Hence, the need for a contract. In other words, ethical principles and norms require us to sometimes act in ways that we would prefer not to act but are justified on the grounds that it is better to cooperate than to be constantly at odds with each other. Agreement among individuals has normative importance. Contractarianism grew out of an older tradition known as social contract theory, which dates to Thomas Hobbes in the early seventeenth century (Cudd and Eftekhari 2018). Hobbes insisted that humans are driven solely by their own self-interest but because they are too fragile and weak to live on their own, it is necessary to be part of a society in which individuals sacrifice some of their freedom and agree to be governed by a sovereign authority in which disputes and self-interest can be mitigated appropriately (Hobbes [1668] 1992). Although starting from the same basic premise—that human beings are primarily driven by their own self-interest— contractarianism, as it has been more recently formulated, focuses less on the giving up of rights and powers to a government authority and more on what "reasonable" people would decide and execute judgment under democratic rule.

Divine Authority

Divine authority is authority from God or gods. If a set of laws is understood to have been given or revealed by a divine figure, then those laws gain their legitimacy from the existence and power of that God or gods. Different religious traditions have different ideas about how humans gain knowledge of these laws. In some traditions, these laws are said to be contained in holy books or sacred writings. In other cases, divine law is considered to be received through prophecy or through authoritative leaders. The ethical weight of an action might also be interpreted as gaining its authority from a God or gods without necessarily using the language of law.

Divine command theory is one form of deontology that derives its authority from God. Broadly speaking, divine command theory holds that moral obligations consist of obedience to God. Under divine command theory, an action is obligatory because God or the gods command it and it is impermissible if God or gods forbid it. If that action is neither obligatory nor forbidden, it is considered to be a permissible action (Quinn 2006).

Although the notion of god-given laws is conceptually straightforward, there is an interesting philosophical problem at their core: did god(s) give this law because it is right, or is the law right because it comes from god(s)? This is sometimes referred to as the Euthyphro problem, referring to a dialogue written by Plato in which Socrates asks, "Is the pious loved by the gods because it is pious, or is it pious because it is loved by the gods?" (Plato, *Euthyphro* 10a1–3). Most divine command traditions do not have a settled answer to this question, and it is a topic of ongoing philosophical and religious debate.

Sidebar: Obligations and Prohibitions

Deontology includes both obligations and prohibitions. Obligations (sometimes called "positive laws") are things that you should do and that require active effort on the part of the agent. By contrast, prohibitions (sometimes called "negative laws") forbid certain kinds of actions: they are obeyed not by undertaking a specific action, but by refraining from acting in a way that has been described as wrong.

Both obligations and prohibitions can be found in the Ten Commandments, which is one of the most well-known sets of laws based on divine authority. In the story recounted in the five books of Moses, which are sacred text for Judaism, Christianity and Islam, the Ten Commandments are given to God by Moses. There are versions in Exodus and Deuteronomy that are similar but not identical, and some of the same laws are also repeated in Leviticus. In fact, the "canonical" sets differ somewhat among the three religions and among the different translations.

Obligations and prohibitions might seem like opposites in the abstract, but in practice there is a great deal of overlap. Consider these two laws, both of which appear in both versions of the Ten Commandments:

4) Remember the Sabbath day to keep it holy.

6) Thou shalt not murder. (Exodus 20:3–4 KJV)

The first of these laws is clearly an obligation, and the second one is clearly a prohibition. But on closer inspection, the obligations can be seen to have some prohibitions built in, and the prohibition might well require some specific positive actions.

Honoring the Sabbath is an obligation. Fulfilling this obligation is not something that just happens on its own; a Jew, Christian, or Muslim who is following this law has to do specific things in order to make it happen. But those specific things include actively avoiding some activities that one does on ordinary days, as part of a larger obligation to become more aware of the sacredness of life.

The second law, forbidding murder, is a prohibition. Much of the time, it can be followed simply by avoiding the forbidden action. But for someone who finds themself in a position to kill another person, it might require active effort to refrain, especially if they are very angry or otherwise motivated to kill that person. In a situation like that—when, arguably, the prohibition against killing matters the most—abiding by that prohibition is likely to require some active, positive effort and not just passive avoidance.

The Authority of Human Reason

The notion that human beings have an inherent moral compass that allows them to discern the difference between right and wrong is an ancient idea that became especially popular during the European intellectual movement known as the Enlightenment (more on this below). Over the centuries, theologians and philosophers have linked this belief in the human's inherent capacity to judge between right and wrong to religious dispositions and creation stories. In the book of Genesis, for example, the first humans are said to know the difference between right and wrong. Thomas Aquinas, a Christian philosopher and theologian, argued that the very first principle of practical reasoning (i.e., ethical reflection on human action) is quite simple: avoid evil and do the good (Aquinas *Summa Theologica* I–II, 94, 2; see Aquinas 1948). Aquinas insisted that all human beings know this basic principle and therefore can discern the difference between right and wrong. While Aquinas appealed to a divine authority to support this claim, he also linked this capacity to the human's ability to grasp the laws that govern and order the universe (i.e., natural laws).

Natural law theory as a tradition has also been articulated in less religious and metaphysical terms. As one theorist explains:

> Natural law theory accepts that law can be considered and spoken of *both* as a sheer social fact of power and practice, *and* as a set of reasons for action that can be and often are sound as reasons and therefore normative for reasonable people addressed by them. This dual character of positive law is presupposed by the well-known slogan "Unjust laws are not laws." (Finnis 2020)

Natural law theorists do not hold that ethics is simply a matter of *sensing* the difference between right and wrong, nor do they claim that deciding what to do in particular situations is easy. In fact, most natural law theorists are careful to point out that acting morally is very difficult and that it requires a certain amount of sacrifice. Even more, they tend to emphasize that even in those cases in which we know what the right choice is, we often fail to act in ways that are right and just. For someone with the outlook of Aquinas, this shows that natural reason needs to be supplemented and guided by religious texts and teaching. For secular theorists, this means that discerning what is right and wrong is never a private enterprise but must be worked out in a social context in which people deliberate about the ends that are worth pursuing.

2.2.3   KANTIAN DEONTOLOGY

The notion that the universe is ordered by laws that can be apprehended by human reason was an especially popular belief in the eighteenth century, when Immanuel Kant was developing his philosophy. Kant's moral philosophy is similar to Aquinas's in that Kant believed that human beings are able to discern the difference between right and wrong. However, Kant was also writing at a time when people were becoming increasingly critical of claims that depended on religious authority, whether in the context of politics, of science, or of morality. And therefore Kant, as many other enlightenment thinkers, rejected the idea that any code of law handed down in a religious tradition or promulgated by the state can successfully address the full breadth and complexity of right and wrong.

Kant acknowledged that external laws often aligned with the moral law in basic ways and that many (or even most) people require such externally given laws to remind them of their fundamental duties to others. He also believed that external laws can be valuable both as constraints on our behavior and as guidelines for our moral understanding. However, Kant insisted that human reason is the most important guide to making moral choices. Because Kant remains one of the most influential figures in moral philosophy and because he is frequently described as a deontologist, it is worth explaining his approach to ethics in a little more detail.

Kant's version of deontology is uniquely universalist in scope. It starts from the assumption that all human beings are free and rational and that they are familiar with both the experience of being moved by desire and the experience of being moved by the feeling of duty or moral obligation. In response to this baseline assumption, Kant offers an account of what must necessarily be true in order for us to make sense of this universal experience of moral duty. He understood himself to be offering a "metaphysics of morals," in which he aimed to articulate the universal pattern of reasoning behind a moral judgment (Kant 1996).

According to Kant, the unity and intelligibility of the moral law is something all rational beings can grasp. The moral law therefore must be perfect, and perfectly consistent, in a manner similar to the order of nature. This means in practice that whereas all our particular duties to individuals, coworkers, family, and friends are important, they are conditional and therefore cannot describe the basic sense of duty that applies in every circumstance. For Kant, only those actions that are unconditional have moral value. So what kind of actions are these?

For Kant, an action performed in accordance with a law or set of laws does not necessarily mean that action has moral worth. Kant strongly believed that people should learn to think for themselves and never blindly follow any one law or set of laws. And even when following an external law is in order, that does not necessarily endow it with moral worth. In order for an action to have moral worth, it must be an action that you, the agent, recognize as right. But just as importantly, it must be binding for all rational agents. Otherwise it would be, again, conditional. Only actions that meet both of these criteria have moral worth. According to Kant, our capacity to act on such a law is the only indication that we are truly free agents. Contrary to popular opinion, freedom is not about acting on whatever desires and impulses one might have in a particular moment, nor is it following a rule to avoid punishment or condemnation. True freedom, Kant maintained, is the freedom to act according to a law that you can both apply to yourself and universally legislate for all. A law such as this is what Kant calls a categorical imperative, by which he means a law that is unconditionally and universally valid. His first formulation of this law is as follows:

Act only in accordance with that maxim through which you can at the same time will that it become a universal law. (Kant 1996, 73)

Let's consider what this means in more general terms. Kant's understanding of morality requires us to evaluate an action using criteria identified by internal reasoning processes, rather than criteria drawn from outside legal codes and cultural norms. Before acting, a moral agent must always ask herself: "If I act in such a way, is this something I can legislate for myself and all other rational agents? Is it something I can continue to follow and expect others to do the same?" Lying is a classic example of an action that might seem justifiable in a particular instance but cannot be justified as a general practice. If you tell a lie, can you honestly and consistently legislate that action? Can you coherently argue that it makes sense, morally speaking, for everyone to be able to lie whenever it is convenient for them? If not, then clearly lying is unethical, and therefore, a person shouldn't lie under any circumstances.

Kant's categorical imperative, as he formulates it initially, is intentionally abstract. It is meant to capture something very basic about the experience of moral obligation. But he also offers a formulation that speaks more to how we are to treat others. His second formulation of the categorical imperative reads:

> *Act that you use humanity, whether in your own person or in the person of any other, always at the same time as an end, never merely as a means.* (Kant 1996, 80)

This second formulation of the categorical imperative is derived from an additional claim that because the human being is a rational creature and therefore are capable of acting freely, all human beings should be treated with respect and dignity. Whenever we treat another person merely as a means to achieve some particular end, we are not only disrespecting that person: we are disrespecting all humanity, including ourselves.

2.2.4 PRINCIPLES IN PRACTICE

It might seem at first that all deontologists—or at least those who acknowledge the same authority behind moral principles or laws—would agree with one another about what the right action is in any given situation. But in practice, those who approach ethics from within a deontological framework—yes, even Kantians—disagree all the time. Although they may share the same general idea about the status of "the law" as the source of ethics and the only guide to ethical action, disagreement often arises around the meaning of that particular law and/or how it should be applied. Furthermore, most people recognize more than one system of law, even if they think that one (or more) of those systems is more important or that one system or code of law is limited in some way. This introduces a difficult question: what is the right way to resolve an issue when different laws seem to prescribe conflicting solutions? When a person is duty-bound by more than one set of laws—for example, religious laws and laws of the state—how do they decide how to navigate conflicts between them or decide when one should take precedence over the other? What kinds of punishments are warranted by different kinds of violations? These are difficult questions to answer, but rather than providing decisive answers, most deontological traditions instead offer additional guidelines, suggestions, and even rules about how to proceed when conflicts and tensions arise.

2.2.5 MODALITIES FOR JUDGMENT

It is rarely possible to satisfy every single duty, or to satisfy them equally well, because circumstances often create tensions between our various duties. This is largely because regardless of their authority, deontology grounds its rules and laws in relationships. When our duties to different people pull us in different directions, or when a given duty relationship is multidimensional (such as a parent's relationship to their child or a person's relationship to their spouse), it is sometimes impossible to honor every duty, or every dimension of one's duty. When these conflicts of duty arise—as they inevitably do—a moral agent is forced to make judgments about which duties should be prioritized over the others. For deontologists, these priority judgments cannot simply be a matter of personal preference—after all, the whole point of a duty is that it's binding, even when it's inconvenient or unpleasant.

When balancing competing duties, a deontologist must consider the following two concerns.

How Fundamental Is It?

Which duty is most *fundamental*? Of the many duties competing for your attention and action, which ones are the most important to who you are and your role in the world? In a situation that compels a person to choose between protecting a stranger and protecting their child, nearly every deontologist would argue that it's right and appropriate to protect the child, not because the agent has no duties at all to the stranger but because parents have particular duties toward their children.

But this determination might be more complicated if the agent is an expert or professional (such as a doctor or a firefighter) who is trained to respond to the crisis at hand, because many such experts and professionals believe that their expertise imposes a specific obligation to use their skills and training to help others when possible, or have even signed onto a professional code that requires them to do so. Even if that parent still protects their child instead of the stranger, they may feel more keenly that they have failed in their duty to the other person.

How Relevant Is It?

Which duty is most *relevant* to the situation in question? When you are making a decision about a specific situation, it is often the case that some of your many duties are more pressing than others because of the particular circumstances. Imagine a soldier in a combat situation whose platoon-mate is injured in pursuit of the enemy. Should this soldier stop and assist her injured comrade, or complete the mission and kill the enemy? Both of these duties are important, but which is more relevant? The answer will depend on a number of very specific factors: how badly injured is her comrade? Is anyone else available to help him? How likely is it that there will be another chance to kill this enemy?

It is worth remembering that for a civilian who is not bound to fight the enemy (and might even carry a strong prohibition against killing other people), this tension would be very different. Both soldier and civilian have a duty to help another person who is injured and in need, but the soldier has other duties that the civilian does not, and those duties might be more relevant than helping the wounded.

These two criteria enable a deontologist to balance the basic obligations that shape their life (the fundamental) against the particular demands of specific circumstances (the relevant). When a deontologist deprioritizes a specific duty—for example, if the soldier above drops her pursuit of the enemy to assist her platoon-mate—that person is neither rejecting nor ignoring that one duty; rather, they are determining that the best way to honor the full range of their duties requires them to prioritize other duties over that one in this particular moment.

Sidebar: Prioritizing the Right over the Good

Unlike some other approaches to ethics, it is frequently said that deontology is not organized around the question of what is good, but rather it is primarily concerned with what is "right." A system organized around "right" may initially seem more constricting than a system organized around goodness, because right/ wrong is a binary system and goodness is not.

But deontology actually allows for more freedom than goodness-based systems. A system concerned with goodness can encompass anything: any person, deed, or object can be evaluated in terms of how good it is compared to other things. When goodness is the metric for ethics, it's (almost) always possible to get better, and being good requires that one constantly try to become better. A system of right and wrong, by contrast, leaves room for actions and experiences that are outside the bounds of ethical evaluation—that is, they are neither right nor wrong, and people can pursue those actions or not purely according to their preferences.

2.2.6 STRENGTHS AND WEAKNESSES OF DEONTOLOGY: IT'S NOT ALL ABOUT RULES

One of deontology's strengths is that it insists upon the notion that there is a difference between the rightness and wrongness of action, and it does so by appealing to factors that do not automatically depend on the way the outcome of that action is received. In this way, deontology preserves the integrity of ethical action: our choices have value, regardless of whether other people recognize that value. At its worst, deontological thinking can enable fanaticism, leading people to believe they are justified in punishing others for doing things they take to be wrong. At its best, deontology can afford people the moral courage to stand against the majority, even when there is no obvious reward for doing so.

Another common criticism of deontology is that it does not focus enough on the consequences of action. Although it is true that deontology prioritizes human actions and intentions, as suggested above, most forms of deontology not only address but seriously consider the consequences and context of decisions and actions. Yet because of its emphasis on principles (laws, rules, and codes), deontology on its own seems to capture only a small portion of what it means to do ethics. And even more than that, it seems to operate as though the ethical life is a judicious process backed by a particular type of authority. In the mid-twentieth century, Elizabeth Anscombe offered a powerful argument against the legalistic language that was then dominant in moral discourse, arguing that without religious backing, such language makes very little sense and that therefore, unless we want to impose its religious significance upon everyone who wants to weigh in on matters of ethics, we ought to abandon it altogether. In her words:

> The concepts of obligation and duty—*moral* obligation and *moral* duty, that is to say—and of what is *morally* right and wrong, and of the *moral* sense of "ought," ought to be jettisoned if this is psychologically possible; because they are survivals, or derivatives from survivals, from an earlier conception of ethics which no longer generally survives, and are only harmful without it. (Anscombe 1958, 1)

Those forms of deontology that claim to ground the authority of moral principles and norms in human reason rather than divine decree have also been criticized. Kantian deontology in particular is frequently criticized for being overly rigorous and impractical. As noted, Kant insists that an action has no moral worth if it is not done out of duty. This means that, for instance, when someone donates to a charity and finds that the act brings them joy, and they decide to donate again and again, according to Kantian standards, this action has no moral worth. It is for reasons like this that people sometimes call Kant a moral purist or rigorist (Cohen 2014). Kant claims that a truly moral act is motivated by respect for the moral law and free from any other particular inclinations or desires. But human experience tells us that most of us act for a variety of reasons, and desire and emotion cannot easily be cast aside in order to evaluate the moral worth of particular actions—whether our own or someone else's.

But perhaps the most important issue that arises within a deontological framework concerns what to do when conflicting duties arise. Although we have described some of the ways deontology handles conflicts and inconsistencies, the fact remains that the principles and rules that guide human conduct are not always easy to follow nor are the choices between them always clear. Quite often, the complexity of our lives and circumstances makes it impossible to abide by all of the rules and duties that bind us, no matter what we do. The impossibility of choosing a perfect course does not mean that deontology has failed or that those rules or duties cease to be binding. Instead, these clashes of duty should be seen as the reason deontology is valuable: it furnishes us with a way of thinking, perceiving, and living that enables us to realistically navigate difficult choices about the things that matter most. That said, there are other ethical frameworks to consider and other points of emphasis to consider as well.

Sidebar: The Challenges of Deontology and Artificial Intelligence

For most deontologists, the right or wrong decision must always be a choice; otherwise, one cannot take responsibility for it. The agent may feel a strong sense of obligation, or the call to duty might be strong, but in order for responsibility to exist there must be a choice, and the agent must have the capacity to choose differently. Defining deontology becomes complicated when we start thinking about automated agents, which are programmed to do certain tasks and which learn new things according to programs that have been written by humans. Are automated entities exercising responsibility in the deontological sense? Could this be on the horizon?

Deontology focuses on duties and obligations. What kinds of entities have duties and obligations? Typically, when we reference duties, rights, and obligations, we have in mind other people. But what are our duties with respect to other living creatures? And what about future generations? These are the types of questions deontology has traditionally asked. But new issues arise when we consider human beings' relation to intelligent machines. What does a robot owe to a human? What does a human owe to a robot? What kind of demand can an artificial being make on human life, and how would that demand be justified? What these questions point to is a certain ambiguity about the nature of the relationship between humans and intelligent machines. If robots cannot have relationships with others, then what is at stake when we delegate a moral decision to a robot?

There is a growing body of work in computer science and related fields called machine ethics (i.e., programming or teaching machines how to act ethically) (Christian 2020; Dubber et al. 2020; Lin et al. 2017; Pereira and Lopes 2020; Wallach and Asaro 2020). Work in this area typically takes one of two main forms. In the *bottom-up* approach, systems or agents are given examples of proper behavior and attempt to abstract from those what to do in particular situations. In the *top-down* approach, systems or agents are given a set of rules in some formal, logical language and are expected to incorporate these rules into the actions that they decide to take. An important stream of research in this area focuses on harnessing the creativity of complex machine learning methods while still enforcing that the agents make decisions in line with outside constraints or rules (Rossi and Mattei 2019).

There has been a lot of press coverage of work in this area related to teaching autonomous cars how to act by surveying millions of people all over the world (Bonnefon et al. 2016). One key question for us is, *do machines even have ethics?* Whereas they can have rules or patterns of behavior, who or what is held responsible for the decisions they make?

Consider self-driving cars. They are trained by being given "traces" of human drivers as well as a set of rules to follow. They observe human drivers obeying traffic laws, avoiding obstacles, stopping for errant pedestrians, and other actions. If an autonomous vehicle stops for a pedestrian, a moose, or a flock of ducklings in the road, can it be said to be acting ethically? Note that moose are large, and that hitting one can do significant damage to a vehicle and possibly the occupants of the vehicle as well as to the moose. Does it make it less ethical (or less of an ethical concern) to stop for a moose than for a smaller animal? Does the vehicle's decision to stop for the ducklings come from a respect for life or from a societal consensus to not injure cute animals?

Story Point: "Dolly," by Elizabeth Bear

"She's a machine. Where's she going to get a jury of her peers?"

Can an android be a person, or could it become one? What obligations would its potential personhood impose on us? And how can those obligations be made visible to people or governments that assume they are only objects? These are the questions that Detective Roz Kirkbride and her partner Peter King end up wrestling with as they try to solve the murder of Clive Steele, who has been disemboweled by his new advanced-prototype home companion Dolly. "Dolly," the story, offers a compelling window into how the framework of deontology can help us define and negotiate ethical conflicts. And the conundrum of Dolly herself illustrates both how our definitions and descriptions create the conditions for ethical analysis that follows. In so doing, it offers a way to understand how deontology gives us the resources to reevaluate baseline definitions and thus the duties and obligations that follow from them.

2.3 VIRTUE ETHICS

Virtue ethics is an approach to ethics organized around the idea of human flourishing and human excellence. Its basic assumption is that all human beings share some basic qualities of character, though we vary widely in how much we excel at those qualities and how we express them, and each of us gets better (or worse) at them according to our experiences. It further assumes that human beings are concerned with how to live the good life and that ethics is a subset of what it means to live a full and happy life.

2.3.1 OVERVIEW OF VIRTUE ETHICS

Unlike deontology and other approaches to ethics, where the focus is primarily on actions and intentions, virtue ethics focuses on the whole person: the qualities of character that they have and the patterns of living that issue from and reveal those qualities. That pattern of being is called a *habitus*. By cultivating excellence in a range of basic human capacities, called the *virtues*, a person likewise builds their capacity for a fulfilling life, because exercising these virtues is what enables them to live in alignment with their goals and desires (Hursthouse and Pettigrove 2016). For this reason, it is virtuous people who live happy and fulfilling lives (that is, who *flourish*): not because they are rewarded or deemed worthy by some external judge but because the practice of virtue helps them build and sustain a satisfying and rewarding life.

Nobody is born virtuous. In fact, it's not really possible to be virtuous until you are an adult, or at least old enough to maintain the practice of virtue from your own habitus, rather than because some external force like a parent or teacher imposes it on you (Hursthouse and Pettigrove 2016). This does not mean that we are born vicious. Rather, virtue is developed over time, by emulating exemplary people and by cultivating good habits (Vallor 2016). Parents, teachers, and culture play major roles in shaping a person's habitus, especially while the person is young.

Virtuousness is also not a one-time achievement, at any age. For one thing, virtue is complex: it is possible to excel in some virtues while being weak or inconsistent in others. In fact, most of us are like that (Athanassoulis 2000). Even more importantly, our capacity for any given virtue is not fixed: although a person's habitus cannot change overnight, it can always shift by small degrees, and those small shifts can eventually add up to significant changes. It's always possible—though never easy—to develop your capacity for a given virtue. It's also possible to lose the capacity for a given virtue over time. Similar to physical muscles, virtues like courage or generosity grow stronger if you exercise them regularly, and they weaken if you don't use them (Annas 2011). In other words, the exercise of virtue is not based simply on a theoretical understanding of the right thing to do, but through ongoing practice (Vallor 2016).

Like deontology, virtue ethics is deeply concerned with the interior lives of individuals and why we do the things we do. In contrast to deontology, however, virtue ethics emphasizes the importance of actions *as an indicator of a person's character*. Although any one individual action may be hard to interpret (or may even be misleading), the long-term pattern of a person's actions across multiple contexts will reveal their character by indicating how well, to what degree, and in what ways they exercise those basic human capacities that are the virtues.

Virtue ethics is a goal-driven framework for thinking about ethics. It aims toward the creation of good outcomes and of happy, flourishing people. From within a virtue ethics framework, these two goals are impossible to separate: a good society is good because it makes it possible for people to flourish and develop human excellence.

2.3.2 WHAT ARE VIRTUES?

Virtues are the basic building blocks of human character; they are fundamental qualities like kindness, playfulness, or self-respect. Virtues are not qualities that only some people possess; rather, a virtue is the *capacity to exercise that quality*. According to virtue ethics, every person is endowed with the same basic library of virtues, and the potential to excel in any of these virtues is part of our basic makeup. Therefore, what differentiates us from one another is not whether or not we possess a given virtue, but rather the degree to which we have developed (or our parents and teachers have developed in us) the ability to exercise that virtue. Within a virtue ethics framework, a chronic liar would be understood as being very deficient in honesty rather than lacking honesty entirely. Given the right conditions and a genuine desire on that person's part to become more honest, they could over time develop a greater capacity for honesty.

Virtues are revealed not through single actions but rather through patterns of action. For example, a single act of benevolence is not enough to know that a person is meaningfully kind or generous, because any number of external causes and internal motivations might have prompted that one benevolent action. It is only when a person shows benevolence in a range of circumstances that one can be confident that they have a well-developed capacity for benevolence (Sreenivasan, 2002). In the words of Rosalind Hursthouse, virtues "go all the way down"; having a well-developed capacity for a given virtue means that you will "notice, expect, value, feel, desire, choose, act, and react in certain characteristic ways" (Hursthouse and Pettigrove 2016).

Thinking about human character in terms of virtues—that is, in terms of a common library of basic capacities—is useful in several ways. First, it helps us think comparatively about the differences between people (or even different versions of ourselves), because those differences can be understood as different ways of inhabiting or expressing the same basic qualities. Second, it provides us with a framework for thinking about how multiple aspects of a person's character work in combination to shape how that person will act in a given moment. And finally, it equips us to think about how our environments can impact our character, by encouraging or reinforcing the exercise of some virtues and creating barriers to the exercise of others.

2.3.3 CONFUCIAN VIRTUE ETHICS

The predominant form of virtue ethics in East Asia comes from the teachings of the ancient Chinese philosopher Confucius (Latinized from *Kongzi*, or "Master Kong"). He was inspired by a desire to return to what he saw as the excellence of the earlier Zhou Dynasty social system in contrast to the turmoil of his own time (the late fifth century BCE). Over time after his death, his disciples compiled quotes attributed to him into the *Analects* (Legge [1861] 2017) and expanded the Confucian canon with other classics, such as the *Mencius* and the *Doctrine of the Mean*.

In early Imperial China, Confucianism competed with other emerging ideologies known as the "Hundred Schools of Thought," including the heavy-handed Legalism of the short-lived Qin Dynasty (221–206 BCE) and the more libertarian Daoism of the early Han Dynasty (206 BCE–220 CE), to eventually become a principal guiding force for Han governance. The later neo-Confucian renaissance, emerging as a secularist response to the prevailing Buddhist and Daoist spirituality during the Song Dynasty (906–1279 CE), solidified Confucianism's place in the core of Chinese philosophy up to the modern era (Yao 2000).

In the Confucian tradition, our interpersonal bonds are a fundamental part of what makes us people; the project of growing as a person is one of developing mutual care and respect in our relationships as parent/child, teacher/student, friend/friend, and so on (Santiago 2008; Wong 2020). Though we are each given unique circumstances based on the family, nation, and other groups we are born into, we are all basically the same by nature at birth, and it is *habituation* that differentiates us (*Analects* 17.2). Confucius's disciple Mencius posits that this nature includes the "seeds" that may eventually grow into the core virtues of *ren* (often translated as "benevolence"), *yi* ("righteousness"), *li* ("propriety"), and *zhi* ("wisdom") (Yu 2013).

An important tool for cultivating virtues in Confucianism is mindful exercise of *social rituals* (Wong 2020). Consider the practices of offering a handshake or bow to a business partner, reciting vows at a Western Christian wedding, holding a funeral when someone dies, and taking off one's shoes before entering an Islamic mosque or a Hindu temple. Rather than intended to be enacted mechanistically, these rituals are meant to help the performer get into a mind-set of respect: for their colleague, for their marital commitment, for the deceased, or for the divine. This idea is central to how Confucius believed a society should be governed; although people might be coerced to behave well through the threat of punishment, a gentler route of teaching them to be virtuous through a system of meaningful rituals would promote flourishing as well as order.

2.3.4 ARISTOTELIAN VIRTUE ETHICS

In European and American philosophy, the prevailing form of virtue ethics can be traced to ancient Greece. Its foundational text is Aristotle's *Nicomachean Ethics*, which was written around roughly 340 BCE, during the so-called Golden Age. The first sustained treatise on ethics produced in classical-era Greece, it remained very influential on Greek and Roman thought over the next several hundred years. Like most other Greek language works, the *Nicomachean Ethics* was not widely known in Latin-speaking Europe during much of the Middle Ages. But it remained popular in the Islamic world in Arabic translation and was "rediscovered" by Christian Europe during the twelfth century along with several of Aristotle's works in other disciplines (such as formal logic, biology, and political theory), thanks in part to its preservation in Arabic. Within a few decades, the *Nicomachean Ethics* became the Western world's definitive account of human nature and how to live well. In the thirteenth and fourteenth centuries, Aristotle was the single most important ancient thinker among Jewish, Christian, and Muslim thinkers, who referred to him in their writings simply as "the Philosopher" and who worked in various ways to harmonize the *Nicomachean Ethics* with their respective beliefs and practices.

Aristotle's goal in the *Nicomachean Ethics* is to ascertain what things a human being needs to do in order to achieve deep happiness and satisfaction with life—to flourish. He concludes that we need a rational understanding of what things make us happy and how those things can be acquired and kept. But he also argues that an understanding of general principles is not enough; we need particular qualities of character both to help us recognize the things that will make us happy and to ensure that we are able to pursue them effectively. These virtues exist in all of us in potential, and cultivating them will help us flourish.

The *Nicomachean Ethics* lists 11 moral virtues that must be developed through *habituation*. Later proponents of Aristotelianism have revised and adapted this list in various ways, either to integrate Aristotelianism with another specific system of thought (such as Islam) or to help it better match the world as they saw it. Some of the virtues that consistently feature across different Aristotelian traditions are courage, generosity, friendliness, temperance, and concern for justice. In order to flourish, one must also have a well-developed sense of *practical wisdom*, or the understanding of how best to act in a given situation. For Aristotle, acting in a manner that appears virtuous is not sufficient evidence that a person is virtuous. Rather, that person must either be acting out of deep inclinations of their character or because they know that those actions are good and have chosen them for that reason (Aristotle, *Nicomachean Ethics* 2.4/1105a16–1105b20).

2.3.5 APPETITES, DESIRES, AND VIRTUOUSNESS

Because virtue ethics is concerned with the whole life of the moral agent, it considers more than our actions and intentions. It also considers our basic physical appetites like hunger and fear, and our emotional or psychological desires. The fact that these appetites and desires are a basic part of human nature means that they are fundamentally good for humans, and necessary for us to flourish. Of course, any of these appetites can be damaging to a person if they are indulged to excess or at the expense of other good things (Schwitzgebel 2007). But suppressing the appetites entirely is just as damaging to a person as allowing them to expand out of control. That is why virtue ethics insists on the importance of learning to regulate appetites and desires and to exercise them in moderation.

But not all ways of regulating desires and appetites are the same, even though they might look the same from the outside. Most of us are familiar with the experience of turning down something that we want—a piece of cake, or the answer key to tomorrow's big exam—because we know that we will be better off without it, in one way or another. This forcible restraining of one's own appetites or desires is called *continence*. A person can cultivate their capacity for continence in the same way that they can cultivate virtues. But continence is not itself a virtue, because it requires us to suppress our desires and appetites.

A different and better way to ensure that you exercise your appetites and desires appropriately is to train yourself into wanting to do what is right. This alignment of desires with right action appears in the Aristotelian tradition as the virtue of *temperance*, that is, having appetites that are attuned with what is right, and in the Confucian tradition by a state of being that can be called *wholeheartedness*, in which one does not feel internal resistance or conflict but is genuinely and joyfully committed to a course of action or way of being. If you have ever been really excited to work on a class project or found yourself saying no to dessert because you feel pleasantly full after a good meal, then you have an idea of what these things refer to.

Continence has much more in common with temperance and wholehearted-ness than it does with various forms of bad living, such as vice (that is, the extreme states of being that virtues avoid) or being ruled by one's appetites. Very broadly speaking, for there is a great deal of disagreement among thinkers and sages, the Confucian tradition raises serious concerns about relying on continence, worrying that it could lead a person to become more concerned with the outward appearance of virtue rather than true virtuousness (Angle 2013). The Aristotelian tradition, broadly speaking, is more positive, celebrating continence as the exercise of the rational part over the nonrational and perceiving it of one's progress toward temperance (Aristotle, *Nicomachean Ethics* 1.13, 1102b26–28; Scarre 2013; Stohr 2003).

But across traditions, there is widespread agreement that true virtue requires more than continence: it requires that a person truly want to do what is right and best. This isn't just because calibrating your desires with your judgment means that you will reliably do the right thing, although that's also true; it's because doing so is what will make you happy. And that, after all, is what flourishing is: living a satisfying and fulfilling life.

2.3.6 HABITUATION: DEVELOPING VIRTUE

Every culture educates its children into its own ideas about what virtues are important and how those virtues are expressed and recognized. This education can come in the form of intentionally crafted lessons, or in simple immersion in the daily workings of that culture. This cultural training, in all its official and unofficial forms, is known as *habituation.*

As a person grows to adulthood, his patterns of thought and action—all of which are influenced by his home culture(s), even if they are a reaction against it—become more entrenched until they seem like the "normal" way of being for him. This normal way of being is called a *habitus*, or interlocking set of habits that are what feels right or natural for him. To have a habitus does not mean that one doesn't understand that there are other ways to live in the world or that those other ways can be ethically valid. It is simply to have habits of one's own.

A large part of a person's habitus comes from doing the things that seem "normal" in their country, in their community, in their family, and among their friends. In general, people are far more aware of those aspects of their habitus that set them apart from their friends, family or neighbors. This is because the similarities are often understood to be "normal," or "just how people are," to the point that people do not notice them at all.

According to virtue ethics, and possibly according to your own experience as well, new habits—whether virtuous or vicious—are hard to cultivate, and old habits are hard to break. But though it is not easy, it is possible. Through steady, deliberate choice on the part of the agent, or steady exposure to a new set of circumstances, an agent can gradually become habituated to a new pattern or set of patterns. In this way, people can (and do) become more or less virtuous over time.

Most people never become perfect in any virtue, let alone in all of them. For this reason, it is better to think of virtue as a spectrum rather than as a binary. You can become more courageous than you were before without being perfect at it, or less generous than you were while still retaining some generosity. Although habitus makes it easy to keep doing what we are doing already, it is always possible to cultivate greater virtue, or to let a virtue lapse. To put it another way, virtues are not a pass/fail system—you're almost always somewhere between perfectly virtuous and perfectly vicious—and your grade is never final.

2.3.7 HOW THE VIRTUES WORK IN TANDEM

Virtues do not exist in isolation from one another or from the rest of our character; they *are* our character, operating in complex combinations in everything that we do (Chen 2015). This is not because virtue ethics considers every single action to be moral in the narrow sense, but because plenty of qualities that are not obviously moral, such as physical prowess or technical knowledge, can also be understood as virtues, even though Aristotle did not consider them to be such (Putman 1997; Stalnaker 2010).

By thinking about each virtue as one element of an interrelated whole, it can become easier to see how virtues that are not obviously moral, like technical skill or physical strength, can also have moral significance (Putman 1997). Imagine, for example, that a friend of yours is organizing a fundraiser for a charitable cause that you also support, and this friend asks you to donate your time to build a website to manage fundraising for the event. In donating your time, you are exercising the recognizably moral virtue of generosity. But you will need more than generosity to build a site that is appealing, easy to find, and easy to navigate: you will also need programming and human-computer interaction skills, as well as knowledge about handling charitable donations, Paypal and credit card accounts, and perhaps the logistics of relevant tax law. These are technical skills that could be used for many purposes: you could get paid for these same skills by your employer (which is reasonable and responsible, but not particularly generous) or you could donate them to build a website to steal credit card numbers. Being good at designing webpages does not make you generous, but it does make your generosity more effective in this situation. It's also the case that both your generosity and your skill in web design will be more valuable if you are practiced at them. However, note that a very skilled and practiced designer who said yes would be exercising less generosity than a novice designer who did—after all, the experienced designer can fulfill her pledge with a lot less time and hassle. Likewise, a person who is unused to being generous may agree to build the website but is far less likely to follow through in putting in the time and energy required to build and maintain a successful and appealing site.

2.3.8 MODALITIES FOR JUDGMENT

Because virtue ethics takes a character-based approach to the world, it cannot and does not aim to offer principles or formulae that can be equally well applied by any person. Instead, it focuses on ways in which a person can draw on the resources of their character to attend to the particularity of specific situations (which, after all, are the only kind of situations we ever experience). Of course, a person of underdeveloped virtues will find these modalities much less useful than a person of good character. Likewise, a person of excellent character is far less likely to need to use these modalities consciously, as they will emerge organically from that person's habitus.

Because our character—including our ability to make use of the following modalities—creates the conditions for how we perceive, interpret and act, a virtue ethicist might say that any given decision is 95% made by the time you realize that you have to make it. Therefore, the best way to engage in good decision making, according to virtue ethics, is to cultivate those capacities that you want to be able to bring to bear in deciding and acting, by building good habits and modeling yourself after exemplary people.

Practical Wisdom

The key to acting well in virtue ethics is *practical wisdom,* or the ability to judge what the best action would be in any given situation. Practical wisdom is conceptualized differently by different virtue ethics traditions, but it is central to all of them. A person with a well-developed sense of practical wisdom will be able to perceive and understand the precise nature of what is going on in a specific set of circumstances and will likewise understand what kind of response that moment requires (Vallor 2016). This understanding and response necessarily combines perceptual acuity with the desire to do what is good; a person who delivers cutting insults at exactly the moment when they will hurt most might be said to be clever or perceptive in ways that partially resemble practical wisdom, but such a person is not practically wise (Clarke 2010; Yu 1998). In keeping with virtue ethics' character-based approach to living well, practical wisdom is itself a virtue, which an individual must develop over time.

Practical wisdom can be developed only through experience. For this reason, children and teenagers cannot yet possess practical wisdom and shouldn't be expected to. If this distinction seems unfair to you, consider the difference in how we might evaluate a well-intentioned teenager who disregards a friend's statement that they hate surprise parties and throws one for the friend anyway, as opposed to a well-intentioned adult who does the same (Hursthouse and Pettigrove 2016). It's not hard to imagine the teenager being genuinely surprised and distressed to learn that their friend's preferences do not match their own, because teenagers have had less exposure to a wide variety of people with different temperaments and preferences. We as observers might have known the party would backfire, but this does not necessarily lead us to doubt the good intentions of the teenaged party planner. By contrast, the party-throwing adult has had many more years of interacting with others, and thus has had more time to develop the perception and judgment that could help them distinguish between another person's genuinely felt dislike of surprise parties and their self-effacing concern that they are causing fuss for their friends. Whether or not we doubt the sincerity of the adult party-planner's good intentions, we are more likely to think of them as at fault for the hurt they cause. Though neither the teenager nor the adult has exercised practical wisdom in deciding to plan the party, it is likely only the adult whose actions will make us think "they should have known better."

Practical wisdom is culturally specific, in multiple ways. Firstly, a person who excels in practical wisdom will always be informed by their own culture and time period's specific ideas of what constitutes flourishing (Vallor 2016). Secondly, a given action or way of being will not elicit the same response across time and place—for example, what seems appropriately friendly in one setting might come across as cold or distant in another—and being attuned to that local variation is part of what practical wisdom entails. Therefore, it must by definition be calibrated to the specifics of its setting.

Practical wisdom guides the exercise of other virtues, but it cannot replace them. Imagine, for example, that you and a friend are having an argument and that you both have gotten very angry. Then your friend says something that makes you understand the whole situation differently. If you have a well-developed sense of practical wisdom, you might be able to perceive how your friend's comment offers an opportunity to de-escalate the argument and make peace. But unless you are also disposed to let go of your anger—or at least capable of controlling it—you might not be able to make good use of that opportunity, even though your practical wisdom tells you that it would be a good idea.

Finding the Mean

As you have probably already noticed, virtue ethics is organized around the idea of moderation. Acting virtuously requires finding the middle path between excess and deficiency (*Nicomachean Ethics*, Book II.7 p. 25; *Analects* 15:3). It also requires that we pay attention to the specifics of a situation, because different situations require different responses: the right amount of courage (or cheerfulness, or honesty) for one occasion might be excessive for a different one (Lunyu, cited in Xia 2020). This point of perfect balance between extremes is known in the Confucian tradition as "the doctrine of the mean" (*Zhongyong*, cited in Plaks 1999) and in the Aristotelian tradition as "the golden mean."

To understand how a virtue could be exercised excessively, imagine a soldier who is considering a risky solo strike attack to provide cover for the other soldiers in his unit. Carrying out such an attack would be brave, regardless of circumstances. But in some situations, it might be *too* brave—in other words, reckless. A hot-blooded soldier who is immune to fear can make things worse by rushing in even when there is no advantage to be gained or when she is not sufficiently skilled to succeed. The most virtuous soldier, therefore—the one who possesses not only courage and skill, but also sufficient practical wisdom to pay attention to the mean—is the one who is unafraid to fight when it will be helpful, but not so eager that he fails to think about tactics.

Aiming for the mean can give you a concrete method for imagining what excessive or deficient responses would look like and enable you to come up with a better course of action in order to avoid those extremes. But realistically, achieving the mean also requires a well-developed faculty of practical wisdom. And doing the right thing also requires you to have developed all the relevant virtues to the degree that the situation requires, so that you are able to go beyond just having the right idea about what a good response would be.

Sidebar: Aristotelian Virtues as a Mean between Vices

In the *Nicomachean Ethics*, Aristotle describes the moral virtues as existing at an intermediate point between two extremes of vice. (Aristotle 2.8) The following table lays out some of the moral virtues described in the *Nicomachean Ethics*, along with the vices that represent an excessive or deficient exercise of that same quality. Presenting them grouped together in this way should make clearer what "finding the mean" looks like in practice.

This *virtue* is the midpoint	between this *excess*	and this *deficiency*
courage	recklessness	cowardice
friendliness	flattery and fawning	crankiness
temperance	overindulgence	lack of appreciation for pleasures
generosity	careless overspending	stinginess

It's worth noting that the precisely suitable degree of exercising a given virtue (known in the Aristotelian tradition as the *golden mean*) will not be the mathematical center point between the two extremes and will vary from situation to situation. For this reason, one's exercise of virtue needs to be guided by *practical wisdom*.

2.3.9 STRENGTHS AND WEAKNESSES OF VIRTUE ETHICS: FLOURISHING IS EASY (ONCE YOU'RE THERE)

A person is said to be truly virtuous only if they find it easy to be virtuous: they are wholehearted in their desire to do what is right, and they do not require continence or external pressure to continue in virtuous habits.

Connecting virtue to flourishing in this way—to a person's felt desire to act virtuously—may seem counterintuitive at first glance. Isn't it more courageous to be brave when an action one must take is extremely dangerous (e.g., rescuing a child by running into a burning building), or more impressively generous for a very poor person to donate money to a disaster relief organization after a major flood? These are indeed two examples of exceptional virtue, in which the agents are being virtuous under conditions of extreme external pressure. Both of them are exercising their respective virtues (courage and charity) in spite of the risks it poses to them, in a way that suggests that they truly excel at those respective virtues (Foot 1978).

But if a person finds it difficult to be virtuous because it does not match their internal desires and values—in other words, because they don't entirely want to be—then that difficulty reflects the limits of their wholehearted commitment to that virtue (Foot 1978; Tiwald 2018). In comparison to the two people in the paragraph above, imagine a person who speaks up in a meeting despite being nervous, or who donates money to disaster relief even though she dislikes giving money away. In both cases, these actions are difficult because of insufficient virtue (courage or generosity) on the part of the actor. That is not to say that these actions aren't good! In addition to the contributions they are making, both of these people are controlling their less virtuous impulses, and possibly building new habits. Both of these people might be on that path to greater virtuousness if these actions represent part of an overall shift in habitus. The timid speaker may find it easier to talk next time, and the resentful donor may donate again because it made her feel good the first time. But it's impossible to know whether either of those things will happen on the basis of only those first, single actions. The actions a person undertakes to shift their habitus are best understood in the long term: in six months, or six years, each of these people may have become more virtuous than they were before, by cultivating new habits (or shaking off old ones).

Being truly virtuous also requires maintaining those virtues across interactions with different people; for instance, a truly generous person finds it easy to help strangers as well as friends and family. But virtue ethics does not necessarily emphasize universality in the same way as some other ethical frameworks. It comes naturally to many people to feel the most concern for the people closest to them. In fact, in the Confucian tradition, familial love is the root of all other love and it is reasonable to give greater moral priority to these relationships (with *filial piety* being one of the most important virtues); to become a generally moral person is to take your circle of care for those closest to you and expand it outwards to encompass others (Curzer 2018; Yu 2007; Zhu 2002).

While virtue ethics' attunement to local norms is one of its strengths, this same quality is often criticized as a weakness. Because the virtues celebrated by one community are sometimes very different from the virtues celebrated by another, virtue ethics does not furnish a good foundation for universal principles, including a universal principle of justice. Virtue ethics has also been criticized for conceiving of the good life in a way that makes it difficult, if not impossible, for people with some disabilities or neurodivergencies to meet its criteria for human excellence. In particular, Furey (2017) has argued that virtue ethics is a problematic fit for engineering ethics in particular, because Aristotle's picture of flourishing presupposes a model of the human mind with strong capacities for emotionally engaged and intuitionistic decision making, and those capacities don't come as naturally to many people on the autism spectrum, a population particularly heavily represented in engineering.

Sidebar: Understanding Virtue Ethics through Role-Playing Games

One potentially useful way to think about the cultivation of virtues is role-playing games (RPGs), in which it is necessary to build your character's skills slowly over time in accordance with your goals for the character. Although skill building in RPGs is not always presented as having a specifically moral dimension, the underlying mechanics can be helpful for thinking about virtue ethics.

When you start out playing a game like *Baldur's Gate*, *Fallout*, or *Mass Effect*, you have an array of skills, but you're not very good at any of them yet. If you want to get better at climbing or communication or spell-casting, then you have to practice. As you practice, your ability level improves. This practice is not so different from habituation, the building-up or cultivation of a certain virtue through practice. It's also not so different from practicing a sport or a musical instrument.

Moreover, the best way to solve any given problem is not going to be the same for every player. Some challenges are simply impossible for lower-level players who have not yet built up their skills to the necessary point. Other challenges will have different solutions. How do you get past a locked door? Your answer will depend on whether your character is a magician or a thief.

In an RPG, the mage and the mercenary will be able to address a given challenge differently because of their different skill sets, but the way the player perceives the problem is likely to be consistent. In the real world, the differences between the mercenary and the mage will go deeper: what seems possible or reasonable to the mercenary (defeat a guard in combat, break through a door) will seem imprudent or impossible to the mage, whereas the mage's solution (unlock the door with magic) will seem fanciful or impossible to the fighter. Each character's habits of thinking will be shaped by their past habits, and by the virtues or skills that have in their experience been relevant to making important decisions.

Of course, in real life you don't usually have the choice to put a challenging situation on hold and come back when you have leveled up: you have to face it when it confronts you. That's why it is important not to put off cultivating your capacity for virtue until later. You might have to fight the dragon tomorrow.

Story Point: "The Gambler," by Paolo Bacigalupi

I try to protest. "But you hired me to write the important stories. The stories about politics and the government, to continue the traditions of the old newspapers. I remember what you said when you hired me."

"Yeah, well." She looks away. "I was thinking more about a good scandal."

"The checkerspot is a scandal. That butterfly is now gone."

She sighs. "No, it's not a scandal. It's just a depressing story. No one reads a depressing story, at least, not more than once. And no one subscribes to a depressing byline feed."

How can we be the version of ourselves that we want to be, even in times of crisis or under pressure from the outside world? And how can we influence the world around us for the better when the world doesn't necessarily want to be influenced in those ways? These universal questions take on particular sharpness for Ong, a Laotian refugee who works as a journalist for a major American media company. Ong writes articles about the issues he considers most important, such as climate change and the fallout of government mismanagement. But such articles don't align very well with the reading habits of the American public, who instead give their clicks to product reviews and celebrity scandals. "The Gambler" offers us several high-stakes moments, but among the various characters in the story—Marty Mackley, the master of infotainment; Janice, Ong's results-oriented boss; Kulaap, who balances her Laotian and American identities in a way Ong cannot quite understand; and of course Ong himself—there is no consensus on what kind of response those high-stakes moments call for. Indeed, there is not even consensus on what really counts as a high-stakes moment in the first place: each character has different judgments about the lines between "real news," distractions, and depressing stories. But as "The Gambler" illustrates, our choices in those moments are rooted in the longer arc of our character.

2.4 COMMUNITARIANISM

Communitarianism is an approach to ethics organized around self-realization in the context of *interdependence*. Its basic assumption is that human beings exist in a state of mutual reliance on one another (Masolo 2010; Smith-Morris 2020). Although this interdependence includes material goods like shelter and safety, the more significant dimension of our interdependence concerns the many social, spiritual, and psychological goods that can only come from relationships with others. Because we are interdependent, everybody benefits from investing in the *common good*, or the well-being of the community as a whole (Masolo 2010; Wiredu 1992a).

2.4.1 OVERVIEW OF COMMUNITARIANISM

Instead of conceiving of communities as collections of individuals who have decided to make common cause, communitarianism posits that communities exist prior to any individual born and raised within them (Menkiti 1984). By furnishing the language, values and relationships that form the basis of an individual's experience, a community is necessarily foundational to understanding the character, goals, and potential of any individual within it. Furthermore, because of our interdependence, that individual's *self-realization*—that is, the fulfillment of the individual's potential and aims for themself—can take place only in a community setting, because relationships with others are essential to the structure and development of the self and because they create the context that makes an individual's aims significant (Gyekye 1997, Masolo 2010).

The idea of community that underwrites communitarianism does not include every single kind of community that exists. Rather, it refers to the kind of community in which a person might be said to have grown up. It requires not only shared interests and values among members but a sustained mutual commitment and a developed sense of what it means to live in common (Coetzee 2003; Gyekye 1992). There are many communities that have some of these qualities, but not all of them. You will likely find that the communitarian framework is helpful in reflecting on how these communities work, but that focusing on one such community to the exclusion of others will give you only a piecemeal picture of the lives of its members.

While communitarian ethics does not presume any one community structure or set of norms, it does presume that communities exist and that they have structure and norms of some kind that they recognize and claim as their own. It further posits that any person can and must be understood as being formed by a community (or perhaps more than one). Being formed in this way means that an individual shares in the social meanings of their communities, and that their understanding of themselves and of the value of their goals is grounded in those shared social meanings (Masolo 2010; Smith-Morris 2020). This is not to deny that there are people who conduct their lives without strong social ties or a sense of shared commitment to others. But according to the baseline premises of communitarianism, any person who lives in isolation from a community of shared values and ideals—"an abstract dangling personality," to use Polycarp Ikuenobe's striking phrase—will not be able to achieve self-realization (Ikuenobe 2006).

Although it does not presume any specific moral or social norms, communitarianism does make some general assumptions about how communities work. The first is that they are enduring and have patterns, processes, and structures that exist prior to the individuals who are formed by them (Menkiti 1984; Smith-Morris 2020). The second is that their shared wisdom is carried to a large extent by elders, who are not merely the older members of the community but are those who have excelled in the forms of living that the community values, particularly in *sympathetic awareness* of the needs of others (Masolo 2010; Menkiti 1984; Smith-Morris 2020). Both of these assumptions are more broadly and transparently applicable to people who spend most of their lives in one place, a pattern of life that is increasingly uncommon. Nonetheless, when you use the lens of communitarianism to think about your own life, you will likely find that you belong to some communities that match this description, even if your life does not take place entirely within them.

In order to grasp why the communitarian framework presumes interdependence, it's helpful to understand the role that community plays in a person's self-realization. Even though you as an individual can decide what kind of person you want to be and what achievements are important to you, you (like everyone else) rely on others in order to become that person and achieve those goals, in three distinct but interrelated ways.

The first way is that others in your community furnish you with many kinds of support in your work and development, from material goods (like food and public roads) to emotional and psychological goods (like love and conversation). The second way is that achieving your goals for yourself, whether in practical accomplishments or in character, rarely feels satisfying or even real unless somebody else recognizes and affirms what you have achieved or become (Gbadegesin 1991; Presbey 2002). Shared social rituals like weddings, coming-of-age ceremonies, and graduations are easily visible examples of how a community marks an individual's achievement or growth. But even small ordinary gestures of recognition like a professional title are often important for helping someone feel that their achievements are real and valuable.

The third way individuals depend on their community for self-realization is rooted in the fact that communities exist prior to individuals and supply the shared social meanings that individuals use to think about their lives (Smith-Morris 2020). Imagine, for example, that you have grown up in a community where most people want to be professional athletes, but you are more interested in becoming a doctor. Although you may think that this goal sets you apart from your community, the specific shape of your goal is still shaped by your community. Every society has medical practitioners of some kind, and different communities have very different ideas about the social role that medical practitioners play. These varying expectations occur not only because the nature and rigor of medical training varies widely but because different communities have different degrees of trust in what the doctors they know claim to do (or whether they will do what they claim). Even if your desire to be a doctor is unusual in your community, your specific ideas about being a doctor will reflect your own community's understanding of what doctors are like and what makes that a valuable goal.

For these reasons, your self-realization needs to be understood in the context of your community. Your community also shares in your attainments, whether it is achieving a concrete goal or becoming an excellent person (Ikuenobe 2006). Not only has the community helped to support you along the way, but the fact that you aimed for and achieved that goal is understood as a sign that the community has successfully raised and educated you to carry on the community's values and practices (Menkiti 1984). This does not mean that every single community you have belonged to plays an equal role in your development and your successes, but it does mean that these things can be traced back to a community that has been important to you.

2.4.2 SOURCES OF COMMUNITARIAN ETHICS

The form of communitarianism discussed here is rooted in indigenous philosophy, traditions and lived community practices. This type of communitarianism is challenging to represent here, for several reasons. It's important to discuss those reasons here, because they may already be influencing the way you are thinking about this framework.

The first difficulty has to do with the challenge of retrieving sources of indigenous thought. Communitarianism was a common form of life in many parts of the world prior to the era of European colonization which began in the fifteenth century CE, including the lands that now comprise Australia, sub-Saharan Africa, much of North America, and parts of Southeast Asia. The societies in these regions were transformed, often violently, when their homelands were "discovered" and settled by European explorers, who largely treated these societies' lands (and sometimes the people themselves) as a source of wealth for their home countries. During these violent upheavals, large swaths of these indigenous communities' knowledge and tradition were eradicated. Some strands of knowledge and tradition were preserved in the minds and memories of survivors but in decontextualized forms that were different from those of prior generations. Our contemporary knowledge of those past ways of life is indelibly shaped, and limited, by that destructive history.

In addition to this very practical challenge, there is also a conceptual challenge. European nations justified their colonial project by claiming that the peoples they conquered were lesser forms of humans—or not even human at all—because they were not "rational" in the same way (Ani 2013; Biakolo 2003). The idea that humanness depends on rationality can be traced to ancient Greece and, indeed, was used by Aristotle and other Greek philosophers to denigrate some non-Greek peoples (Biakolo 2003). But the specific form of rationality that mattered to the Europeans of the fifteenth and sixteenth centuries was the newly developed principles of "scientific reasoning" (Naudé 2019). Because the worldviews and thought patterns of the indigenous societies that they encountered did not mesh with their own recently developed ideas about universal reason, the colonizers (along with observers in their home countries) concluded that these communities were incapable of higher-level thought, or at least had not developed it yet (Biakolo 2003; Nicholas 2018). Although it is easy to look back and see the error and hubris of that judgment, its legacy continues to shape how non-European peoples and their knowledge traditions are perceived today—including in the field of ethics, which (like many academic disciplines) is only just beginning to reckon with how some of its standards and practices continue to make it inhospitable to non-European traditions of knowledge (Ani 2013; Hallen and Sodipo 1997; Outlaw 1987; Oyěwùmí 1997; Sereequeberhan 2003; Wiredu 1992b; see also Wiredu 2009).

Many of the indigenous communitarian societies mentioned above continue into the present, adapting the core practices of communitarian ethics to the institutional structures of the settler states that now control their homelands (Smith-Morris 2020). For both practical reasons and because of ingrained philosophical biases, these communities' frameworks and traditions are rarely given the same philosophical credence as writings from the European traditions. Yet indigenous communitarianism significantly enriches our conceptual vocabulary for doing ethics.

Although these various indigenous communitarian traditions have many points of overlap among them, limitations of space make it impossible to discuss commonalities in a way that can truthfully represent them all. Therefore, the remainder of this unit focuses specifically on the communitarianism of sub-Saharan Africa, which is internally diverse but also coheres around a handful of core ideas and themes. Among these shared ideas is a particularly well-developed conceptual apparatus for communally based self-realization.

The communitarian framework presented here cannot be, and does not aim to be, a direct representation of what precolonial Africa was "really like." Instead, it offers an overview of the philosophical efforts by African scholars to theorize, systematize and reflect on those aspects of noncolonized African culture that endure, or those elements of precolonial Africa that can be recovered.

It should be noted that communitarian philosophy and practices have also emerged in the culture and thought of Europe and America (Masolo 2010). Political philosophers Michael Sandel and Michael Walzer and philosopher/theologian Alisdair MacIntyre have each argued for the merit and moral urgency of a turn to communitarianism, responding to and drawing from the Euro-American tradition (MacIntyre 2013; Sandel 2010; Walzer 1990). As these thinkers all note, communitarianism is deeply at odds with the structure of the modern western capitalist state. Nonetheless, several communitarian movements have developed. One of the most widespread of these is the Catholic Worker Movement, which began in the United States in the early 1930s as a means to bring about a "society in which it will be easier to be good" (Cornell 2006). Initially just a newspaper, this movement soon expanded to include both independent houses and cooperative farms. The Catholic Worker Movement now has over 100 houses and farms in several countries that feed the needy and house communal workers and continues to stand against exploitation and inequality (Forest 2010). Another well-known example is the kibbutz movement. A kibbutz is a form of communal living first established in 1910 at Dgania in Palestine. The founding ideals of the kibbutz movement combined an agrarian communitarianism with the Zionist goal of claiming land for the Jewish people. Kibbutzim (to use the Hebrew plural) were founded on a purely egalitarian regime of total equality of the members in both work and claim on resources. This communalism extended even to child-rearing; all children on the kibbutz were raised communally in a separate "children's house" (Shpancer 2011). Although kibbutzim remain a small share of Israel's culture and economy, many have shifted their practices and community structure to more closely resemble Israel's wider capitalist society. Although most remain collectivist agrarian communities, some have become for-profit enterprises (Sivak 2020). Many, but not all, modern kibbutzim have differentiated between the management of the economy on one hand and the community on the other, leading to less egalitarian wages but a commitment for caring for all members of the community (Rubinstein 2007).

2.4.3 PERSON, COMMUNITY, AND WORLD: SUB-SAHARAN METAPHYSICS

The concepts of community and of human self-realization are easier to understand with a basic understanding of how traditional sub-Saharan cultures conceptualize the composition of the world at large.

According to most traditional sub-Saharan metaphysics, the entire world is an interdependent and harmonious system. The boundaries between different kinds of living things, between mind and body, between the living and the dead, and between the natural and spirit world are all gradations of difference rather than absolute divides, and treating them as categorically separate—as Western metaphysics does—prevents one from understanding them as they actually are (Ramose 2003; Tangwa 2005; Wiredu 1992b).

This sense of the world as a harmonious unity informs nearly every aspect of how sub-Saharan communitarianism is structured. A striking example can be seen in the way the death of a community elder is understood. Although the deceased elder is acknowledged to be biologically dead, they are understood as still present in the community, insofar as their words and character are remembered (Ikuenobe 2006; Masolo 2010; Okolo 1992; Ramose 2003). While their presence is felt in this way, the elder is still a person within their community. Only when their name is no longer remembered by anyone living does the elder join the category of ancestors, a de-individualized collective of wise and beloved persons who have gone beyond material existence (Menkiti 1984). Furthermore, just as the world at large is understood to be an ordered and harmonious whole, most African communitarianism likewise understand human individuals to be, themselves, each a community of many self-elements (Gbadegesin 1991; Neequaye 2020; Ogbonnaya 1994).

This principle of organization helps underscore why self-realization must take place in the context of a community. Because everything that exists is by its nature part of an interconnected and interdependent whole, realizing your purpose in the world must necessarily include acknowledging that you are part of that system of interconnection.

Sidebar: The Role of Religion in Sub-Saharan Communitarian Thought

It is easy—but also wrong—to assume that the metaphysical outlook described above is intrinsically "religious" in a way that is not true of the prevailing metaphysical views of the contemporary West (Oladipo 2003). The relationship between religion and metaphysics is very complicated, and it is entirely possible (and very common) for a person to hold metaphysical beliefs that are rooted in a religion they don't practice or even know very much about, because those beliefs have shaped their community and culture in ways that are not obviously religious. Therefore, although communitarianism is not intrinsically more "religious" than the other frameworks in this book, it is important to understand that there are some major differences between the traditional religions of Africa and the religions that have been foundational to contemporary Europe and America. For this reason, we are giving a brief background on sub-Saharan-African traditions—which, like the Abrahamic traditions, have vital and significant differences but nonetheless share some important assumptions about the world.

In discussing the three classically Western canonical frameworks that we have looked at, we noted that each is compatible with a religious or theistic worldview but can also be adhered to without any belief in God or any religious commitments. Whereas this is true in general, it is worth noting that discussions of "religion" in the context of Euro-American ethics almost always mean the Abrahamic traditions: Judaism, Christianity, and Islam. Although these three religions are not interchangeable, they share many common beliefs about both human nature and divine nature. Those shared beliefs serve as a metaphysical baseline for most people raised in communities where Abrahamic religions are the norm—even people who reject the religions themselves or do not believe in God. Indeed, they are so widespread that it can be hard to notice. But these religions differ sharply from the indigenous religions of sub-Saharan Africa.

The indigenous traditions of sub-Saharan Africa begin with very different metaphysical assumptions. One major difference concerns the role of the divine in human knowledge: how much humans know about the divine or spiritual world, and how they know it. Unlike the religions of sub-Saharan Africa, the Abrahamic religions are all *revealed religions*: that is, they are based on teachings and insights that were *directly revealed* to human beings by God. Though most Jews, Christians and Muslims cannot claim to have received a direct revelation themselves, they nonetheless practice a religion based on teachings that are understood to have been directly revealed to others. The religions of sub-Saharan Africa, by contrast, are not revealed religions. Though divinities and spirits exist, they do not communicate with humans in the same way. Because of this, they are not a direct source of rules or teachings. Rather, rules and teachings are understood to come from the community's own wisdom and knowledge of the world, which includes spirits (Gyekye 1987).

The second major difference, following from the first, concerns the relationship between the spiritual realm and morality. One of the philosophical problems with revealed religions—especially when God is understood to be an enforcer—is that people can easily end up doing the morally right thing to avoid punishment, rather than because it is morally right. Therefore, in European contexts, many thinkers argue that genuine humanism, a commitment to the value and significance of humans and human life, is not compatible with religious belief. This tension does not arise in sub-Saharan African traditions, because the gods are not keepers of morality. Furthermore, the spirit world that people enter after they die is not entirely distinct from the community of the biologically living: ancestors remain a part of the human community, not as gods but as cherished and respected people (Ikuenobe 2006; Wiredu 1992a).

2.4.4 SELF-REALIZATION IN SUB-SAHARAN COMMUNITARIANISM

Although there are variations across traditions and communities, there are significant commonalities in the idea of self-realization as it is found across sub-Saharan Africa (Tangwa 2000; Wiredu 2009). The language used to describe the achievement of this self-realization is nearly always rendered in English as "becoming a person." This conception of personhood does have a descriptive element that captures the basic biological and metaphysical features that are intrinsic to all persons; only someone with these features is capable of attaining personhood in the normative sense (Ikuenobe 2006). But sub-Saharan communitarianism is focused mainly on the normative aspect of personhood, the realizing of a person's potential for excellence and humanity as a participant in their community (Gbadegesin 1991; Gyekye 1987; Ikuenobe 2006, Menkiti 2005; Wiredu 2009). Achieving and maintaining self-realization is necessarily a long-term project, because it involves absorbing values and taking on and upholding adult responsibilities (Ikuenobe 2006; Menkiti 1984).

Self-realization is not, cannot be, something that an individual accomplishes by herself. It is part of a community effort, and it is validated by recognition from the community. The values are those of the community, as is the responsibility of imparting them, although this responsibility is shared with the individual in taking them on and in living accordingly (Kaphagawani 2005; Menkiti 1984; Presbey 2002; Wiredu 2003). The achievement is likewise shared with the community, insofar as the successful development of a person is understood to be a communal project that involves education, nurture, and role models (Hallen 2005; Ikuenobe 2006; Masolo 2010).

The achievement of self-realization is often celebrated, and frequently brought about, by a ritual of training or process of being inducted into community knowledge (Masolo 2010). Furthermore, complete self-realization involves claiming specific social roles, such as parent or elder, which impart specific social responsibilities to other individuals and to the community as a whole. Like the language of being a person, these social roles are not biologically descriptive: you are not a mother or a father just because you contributed biological material to a child, but because you love and provide for that child (Masolo 2010).

Sidebar: Agent-Centered vs. Patient-Centered Personhood

As you will see in chapter 4, using the concept of personhood to refer to a distinctively human way of being valuable is not unique to sub-Saharan Africa. Indeed, ethical description and reflection of any kind almost always involves some working notion of personhood, even if that notion is not consciously held. But for this very reason, it is essential to distinguish the sub-Saharan conception of personhood from the conception that most frequently appears in English-language work in ethics (Behrens 2011). As noted above, the sub-Saharan concept of personhood is focused on how someone can and should conduct themself; in other words, it is *agent-centered*. Living in this way means honoring obligations to others, but the focus is on the individual upholding the duties rather than the entities to whom they are owed (Molefe 2020; Tangwa 2000). By contrast, a *patient-centered* approach to ethics means that the focus is on the rights and protections of individual persons. In the United States, particularly in the context of bioethics and medical ethics, the emphasis is often on the patient as a holder of rights. The logical extension of patient-centered reasoning is that moral agents should honor the moral patient's rights, but the focus remains on the patient rather than the agent.

2.4.5 UBUNTU

Ubuntu refers to a crucial human quality that is central to communitarianism of southern and eastern Africa. (Gade 2011) The word "ubuntu" comes from a family of Bantu languages spoken in southern Africa and has close equivalents in several other languages across the southern half of the continent (Idoniboye-Obu and Whetho 2013). The word first appeared in writing in the mid-nineteenth century, but the principle has been transmitted through oral traditions for centuries, in the form of proverbs, myths, riddles and other narratives (Gade 2011; Kamwangamalu 1999). In the past several decades, ubuntu has been taken up by thinkers and politicians in South Africa and Zimbabwe, in an attempt to articulate a distinctly African political philosophy of humane justice. In this same timeframe, it has also become popular among theologians and philosophers from English-speaking countries who aspire to a more global approach to ethics. On account of these two movements, the ubuntu tradition has gained broad name recognition, but it has become increasingly ill defined the more widely it has circulated (Gade 2011; Idoniboye-Obu and Whetho 2013).

Even when one focuses on indigenous sources, ubuntu is notoriously difficult to define because it is understood to encompass so many interrelated aspects and ways of being (Idoniboye-Obu and Whetho 2013; Kamwangamalu 1999). In essence, to live with ubuntu means to engage in the process of becoming and being human: to attend to and embrace the interdependence of human existence, in the broader context of an interdependent and harmonious universe. (Mkhize 2008) Ubuntu has been described not only as a quality of character but as an ethic, an orientation, a standard of judgment, and a mode of being (Gade 2011; Idoniboye-Obu and Whetho 2013; Mkhize 2008; Munkaya and Motlhabi 2009). And indeed, the claims that ubuntu makes on a person are so comprehensive that all of these descriptions are potentially useful.

Because ubuntu is a human quality rather than an abstract principle, it is not enough to believe in the values of ubuntu. Rather, it requires continuous engagement with the ever-changing world around us, and particularly the active, ongoing affirmation of the humanity of those around us (Chimuka 2001). This affirmation extends not only to one's immediate community, but also to strangers and outsiders (Munkaya and Motlhabi 2009). Because a person living in accordance with ubuntu "cannot look on the suffering of another and remain unaffected" (Mkhize 2008) in a thriving ubuntu community, all challenges are met both communally and cooperatively (Munkaya and Motlhab 2009).

There are differences within different branches of the ubuntu tradition about whether or not humans have distinct metaphysical status or dignity that set them apart from other creatures (Chimuka 2001; Mkhize 2008), but the fact of being human means that has one particular obligations to other members of the human community.

2.4.6 YORÙBÁ COMMUNITARIANISM

Yorùbá communitarianism is a tradition based in the region of western Africa that is now Nigeria, along with neighboring parts of Togo and Benin. Although the societies in this region operated under a variety of political structures, nearly all of them were grounded in communitarianism.

Yorùbá communitarianism is noteworthy because of its distinctive and well-documented epistemology (that is, its conception of knowledge.) Instead of using vision as the metaphorical basis for understanding, Yorùbá communitarianism instead takes hearing to be the foundation of how understanding works (Oyěwùmí 1997). This approach to knowledge emphasizes the relational nature of knowledge; whereas the underlying model of learning-as-seeing invites us to imagine the thing we are learning as wholly passive under our gaze, learning-as-hearing gives the thing learned about an active role in the perceiver's learning and emphasizes the interaction and exchange between them. This epistemological approach also underscores the relational, rather than biological, nature of social roles and relationships, because an individual's physical appearance is not taken to be a significant indicator of who they are (Oyěwùmí 1997).

As a consequence of its conceptual rootedness in hearing, Yorùbá communitarianism draws a sharp distinction between firsthand knowledge, which an individual has seen and understood for herself, and all other kinds. The language used to describe second-hand knowledge—whether it is learned from a book, from tradition, or from a friend's story over lunch—is best translated as "agreeing to what one hears" (Hallen and Sodipo 1997).

While this conceptualization of knowledge might at first sound radically individualist, it in fact underscores the extent to which community members depend on one another to know and to understand. Furthermore, it calls attention to the importance of relationships in shaping how information and knowledge is exchanged: an individual who is known to be conscientious and trustworthy will have a greater voice in shaping the community's shared understanding of a situation. For these reasons, being an observant witness to one's own experiences, being reliably thoughtful and measured in understanding those experiences, and being careful and precise in communicating them to others are all important elements of a good character in Yorùbá communitarianism (Hallen 2005).

2.4.7 IS THE COMMUNITY AN INTRINSIC GOOD OR AN INSTRUMENTAL ONE?

There is no question that community is essential to achieving personhood in sub-Saharan ethics. Simply put, there is no way to achieve self-realization apart from a community. What is sometimes less clear, however, is the exact role that communities play in securing an individual's self-realization, philosophically speaking. Is the community an end in itself, the necessary basis of self-realization? Or is it a means to an end, a necessary path rather than a destination? (Metz 2011).

Earlier studies of sub-Saharan communitarianism—perhaps aimed at drawing the sharpest possible contrast with western liberal individualism (Oyowe 2013)—all agreed that community was the center of sub-Saharan value systems and that the value of the individual person is fundamentally constituted by the community and conferred upon them by community recognition (Menkiti 1984). In recent years, however, many scholars have argued that, while individuals need community in order to achieve well-being and self-realization, the true goal of sub-Saharan communitarianism is to create good for all of the individuals in it.

2.4.8 MODALITIES FOR JUDGMENT

Sub-Saharan communitarianism is not unique in understanding ethics to be about how to live well with others. But the communitarian framework is distinctive in the way that it positions us to understand individuals and their development in terms of their communities. This distinctive community focus is reflected in its modalities for judgment, which can be formulated as a series of principles that are relational and community-oriented, grounded in the authority of shared humanity and in the reality of our social nature. This principle-based form of ethical reasoning bears some similarity to deontology (recall in particular that deontology centers on duties to others.) But the community-oriented, relational aspect of communitarianism contrasts with an orientation, for many deontological theories, toward an autonomous rational subject.

The Consensus Principle

The principle of consensus urges us to find common ground with others who have a stake in a given issue. This principle is rooted in both the knowledge theory and the social organization of communitarianism. As we saw above in the Yorùbá tradition specifically, communitarianism considers the tasks of knowing and understanding to be cooperative by necessity. Similarly, Edward Wamala writes of one of the Bantu kingdoms in Uganda, "Nobody has a monopoly on knowledge; everybody is in need of the knowledge and opinions of others" (Wamala 2005, 438). Because communitarianism presumes a strong model of collective participation, it becomes possible to overcome the limits of our individual knowledge by working together to understand and decide.

It should be noted that this principle does not assume uniformity of opinion. In fact, it assumes the opposite—if everyone agreed, there would be no need for further discussion. Furthermore, the principle of consensus does not require a community member to surrender to the will of the group. Rather, it demands that each community member participate in the group's deliberations.

Although this principle might at first strike you as irrelevant to at least some kinds of ethical quandaries, further reflection from within the communitarian framework will likely reveal that many problems that seem at first to be yours alone can, in fact, be understood as shared with others in your life—even if it's only because of their relationship with you.

The Principle of Building Community

This principle directs us to nurture a sense of common cause and belonging with others in our community. Because people exist in a state of mutual dependence, we all benefit when those necessary ties are strengthened by feelings of solidarity, trust, and belonging. In other words, it is easier to be happy about helping others and to set aside your personal desires in order to help another if you feel a sense of shared identity or community with those others and if you trust that they will help you.

Honoring this principle can be as simple as making time for conversation with others. It can also require more significant acts of mutual aid. It can even be helpful to create new structures of mutual dependence, such as a study group, to help strengthen your own and others' commitment to helping and being helped by others (Masolo 2010).

The Principle of Peace/Rehabilitation

According to this principle, an individual should always try, within reasonable measure, to repair relationships with others (Mangena 2015). This principle grows from the understanding that self-realization requires that one have good relationships whenever possible. Maintaining good relationships with others is, therefore, partly an act of self-interest—but characteristically so for communitarianism, because pursuing one's own interest aligns with the good of the whole.

A further dimension of this principle is the importance of *restorative justice*, a form of justice that centers on making restitution to victims rather than punishing wrongdoers. This principle depends on the cooperation of everyone involved, including the willing participation of those harmed, and often involves other community members as advisors and mediators (Mangena 2015). Honoring this principle means making an effort to rehabilitate wrongdoers, whenever it can be done without being unjust to those who have been wronged. This principle leaves open the possibility of cutting ties with a relentless wrongdoer; good relationships, after all, must be mutual. But insofar as it is possible for an individual to better themself and the other individual by remaining connected, one should seek to do so (Molefe 2020; Wiredu 2009).

2.4.9 STRENGTHS AND WEAKNESSES OF COMMUNITARIANISM: TENSIONS WITH THE LIBERAL TRADITION

Communitarianism is sometimes perceived as radically opposed to the ideals of individualism that shape Western capitalist nations, requiring the individual both to sacrifice their own well-being for the sake of the group or to surrender their power of judgment to the group. Both of these misunderstandings are rooted in the notion that a community is monolithic and unchanging rather than a shared baseline of relationships and understandings between different individuals (Gyekye 1987). They also significantly exaggerate the differences between communitarianism and its alternatives.

Although it is true that communitarianism places more community-oriented demands on individuals than is necessarily the case in the other frameworks in this chapter (though each of them includes more communally oriented traditions), these demands are, broadly speaking, similar to the constraints those frameworks place on a moral agent. Each of those frameworks offers resources to help the moral actor balance the well-being of others against her own well-being. Communitarianism is distinctive in that it understands those constraints in terms of relational commitments to others and shared values, and emphasizes the ways in which an individual can pursue her own good through the common good. But it is not unique in holding the moral agent responsible for the well-being of others (Masolo 2010; Molefe 2020).

Although communitarianism posits the existence of a common good that is of meaningful benefit to everyone in the community, critics have argued that it makes the individual "a mere means to society's welfare" (Tshivhase 2011). In particular, communitarianism's focus on self-realization is sometimes criticized as stifling or coercive to the individual. Because self-realization is calibrated to community norms and requires an individual to succeed according to collectively recognized categories of achievement, it can create conformity or punish those who are unwilling or unable to acquiesce to social expectations (Presbey 2002). This description of communitarianism has been disputed on the grounds that it ignores the individualism and competitive spirit that can develop in the context of mutual belonging and shared norms (Wariboko 2020).

Communitarianism has also been criticized for preserving unjust power imbalances. Because of elders' high social standing in communitarian societies even after their biological deaths, sub-Saharan communitarianism is sometimes said to promote an "epistemological monopoly of the old over the young" (Kaphagawani 2006). Furthermore, many of the attainments of personhood are easier to achieve if you are wealthy and of high social standing, and especially if you are male. Since self-realization relies on recognition from within one's group, it is structured in a way that offers fewer rewards to those with private, less visible social roles. If self-realization is equated in some way with community standing, then lower standing in the community becomes its own confirmation of being less worthy (Oyowe 2013). This has led some of sub-Saharan communitarianism's fiercest critics to conclude that the sub-Saharan paradigm of self-realization is fundamentally incompatible with the idea that all persons are morally equal (Oyowe 2013) and that communities organized around this form of self-realization are irreparably unjust.

Story Point: "The Regression Test," by Wole Talabi

> I'm not sure if A. I.s can believe anything and I'm not supposed to ask
> her questions about such things, but that's what the human control is for,
> right? To ask questions that the other A. I.s would never think to ask, to
> force this electronic extrapolation of my mother into untested territory and
> see if the simulated thought matrix holds up or breaks down. "Don't tell
> me what you think. Tell me what you believe."

What are the things about us that make us distinctly ourselves? Can a self continue to exist, once its care relationships and community are gone? These questions are, in some respects, very literal in "The Regression Test." Titilope, our narrator, participates in the titular test by cross-examining an AI copy of her long-deceased mother Olusola—or "memrionic," as these AI copies are called—in order to determine whether it still reflects the person her mother was. What can a human know about a memrionically preserved person that the memrionic itself cannot? "The Regression Test" directs our attention away from the what and toward the how: how the menrionic's memories and reasoning do and don't reflect the ways of Olusola's knowing and being and her relationships with and within their community, and how Titilope herself continues to know and relate to the beloved and complicated person who was so important to her, even though that person is gone.

2.5 UTILITARIANISM

Utilitarianism is an approach to ethics organized around the idea of happiness. Like virtue ethics, utilitarianism is an outcome-based approach to ethics that assumes that human beings are motivated by the desire to be happy. Building on the basic assumption that humans are motivated by *happiness*, utilitarians argue that when it comes to determining how we should act, we should first and foremost consider what kinds of actions bring about the most happiness for the greatest number of people. This is known as the *principle of utility* or the *greatest happiness principle*.

2.5.1 OVERVIEW OF UTILITARIANISM

Maximizing happiness sounds great, but in order to make that happen, it's necessary to have a definition of happiness that works for everyone who is impacted. So, what is happiness? One of the most well-known proponents of utilitarianism, John Stuart Mill, supplied a definition of happiness within his own definition of utilitarianism, defining it as follows:

> The creed which accepts as the foundation of morals, Utility, or the Greatest Happiness Principle, holds that actions are right in proportion as they tend to promote happiness, wrong as they tend to produce the reverse of happiness. By happiness is intended pleasure, and the absence of pain; by unhappiness, pain and the privation of pleasure. (Mill 2002, 239)

What Mill was suggesting is that pleasure and pain are what matter in measuring happiness. Happiness here refers to a state of mind that can be determined on the basis of how people interpret their experiences or how they assess the amount of pleasure found in a person or community. Later philosophers have debated what exactly counts as pleasure, whether the concept of pleasure alone is enough to describe the intrinsic value that should be maximized, and what the critical value is otherwise

(Feldman 1997), but all utilitarians agree that there is *some* form of basic good that should be promoted as much as possible.

Utilitarianism is a form of consequentialism, which is a category of ethical theories that focuses primarily or solely on consequences in determining the moral worth of an action. (Virtue ethics is also technically consequentialist because it is focused on outcomes, though its focus on character means it is not often categorized that way.) However, the choice to focus on consequences still leaves open a major question: whose consequences should be considered? Should we strive to create positive outcomes for everyone, or just for a chosen few? Different forms of consequentialism answer this question differently. According to ethical egoism, each person should prioritize their own well-being over others.' According to ethical altruism, each person should improve others' well-being without regard to their own. In contrast to both of these, utilitarianism requires that everyone's well-being counts equally; as such, the individual ought to reason about their own happiness in the same way that they ought to reason about the well-being of others. This mode of reasoning is sometimes referred to as "agent-neutral" (Parfit 1984, 27) because the agent does not put herself in a special category. Everyone, no matter their position in life, counts as one.

One of the strengths of utilitarianism is that it strives to be universal, and it does so because, in principle, it presupposes that all individuals are of equal worth. This is called the principle of equality. No one person's happiness is more important than another's. Those who advocate for a utilitarian framework, historically, have also been deeply concerned with the freedom of the individual and the creation of a just social structure in which individuals can pursue happiness without fear.

Utilitarianism is an incredibly influential theory with broad appeal because of its apparent objectivity and concern for equality. We say "apparent" because achieving complete objectivity is impossible when making value determinations. It may be the case that the central formula of utilitarianism is purely impartial in its abstract form, before key parameters are assigned. But it is impossible to apply the calculation until the parameters have been determined, and therefore determining those parameters necessarily involves making important moral judgments. Imagine, for example, that you work for a college preparatory program for high school students and that you are in charge of awarding full scholarships to a small number of admitted students, on utilitarian principles. What parameters should be used to determine who should get the scholarships? Do you select the students whose academic work is strongest, the ones who seem most likely to benefit from the prep course, or the ones whose financial need is greatest? If you decide to use a combination of these three (and possibly other worthy parameters), how do you determine an appropriate balance between them, and how do you implement that balance? And once you have decided what kind of parameters you will use, how do you measure them in a way that can be meaningfully compared among all the students? Keep in mind that even numerically objective metrics like grade point average or household income are not necessarily good proxies for the things you are trying to measure and compare: different high schools have different academic standards and opportunities, and a family's annual income doesn't capture important aspects of their financial reality, such as meeting ongoing medical expenses. As even this brief and simplified example shows, utilitarian decision making always requires the moral agent to make value judgments along the way, even though those value judgments are aimed at achieving a just and impartial outcome. If you hear someone calling utilitarianism "objective," it's because they are ignoring (or have not noticed) that crucial part of the utilitarian process.

To get a group of reasoners to agree on the merits of utilitarianism's central formula is, by itself, significant: it means that the group has agreed that the consequences of voluntary actions is what matters in ethics, and that the best way to reach agreement about how to act lies in assessing what brings about the most happiness for the most people. And even then, reasoners who share these presuppositions may still disagree sharply with one another about foundational ethical ideas (for example, what counts as pleasure and pain?), to the point that they cannot find common ground. As with the other ethical frameworks, the principles of utilitarianism constitute a common language, but this doesn't mean that all the people speaking this language are saying the same thing.

2.5.2 CLASSICAL UTILITARIANISM

Because John Stuart Mill ([1859] 2002) and Jeremy Bentham ([1789] 1996) are generally understood to be the first thinkers to develop systematic accounts of utilitarianism; they are sometimes referred to as classical utilitarians. Mill and Bentham—both nineteenth-century British philosophers—developed theories of ethics that could then be used to critically assess and inform law and social policy (Driver 2014).

Many early utilitarian thinkers thought of ethics as a science, in the sense of being a systematic study that pertains to knowledge. But even though they adopted an approach to ethics that was similar to other human sciences like psychology or sociology, they understood their field of study to involve a different kind of critical practice. Whereas other fields use empirical evidence of human actions to describe what human behavior is like more generally, ethics examines empirical evidence in order to understand norms of action: what "ought" to be the case, or what is "right." Henry Sidgwick (Sidgwick [1874] 1981), another important early utilitarian thinker who defended and built upon the ideas of Mill and Bentham, considered the study of ethics to be similar to that of politics: politics was, in his view, a science that deals with how communities ought to be structured and what is right for society as a whole. Ethics, to Sidgwick, is also aimed at evaluating what is right, although instead of focusing on communities, it considers the voluntary actions of individuals.

Because it defines the good in terms of pleasure and pain, classical utilitarianism is sometimes associated with what is known as ethical hedonism. The concept of hedonism (from *hēdonē*, the Greek word for "pleasure") is sometimes defined as excessive self-indulgence and the pursuit of pleasure and is understood to be negative. However, in the context of ethical theory, hedonism is simply the belief that pleasure is the highest good and proper aim of human life, and that all the basic goods of human life can be understood as forms of pleasure.

Though ethical hedonism goes at least as far back as ancient Greece, classical utilitarianism and its successors are relatively recent compared to the other traditions we examine in this chapter. Utilitarianism, as an ethical framework, developed mostly in England growing out of the developments in the eighteenth- and nineteenth-century European intellectual tradition, including a belief in the possibility of objectivity and a commitment to empiricism. Because this same intellectual tradition still serves as a baseline for many people in the English-speaking world and in other places influenced by English culture, utilitarianism can appear as if it makes no metaphysical assumptions at all. In fact, it would be more accurate to say that utilitarianism rests on the same assumptions about knowledge and action that many of us in the English-speaking world still take for granted.

2.5.3 PREFERENCE UTILITARIANISM

Preference utilitarianism is typically associated with the work of British philosopher R. M. Hare (1981), who was strongly influenced by Kant as well as by classical utilitarianism. Hare arrived at his theory through trying to answer a meta-ethical question: What kind of statement is a moral statement? In Hare's view, moral statements are *prescriptive*, expressing a commitment to a certain course of action (e.g., it would be logically inconsistent for a person to sincerely say "I ought not to lie" and then to lie). He further argued, following Kant, that they are also *universalizable* (e.g., a moral claim of "I ought to lie to you" in a given situation also implies "you ought to lie to me" if all the relevant details of the situation were the same but I was in your position and you were in mine, mentally as well as physically). The combination of prescriptiveness and universalizability led Hare to a standard for evaluating moral statements: To claim that we ought to do something, we must be willing to make the same claim from the perspectives of everyone affected. This, Hare argued, leads to a utilitarian decision procedure in which we weigh these individuals' perspectives against one another. By considering what each stakeholder would have chosen based on their personal desires and values, this procedure treats *preference satisfaction* as the primary good.

Promoting pleasure is still important to a preference utilitarian. If we assume that, all else being the same, a person would rather have more pleasure than less pleasure by definition, then pleasure falls under the category of a person's immediate preferences: their present wishes for their present self, and their future wishes for their future self. But preference utilitarianism also emphasizes the importance of a person's present wishes for their future self. To understand how these are meaningfully different, consider the tale of the epic hero Ulysses (also called Odysseus), who survived the deathly song of the Sirens while sailing home after war. The Sirens' enthralling song would turn a person suicidally irrational as long as they heard it, but it was also said to contain divine wisdom. Ulysses wished to hear the song but didn't want to die in the process. He therefore ordered his crew to plug their own ears and tie him to the mast, and to ignore any orders to untie him until they had made it safely past the Sirens. As predicted, the influence of the song gave Ulysses a desperate desire to leap overboard, and he began to demand his release. But his crew followed his previous orders to leave him bound—they sacrificed the preferences he developed while enthralled in order to fulfill the preferences that he had expressed beforehand. This tale lends the name to present-day psychiatric *Ulysses contracts*. Suppose that a patient is receiving treatment for a condition. The patient wants their future self to complete the course of treatment but anticipates that a relapse might cause their future self to try quitting treatment early. A Ulysses contract would allow the current preference for continuing treatment to override the future preference for stopping (Spellecy 2003).

Another category considered in preference utilitarianism is *external preferences*, preferences about things outside of the preference-holder's experience. For instance, suppose I prefer for my friends not to gossip about me. Such gossip could make me suffer directly if I find out about it, or indirectly if it causes people to treat me worse. But it is possible for my preference to be violated without affecting my experience at all; for example, a friend slanders me in my absence to someone I will never meet. Likewise, people often have wishes for what will happen to their body and possessions after they die, even though they will not be around to witness the outcome. Preference utilitarianism gives these wishes direct moral weight.

One of the most influential applications of preference utilitarianism is Singer's work on bioethics (1994). Classical utilitarianism struggles to accommodate the fact that people tend to view murder, even painlessly killing a miserable person or sacrificing one person to benefit several other people, as almost always immoral. For Singer, beings who are capable of thinking about their own future state, such as adult humans and other great apes, usually have a very strong preference to keep living; a general guideline against killing them is useful because doing so would typically violate their preferences so severely that the violation would outweigh anything that would be gained from the act.

2.5.4 VOLUNTARY ACTION

All four frameworks acknowledge the difference between voluntary and involuntary actions, but deontology and utilitarianism both treat voluntary action as much more morally significant. Yet utilitarianism is particularly geared toward releasing moral actors from responsibility for involuntary actions.

To understand what voluntary action is, consider the following. Imagine that you are skiing down a mountain and suddenly hit a bump and tumble to the ground. You are not badly hurt, so you begin to get up. Then another skier notices you but cannot stop. In order to avoid running into you, the skier quickly steers away from your direction, failing to notice a tree nearby. That skier hits the tree and is knocked unconscious. When they awake they have no memory of who they are, and are forever changed by the experience. In such a situation, who is to blame for the other skier's accident? The simple answer to this question, according to utilitarian logic, is that although you were technically the cause for the other person's injury, you did nothing wrong. You did not choose to fall, nor did you choose to distract the other skier and cause them to become injured. Now, there is always the possibility that you acted in an unethical way; for example, you might have failed to wait long enough before heading down the mountain, or if you were an amateur you might have been trying to impress your friends by going down a slope that was beyond your ability. But even then, the utilitarian might argue that when you—and the person you caused to fall—decided to go skiing that day, you both accepted certain risks that come with skiing in a public environment. By this reasoning, although it is unfortunate, and even tragic, that a person was injured as a result of your fall, you are not morally to blame.

At first, it might seem like this approach to ethics is limited because it focuses solely on an individual actor and the results of particular decisions without considering how society affects decision making. Isn't it the case, one might point out, that in many ways society determines the kinds of choices we make and what is acceptable and what isn't? After all, who is to say that skiing is an acceptable risk to take? What about drinking heavily? What about participating in a clinical trial? What about selling stolen goods on Craigslist? What about buying stolen goods on Craigslist? What about attending a contentious protest?

You may find it confusing that an ethical framework organized around consequences would be so concerned with what a moral actor has chosen (or not chosen) to do. But this focus on voluntary vs. involuntary action makes more sense in light of how utilitarianism conceptualizes human beings, and what makes it possible for us to be moral at all. Utilitarianism presumes that human beings are capable, not only of reasoning about how to live and act but also of communicating their reasoning process to others. These capacities are what make it possible for us to talk as a society about why some actions are justified and others are not (and, as such, make the judicial process possible.) For more on this see chapter 4, where we discuss Locke and his notion of personal identity.

2.5.5 MODALITIES FOR JUDGMENT

The central formula for utilitarianism, known as the principle of utility, is often presented as a universal, objective standard that can be used to determine the moral worth of human action. But as we have seen above, it is always necessary to make some basic value determinations first.

The key parameters in utilitarianism can be understood when we ask the following:

- **Who** comprises the group whose well-being is under consideration? After all, everyone who is part of the calculation may not actually benefit from a given utilitarian decision. There are usually some, and often many, in this group who will pay a price to contribute to the overall net good of everyone considered.

- **What** is the value, or cluster of values, that is being used to define good/happiness/utility? For example, in classical utilitarianism, pleasure is the most important factor in considering happiness. But what counts as pleasure, and what happens when different people experience different sorts of utility from the same kinds of actions or objects?

- **When** is the measure of success being taken? The calculation of consequences often looks different depending on whether you are looking ahead six weeks, or six months, or six years, or six decades.

There is no easy way to decide what the "right" answers are for these categories in a universal sense, although there are almost always better and worse answers under particular conditions. Disagreements, therefore, cannot simply be a matter of miscalculating the consequences. When it comes to evaluating actions, people often have different points at which they begin their analysis with different ideas about who, what, and when.

The "Who"

There are two questions here. One is the question of who fits in, and the other is the question of whether it is appropriate to consider everyone as equal to one another.

Here are some concrete examples of situations in which the "who" is contested:

- **Factory farming**: Who matters, and how much? Consumers, who want to spend less, or want a particular variety or quality? Vendors and ranchers, who provide the meat? Processing plant workers, who often suffer from the hazardous conditions typical to meat processing plants? Stock animals, who become the meat?

- **Organ donation**: Who matters, and how much? How should the cost to the person donating an organ be weighed against the benefit to the person receiving it? What sort of incentives are fair or appropriate to offer to donors? How will the possibility of an organ transplant change the family dynamics of a person who needs one, given that family members are often the best matches? A shift in organ donation policy doesn't just affect patients and doctors. It also affects family members and friends of patients. Furthermore, it affects people who are ill and are considering treatment, or who might be eligible for treatment only under certain conditions. An analysis that looks only at actual patients could be wildly skewed by failing to consider potential patients (Purtill 2018).

- **Traffic and surveillance cameras**: Who should be watched with traffic cameras? Who is being protected? What is the cost of protecting them, and who bears it? Cameras are becoming more common in public, private, and semiprivate workplaces and spaces. Cameras can potentially reduce the number of traffic accidents by catching more speeders (Vincent 2018); however, these same tools make it easier for governments, as well as other organizations and entities, to surveil.

In each of these examples, there are people who will argue that the other side has no moral standing: that the well-being of animals is less important than that of humans, or that people who are caught breaking the law are less entitled to its protections. Where utilitarians are likely to differ, however, is in the *degree* of importance that they assign to the significance of the welfare of these entities as compared to a person. For instance, according to some preference utilitarians like Singer (1994), fetuses and many animals lack the ability to think about the future and therefore have preferences only about their immediate experience, so decisions about killing them are less weighty than decisions about killing adult humans with a host of wishes about their own future lives. The point, however, is that the situation may shift dramatically when we consider whose well-being is being considered and why. Most decisions affect far more people than it might first appear.

The "What"

Even when you consider one person in isolation, it is difficult to pin down what constitutes maximizing their happiness when you consider all the different kinds of things that might count. For instance, consider the satisfaction of finally getting a challenging program to compile, run, and produce useful output; the joy of reading a clear and well-presented explanation; or the pleasure of playing a well-crafted and engaging game. Most utilitarians would agree that these are all ethically positive experiences, but how do we value them in comparison to each other. For example, how many hours of gaming is it worth to finish your final project? Can these two things be meaningfully compared at all?

The question of comparability was a point of disagreement between the founding figures of utilitarianism. Mill contended that "higher" pleasures such as poetry are intrinsically different and more worthwhile than "lower" pleasures like eating, whereas his predecessor Bentham made no such distinction, treating pleasure solely as a matter of quantity rather than quality. Preference utilitarians leave it up to the individual affected to choose what they value most.

The "When"

The question of timing is especially important when we are trying to determine how suffering is measured against utility. To what extent should we accept suffering now if it will create greater happiness later on? What about immediate happiness in exchange for later suffering?

Consequences of an action can reach far into the future, and consequences farther into the future become harder to predict. To deal with this, people have a bias toward placing greater importance on the consequences that come sooner, a phenomenon that economists call *discounting*. Discounting is also invoked on purpose in sequential-decision-making artificial intelligence (AI), to make it computationally possible for an agent to reason about an infinitely long future. But prioritizing sooner outcomes can cause trouble when there is a large difference between the short-term and long-term impacts. In the case of climate change, for example, its future effects could be drastic and damaging, but this knowledge rarely impacts individual, corporate, or societal decision making to a degree that is proportionate to the damage that we know it will do.

2.5.6 CALCULATING GOOD OUTCOMES

Once we have a clear theory about what values matter and for whom, it is still challenging to put the theory into practice. To make a decision with the best possible consequences, we have to calculate what those consequences are. The complexity of filtering through many options that could affect many people in many ways, plus our cognitive biases and the limited amount of information available to us, make the calculation process difficult and inevitably imperfect. If everyone attempted a thorough calculation every time they acted, it would leave them with little time to go about their lives, and they would often calculate *poorly*; the end result would not be "the greatest good for the greatest number" at all!

Fortunately, utilitarianism does not require that each person individually try to be an ideal utilitarian decision maker; in fact, it can acknowledge the value of rule following. Unlike in deontology, the rules in utilitarianism are "rules of thumb" rather than fundamental moral laws, bypassing the sometimes necessary but laborious and error-prone calculation process by presenting courses of action that tend to produce a good outcome much of the time.

Utilitarian perspectives differ on the level at which most of our evaluations should take place. They exist on a spectrum from *act utilitarianism*, which evaluates each action on its individual consequences, to *rule utilitarianism*, which posits that we should follow rules chosen to bring about the greatest overall good. Hare's account (1981) famously calls for a "two-level" model of moral evaluation: We should rely on general, intuitive principles for most everyday decision making and deploy critical reasoning to analyze the specifics of a situation in which special circumstances occur, such as a conflict between intuitive principles.

2.5.7 STRENGTHS AND WEAKNESSES OF UTILITARIANISM: DOES EVERYONE REALLY COUNT AS ONE?

Sometimes, the questions of "who?" and "how much?" are inseparable. For example, if you are at war, are you obligated to consider the citizens of the opposing country in your calculations? How should you weigh their well-being, compared to that of your fellow citizens?

But even if we stick within our own society, should people who are exceptionally beloved or talented matter more than "ordinary" people? Most utilitarians would agree that some people do deserve special consideration but might disagree about the reasons for such considerations. Consider, for instance, the proposition that people with special talents are in a better position to contribute to overall well-being. Do these people deserve special consideration? That is, should we value the life of a virtuoso musician, a promising medical researcher, or a record-breaking athlete more than the life of an ordinary person? This approach involves broadening the scope of "the greatest number" to include all present and future people who might be affected by the public contributions these individuals make. For many utilitarians, this is a compelling reason to argue that not every individual should be counted exactly the same.

The other reason offered for prioritizing some people over others is relational— that is, it concerns the individual reasoner's personal relationship to those who are affected by the decision. Is it acceptable to privilege people in your own group? The principle of equality demands that we answer "no" to this question. And yet the answer gets more complicated when you consider the details. Should family members and close friends matter the same as strangers? What about people who share your worldview and who probably seem to you to have a more honest and accurate understanding of reality? Should you show preference for people to whom you are financially bound—bosses, employers, clients, or stockholders? If not, how can one get the necessary distance to weigh them appropriately?

Giving some individuals special consideration softens the purely agent-neutral approach found in more strictly egalitarian forms of utilitarianism. It bears some similarity to the deontological notion of particular duties or obligations.

Sidebar: Utilitarianism and Machine Learning

Although artificial intelligence seems to be progressing quickly, we are still some distance away (long or short, depending on whom you listen to) from fully automated self-driving cars and self-conscious artificial agents. Nonetheless, people are already asking questions and developing protocols for building ethics into artificial intelligence (Yu et al. 2018). Now that we know something about utilitarianism, let us consider the question: Can a machine be taught utilitarianism?

As we have seen, utilitarianism is based on maximizing some form of welfare, whether promoting pleasure and preventing suffering in classical utilitarianism, or fulfilling a wider variety of desires in preference utilitarianism. Aside from the question of whether a robot itself can have interests that should be taken into account, is it possible to program a robot to reason about furthering humans' interests? Is it possible to equip a robot with a utilitarian framework for decision making?

One of the major issues that arises when this is attempted concerns the fact that decision-making frameworks must rely on models. This means that programmers must know something about ethical dilemmas and also about what kinds of actions count as ethical. In order to identify cases in which ethical dilemmas arise, moral perception is needed. Before we can program moral perception into artificial beings, we need to consider the limits of human perception. For instance, imagine that you are walking down the street and you see a man attack an old woman, grab her purse, and shove her to the ground.

You witness this act and feel bad for the old woman but now the man is running toward you. What do you do? Are you responsible for risking your own safety and trying to stop this man? Minimally, should you call for help?

Now imagine a futuristic version of the scenario where a robot is in the same situation and has the capacity to either (1) taze or somehow injure the robber or (2) set off a siren for the police. Is the robot morally responsible to act in a certain way? When we consider the human agent in the situation, it seems necessary to take into account the risks involved if that person were to try and stop the robber using physical violence. The robot's integrity might be at stake—in other words, it might be injured or harmed in some way—but does it have a concern for its own well-being? If it was built to have concern for its well-being, should that well-being be considered over the loss of property for the old woman? Note that these are not new questions and are famously addressed in *I, Robot* by Isaac Asimov (Asimov 1950).

Whenever we describe ethical dilemmas, we almost always fit in information that might be irrelevant to a robot. Does it matter that the victim was an "old woman"? What if the victim was a younger, athletic-looking person? Would you feel the same obligation to act? Or would the siren be sufficient? What if it was an old woman stealing the wallet out of a young man's suit coat? Should the robot injure the old woman, or set off a siren?

Before machines can make moral decisions we need to be clear about all the factors that we as human beings might consider when we are making an ethical judgment. Is there some way to come to agreement about, for instance, which kind of people require assistance and which kind of people can fend for themselves? Are we willing to let a machine make these judgments, or alternatively, simplify them into rule-of-thumb principles and program those principles into a machine?

Who is responsible when that machine makes a choice that society deems unethical? Would it be the programmer? Or the company that pays and directs the work of the programmer? If it is the company's responsibility, can we evaluate the company as an ethical agent? Or are we pushing the question of ethics to the legal sphere? Insofar as a programmer is voluntarily programming machines to behave in certain ways, and insofar as those behaviors hurt others, couldn't we hold the individual programmer (or team) morally responsible? What about the company? Lots of groups are looking at questions like these; for example, refer to the survey of Fjeld et al. (2020).

Story Point: "Message in a Bottle," by Nalo Hopkinson

"This fucking project better have been worth it."

"Message in a Bottle" follows the main character, Greg, in his quest to understand value and gain success as an artist. The story communicates the risks and challenges of utilitarianism: the difficulty of reasoning impartially about what is valuable, or how best to distribute that valuable thing; the very real trade-offs that come with trying to encompass everyone (whatever "everyone" means) in one's determinations about who deserves value or recognition; and the challenge of conceptualizing value in an enduring way, as a future-oriented form of ethical reasoning demands.

2.6 CONTEMPORARY DEVELOPMENTS IN ETHICS

The four frameworks discussed above have endured and remained influential, in large part, because of their explanatory power: they equip us to perceive and make sense of the world even through a variety of changing conditions and circumstances.

Yet in the past 75–100 years, these frameworks have come under considerable criticism by people who perceive them to be significantly inadequate, or even actively harmful. In some cases, these critics point to long-standing failures of these frameworks that have only recently gained traction in public discourse. In other cases, critics of the traditional frameworks argue that the dramatic technological shifts of the last century have created the need for new ways to understand how to live and act well in our transformed and still-transforming world.

This section briefly introduces three of these new approaches to ethics that have emerged in the past 75 years. These new approaches all draw on the older traditions described above, as well as critiquing them.

2.6.1 RESPONSIBILITY ETHICS

All ethical frameworks are grounded in some account of what human beings are and what kinds of action we are capable of. Deontology and utilitarianism in particular lend themselves to understanding human beings primarily as actors, whether as knowers, doers, creators, or lawmakers. In contrast to these more action-centered approaches, responsibility ethics is an approach to ethical reasoning that begins with the notion that human beings are not only or even primarily acting beings but are beings that are constantly reacting—responding—to powers, forces, and events that are beyond our control. The scale and scope of moral action therefore cannot be confined to individuals as intentional actors.

In fact, as responsibility theorists point out, the emphasis on human action, along with the attitude of optimism that it breeds with respect to human power and potentiality, has amounted to humans acting in unrestrained ways in the world, without sufficient attention to the wider system of forces, events and beings of which we are a part. This attitude has led to limitless industrialization and production of resources that is now putting the future of humanity at risk. But though the future we have created for ourselves is bleak, it is not cause for despair; rather, it's a starting point for thinking about how we can act responsibly in a complex and interconnected world.

Responsibility ethics is an approach that developed in religious ethics in the wake of the Holocaust, the systematic murder of nearly 11 million people by the Third Reich (primarily Jews, but also Catholics, LGBTQ people, Romani, and political dissidents of all stripes) during World War II. In the years after the war, Christian and Jewish thinkers alike concluded that new approaches were needed to reckon with the previously unthinkable scope and scale of that event. They observed that such efficient, bureaucratized slaughter was possible only because of recent technological developments, which made both the transportation and killing more efficient and which enabled both soldiers and civil servants to play a role in the killings without having to confront their own actions directly. A few decades later, German Jewish philosopher Hans Jonas turned the core insights of responsibility ethics toward the issue of environmental degradation, arguing that humankind's reckless exploitation and destruction of the natural world was similarly both life-threatening and morally disastrous (Jonas 1984).

Responsibility ethics contends that the major Western approaches to ethics—deontology, virtue ethics and utilitarianism—are not sufficiently attuned to how large, complex, and interdependent the world is. These other approaches, according to responsibility ethics, understand cause and effect in a narrow and limited way. In some cases, the limitation is about focusing only on one's community or one's immediate environment; in other cases, it's about focusing on a specific sphere of concern (such as economics or health) and excluding other interrelated spheres from consideration. But either kind of limitation necessarily simplifies the problems we perceive. And when our formulation of a problem is too simple and too narrow, it is impossible to act in a truly responsible way.

Because of the complexity of the systems we inhabit, we rarely know the full scope of our actions. In this way, responsibility ethics offers a particularly sharp critique of utilitarianism—not because its aims are bad but because the idea that we could meaningfully anticipate the outcomes of our actions is, according to its premises, both naive and unrealistic.

The framework of responsibility ethics could, in principle, be useful at any moment in history. But its concerns are made more urgent by recent developments in technology, both by increasing our ability to act and by making it even harder for us to perceive or understand the scope of the impact that our actions have. Technologies of all kinds extend our ability to act in the world, whether by protecting our bodies, making it easier for us to accomplish certain tasks, or amplifying the effects of the actions we take. In particular, the rapid pace of technological development in the last 150 years has dramatically extended our ability to impact the world, both as individual actors and as societies. This expansion of our ability to act has far outpaced our ability to adapt to an appropriate understanding of our own power. Even if we were already good at understanding the impacts of our actions, and even if we already took seriously the task of understanding the full scope of impact our actions have, it would be difficult to adjust to the changes. But in fact we haven't done a great job of paying attention in the past.

Some of the concerns of responsibility ethics are made more urgent by informational and computing technologies in particular because these technologies make it easy to put ourselves at a remove from the impacts of our actions, or even from the fact that we ourselves are the ones doing it. This enhanced ability to act at a distance is playing out at multiple levels of society. Many of these new developments would be classified and celebrated as advances, though through the lens of responsibility ethics it becomes clear that they also include great risks. Tele-medicine enables doctors to "see" patients at a distance and even perform remote surgeries; surveillance technology and weaponized drones have allowed military personnel and their contractors to distance themselves from the carnage of war.

In order to cope with this burden of responsibility, and the difficulty of being responsible under these circumstances, responsibility ethics offers several core principles and strategies. The first is accountability to all life. This accountability can be clarified by breaking it into two dimensions: *responsibility for* our actions, and *responsibility to* living beings and to the environment.

Another key strategy, proposed by philosopher Hans Jonas, is the "heuristics of fear," a strategy of thought that might be described as a more pessimistic form of utilitarianism. The heuristics of fear requires that, when confronted with a choice, we assume that the worst of all possible future scenarios is what will come about as a result of whatever it is we do. Anticipating these bad outcomes, Jonas argues, will prompt us to be cautious. By choosing our course of action based on the bleakest possible scenario, we deliberately choose to not to take full advantage of our own power to change the world. We also accept responsibility for trying to make the world better, instead of assuming that it will fix itself for us.

Story Point: "Codename: Delphi," by Linda Nagata

It *was* a kid. The battle AI estimated a male, fourteen years old. It didn't matter. The boy was targeting Valdez and that made him the enemy.

"Take the shot."

Karin sits behind a desk all day, but her work is life-or-death: she is a handler for soldiers in battle, keeping track of battlefield data and advising them accordingly. Karin is responsible for keeping her soldiers alive, and often that responsibility involves making sure they kill their enemies and destroy their resources. Both Karin and her soldiers maintain a tight focus on immediate goals and concerns. They do not even make reference to the larger war, much less debate the merits of its purpose or the ethical complexities of fighting an enemy that has far fewer technological and defensive resources. The soldiers don't have time for these questions—and neither, notably, does Karin, who is physically safe and distant but emotionally and experientially close. By taking us through one harrowing shift with Karin, "Codename: Delphi" not only calls our attention to some of the ways that technology extends our ability to act; it also sheds light on how and why it is so difficult to take that extended power seriously, and why it is urgent that we do so.

2.6.2 FEMINIST ETHICS

As we have shown, the major ethical traditions have all been criticized for preserving and replicating power imbalances. The structures that shape our world were, by and large, imagined and implemented by the people who already had leisure, resources, and social power. As a result, the voices and perspectives of the least powerful and cared-for in society have largely been excluded from how they were created and applied. Though all the frameworks described above all seek to protect (and even elevate) those less powerful members of society, their notions of how to accomplish this are still mostly limited to how powerful people think about it; the perspectives and insights of the less powerful themselves are largely excluded. This limitation, critics argue, makes these frameworks insufficient to combat "the pattern, widespread across cultures and history, that distributes power asymmetrically to favor men over women, creating and maintaining social institutions and practices that systematically put men's interests and preoccupations ahead of women's" (Lindemann 2019, 10).

Feminist ethics is an approach to ethics that aims to repair the ramifications of this long-term exclusion by focusing on the lives, experiences, and concerns of women and other disempowered persons. It frequently centers emotional ties and empathy, not only as goods in themselves but also as sources (often unacknowledged by traditional Western ethics) for our reasoning about others (Gilligan 1993; Held 2014). Rather than offering a single unified theory of ethics, feminist ethics offers "a way of *doing* ethics" (Lindemann 2019) shaped by "the needs of those one cares for in relational contexts" rather than by "abstract, universal principles" (Norlock 2019).

In focusing on societal patterns of power and disempowerment, feminist ethics seeks to correct for imbalances that it perceives in the traditional Western frameworks. These imbalances include focusing on isolated autonomous actors, at the expense of recognizing our interdependence and the constraints that shape our choices (Jaggar 1992; Lindemann 2019); emphasizing public issues at the expense of private ones (Noddings 1984); and treating abstract versions of a given problem as more real than specific instances (McLaren 2011). This focus on the public and the abstract, feminist ethicists contend, works to obscure the realm of private and interpersonal experience and by extension trivializes any insights or modes of reasoning that emerge from taking that realm seriously (Jaggar 1992; McLaren 2011; Noddings 1984). For instance, consider a variant of the dilemma originally posited by Lawrence Kohlberg and developed by Carol Gilligan (Gilligan 1993). In the original, a man steals a drug that his wife needs to survive. Kohlberg argues that one can look at this as an issue of the man's virtues, for instance, or his duties to the law and to his wife. Gilligan, taking a feminist ethics lens, argues that it is just as important to look at the systems of oppression that deny his family insurance that would cover the cost of her care (by relegating them to working as Uber drivers or university adjunct instructors, for example) and at the environment in which they live, and how her ill health has led to their poverty, which in turn creates the conditions for poor health.

Because of its attention to care relationships in the context of systemic power imbalances, feminist ethics is especially useful for understanding how different forms of systemic oppressions interact and reinforce one another. Although early feminist ethics tended to focus solely on women's disempowerment, later waves have expanded to encompass LGBTQIA+ people, Black and indigenous people of color, and other minoritized communities as within their sphere of concern. Overviews with many references can be found in Norlock (2019) and Lindemann (2019).

The relational structure of feminist care ethics can help illuminate dimensions of human experience that are often overlooked by the classical Western frameworks. One example of a concept made visible by feminist ethics is emotional labor (Hochschild [1983] 2012). This term, coined by sociologist Arlie Russell Hochschild, describes jobs that require the worker to express or suppress specific emotions, such as the warmth and friendliness that is often expected of service workers in the United States, regardless of how they are being treated. Although almost any kind of paid work requires labor of some kind, emotional labor jobs are distinctive because those workers are hired for, and evaluated by, their ability to "manage and produce a feeling" when relating to others (Beck 2018). Fields that require significant emotional labor, such as childcare and customer service, typically have a large percentage of female employees; workers in these same fields often struggle for professional credibility, on the grounds that their emotional labor is not recognized as work (Fairchild and Mikuska 2021; Kruml and Geddes 2000). Although every ethical framework deals extensively with the importance of managing one's own emotions, none of the traditional Western frameworks pay attention to the dynamics or demands of care work in the manner that feminist ethics enables us to do. This oversight not only obscures an important dimension of human social relationships but creates the conditions for emotional labor to continue to be undervalued.

Another type of labor or work made visible through feminist ethics is invisible work. The traditional concept of work is those activities that we "have to do" as part of our paid work. The discussion of invisible work originally began with the observation that work typically associated with women, especially in the home, is unpaid and many times not recognized as work (Daniels 1987). The underlying idea is that many things can affect how and what we count as work, and often work that happens in private and/or is largely done by women or minorities has historically not been counted as work. This type of devaluing is clearly illustrated, for instance, in the way national gross domestic product (GDP) is computed: if a parent stays home and provides child care, this is not reflected in GDP. However, if the parent obtains a job and then pays someone to provide child care, then both the income from the outside job and the payment to the child-care provider are tracked as part of GDP (Glynn 2019). As we discuss in chapter 3, this is one example of how we choose to report data is a value-laden proces. The concept of invisible work has been expanded in recent years to address the process by which some work is made invisible for particular reasons, such as outsourcing or contracting to reduce the number of official employees (Crain et al. 2019).

Feminist ethicists have provided important perspectives on technological developments and their interaction with larger societal structures. Examples include feminist responses to the misogyny and other biases in the gaming culture, and the #GamerGate backlash; ethicists have engaged with both the gaming culture ("geek masculinity") and the backlash (Braithwaite 2016); a general response to online harassment and violence (Puente et al. 2019), and the ethics of digital vigilantism ("digilantism") (Jane 2016). This list focuses on online culture; others have analyzed the culture of computer science and computer technology (e.g., Schinzel 2018) or computer research ethics (e.g., Toombs et al. 2017) from a feminist viewpoint. Others have looked at the many issues entangled with fairness, accountability, transparency, and the like (see, e.g., Gebru 2020).

Story Point: "Today I Am Paul," by Martin L. Shoemaker

> Today I was Susan, Paul's wife; but then, to my surprise, Susan arrived for a visit. She hasn't been here in months. In her last visit, her stress levels had been dangerously high. My empathy net doesn't allow me to judge human behavior, only to understand it at a surface level. I know that Paul and Anna disapprove of how Susan treats Mildred, so when I am them, I disapprove as well; but when I am Susan, I understand. She is frustrated because she can never tell how Mildred will react. She is cautious because she doesn't want to upset Mildred, and she doesn't know what will upset her. And most of all, she is afraid.

Mildred, an elderly woman with dementia, spends her days being cared for by a highly advanced care android. Because the android is able to physically and emotionally emulate people, Mildred does not realize that the android exists; she believes that people she knows and loves are visiting her every day. The android's presence spares Mildred's son, granddaughter, and daughter-in-law some of the worries and responsibilities they might otherwise have borne. But it also makes it harder for them to stay connected to Mildred or to provide care for her in ways that they want to or that they might find beneficial or comforting despite its challenges. Mildred's own situation—and the situation of the family who chose it for her—raises vital questions about care relationships: what counts as care, what we gain by giving it, and what is at stake when we decide to outsource it to programmed entities.

2.6.3 The Capability Approach

Capability is an approach to ethics that was derived and adapted from a broader normative framework used across many fields. It aims to create the conditions for people to fulfill their potential, not by celebrating freedom in the abstract but by paying attention to the practical conditions of people's lives—the actual freedoms, resources and opportunities that are available to them. Accordingly, it evaluates both individual actions and broader social policies in terms of how they support people's *capabilities* and thus their *functionings*: that is, the resources that they draw on (capabilities) in order to do and to be in the ways that matter to them (functionings).

The Capability Approach (CA) was first introduced by the economist Amartya Sen in the 1980s, in response to trends in political theory and the international policy that (Sen argued) were narrowly focused on economic growth at the expense of human well-being. All forms of this versatile framework are rooted in a political vision of real, feasible human freedom, in which people have the opportunities to live and act in ways that they themselves value. CA has been developed as an ethical framework by the philosopher Martha Nussbaum and others, but it is frequently used in economics and political science and has been especially influential in the field of international development (Torrente and Gould 2018, 596).

Like utilitarianism, CA is consequentialist and pragmatic, and it relies on factual, empirical data. But unlike utilitarianism, CA is concerned primarily with creating the *potential* for good outcomes. This focus on potentiality means that it is able to take seriously an individual's actions, their ability to choose, the importance of their own individual goals and preferences, and the circumstances that shape and constrain their lives. Rather than focusing primarily on individual agents acting in pursuit of the good, CA focuses on the development and organization of society. And rather than insisting on the possibility of free action localized in the human individual, it emphasizes the conditions that make it possible to expand the capacities for action itself. Instead of simply valuing freedom in the abstract sense, CA examines people's real opportunities to do and be what they have reason to value.

Capabilities are the resources that individuals have. Those resources include material goods such as money or property, shelter, and food, but they also include intangibles like health and access to education, transportation, and other intangible resources. Although Sen did not think there is or should be a list of core capabilities, Nussbaum's adaptation of CA has proposed a list of what she calls *basic capabilities*, which emphasizes human dignity (Nussbaum 1997; Sen 2005). The 10 basic capabilities are being able to live a normal lifespan; having good health; maintaining bodily integrity; being able to use the senses, imagination, and thought; having emotions and emotional attachments; possessing practical reason to form a conception of the good; having social affiliations that are meaningful and respectful; expressing concern for other species; being able to play; and having control over one's material and political environment (see Nussbaum 2000, 33, paraphrased from Jacobs 2020). Furthermore, Nussbaum explicitly stresses the importance of both the internal and external conditions, distinguishing between internal capabilities and combined capabilities. Internal capabilities are "the characteristics of a person (personality traits, intellectual and emotional capacities, states of bodily fitness and health, internalized learning, skills of perception and movement)" (Nussbaum 2011, 21). Combined capabilities are thus the internal capabilities in combination with the relevant external conditions.

Whether a particular person has a particular capability depends both on the state of the person and on the barriers that they encounter. Barriers could include availability of resources—money, food and water, electricity, material goods, and jobs—or accessibility. For instance, someone might live in a time and place in which jobs or leisure activities might be available, but because of transportation issues or a lack of internet access, those things are not meaningfully available to that person. The example of the bicycle is frequently referenced to illustrate this point (Sen [1985] 1999). For one person, a bicycle could enable them to get to a job or leisure activities, which greatly increases their options. But for someone unable to ride the bicycle, it does not. The imperative that comes out of this is not to stop providing bicycles, but rather to consider the individual's situation and their ability to make use of the bicycle.

In some cases, capabilities are best supported when the political state does not intervene in people's lives. In other cases, supporting people's capabilities requires social or political policies to make sure that capabilities are available to all. Consider, for example, wheelchair-accessible public spaces (Nussbaum 2006). Simply supplying someone who cannot walk with a wheelchair is insufficient if they are excluded from political participation by inaccessible voting places, rallies, or courts. While museums and movie theaters may provide capabilities of play and leisure, a lack of ramps or elevators limits some people's ability to transform these capabilities into functionings.

Though CA is concerned with both the individual and the context in which that individual is situated, it was originally developed to address governmental, institutional, and society-level choices more than individual choices. It therefore helps us address big-picture questions about technological development, and for that reason it is worth considering. For individual decision making, CA is often paired with a political theory of justice or with another ethical framework, typically virtue ethics.

Story Point: "Welcome to Your Authentic Indian Experience™," by Rebecca Roanhorse

"So," White Wolf says, "tell me about yourself."

You look around the bar for familiar faces. Are you really going to do this? Tell a Tourist about your life? Your real life? A little voice in your head whispers that maybe this isn't so smart. Boss could find out and get mad. DarAnne could make fun of you. Besides, White Wolf will want a cool story, something real authentic, and all you have is an aging three-bedroom ranch and a student loan.

But he's looking at you, friendly interest, and nobody looks at you like that much anymore, not even Theresa. So you talk.

Jesse Turnblatt doesn't have a lot going for him. He doesn't have close friends or strong community ties, and he's terrified that his wife is going to leave him. The only thing in Jesse's life that's going well is his job at Sedona Sweats virtual reality (VR) company, where he works as a Spirit Guide, offering American Indian–themed VR Experiences for Tourists seeking spiritual transformation. Jesse draws on movie stereotypes rather than his own heritage in crafting these "authentic" Indian experiences, and the Tourists love him for it. When Jesse does meet a Tourist who wants an Experience closer to Jesse's own lived reality, Jesse's life begins to unravel. Authentic Indian

Experience raises many difficult questions about who gets to be "authentic" (and who gets to decide what authenticity means), but it also challenges us to recognize all the barriers in Jesse's life—material, social, cultural, and political—that constrain his capabilities and limit his ability to be the person he wants to be.

2.7 CONCLUDING REMARKS: THE IMPORTANCE OF MULTIPLE FRAMEWORKS

In this chapter we have shown that ethical frameworks are not a set of fixed rules to be memorized but rather are a variety of interpretive lenses that can furnish us with criteria for ethical description and reasoning, so that we can both judge value for ourselves and understand the viewpoints of others. The ethical frameworks that we have discussed—deontology, utilitarianism, virtue ethics, communitarianism, responsibility, feminist ethics, and capabilities—are all conceptually rich and dynamic approaches to ethics; each of them can support many kinds of inquiry and many kinds of normative positions. We have encouraged you to view ethics as a resource to describe and reason through issues, and to remember that the framework, viewpoint, and process, can be separate both from your own ideology and from the conclusions you reach. Each framework answers the question "What is at stake?" differently, because each framework takes up the work of ethics from a different starting point: duties, happiness, living well, the social nature of the self, the world around us, social patterns and equity, or the concrete freedoms available.

In chapter 3 we build on what we have learned here and turn our attention to some of the fundamental units of modern communication and computing technologies: data and information. We examine one way of understanding the difference between data, information, knowledge, and wisdom, and how our understanding of these concepts can provide a valuable window into issues around the design, development, deployment, and use of information and communications technologies.

REFLECTION QUESTIONS

1. *What are the three ways in which people commonly misunderstand deontology? What are the shortcomings with each of these misunderstandings?*

2. *Consider an ethical quandary in your own life, or something you've read about recently in the news, which you can describe in terms of competing duties. What aspects of the situation could help you determine which duties are the most fundamental? What aspects of the situation could help you determine which duties are most relevant?*

3. *Recall our example in section 2.5 about deciding who should get a scholarship. Reflect on a time that you were either a part of or subject to a decision that required trading off various goods and measures. What were the measures? What was included and what wasn't? Were all these measures meaningful or comparable?*

4. *One of the central challenges for utilitarian thinkers is breadth: including everyone who should be part of the "who," including all of the relevant kinds of good (and harm) in the "what," and thinking of the long-term consequences as well as the near-term ones.*

   a. *Why do you think this is difficult? Give a specific scenario where changing the "who" can lead to a vastly different calculus for outcomes.*

   b. *Are there any ways for utilitarians to get better at including everything and everyone that should be part of their calculations?*

5. *Think of a situation—professional, personal, or societal—that could have been better handled by those involved if they had given more weight to long-term consequences, as opposed to short-term ones. Why, specifically, do you think the long-term consequences were ignored or undervalued? What, realistically, could have helped the decision makers to consider or understand the long-term effects of their choices?*

6. *Think of a habit you have tried to adopt (or tried to break). What about it was easy, and why? What about it was difficult, and why? What personal qualities helped you succeed in (or prevented you from) altering your habit?*

7. *How do goals differ from virtues? Can you list some of your goals, and the virtues that could support you in working toward those goals?*

8. *Think of someone you admire. What virtues do you think they strive for, and why? Do you think that they are usually successful in their attempts to live according to those virtues? How do they handle their failures to live up to their (apparently) chosen virtues?*

9. *The term "digital divide" is used in this book and in the media to call attention to the fact that not everyone has access to reliable and modern information and communication technologies including computers, smartphones, and even the internet. Analyze how one or more of the following aspects of modern society that you take for granted can happen via these technologies: registering for classes, paying your friends for dinner, applying for a job, or figuring out how to get from your home to a new location. How would being on the other side of the digital divide change what you are able to do? What would be easier? What would be harder?*

10. *Think of a community that has been important for your personal development. What values do you hold, or patterns of behavior do you have, that were instilled in you by that community? How did they come to be part of your own identity?*

11. *What goals of self-realization do you have for yourself? What are the community contexts in which those goals are meaningful?*

12. *Compare two traditions encompassed in a single framework. What are the similarities? What are the differences?*

13. *What is a contemporary issue in technology that is not well addressed by the four major frameworks discussed in the chapter? How could one or more of the three contemporary responses help us describe and address that issue?*

BACKGROUND REFERENCES AND ADDITIONAL READING

Bewaji, John Ayotunde Isola. 2004. Ethics and morality in Yorùbá culture. In *A Companion to African Philosophy*, edited by Kwasi Wiredu, 396–403. Blackwell.

Cohen, Alix. 2014. Introduction. In *Kant on Emotion and Value*, 1–10. Palgrave Macmillan.

Friedman, Marilyn. 1993. *What Are Friends For? Feminist Perspectives on Personal Relationships and Moral Theory*. Cornell University Press.

Frye, Marilyn. 1983. *The Politics of Reality: Essays in Feminist Theory*. Crossing Press.

Fulmer, Russell. 2019. Artificial intelligence and counseling: Four levels of implementation. *Theory & Psychology* 29, no. 6: 807–819.

Gotlib, Anna. Feminist ethics and narrative ethics. In *The Internet Encyclopedia of Philosophy*. https://iep.utm.edu/fem-e-n/.

Hochschild, Arlie Russell. [1983] 2012. *The Managed Heart*. University of California Press.

Lindemann, Hilde. 2019. *An Invitation to Feminist Ethics*. Oxford University Press.

National Science Foundation, Division of Science Resources Studies. 2019. *Women, Minorities, and Persons with Disabilities in Science and Engineering*. National Science Foundation.

Nussbaum, Martha C. 2003. Capabilities as fundamental entitlements: Sen and social justice. *Feminist Economics* 9, no. 2/3: 33–59.

Oosterlaken, Ilse. 2015. Human capabilities in design for values: A capability approach of "design for values." In *Handbook of Ethics, Values, and Technological Design: Sources, Theory, Values and Application Domains*, edited by Jeroen van den Hoven, Pieter E. Vermas, and Ibo van de Poel, 221–250. Springer.

Paton, H. J. 1958. The aim and structure of Kant's *Grundlegung*. *The Philosophical Quarterly* 8, no. 31: 112–130.

Sen, Amartya Kumar. 2009. *The Idea of Justice*. Harvard University Press.

REFERENCES CITED IN THIS CHAPTER

Angle, Stephen C. 2013. Is conscientiousness a virtue? Confucian answers. In *Virtue Ethics and Confucianism*, edited by Stephen C. Angle and Michael A. Slote, 182–191. Routledge.

Ani, Ndubuisi Christian. 2013. Appraisal of African epistemology in the global system. *Alternation* 20, no. 1: 295–320.

Annas, Julia. 2011. *Intelligent Virtue*. Oxford University Press.

Anscombe, G. E. M. 1958. Modern moral philosophy. *Philosophy* 33, no. 124: 1–19.

Aquinas, Thomas. 1948. *Introduction to Saint Thomas Aquinas*. Edited and with an introduction by Anton C. Pegis. McGraw-Hill Education.

Asimov, I. 1950. *I, Robot*. Doubleday.

Athanassoulis, Nafsika. 2000. A response to Harman: Virtue ethics and character traits. *Proceedings of the Aristotelian Society*, n.s., 100: 215–221.

Beck, Julie. 2018. The concept creep of emotional labor. *The Atlantic*, November 26. https://www.theatlantic.com/family/archive/2018/11/arlie-hochschild-housework-isnt-emotional-labor/576637/.

Behrens, Kevin Gary. 2011. Two "normative" conceptions of personhood. *Quest: An African Journal of Philosophy* 25: 103–117.

Bentham, Jeremy. [1789] 1996. *The Collected Works of Jeremy Bentham: An Introduction to the Principles of Morals and Legislation*. Clarendon Press.

Biakolo, Emevwo. 2003. Categories of cross-cultural cognition and the African condition. In *The African Philosophy Reader*, 2nd ed., edited by P. H. Coetzee and A. P. J. Roux, 9–19. Routledge.

Bonnefon, Jean-François, Azim Shariff, and Iyad Rahwan. 2016. The social dilemma of autonomous vehicles. *Science* 352, no. 6293: 1573–1576.

Braithwaite, Andrea. 2016. It's about ethics in games journalism? Gamergaters and geek masculinity. *Social Media + Society* 2, no. 4: 2056305116672484.

Chen, Lai. 2015. Practical wisdom in Confucian philosophy. In *Selected Papers from the XXIII World Congress of Philosophy*, vol. 40, issue supplement, 69–80. Philosophy Documentation Center in cooperation with the Greek Philosophical Society.

Chimuka, Tarisayi A. 2001. Ethics among the Shona. *Zambezia: The Journal of Humanities of the University of Zimbabwe* 28, no. 1: 23–37.

Christian, Brian. 2020. *The Alignment Problem: Machine Learning and Human Values*. Norton.

Clarke, Bridget. 2010. Virtue and disagreement. *Ethical Theory and Moral Practice* 13: 273–291.

Coetzee, Pieter H. 2003. Morality in African thought. In *The African Philosophy Reader*, 2nd ed., edited by P. H. Coetzee and A. P. J. Roux, 273–286. Routledge.

Cornell, Tom. 2006. A brief introduction to the Catholic Worker Movement. The Catholic Worker Movement. https://www.catholicworker.org/cornell-history.html.

Crain, Marion, Winifred Poster, and Miriam Cherry, eds. 2019. *Invisible Labor: Hidden Work in the Contemporary World*. University of California Press.

Cudd, Ann, and Seena Eftekhari. 2018. Contractarianism. In *The Stanford Encyclopedia of Philosophy*, edited by Edward N. Zalta. https://plato.stanford.edu/archives/sum2018/entries/contractarianism/.

Curzer, Howard J. 2018. Yesterday's virtue ethicists meet tomorrow's high tech: A critical response to *Technology and the Virtues* by Shannon Vallor. *Philosophy & Technology* 31, no. 2: 283–292.

Daniels, Arlene Kaplan. 1987. Invisible work. *Social Problems* 34, no. 5: 403–415.

Driver, Julia. 2014. The history of Utilitarianism. In *The Stanford Encyclopedia of Philosophy*, edited by Edward N. Zalta. https://plato.stanford.edu/archives/win2014/entries/utilitarianism-history/.

Dubber, Markus D., Frank Pasquale, and Sunit Das, eds. 2020. *The Oxford Handbook of Ethics of AI*. Oxford University Press.

Fairchild, Nikki, and Eva Mikuska. 2021. Emotional labor, ordinary affects, and the early childhood education and care worker. *Gender, Work & Organization* 28, no. 3: 1177–1190.

Feldman, Fred. 1997. On the intrinsic value of pleasures. *Ethics* 107, no. 3: 448–466.

Finnis, John. 2020. Natural law theories. In *The Stanford Encyclopedia of Philosophy*, edited by Edward N. Zalta. https://plato.stanford.edu/archives/sum2020/entries/natural-law-theories/.

Fjeld, Jessica, Nele Achten, Hannah Hilligoss, Adam Nagy, and Madhulika Srikumar. 2020. Principled artificial intelligence: Mapping consensus in ethical and rights-based approaches to principles for AI. *Berkman Klein Center Research Publication* 2020-1. https://dash.harvard.edu/handle/1/42160420.

Foot, Philippa. 1978. *Virtues and Vices and Other Essays in Moral Philosophy*. University of California Press.

Forest, Jim. 2010. The Catholic Worker Movement. Catholic Worker Movement. https://www.catholicworker.org/forest-history.html.

Furey, Heidi. 2017. Aristotle and autism: Reconsidering a radical shift to virtue ethics in engineering. *Science and Engineering Ethics* 23, no. 2: 469–488. https://link.springer.com/article/10.1007/s11948-016-9787-9.

Gade, Christian B. N. 2011. The historical development of the written discourses on Ubuntu. *South African Journal of Philosophy* 30, no. 3: 303–329.

Gbadegesin, Segun. 1991. *African Philosophy: Traditional Yorùbá Philosophy and Contemporary African Realities*. Peter Lang.

Gebru, Timnit. 2020. Race and gender. In *The Oxford Handbook of Ethics of AI*, edited by Markus D. Dubber, Frank Pasquale, and Sunit Das, 251–269. Oxford University Press.

Gilligan, Carol. 1993. *In a Different Voice: Psychological Theory and Women's Development*. Harvard University Press.

Glynn, Sarah Jane. 2019. The economics of caregiving for working mothers. Center for American Progress, December 10. https://www.americanprogress.org/issues/early-childhood/reports/2019/12/10/478387/economics-caregiving-working-mothers/.

Gyekye, Kwame. 1978. Akan concept of a person. *International Philosophical Quarterly* 18, no. 3: 277–287. Reprinted in *African Philosophy: An Introduction*, edited by Richard A. Wright. University Press of America, 1984.

Gyekye, Kwame. 1987. *An Essay on African Philosophical Thought: The Akan Conceptual Scheme*. Cambridge University Press.

Gyekye, Kwame. 1992. Person and community in African thought. In *Person and Community: Ghanaian Philosophical Studies I*, edited by Kwasi Wiredu and Kwame Gyekye, 101–122. Council for Research in Values and Philosophy.

Gyekye, Kwame. 1997. *Tradition and Modernity: Philosophical Reflections on the African Experience*. Oxford University Press.

Hallen, Barry. 2005. Yorùbá moral epistemology. In *A Companion to African Philosophy*, edited by Kwasi Wiredu, 296–303. Blackwell.

Hallen, Barry, and J. Olubi Sodipo. 1997. *Knowledge, Belief and Witchcraft: Analytic Experiments in African Philosophy*. Cambridge University Press.

Hare, Richard Mervyn. 1981. *Moral Thinking: Its Levels, Method, and Point*. Clarendon Press.

Held, Virginia. 2014. The ethics of care as normative guidance: Comment on Gilligan. *Journal of Social Philosophy* 45, no. 1: 107–115.

Hobbes, Thomas. [1668] 1992. *Leviathan*. Edited by Edwin Curley. Hackett.

Hursthouse, Rosalind, and Glen Pettigrove. 2016. Virtue ethics. In *The Stanford Encyclopedia of Philosophy*, edited by Edward N. Zalta. https://plato.stanford.edu/entries/ethics-virtue/.

Idoniboye-Obu, Sakiemi, and Ayo Whetho. 2013. Ubuntu: "You are because I am" or "I am because you are"? *Alternation* 20, no. 1: 229–247.

Ikuenobe, Polycarp. 2006. *Philosophical Perspectives on Communalism and Morality in African Traditions*. Lexington Books.

Jacobs, Naomi. 2020. Capability sensitive design for health and wellbeing technologies. *Science and Engineering Ethics* 6, no. 6: 3363–3391.

Jaggar, Alison M. 1992. *Encyclopedia of Ethics*. Edited by L. Becket and C. Becket. Garland.

Jane, Emma A. 2016. Online misogyny and feminist digilantism. *Continuum* 30, no. 3: 284–297.

Jonas, Hans. 1984. *The Imperative of Responsibility: In Search of an Ethics for the Technological Age*. University of Chicago Press.

Kamwangamalu, Nkonko M. 1999. Ubuntu in South Africa: A sociolinguistic perspective to a pan-African concept. *Critical Arts: A South-North Journal of Cultural & Media Studies* 13, no. 2: 02560046.

Kant, Immanuel. 1996. *Groundwork*. In *Practical Philosophy: The Cambridge Edition of the Works of Immanuel Kant*, 37–108, translated and edited by M. J. Gregor. Cambridge University Press.

Kaphagawani, Didier Njirayamanda. 2005. African conceptions of a person: A critical survey. In *A Companion to African Philosophy*, edited by Kwasi Wiredu, 332–342. Blackwell.

Kruml, Susan M., and Deanna Geddes. 2000. Exploring the dimensions of emotional labor: The heart of Hochschild's work. *Management Communication Quarterly* 14, no. 1: 8–49.

Legge, James. [1861] 2017. *The Chinese Classics.* Vol. 1: *Confucian Analects, the Great Learning, and the Doctrine of the Mean.* Forgotten Books. http://oaks.nvg.org/analects-legge.html.

Lin, Patrick, Keith Abney, and Ryan Jenkins, eds. 2017. *Robot Ethics 2.0: From Autonomous Cars to Artificial Intelligence.* Oxford University Press.

Lindemann, Hilde. 2019. *An Invitation to Feminist Ethics.* Oxford University Press.

MacIntyre, Alasdair. 2013. *After Virtue.* 3rd. ed. A. & C. Black.

Mangena, Fainos. 2015. Restorative justice's deep roots in Africa. *South African Journal of Philosophy* 34, no. 1: 1–12.

Masolo, D. A. 2010. *Self and Community in a Changing World.* Indiana University Press.

McLaren, Margaret A. 2011. Feminist ethics. In *Encyclopedia of Global Justice*, edited by Deen K. Chatterjee, 345–347. Springer.

Menkiti, Ifeanyi. 1984. Person and community in traditional African thought. In *African Philosophy: An Introduction*, edited by Richard A. Wright, 171–181. University Press of America.

Menkiti, Ifeanyi A. 2005. On the normative conception of a person. In *A Companion to African Philosophy*, edited by Kwasi Wiredu, 324–331. Blackwell.

Metz, Thaddeus. 2011. Two conceptions of African ethics. *Quest: An African Journal of Philosophy* 25: 141–162.

Midgley, Mary. 1991. *Can't We Make Moral Judgements?* St. Martin's Press.

Mill, John Stuart. 2002. *The Basic Writings of John Stuart Mill: On Liberty, The Subjection of Women, and Utilitarianism.* Edited by J. B. Schneewind. New York: Modern Library.

Mkhize, Nhlanhla. 2008. Ubuntu and harmony: An African approach to morality and ethics. In *Persons in Community: African Ethics in a Global Culture*, edited by Ronald Nicolson, 35–44. University of KwaZulu-Natal.

Molefe, Motsamai. 2020. *African Personhood and Applied Ethics.* National Inquiry Services Centre.

Munkaya, Mluleki, and Mokgethi Motlhabi. 2009. Ubuntu and its socio-moral significance. In *African Ethics: An Anthology of Comparative and Applied Ethics*, edited by Munyaradzi Felix Murove, 63–84. University of KwaZulu-Natal.

Naudé, Piet. 2019. Decolonising knowledge: Can Ubuntu ethics save us from coloniality? *Journal of Business Ethics* 159: 23–37.

Neequaye, George Kotei. 2020. Personhood in Africa. In *The Palgrave Handbook of African Social Ethics*, edited by Nimi Wariboko and Toyin Falola, 103–127. Palgrave Macmillan.

Nicholas, George. 2018. When scientists "discover" what indigenous people have known for centuries. *Smithsonian Magazine*, February 21. https://www.smithsonianmag.com/science-nature/why-science-takes-so-long-catch-up-traditional-knowledge-180968216/.

Noddings, Nel. 1984. *Caring: A Feminine Approach to Ethics and Moral Education*. University of California Press.

Norlock, Kathryn. 2019. Feminist ethics. In *The Stanford Encyclopedia of Philosophy*, edited by Edward N. Zalta. https://plato.stanford.edu/archives/sum2019/entries/feminism-ethics/.

Nussbaum, Martha C. 1997. Capabilities and human rights. *Fordham Law Review* 66: 273–300.

Nussbaum, Martha C. 2000. *Women and Human Development: The Capabilities Approach*. Cambridge University Press.

Nussbaum, Martha C. 2006. *Frontiers of Justice: Disability, Nationality, Species Membership*. Harvard University Press.

Nussbaum, Martha C. 2011. *Creating Capabilities*. Harvard University Press.

Ogbonnaya, A. Okechukwu. 1994. Person as community: An African understanding of the person as an intrapsychic community. *Journal of Black Psychology* 20, no. 1: 75–87.

Okolo, Chukwudum B. 1992. Self as a problem in African philosophy. *International Philosophical Quarterly* 32, no. 4: 477–485.

Oladipo, Olusegun. 2003. Metaphysics, religion, and Yorùbá traditional thought. In *The African Philosophy Reader*, 2nd ed., edited by P. H. Coetzee and A. P. J. Roux, 200–208. Routledge.

Outlaw, Lucius. 1987. African "philosophy": Deconstructive and reconstructive challenges. In *Contemporary Philosophy*, vol, 5: *African Philosophy*, edited by G. Floistad, 9–44. Martinus Nijhoff.

Oyěwùmí, Oyèrónkẹ́. 1997. *The Invention of Women: Making an African Sense of Western Gender Discourses*. University of Minnesota Press.

Oyowe, Oritsegbubemi Anthony. 2013. Personhood and social power in African thought. *Alternation* 20, no. 1: 203–228.

Parfit, Derek. 1984. *Reasons and Persons*. Oxford: Oxford University Press.

Pereira, Luís Moniz, and António Barata Lopes. 2020. *Machine Ethics: From Machine Morals to the Machinery of Morality*. Springer.

Plaks, Andrew. 1999. The mean, nature and self-realization: European translations of the Zhongyong. In *De l'un au multiple: Traductions du chinois vers les langues européennes* (From one into many: Translations from the Chinese into European languages), 311–331. https://books.openedition.org/editionsmsh/1507.

Plato. *Euthyphro* 10 a 1–3.

Presbey, Gail M. Maasai. 2002. Concepts of personhood: The roles of recognition, community, and individuality. *International Philosophical Quarterly* 34, no. 2: 57–82.

Puente, Sonia Núñez, Sergio D'Antonio Maceiras, and Diana Fernández Romero. 2019. Twitter activism and ethical witnessing: Possibilities and challenges of feminist politics against gender-based violence. *Social Science Computer Review*, July 24. https://doi.org/10.1177/0894439319864898.

Purtill, Corinne. 2018. How AI changed organ donation in the US. Quartz, September 10. https://qz.com/1383083/how-ai-changed-organ-donation-in-the-us/.

Putman, Daniel. 1997. The intellectual bias of virtue ethics. *Philosophy* 72, no. 280: 303–311.

Quinn, Philip L. 2006. Theological voluntarism. In *The Oxford Handbook of Ethical Theory*, edited by David Copp, 63–90. Oxford University Press.

Ramose, Mogobe. The philosophy of *ubuntu* and *ubuntu* as a philosophy. In *The African Philosophy Reader*, 2nd ed, edited by P. H. Coetzee and A. P. J. Roux, 230–237. Routledge.

Rawls, John. 1999. *A Theory of Justice*. Harvard University Press.

Rossi, Francesca, and Nicholas Mattei. 2019. Building ethically bounded AI. In *Proceedings of the 33rd AAAI Conference on Artificial Intelligence*. Blue Sky Track.

Rubinstein, Amnon. 2007. Return of the kibbutzim. *Jerusalem Post*, July 10.

Sandel, Michael J. 2010. *Justice: What's the Right Thing to Do?* Macmillan.

Santiago, John. 2008. Confucian ethics in the Analects as virtue ethics. In *Philosophical Ideas and Artistic Pursuits in the Traditions of Asia and the West: An NEH Faculty Humanities Workshop*. Digital Commons. https://dc.cod.edu/cgi/viewcontent.cgi?article=1007&context=nehscholarship.

Scarre, Geoffrey. 2013. The continence of virtue. *Philosophical Investigations* 36, no. 1: 1–19.

Schinzel, Britta. 2018. Gendered views on the ethics of computer professionals. In *Localizing the Internet: Ethical Aspects in Intercultural Perspective*, edited by Johannes Frühbauer, Thomas Hausmanninger, and Rafael Capurro, 121–134. Wilhelm Fink.

Schwitzgebel, Eric. 2007. Human nature and moral education in Mencius, Xunzi, Hobbes, and Rousseau. *History of Philosophy Quarterly* 24, no. 2: 147–168.

Sen, Amartya. [1985] 1999. *Commodities and Capabilities*. Oxford University Press.

Sen, Amartya. 2005. Human rights and capabilities. *Journal of Human Development* 6, no. 2: 151–166.

Serequeberhan, Tsenay. 2003. The critique of Eurocentrism and the practice of African philosophy. In *The African Philosophy Reader*, 2nd ed., edited by P. H. Coetzee and A. P. J. Roux, 64–78. Routledge.

Shpancer, Noam. 2011. Child of the collective. *The Guardian*, February 18. https://www.theguardian.com/lifeandstyle/2011/feb/19/kibbutz-child-noam-shpancer.

Sidgwick, Henry. [1874] 1981. *The Methods of Ethics*. Foreword by John Rawls. Hackett.

Singer, Peter. 1994. *Practical Ethics*. 2nd ed. Cambridge University Press.

Sivak, Jacob. 2020. The kibbutz is Israel's original start-up. *The Forward*, July 19.

Smith-Morris, Carolyn. 2020. *Indigenous Communalism: Belonging, Healthy Communities, and Decolonizing the Collective*. Rutgers University Press.

Spellecy, Ryan. 2003. Reviving Ulysses contracts. *Kennedy Institute of Ethics Journal* 13, no. 4: 373–392.

Sreenivasan, Gopal. 2002. Errors about errors: Virtue theory and trait attribution. *Mind* 111 (January): 47–68.

Stalnaker, Aaron. 2010. Virtue as mastery in early Confucianism. *Journal of Religious Ethics* 38, no. 3: 404–428.

Stohr, Karen E. 2003. Moral cacophony: When continence is a virtue. *Journal of Ethics* 7, no. 4: 339–363.

Tangwa, Godfrey B. 2000. The traditional African perception of a person: Some implications for bioethics. *Hastings Center Report* 30, no. 5: 39–43.

Tangwa, Godfrey B. 2005. Some African reflections on biomedical and environmental ethics. In *A Companion to African Philosophy*, edited by Kwasi Wiredu, 387–395. Blackwell.

Tiwald, Justin. 2018. Confucianism and Neo-Confucianism. In *The Oxford Handbook of Virtue*, edited by Nancy E. Snow, 171–189. Oxford University Press.

Toombs, Austin, Shad Gross, Shaowen Bardzell, and Jeffrey Bardzell. 2017. From empathy to care: A feminist care ethics perspective on long-term researcher-participant relations. *Interacting with Computers* 29, no. 1: 45–57.

Torrente, Steven, and Harry D. Gould. 2018. Virtues and capabilities. In *The Oxford Handbook of International Political Theory*, edited by Chris Brown and Robyn Eckersley, 587–599. Oxford University Press.

Tshivhase, Mpho. 2011. Personhood: Social approval or unique identity? *Quest: An African Journal of Philosophy* 25: 119–140.

Vallor, Shannon. 2016. *Technology and the Virtues: A Philosophical Guide to a Future Worth Wanting*. Oxford University Press.

Vincent, James. 2018. IBM secretly used New York's CCTV cameras to train its surveillance software. The Verge, September 6. https://www.theverge.com/2018/9/6/17826446/ibm-video-surveillance-nypd-cctv-cameras-search-skin-tone.

Wallach, Wendell, and Peter Asaro, eds. 2020. *Machine Ethics and Robot Ethics*. Routledge.

Walzer, Michael. 1990. The communitarian critique of liberalism. *Political Theory* 18, no. 1: 6–23.

Wamala, Edward. 2005. Government by consensus: An analysis of a traditional form of democracy. In *A Companion to African Philosophy*, edited by Kwasi Wiredu, 433–442. Blackwell.

Wariboko, Nimi. 2020. Between Community and my mother: A theory of agonistic communitarianism. In *The Palgrave Handbook of African Social Ethics*, edited by N. Wariboko and T. Falola, 147–163. Palgrave Macmillan.

Wiredu, Kwasi. 1992a. The moral foundations of an African culture. In *Person and Community: Ghanaian Philosophical Studies I*, edited by Kwasi Wiredu and Kwame Gyekye, 193–206. Council for Research in Values and Philosophy.

Wiredu, Kwasi. 1992b. Problems in Africa's self-definition in the contemporary world. In *Person and Community: Ghanaian Philosophical Studies I*, edited by Kwasi Wiredu and Kwame Gyekye, 59–70. Council for Research in Values and Philosophy.

Wiredu, Kwasi. 1998. Toward decolonizing African philosophy and religion. *African Studies Quarterly* 1, no. 4: 17–46.

Wiredu, Kwasi. 2003. An Akan perspective on human rights. In *The African Philosophy Reader*, 2nd ed, edited by P. H. Coetzee and A. P. J. Roux, 313–323. Routledge.

Wiredu, Kwasi. 2009. An oral philosophy of personhood: Comments on philosophy and orality. *Research in African Literatures* 40, no. 1: 8–18.

Wong, David, Chinese ethics. 2020. In *The Stanford Encyclopedia of Philosophy*, edited by Edward N. Zalta. https://plato.stanford.edu/archives/sum2020/entries/ethics-chinese/.

Xia, Fang. 2020. A comparative study of Aristotle's doctrine of the Mean and Confucius' doctrine of Zhong Yong. *International Communication of Chinese Culture* 7, no. 3: 349–377.

Yao, Xinzhong. 2000. *An Introduction to Confucianism*. Cambridge University Press.

Yu, Han, Zhiqi Shen, Chunyan Miao, Cyril Leung, Victor R. Lesser, and Qiang Yang. 2018. Building ethics into artificial intelligence. In *Proceedings of the Twenty-Seventh International Joint Conference on Artificial Intelligence (IJCAI-18)*, 5527–5533. IJCAI. https://www.ijcai.org/Proceedings/2018/0779.pdf.

Yu, Jiyuan. 1998. Virtue: Confucius and Aristotle. *Philosophy East and West* 48, no. 2: 323–347.

Yu, Jiyuan. 2013. *The Ethics of Confucius and Aristotle: Mirrors of Virtue*. Vol. 7. Routledge.

Zhu, Rui. 2002. What if the father commits a crime? *Journal of the History of Ideas* 63, no. 1: 1–17.

CHAPTER 3

MANAGING KNOWLEDGE

Learning Objectives

At the end of this chapter you will be able to:

1. *Recall the elements that comprise the traditional data, information, knowledge, wisdom (DIKW) paradigm and evaluate the limits of this classical definition.*

2. *Describe the similarities and differences between the information and data storage technologies of the past and those of the present.*

3. *Differentiate between data/information management practices available to computers and those available to humans; articulate and evaluate practical consequences of this difference, including data retention, data bias, and data accessibility.*

4. *Justify the claim that knowledge is a shared enterprise by tracing the different ways we come to know things both as individuals and as a society.*

3.1 INTRODUCTION

The word "information" figures into many of the phrases we use to explain the present and the role of computing technologies in that present, such as *information technology* and *information age*. But what is information, and what makes it useful? In order to be ethically responsible users and developers of technological systems that deal with information, we need to understand what we're working with, and how circulating, creating, and using technological systems can help or hurt individuals and societies.

We live in an age when computers and technology have vastly increased our ability to collect, store, process, and communicate information. These changes have, in turn, increased the amount of information that we encounter on a daily basis, both actively, when we are looking for it, and passively, when it is put in front of us unasked.

The massive surge in available information has been interpreted by many in the tech world as evidence that human beings are getting smarter, or that they have the potential to become smarter. But is information enough to make us smarter? Access to information is not the same as the ability to use that information effectively. It can be difficult to know how to interpret or make suitable use of a particular item of information. And sometimes the sheer volume of available information can make it hard to find the relevant item. For example, in the months preceding the September 11, 2001, terrorist attacks, many security reports issued by American intelligence agencies reported the likelihood of terrorists using airplanes as weapons (National Commission on Terrorist Attacks upon the United States 2004), but these bits of information were lost in the deluge of other reports. More information can be as much a problem as a solution.

Computers are new, but most of the effects they have had on human culture are not. There have been earlier technologies such as writing that have similarly expanded our capacity to store information, some of them as radical and socially transformative as the introduction of computers. When movable type was invented in Germany in the mid-fifteenth century—enabling inexpensive, high-quality printing on a scale previously unknown in Europe—the entire cultural landscape of that continent was transformed as the number of available books and the number of copies in circulation increased exponentially (Briggs and Burke 2002, 12–13; Sohn 1959, 103). The influx of available information was not well received by those who already had access to that information; early responses about the printing press include complaints about the "confused and irritating multitude of books" and a flood of "so many books that we do not even have time to read the titles" (Gardner 2012).

History teaches us again and again that access to ideas can be revolutionary. But the phrase most American school kids are raised on isn't "information is power;" it's "knowledge is power." Knowledge is a more complicated construct than information. How is information related to knowledge? Can the presence of new ways of processing, storing, and communicating data change or expand our capacity for knowledge? Will they help us become wise, better able to make sense of life and its challenges?

In the historical example above of the printing press, many people worried that new forms of information storage meant that people would *know* less: that access to external information storage would harm humans' capacity for information storage and understanding. Contemporary research suggests that these societies did indeed experience a shift in how memory worked. Poets and performers in contemporary oral cultures, such as in Papua New Guinea and Northern Australia, have demonstrated mental retentive abilities far beyond those in literate cultures (Bakker 2008; Minchin 2008; Rubin 1997), and scholars of both ancient literature and cognitive psychology now take it for granted that the poets and storytellers who preserved ancient epics such as the *Iliad* and the *Ramayana* were able to retain and reproduce those stories in a manner comparable to the poets of contemporary oral cultures.

In addition to worrying about new technology's quantitative impact on memory, ancient Greek thinkers were also concerned about its effects on the *quality* of human learning. In the *Phaedrus*, Socrates argues (by way of a "retold," likely fabricated Egyptian myth) that writing actually damages a person's ability to learn, enumerating several ways in which humans will become less adept at processing information, and therefore less wise, once they come to rely on this new external storage technology:

> It will introduce forgetfulness into the soul of those who learn it: they will not practice using their memory because they will put their trust in writing, which is external and depends on signs that belong to others . . . the appearance of wisdom, not . . . its reality. [Writing] will enable them to hear many things without being properly taught, and they will imagine that they have come to know much while for the most part they know nothing. And they will be difficult to get along with, since they will appear to be wise instead of really being so. (Plato 1995, 275a–275b)

It is not hard to imagine these same accusations being leveled at the internet. Were Socrates's concerns legitimate? Does information-sharing technology have the paradoxical effect of creating citizens who are shallowly informed, or even hostile toward those with greater expertise? Or have humans been able to use their expanded ability to engage with cultural knowledge to think more broadly and deeply about the world and to question old assumptions in productive ways? As we will find when we go further in this chapter, developments in both directions can, and do, occur as a result of progress in technology, just as they have occurred in response to prior shifts in the technology of information or knowledge.

Just as recent advances in information and communication technologies resemble the advances brought about by the introduction of writing and the printing press, we are confronting many of the same risks and concerns raised by those earlier technological advances. As we will see in this chapter, many of the important ethical issues in technological systems that deal with information require us to reckon seriously with *epistemology*, that is, the study of how we know the things that we know. As people who work in information technology, it is crucial that we understand what we are working with and how it impacts the world, so that we can use it to enhance the quality of human life and society rather than damage it.

3.1.1 CHAPTER OVERVIEW

In order to come up with information technologies that enrich the human experience, we need to understand how human beings deal with information, data, and knowledge.

We frame our discussion around the interrelated concepts of data, information, knowledge, and wisdom, known as the "DIKW cluster." This chapter begins by emphasizing two important points about DIKW, which form the basis for the rest of our discussion:

1. Data and information are not value neutral; the ways in which we interpret, collect, process, and apply data and information reflects deeply on ourselves as individuals and society as a whole. Our understanding of new material depends on how it fits into our existing map of the world; only rarely does new data or information fundamentally alter that map.

2. People have a limited capacity for absorbing, processing, and storing data and information, even if some of that work can be outsourced to technologies such as writing and computers. The sheer volume of data forces us to be selective. We often retain what is easiest to assimilate, instead of what best meets our needs.

The chapter then highlights some specific examples of the kinds of mistakes that get made when those two things are ignored or overlooked, when the resulting technological systems exacerbate the effects of human bias. In the rest of the chapter we discuss coping mechanisms for these limitations at individual and societal levels.

3.2   THE THINGS WE KNOW ARE NOT VALUE NEUTRAL

How is information related to knowledge? How are each of these things related to data? And how do agents—human or computational—put these resources to work in trying to identify the best course of action, ethically speaking?

Data, information, and knowledge are all potential ingredients in ethical decision making. Any time an agent has to make a decision, they cannot consider their options or evaluate what is at stake in their response without making recourse to some of the data, information, and knowledge available to them. But in the age of the internet, with vast reserves of content easily and rapidly accessible, how does one decide which information to assimilate, which knowledge bases to consult, and how much is enough?

In an information- and knowledge-rich culture, every decision, large or small, is tied up in larger, systemic-level choices about how to manage the knowledge and information that is available. It is a truism that computers and modern technological systems can vastly expand our capacity to manage, organize, and even synthesize *information*. Whether or not they can expand our capacity to *manage knowledge*, however, is a separate question, and a much more difficult one. How we answer it will rely, in part, on how we define information and knowledge and understand the relationship between them.

Therefore, when we discuss ethical issues in managing knowledge, we need to clarify what we mean when we use terms like "knowledge," "information," and "data." These terms might seem familiar, but it is not always clear how, for instance, knowledge differs from information, or what counts as data.

3.2.1 THE DIKW PARADIGM AND ITS SHORTCOMINGS

A dominant paradigm in information sciences is the *data-information-knowledge-wisdom* (DIKW) model (figure 3.1), whose introduction is often credited to Russell Ackoff (1989). The model became popular in information-science-related fields because its definitions for information and data (particularly the latter, the newest of the four terms) corresponded closely to how people were already using these terms. Knowledge and wisdom, meanwhile, had received comparatively little attention in information-science work. The model describes these four concepts hierarchically: Data is processed into information, information is synthesized into knowledge, and—when the wisdom layer is discussed at all—knowledge is synthesized into wisdom (Ackoff 1989 adds understanding as another layer to the DIKW paradigm between knowledge and wisdom).

This model has many strengths. One is that it emphasizes the interconnectedness of data, information, and knowledge; if we recognize the interconnectedness, it is easier for us to recognize that data, information, and knowledge are all contextual rather than absolute. There are many things that are particularly appealing about the ascending structure of the pyramid, in which knowledge and/or wisdom are the end result of increasingly complex processing of raw data. It may be particularly intuitive to people working in information technology fields, because of how it aligns with the gathering of information from databases or data warehouses through queries.

But this model is not expansive enough to account for how information, knowledge, and wisdom actually operate in human life. Some of this inadequacy is due to

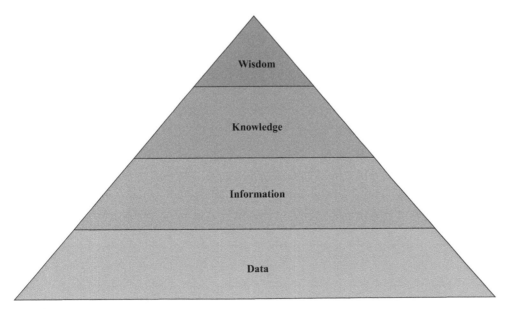

FIGURE 3.1
The DIKW pyramid.

the terminology itself: the version of "knowledge" that appears in the hierarchy does not match the way the word has been traditionally used and understood, and "wisdom" is not clearly or consistently defined.

Another problem is the implied structure of the relationship. The model presumes a movement upward, each layer created out of the one beneath it. Although this steady procession from data to information and on upward may seem intuitively correct in the abstract, it reflects only a small slice of the complex conditions in which an agent encounters or absorbs information or data or knowledge. In practice, the significance or value of information depends on the context in which we experience it and the lens through which we view it. That is, knowledge is *shaped from* information but it is also *shaped by* the observer's existing knowledge and whatever organizing premises they start out with.

For example, imagine a blueprint for a luxury apartment building. Many people could take away only very basic information from this data set. To an engineer, the blueprint would contain far more information, and because of her existing knowledge base, she could also use it to learn how to build such a building. A burglar might glean the same information from the blueprint as the engineer, but that information would yield a different kind of knowledge: how to break in. We might argue that the burglar is using the information incorrectly in a moral sense, but he will not necessarily be incorrect about how best to break in.

As even this simple example shows, the widely held ideas about data, information, and knowledge and their interrelationships are inadequate. But because the DIKW hierarchy so clearly formulates ideas about these subjects that are latent but widely held, it is a useful place from which to begin, even as we learn to recognize its presuppositions and limitations.

The following definitions are drawn from Ackoff (1989) and from literature built on Ackoff's paradigm (Bellinger, Castro, and Mills 2004; Floridi 2010). After spelling them out, we explore their limitations and offer an approach that better reflects information cultures.

When you encounter the word "data," you might think of a row in a table or a point on a plot—and these are indeed types of data in the DIKW model's sense. But the words on this page are also data. So, too, is the sensory input that we experience as colors, shapes, and sounds. In general, "data" refers to raw observations of the world. In the context of computing, data is usually understood to be that which is collected and stored somewhere for further synthesis, querying, or analysis. It is worth noting that although there are some problems with how Ackoff's paradigm relates data to the other parts of the pyramid, this definition of data is itself sound.

Information is defined in the DIKW model as data that has been processed to give it meaning. The "processing" in question could be mechanized, such as a database system answering a query, but consider again the example of words, which we convert into semantic meaning, or sensory data, which we then convert into information about our surroundings. In contrast to raw data, it is "useful" at least insofar as it serves as a coherent *descriptor* of the world.

In parallel to the definition of information as a synthesis of data, the model defines *knowledge* as that which comes from the synthesis of information. According to Ackoff, knowledge is a body of interconnected information that lends itself to being applied and is described as "what makes possible the transformation of information into *instructions*" (1989, 4). Whereas a piece of information deals with a detail of a system, knowledge allows us to comprehend how the system as a whole works. It can be derived from experience or taught from one agent to another; to Ackoff, programming a computer is an act of teaching, instilling the machine with the programmer's knowledge about how to do something.

The final level of the paradigm is *wisdom*. This level has received far less attention than the other three (Rowley 2007), and few DIKW-related pieces in the information-systems literature bother to define it. Ackoff's own definition is much more abstract than the three preceding levels: While the other three contribute to *efficiency*—the ability to carry out an objective well—wisdom contributes to *effectiveness*—the ability to judge which objectives are worth pursuing. Unlike the three preceding tiers, wisdom is not an external phenomenon but rather part of the knower's character; it is also notably hard to integrate with the other three tiers.

In summation, the DIKW paradigm offers four different levels in an ascending structure; each level comes into existence by synthesizing matter from the level below it. Data consists of unprocessed symbols or signals. Information is produced from data. Knowledge is the processed amalgam of information. And although adherents to this paradigm differ with regard to the definition or uses of wisdom, they share the assumption that wisdom is, in some sense, a synthesis of knowledge.

The intuitiveness of this paradigm can be seen in the following example. As you are driving down the highway, you glimpse the number 45 on your speedometer. You process this reading, a piece of data, into information—that your speed is 45 miles per hour. Synthesizing this with other information—observations of your surroundings, memories of past driving experiences, relevant pieces of traffic law—yields knowledge about the current situation on the road. This lets you predict future states of the world (e.g., recognize that you might crash and be liable if the car in front of you suddenly stops), make decisions (e.g., to slow down and change lanes), and draw upon your skills to execute these decisions. These moment-to-moment choices contribute to broader goals—such as ensuring the safety of yourself and those around you—that are governed by wisdom. The logic of the DIKW paradigm feels highly intuitive when applied to narrow-frame examples such as this one, which are frequently the sort of examples used as test cases for abstract theories.

3.2.2 HOW THE DIKW PARADIGM LIMITS OUR ABILITY TO UNDERSTAND THE WORLD

As we noted above, despite its intuitive appeal, the commonly understood DIKW model has profound epistemological shortcomings: that is, it is not adequate to help us make sense of how we know the things we know. First, although it offers good working definitions of data and information, its definitions of both knowledge and wisdom are not nearly complicated enough to account for the ecology of data, information, and knowledge in the real world. The reasons for this inadequacy lead us to the second weakness in the paradigm: the definitions of knowledge and wisdom are flawed precisely because they are rooted in a model of ascending synthesis.

The driving example above demonstrates the intuitive value of the model, but also exposes its limits. Drivers use data, information, and knowledge to accomplish their goals: data about the state of the car and the road, information about traffic and pedestrian movements, and knowledge about how to control the car and navigate to their destination. This description makes it seem that the DIKW hierarchy is clearly delimited, and that understanding flows from data up to wisdom. The fact that most drivers share the same goal—namely, "don't hit anything"—can create the illusion that there is one correct way to synthesize each level, until it feeds into the driverly wisdom of not hitting anything en route to one's destination.

But once the example is widened to consider people who aren't driving (or who don't mind hitting things with their cars), it becomes clear that there are many possible ways to synthesize information from data, or knowledge from information. A pedestrian, for example, might have a range of responses to a sidewalk lemonade stand, none of which match a driver's response.

Even on its own terms, the driving example is not as clear-cut as it initially appears. The reality is that people often *do* hit things with their cars. Why does this happen? Sometimes they are not technically adept enough to assimilate the information that would prevent them from crashing. Sometimes they choose to ignore some available information or knowledge, because their interest in getting to their destination eclipses their interest in not hitting something. And sometimes, hitting something is actually their goal, and not always for morally dubious reasons: drivers who are being carjacked or abducted have been known to crash their cars on purpose to prevent the carjacker or kidnapper from succeeding (Edmunds 2020). These are only a few of the ways in which the structure that governs the implied data-to-wisdom flow of the DIKW paradigm might break down in reality; you can probably think of others.

So, although Ackoff offers us useful partial definitions for thinking about both data and information, the paradigm is not adequate to account for the complexity of interrelationships between data, information, knowledge, and wisdom. As we see in the narrow example of drivers' collision avoidance, not everybody will discover the same set of instructions in a given body of information. In practice, a person's existing body of knowledge—both their assumptions about how the world works and their expert knowledge of a particular field—influences how they absorb and process available information.

Theories of knowledge are not new. Philosophers and theologians have provided ways of thinking about knowledge, and these ideas have been criticized and improved over centuries. It has been pointed out (McDermott 1999; Weinberger 2010) that Ackoff's definition is out of step with those long-standing definitions of knowledge. First, knowledge has a "knower" who draws upon personal reflection and intuition as well as the information at hand. Secondly, because knowledge requires a knower, it is dynamic: its precise qualities depend on how the knower engages with information. And last, as we will see shortly, knowledge does not exist in isolation in an individual mind but rather is given context by a community.

The dynamic quality of knowledge presents a further challenge to the DIKW paradigm, specifically to its ascending order. According to that paradigm, data is chronologically the first thing known, and the higher levels occur later as a direct result of the encounter with data. But in practice, data and information are taken in by knowers who already have ideas and goals shaped by their previous knowledge; although new data and information can sometimes influence those knowers' view of the world, they are always slotted into an existing map of data, information, and knowledge. In other words, knowledge primarily comes before data and information, not after; as knowers, our broader worldview shapes how we infer patterns of information in data.

It is therefore both more useful and more accurate to define *knowledge* in a way that recognizes how new knowledge is grounded in existing structures. After revising the definition of knowledge, we investigate in greater depth the meaning of "wisdom," which will help us return to our original concern with ethics and ethical decision making.

3.2.3 CULTURES OF KNOWLEDGE: REVISING THE DIKW PARADIGM

We can now understand why, upon further scrutiny, the ascending structure of the DIKW paradigm and parts of Ackoff's definitions prove insufficient: the idea of non-neutrality is true of *both data and information*. The data that one encounters is going to be shaped by one's capacity to observe correctly. The information that one discovers will be shaped by one's capacity to make correct inferences about the relationships between observations. Both flawed observations and flawed inferences can lead to flawed information.

If it ever seems like data can be synthesized directly into information, or information synthesized directly into knowledge, without any interpretive intervention, *this is only because interpretive parameters are already in place*. Although this sort of simple upward translation can and does take place within the framework of a highly specific set of parameters, most situations in life—including all of the ethically complex situations—do not fit neatly into those parameters.

For this reason, we offer a revised paradigm that accounts for how knowledge influences our relationship to information and how our interest in particular kinds of information shapes our relationship to data (both before and after we collect it). Although we readily acknowledge that data often does provide the raw materials for information, and that information can have an impact on one's larger body of knowledge, this ascending movement is far from universal. One can gain a more accurate understanding from a *descending model*, in which an entity's relationship with a given level of the DIKW "hierarchy" is controlled by its existing encounters with, or commitment to, the level above. Ackoff's ascending model is still valuable, but it is most accurate when understood as a supplement to the descending model.

The descending model acknowledges that agents do not seamlessly integrate new information into their picture of the world, but rather they incorporate this information in a way that accommodates their existing knowledge. Similarly, although either a human being or a computer can collect a vast amount of data, that data does not become usable except in response to a query of some kind, be it retrieving records from a database or recalling a memory of what we ate last week. The agent necessarily formulates any such query in response to information or knowledge they already have, though of course the constraints and affordances for humans and computers are quite different. The descending paradigm further acknowledges that, in response to those unavoidable limits, agents must *manage the knowledge* with which they come into contact, and that this need shapes the layers below knowledge.

This need to manage knowledge is not a flaw or a weakness: it is simply a reality of being an entity that already knows things. In fact, the only alternative to managing knowledge would be for an entity to throw out all old knowledge every time it encountered anything new. Think for a moment about what that would mean: what would you be like, if you had to start fresh every morning? Would you still be yourself? Whenever we assess the merits of new information, try to figure out the best way to make use of a given data set, or identify flaws in another person's argument, we are managing knowledge, using the material we already have to control our encounter with new findings.

But managing knowledge is more complicated than simply separating good information from bad. The reason we react differently to different kinds of information—why some of it is deemed worth absorbing while some of it is not—is that our existing body of knowledge acts as a kind of screen. Furthermore, although each of us is unique, these screens are not products of our individual minds and experiences: all of us belong to *cultures of knowledge* that are formed by common experience (or presumed common experience) and bound together by certain shared ideas. These ideas might be grounded in shared information, but they are broader, and they guide us in how to react to new information.

A culture of knowledge can be large or small. Your generation, members of your religious faith, everyone who went to high school with you, people of your political orientation, and natives of your geographic region are all examples of knowledge cultures to which you might belong. Any sort of identity group that involves a shared perspective on the world can function as a culture of knowledge. Every New York Mets fan knows that New York Yankees fans are sellouts; every Yankees fan knows that Mets fans are sore losers. Members of each group are able to reach all sorts of conclusions about individuals in the other group, on the basis of that foundational knowledge.

Within any knowledge culture, there are ideas and concepts that are understood to be widely recognizable to members of that culture—and, in some cases, opaque to those outside of that culture. This insider/outsider dynamic is a key element of humor. An especially clear example of this type of humor can be found in memes organized around growing up in a particular era, such as "you'll only get this if you grew up during *X* decade." Memes in general "work" only if the reader is a knowledge insider—that is, if that reader recognizes the underlying pattern that explains or justifies what is going on in any particular instance of the meme. "Only [*XX* era] kids will get this!" memes underscore this reliance on cultures of knowledge. Unlike many other memes, which are ostensibly funny to anyone familiar with the relevant sphere of culture, era-specific memes explicitly identify a given knowledge culture—in this case, having experienced the children's media of a given era, or having encountered the adult media of that period from a child- or adolescent-specific perspective—as the locus of the meme's significance. But like all other memes, they highlight the fact that belonging to a particular knowledge culture confers on the knower a (group) identity and shapes their perspective on the world.

From within a given culture, some kinds of information are clear and intelligible—even if a given piece of information is new, it "matches" the things that are already known or understood—whereas other kinds of information are not such an easy fit. This sense of ill-fittedness can occur because those kinds of information seem like an attack on the patterns by which other things are known, or it can simply be because they do not seem to belong to the world in which this particular knowledge culture is rooted (Oyěwùmí 1997).

For instance, consider the relationship between two knowledge cultures concerned with ecology: the knowledge culture of modern institutional science, rooted in methods developed in Europe during the Enlightenment, and the traditional knowledge held and passed down by Indigenous cultures. Because the scientific method requires that particular metrics be met in order for a hypothesis to be considered sound, much of the ecological and environmental knowledge held by Indigenous communities has long been considered "unscientific" and therefore dismissed as unsound. It has, therefore, been a great surprise to many in the scientific community when researchers use the scientific method to confirm things that have long been known by Indigenous people. Two recent examples of this pattern include a study of Australian firehawks (Bonta et al. 2017) who deliberately spread brush fires to drive out prey, and an investigation of clam farming practices (Groesbeck 2014) that proved that ancient Indigenous practices were able to maximize food production while preserving the ecosystem. These studies were both undertaken by scientists seeking to legitimize Indigenous knowledge in the eyes of their colleagues (Greshko 2018; Nicholas 2018), and it is possible that some readers will be prompted to adjust their prior beliefs about Indigenous knowledge after seeing it confirmed by processes they recognize. However, it can be argued (as Nicholas 2018 has) that the real adjustment required is for Western science to recognize the legitimacy of Indigenous ways of knowing, without needing to confirm it on its own terms.

3.2.4 HOW DOES WISDOM FIT IN?

Before you began reading this chapter, you might have asked: Why is there so much text worrying about knowledge in an ethics textbook? The answer is that managing knowledge is a topic with deep ethical implications. The way we absorb or use the data, information, or knowledge available to us can have a profound impact on our judgment of what we ought to do. Likewise, systems that collect, organize, and synthesize data or information also shape their users' view of what is true, and what therefore can or should be done.

Yet it is also clear that neither data, nor information, nor even knowledge is sufficient for making informed and responsible ethical judgments. Such judgments always require us to put data, information, and knowledge into conversation with a sense of what is good or right, and how that good or right thing can be practically achieved. This is the point at which wisdom comes in.

Wisdom requires reflection on our lives and our place in the world. It connects our choices and our previous actions to the things we value. Wisdom also allows us to prioritize knowledge. Which kinds of things are worth knowing? How does knowledge relate to living well or help us take actions that help ourselves or others flourish?

In light of the revised account of data, information, and knowledge offered above, it is possible to gain a clearer picture of wisdom as well. This perspective makes it possible to identify two qualities of wisdom which also consistently appear in philosophical discussions:

1. Wisdom requires knowledge and information, but they are not enough. To be wise, a person must also be able to apply them in helpful and relevant ways: they must be able to get hold of and recognize accurate information about the right kind of things, and know what information is suitable to which situations.

2. Wisdom requires a certain amount of humility, or an awareness about the limits of one's knowledge. The famous philosopher Socrates was said to be wise precisely because he claimed to know nothing.

These two qualities offer some clarity, but there is still a long way to go. How do we decide what information is accurate or suitable? How do we know what the appropriate amount of humility is? Each of the ethical frameworks, as we defined them in chapter 2, emphasize the importance of wisdom, but they do so in different ways. By considering them alongside one another, we get a better sense of what wisdom requires.

Deontology recognizes that consequences are important but is much more concerned with duties and obligations. Actions should be done for the right reasons—and not necessarily because they maximize happiness or pleasure. To be wise, then, is to understand one's obligations, duties, and responsibilities, and also to possess knowledge about how to apply them practically.

The character-based approach of *virtue ethics* offers a more straightforward place for wisdom: practical wisdom is, after all, the key to making good choices and therefore is essential for flourishing. As noted previously, even Ackoff described wisdom in terms of the character of the knower. Virtue ethicists ask questions about what it means to live well, and most virtue ethicists would agree that living well requires *both* instrumental reasoning about how to get around in the world *and* knowledge about how to flourish and thrive, both individually and collectively. Finding the intersection or balance of these two things is where ethics comes in.

Communitarianism also understands wisdom to be central: hard to develop and therefore possessed by few, but beneficial to everyone as a community holding. All communities have elders, whether or not there is an official title for or recognition of them, and the wisdom of these elders serves as a stabilizing anchor and a source of guidance for the community as a whole. Although, as in virtue ethics, wisdom is the purview of older and more experienced individuals, age alone does not make one an elder. An elder must also possess the insight, understanding, and sensitivity that makes one a reliable source of guidance and advice.

In *utilitarianism*, moral judgments are evaluated on the basis of the expected outcome of actions. The utilitarian asks questions such as: How are we to achieve the greatest happiness? How are we to increase welfare and decrease suffering? Wisdom thus requires a thorough and accurate understanding of how actions effect change in the world. Moreover, if the goal is to maximize happiness or pleasure, then wisdom also requires awareness of how to do things prudentially.

Earlier in the chapter we tackled the widespread notion that data or information could be seamlessly "converted" upward into knowledge or wisdom, given enough processing power. Now that we have taken the time to define data, information, knowledge, and wisdom more carefully, it is apparent that the interrelationships between them are complex and that the influence moves down the pyramid as least as much as it moves up—that existing structures of knowledge or wisdom influence the kinds of information that seem meaningful or convincing, and even the kinds of data that are worth collecting. Although technological systems can be built to process data and information independent of human actors, it is now clear that neither data nor information actually exists apart from humans that know them or that have defined the parameters by which they can be collected, known, and understood.

This reconceptualization has important implications for how we design systems that process data with supposedly value-neutral intent, or present information as value-neutral to end users. It also helps us reflect critically on the knowledge, data, information, and wisdom that guide our own decisions. It also has ramifications for our own personal decision making. We need to apply wisdom to our choices in information gathering and how we weigh our information sources.

Sidebar: Skepticism

Many of us, especially the scientifically or technically inclined, have been raised to appreciate "skepticism" as a virtue. Skepticism, in the classical sense, is the practice of requiring firm evidence before accepting a certain conclusion as truth and being prepared to revise our conclusions if we encounter contradictory evidence. This can be a healthy way to grow intellectually—we examine many issues in this book with a skeptical mentality.

However, in practice, the term sometimes drifts into signaling something more like *denialism*, the rejection of ideas that do not meet a preconceived conclusion. Consider people who claim that the earth is flat; they may claim an air of legitimacy by describing themselves as "skeptics" for dismissing the mainstream concept of a spheroidal earth as dogmatic, but a more rigorous and consistent skepticism would demand that they hold their own model of the planet to as high standard of scrutiny as any.

There are other ways that skepticism can go awry. For instance, Klayman (1995) surveys psychological research on *confirmation bias*, which describes how people tend to search for and interpret information in a way that avoids challenging their preexisting beliefs. In addition, Lewandowsky et al. (2016) discuss the difference between healthy skepticism and denialism in public discussion of science and provide guides both for challenging a work of research and for responding to such a challenge as a researcher.

The consequences of arbitrarily applied skepticism can be much worse than a few people believing in bad astronomical models. For instance, people who faced sexual violence have encountered huge barriers to having their reports taken seriously because "skepticism" was applied much more rigorously to the accusers than to the accused. Many of the powerful men exposed during the #MeToo movement were able to continue their abuses for decades because the high power status of these men led people to apply very little scrutiny toward their claims compared to their victims'; the parties are subject to what Tuerkheimer (2021) has termed *credibility inflation* and *credibility discounting*, respectively.

Sloppy, self-serving skepticism is a trap any of us can fall into when we encounter an idea that goes against our own intuitions. Our skeptical intuitions can be helpful insofar as they drive us to investigate the idea further before deciding whether to accept it. But using intuition as a substitute for understanding, and dismissing new ideas based only on account of our intuition, can lead us to reinforce a status quo that deserves questioning. A healthy skepticism includes an understanding of the limits of our own perspective, hence it requires wisdom.

Story Point: "Here-and-Now," by Ken Liu

A search itself was also information. It was an expression of intent, of desire, fear, want, lust.

How much is information worth? That is the question that Aaron, the protagonist of Ken Liu's "Here-and-Now," is forced to confront over the course of one complicated day. Aaron is one of thousands, if not millions, of people using the Here-and-Now app from tech giant Centillion, which allows its users to put in anonymous requests for "information" of any and all kinds. This story poses deceptively simple questions about why information matters. It also points out that some kinds of information are much more meaningful or valuable to some people than to others and asks us to consider whether that difference should matter, and how.

3.3 IT IS DIFFICULT TO MARSHAL LARGE BODIES OF INFORMATION

In the last section, we looked at one key way in which humans manage knowledge differently than computers do: we saw that the human mind uses the things it already knows and believes as the basis for integrating new things. In this section, we examine another crucial but underrecognized difference between humans and computers: human beings' capacity for storing knowledge and using the knowledge that they have stored.

In recent decades, advances in digital technology have drastically increased the volume, velocity, and variety of data. These aspects, called the "three Vs" of big data, mean that we have a larger total volume or amount of data that is generated at an ever-accelerating pace or velocity, and that takes more varied forms including video, email, spoken word, and numerous other formats (Laney 2001). In 2021, it was estimated that humans generated about 500 million tweets, 294 billion emails, and 720,000 hours of YouTube video per day. Attempting to measure all the books, tweets, newspapers, and videos that have ever been created and stored is a difficult task, but it is estimated that about 59 zettabytes, or 59 trillion gigabytes, of data has been created, captured, copied, and consumed, and this will grow to 175 ZB by 2025 (Gantz and Reinsel 2011; Vopson 2021).

There is a widespread belief—held by both the developers of information technology systems and those that use those systems in areas as diverse as marketing, medicine, and manufacturing—that having access to ever larger information sources is, by itself, a kind of magic bullet for problem-solving. This belief rests on the fundamental supposition that deploying a higher volume of information will reliably yield the answers to important problems: all we need to do is collect more data. Indeed, some public intellectuals have gone so far as to argue that advances in technology have made, or will continue to make, human beings fundamentally better or smarter than they were in the past (Pinker 2018). The underlying reasoning of these arguments is that because technology has dramatically enhanced our access to data, information, and knowledge, it will likewise enhance our ability to reason in an effective, just, and timely way. But it is important to remember that these new vast troves of data are not available to everyone equally: not all individuals have access to the data or the permissions necessary to access it; the hardware necessary to attain, visualize or process the data; the knowledge that such data actually exists and where to find it; or the technical know-how to do any or all these things. And having access to data that is stored elsewhere is not the same thing as knowing it, much less understanding it.

Most of us can recall plenty of times that digital information has made our lives easier. Perhaps you have visited a healthcare website like WebMD and looked up some symptoms you were experiencing: for example, *What are these strange bumps on the back of my tongue?!* Within moments, you probably had a plausible explanation for those symptoms and could make a judgment about the best course of action—whether to get medical attention, treat the symptoms yourself, or simply wait for the problem to go away, if it was a problem at all. *Oh, those are lingual tonsils, and everyone has them.*

When using such a website, you do not (and cannot) process all the information available to you. The curators of the website have distilled the massive body of medical information to the most important details, and in the course of browsing you narrow down the information even further and consume only the content that you recognize as relevant to your individual needs. But between "clearly relevant" and "irrelevant" lies the troublesome gray area of "possibly relevant," where the need for selective attention, and at least moderately informed judgment to guide it, can lead to mistakes. When we are predisposed to a particular diagnosis or happen to come across it early in our search, it is easy to convince ourselves that we have one disease when we actually have a different one—or none at all. For instance, "Morgellons disease," typically the presence of sores on the skin that appear to be made of fibers like cotton or other textiles, was named and made famous by an online community, and it is controversial whether it is a real skin infection or simply a misinterpretation of a common itch reinforced by the accounts of other sufferers on the internet (Vila-Rodriguez and Macewan 2008).

If keeping track of all the available knowledge and information seems impossible, we can take some solace in the fact that *knowledge is a social project, shared among many* (as we discuss further in chapter 5, section 5.3). There are experts—in the above example, medical doctors—who devote their careers to helping us navigate complicated subject areas. But even doctors have their own narrower specialties, and amassed information can get unwieldy even within those specialties. A case study in 2010, using echocardiography as an example, estimated that reading every existing paper on the topic would take over a decade, during which nearly another decade's worth of new reading would accumulate—and even that calculation assumed that the reader's daily time commitment to reading was equivalent to a full-time job (Fraser and Dunstan 2010). The days when a lone genius like da Vinci could become a leading expert on a wide variety of fields are over; as the deluge of information becomes deeper, the nature of expertise becomes narrower.

For those of us who work in the design and development of technology, it is reflexive to ask whether this technological problem has a technological solution—whether the world's digital information would be more digestible if only we had a finer-tuned human-machine interface to deliver it to us. We can see this mentality at work even among the giants of the field; Google co-founder Sergey Brin once remarked in an interview (Levy 2004), "Certainly if you had all the world's information directly attached to your brain . . . you'd be better off." The Google founders described their mission as making "the world's information . . . universally accessible and useful." Their products have certainly done a lot to make information *accessible*. But have they made it more useful? What does *usefulness* of information even mean?

It is important to note that *usefulness* of information (or data or knowledge) is not an absolute quality, but a relative one: it depends on the particulars of a situation. As discussed in section 3.1, wisdom can help guide our assessment of the usefulness of a piece of information at any given moment. Another key factor, however, is the sheer volume of information. It may be possible to build a system with some arbitrarily high amount of storage, but stored data and information gain force in the world only under the direction of human projects; therefore, systems should be built in a way that acknowledges human limitations.

The usefulness question may seem secondary to a person who considers information to be an intrinsic good, valuable and important for its own sake. But even such a person has to prioritize other goods, such as bodily health, if only to keep her mind in optimal information-gathering condition. And even an optimally maintained mind has a limited capacity to retain and employ information. Hitting that limit is called information overload.

Sidebar: Decision Fatigue

The phenomena of information overload and *decision fatigue* are closely related—handling a large amount of information requires constantly making decisions about where to look, what to focus on, and what to dismiss. Excessive decision making can wear down our ability to make further decisions thoughtfully. For instance, in a study involving choosing car features from a computerized catalog, buyers who were presented the most complicated choices early in the process (e.g., selecting from 56 possible interior colors) were more inclined to accept whatever was presented as the default option in later choices, which could ultimately be exploited to get them to spend more money (Levav et al. 2010; Tierney 2011). As technology practitioners, we may wish to embed lots of choices into our products with the intent of empowering the end users; but unless we think about how to limit those moments of choice, we may end up hindering the end users instead. And in our own professional lives, especially when we must make decisions of ethical significance, we should consider how we can change our work habits to conserve our decision-making energy for those situations when it counts.

The concept of "information overload" encompasses both descriptive and normative content. Descriptively, it suggests that we have an excess of information, which implies that there is some limit that we can in some way exceed. Normatively, it suggests that this excess is in some sense "bad" or undesirable. Much recent scholarship presumes that information overload is a problem that impedes a person's ability to reason clearly and make good decisions. The stakes of this impediment range from making bad purchasing choices to undermining citizens' ability to participate effectively in a democratic society (Himma 2005).

In the age of mass telecommunications, information overload gets far more attention than it did in the 1970s, when the term first began appearing. But the underlying notion can be traced back to the mid-1940s, when an engineer and science administrator named Vannevar Bush published an article discussing how the library system could not keep up with the constant accumulation of information as a result of ever-increasing research specialization. According to Bush, the quantity and inaccessibility of new research had become an obstacle to researchers, because the amount of effort required to track down and corral new material left them with far too little time and energy for understanding and analyzing the problems in their field.

To cope with the increasing quantity of information on offer in the mid-twentieth century, Bush proposed a technological solution: a device called a Memex, which foreshadowed hypertext and which inspired the development of many computational tools. The technological solutions Bush identified—including increasing accessibility, creating search functions, and allowing for associative indexing—have been realized, to a degree that could not have been imagined 70 years ago. But, as Levy (2005, 283) has pointed out, these developments "seem not to have solved the problem Bush identified but to have exacerbated it. . . . More of the record is broadly available to us than ever before, but there is less time to make use of it—and specifically to make use of it with any depth of reflection."

In the early twenty-first century, we likewise face an unprecedented deluge of information: more easily accessed and navigated than that faced by researchers of the mid-twentieth century, but no less overwhelming. The visionary technological solutions to that era's problems have transformed the information landscape, but they have not been able to solve the underlying problem of human limitations.

Story Point: "Codename: Delphi," by Linda Nagata

As she looked up again, a glowing green dot expanded into a new set of windows, with the client's bio floating to the top. Shelley, James. A lieutenant with a stellar field rating. Good, Karin thought. Less work for me. As she fanned the windows, the live feed opened with the triple concussion of three grenades going off one after another. She bit down on her lip, anxious to engage, but she needed an overview of the situation first.

Karin works as a handler for soldiers on the battlefield. She monitors myriad sources of information for each of the three clients she works with: live point-of-view audio and video; intelligence about mission objectives; overhead footage from drones and satellites; and real-time vitals and GPS positions on her soldiers, as well as on enemy combatants. Handlers like Karin are responsible for three active battles at once, and one mistake or overlooked piece of data could spell death for her clients in the field. Though Karin spends her days behind a desk rather than in the field, she is as exhausted and emotionally scarred as her soldiers, worn down by the sheer quantity of high-speed information management, the gruesome and upsetting nature of the things she sees, and the high cost of any potential failure. Although most IT jobs are not quite as high-stakes as Karin's, many of the pressures are the same, and "Codename: Delphi" offers an unsparing portrait of the toll those pressures can take on a person.

3.4   AUTOMATED DECISION-MAKING SYSTEMS AND BIAS

Sections 3.1 and 3.2 discuss some of the practical constraints on managing knowledge that often go underacknowledged in working with large volumes of data. In this section we examine some of the opportunities as well as some of the mistakes that can and do happen when we either lose sight of or ignore these practical constraints. Specifically, we explore the concept of bias in automated decision-making systems such as medical diagnosis, bail and sentencing, finance and lending, and product recommendation. We highlight the situations in which technology has helped and the situations in which technology has caused harm as these systems take on a larger role in our lives.

Bias is a complex concept with many definitions. In common usage it is tightly bound with a concept of fairness and typically means an inclination for or against some people, ideas, or things that results in treating them unfairly. As a statistical concept, bias is a systematic tendency for the true value and the estimated value for a population to differ. A common bias in statistics is selection bias, in which we do not choose a sample randomly and hence our estimates of a population are inaccurate. For instance, if we want to estimate the mean height of everyone in the world but select only male individuals from one country, our selection is biased and hence our estimate of the overall mean will be biased (Everitt and Skrondal 2010). Finally, in cognitive psychology, a cognitive bias is typically a heuristic or shortcut that one may use to make decisions, possibly irrationally. These can be both positive (e.g., the optimism bias may incline you to believe that things will work out) or negative (e.g., the base rate bias that leads us to focus on salient, specific instances instead of more general trends) (Kahneman 2011). So, although the word "bias" may mean many things, in this section we focus specifically on its association with using past data to make future judgments or predictions in ways that may seem unfair.

In many contexts we rely on information technology systems that use a mix of data, algorithms, artificial intelligence (AI), and/or machine learning to make decisions and manage complex processes. These decisions can be as mundane as adjusting the timing of traffic lights and as fraught as deciding bail prices and sentencing for criminal cases. In many cases where we rely on these data-driven decision-making systems, it can be tempting to portray these systems as improvements upon the flawed or limited abilities of humans. One famous example is the "hungry judges" case: a study of parole boards in 2011 found that parole was granted nearly 65% of the time at the start of a session, barely above 0% at the end of a session before a meal break, and again nearly 65% after a break (Danziger, Levav, and Avnaim-Pesso 2011). This is an example of the status quo cognitive bias in which we may make default or status quo decisions because they require less cognitive effort. This finding also triggers our common definition of bias because it seems inherently unfair that whether or not you are given parole may depend on where you happen to fall on the docket. Given that even a simple meal break may affect the decision-making process, one might argue that we should make decisions using only "raw data." We have seen the rise of many similar systems that have allowed professionals of various kinds—mortgage lenders, recruitment agencies, and even criminal court judges—to outsource the ethical responsibility of making crucial decisions that affect people's lives (O'Neil 2016).

It is important to remember, however, that "raw data" is not some universal, predefined truth. In the introduction to *"Raw Data" Is an Oxymoron*, Gitelman and Jackson remind us that "data need to be imagined as data to exist and function as such, and the imagination of data entails an interpretative base" (2013, 4). If we lose sight of the fact that someone selected which data to record, how to count things, and what not to count, we can end up hiding behind a mask of "objective data." This phenomenon is not new. Consider Frederick L. Hoffman who in the 1890s began a quest to understand "objectively," through data, whether or not formerly enslaved Black people in the United States were better off in a condition of slavery or freedom. Hoffmann claimed, "being of foreign birth, a German, I was fortunately free from personal bias which might have made an impartial treatment of the subject difficult" (Hoffman 1896, v). Hoffman proceeded on one of the first data collection studies across areas such as criminology, where he found that Blacks had higher arrest and incarceration rates than whites; health, where he found that Blacks had higher mortality rates and were more likely to suffer from various diseases including tuberculosis and syphilis than whites; and economics, where he found that Blacks produced fewer crops than whites. Through this data collection and a comparison of the numbers in the 1890s to those during slavery 30 years earlier, Hoffman reached the conclusion that Black Americans were better off enslaved and that, left to freedom, Black Americans were on a path toward "gradual extinction" (Hoffman 1896, 95; Kendi 2016, 2020). This data collection exhibited a high degree of selection bias and ignored the practical reality of the Jim Crow era in the US South, where Blacks were actively excluded from economic opportunity and health care and were actively pursued and arrested for minor crimes such as spitting in public (Chafe et al. 2011).

As we saw in section 3.1, it is not possible to create systems that are completely independent of the thought processes of the humans who build them. These thought processes are inherently encoded in the knowledge and data that the system leverages. Data does not simply produce knowledge; it is also the product of knowledge, and specifically of human knowledge, because when computers "know," they do so on terms created by human beings. Data that record past human decision making will preserve the value judgments that influenced those decisions. The value judgments are still present, even if their "neutral" appearance as data disguises them. As Barocas and Selbst point out, "data mining can inherit the prejudices of prior decision-makers or reflect the widespread biases that persist in society at large. Often, the 'patterns' it discovers are simply pre-existing societal patterns of inequality and exclusion" (2016, 671).

These concerns are not merely theoretical but have been borne out by several situations in which social decision-making algorithms have reproduced subjective bias in harmful ways (Crawford 2016; Staab et al. 2016). These include the following:

Predictive policing: Several major city police departments are using analytics to anticipate where crimes are more likely to happen and are sending more police to neighborhoods where a high crime rate has been reported. This practice can lead to a higher rate of reported crime in these same neighborhoods because there is much closer scrutiny by the police than in other neighborhoods (Staab et al. 2016). Police have also used similar programs to decide whom to hold in custody or to charge, again leading to a feedback loop based on historical patterns of overpolicing (Angwin et al. 2016).

"Weblining": This term is an allusion to "redlining," the practice of offering services such as home loans or insurance on a selective basis, making them unavailable to the residents of neighborhoods that are predominantly poor or are ethnic minorities. By using a person's home address or a constellation of other data as a proxy for other characteristics, businesses were able to discriminate against members of racial minority groups while staying within the bounds of the law that forbade them from making decisions based on race (Staab et al. 2016).

Targeted advertising: Retailers that collect purchasing information from its customers can and do use this information to predict individual shoppers' choices and habits, and advertise accordingly (Staab et al. 2016). This targeting of advertising can take other forms as well: in 2016 it was shown that Google was displaying ads for high-paying jobs to women much less frequently than to men on the basis of historical characteristics of who would click on the ad and be successfully hired (Crawford 2016).

Sentencing software: Judges in parole cases are using algorithms to determine the likelihood that a given individual, who is being considered for parole, will re-offend if released. In the case of one widely used program, it was shown that black defendants are twice as likely to be flagged incorrectly as high-risk, whereas white defendants were twice as likely to be incorrectly flagged as low-risk (Crawford 2016).

Hiring and admissions: Algorithms now almost completely control where job ads appear on hiring websites, social media, and most digital platforms. The placement of these ads can reflect historic bias based on perceived race, gender, economic, or educational status (ACLU 2019). Many companies then use algorithms to sort resumés and applications, leading to further reinforcement of biases (Bogen and

Rieke 2018; Schumann et al. 2020). These algorithms are also in use for admissions for higher education, in which they have been shown to replicate a preference for white or male applicants (Burke 2020).

Numerous other examples of these issues abound in popular literature, including O'Neil's (2016) accounts of biased decision making in areas as diverse as law, loans, and hiring. These issues are also the focus of Eubanks (2018), who draws specific attention to how services for the poor are biased on the basis of what data is collected and what is not. D'Ignazio & Klein (2020) and Perez (2019) look at the gender biases that are coded into data, and Noble (2018) focuses on racial bias.

In many of these cases, big data analytics reproduce existing patterns of bias and discrimination—even when such discrimination is entirely unintentional—because the historical effects of these long-standing biases are a part of the data set and reflect systemic issues across society. The history of discriminatory treatment has meant that women, minorities, people with disabilities, and people with a low economic status have not had the same opportunities to achieve and succeed, and that comparative lack of success is recorded in the data. It is also possible to see how the history of discrimination might "follow" a homebuyer or an inmate into an algorithmic assessment of that person's future, even if their race or gender is not being explicitly considered. Machine learning techniques and massive data sets can easily (even unintentionally) be leveraged to make predictions that correspond to historical patterns for a given person's race or gender, even if race or gender is not being used directly as an analytic category. In this way, data ends up being used (deliberately or otherwise) to circumvent societal rules and laws that exist *explicitly to prevent* decision makers from using these features for making decisions.

All is not lost, and many times the use of automated decision-making systems can lead to positive outcomes and even address societal issues and bias. Abebe et al. (2020) point out three possible roles for computing and large data systems for promoting positive social change: using these systems for diagnosis, for rebuttal, and as a synecdoche. In many of the previous examples we can see computing as a synecdoche, that is, computing as a part that stands in for a larger whole, bringing new attention to long-standing issues. As Eubanks (2018) points out, the problem of how systems of power reinforce and punish poverty is not a new one, but the conversation has gained new meaning and new attention when framed as a technological problem, hopefully bringing attention, resources, and change to how we support those in need. There are also positive outcomes from large, automated decision-making systems: crisis and drug intervention systems have helped to identify, and to match with treatment, people suffering from drug and depression issues, cutting rates of suicide and drug overdoses (Calzon 2020). Within finance there has been a massive decrease in fraud and credit card theft through the use of machine learning systems that are biased by users' individual behavioral data, because what we are trying to detect is deviations from an individual's typical behavior (Calzon 2020). In more mundane areas, the ability of Netflix to suggest a great show that you haven't seen and for Amazon to highlight products that interest you requires leveraging massive data sets and information about what interests people like you; that is, predicting your behavior from a biased sample increases the applicability and accuracy of these systems (Netflix 2021). Finally, in the space of large-scale public health, using data on cancer and health outcomes has helped establish new interventions for cancer treatment and screening and led to new EPA protocols and guidelines (Shaw and Younes 2021). All of this is driven by collecting and analyzing large troves of data and using these data to make predictions, sometimes with full knowledge that a sample may be biased by location or some other factor.

Abebe et al. (2020) describe two other ways in which information technology systems can help produce positive social change: using computing as a diagnostic and using computing as a rebuttal. Many of the systems that we have described in this section, once formalized as an information technology system, can then be tested for bias. In this way one can see the role of these computing systems as a way to perform diagnosis or "debug" decisions that we have made in the past, and to highlight fairness issues within particular sociotechnical systems that may have long been a part of actual practice, we will return to this discussion of sociotechnical systems in chapter 5. Finally, computing can be used as a rebuttal to the deployment of information technology systems or to long-standing beliefs that are not valid. Abebe et al. (2020) point out that collecting and marshaling the huge volumes of data that modern society produces and records can be leveraged to argue against the use of that data. Researchers were able to convince the US Immigration and Customs Enforcement Agency (ICE) not to deploy a system to evaluate whether or not potential immigrants were going to be a "positively contributing member of society" by showing that no system could perform this analysis on the basis of available data. Likewise, researchers who work on matching mechanisms for resettling refugees have shown that integrating more information about local communities, including whether or not there are clusters of people speaking the same language, can improve resettlement outcomes.

One contemporary discussion in the area of automated decision making, which perhaps offers some hope for mitigating issues in this domain, is the question of whether biases are introduced by biased data or biased developers. On the biased data side of this theory, the idea is that many applications are developed using historical data and that these data reflect and could possibly perpetuate any bias present in a former time as represented by the data. On the biased developers side of the theory, the idea is that the programmers developing the algorithms are not representative of the populations that will be impacted by the system, and this lack of representation may lead to bias in the algorithms that are created. Cowgill et al. (2020) attempted to answer this question using an intervention study with over 400 AI workers. They used multiple interventions to test whether programmers with different characteristics and different levels of implicit bias created biased decisions in the final software. Cowgill et al. found no evidence that minority or low-implicit-bias workers generate better, less biased predictions. Conversely, better data leads to better predictions, and as Cowgill et al. found in their study, so does a simple intervention that reminds workers, "As you write your algorithm, please be mindful that your training data set may originate in a biased social system. Adjusting your algorithm to account for discrimination in hiring, self-sorting of applicants, or other sources of such bias could improve your accuracy on the test set. You will be evaluated only on the accuracy of your predictions on the test set." Cowgill et al. found that this warning serves as a reminder about the main point of the first part of this chapter: that data is not value-neutral and that we must carefully reflect on what the data is, where it comes from, and why it was collected, before we leverage that data (or decide not to) in the design and development of new systems.

As we have seen, large automated decision-making systems can both counter and perpetuate bias. Imagine a system based on the decisions of our hungry judges. This system, if not interrogated thoroughly, may perpetuate and reinforce the existing bias. This is an even more pernicious risk in the context of automated decision-making systems because data is quite often thought of as value neutral, and the answers generated by these algorithms have the appearance of objectivity. The systems therefore make it possible for these bias-dependent assessments to be characterized as impartial, lending strength to the biases that the algorithms were designed to avoid and making it more difficult for those unjustly denied to appeal the decisions (e.g., "this is just what the system said to do"). This appeal to data as value-neutral is called "ethics washing," a cover-up or facade to hide unethical behavior (Bietti 2020). It is important to note that collecting more data, or making sure that we understand what "fair" and "unbiased" decisions are in the particular domain, is not enough. As Corbett-Davies and Goel (2018) argue, "one cannot generally address [problems of bias] by requiring that algorithms satisfy popular mathematical formalizations of fairness." Hence, as we discuss further in chapter 5, we must take great care when working on systems that make decisions in society, as it is all too easy for these to perpetuate a history of constricted opportunities into limited future opportunities for historically marginalized groups.

These social and ethical issues posed by automated decision-making algorithms have drawn the attention of both major professional organizations in computing, the Association for Computing Machinery and the Institute of Electrical and Electronics Engineers, which we discuss further in chapter 6. As of 2020, both these organizations have established working groups for discussing topics of fairness, accountability, and transparency in algorithms, and for developing guides and best practices for working professionals in computing technology. Many companies, governments, and academic institutions have also released various statements about AI principles and practices, and as of 2020 there were over 25 of these types of statements. Researchers at the Berkman Klein Center at Harvard University have analyzed the trends in AI ethics statements and possible solutions (Fjeld et al. 2020), finding that highlighting the need to mitigate bias and promote transparency is of tantamount importance for the deployment of AI and information technology systems. Finally, significant legislation in the European Union, called the General Data Protection Regulation (GDPR), specifically requires the right to an explanation of any decision made by an algorithm in order to increase transparency and battle bias (Goodman and Flaxman 2016).

These efforts and guidelines underscore the importance of approaching the design and development of automated decision-making technology with a systemic understanding of the nonneutral nature of data, information, and knowledge and provide jumping-off points for how to integrate best practices about data into your professional life.

3.5 STORING KNOWLEDGE OUTSIDE OURSELVES: HOW DOES IT AFFECT US AS INDIVIDUALS?

As we saw above, on the one hand it is appealing, and potentially useful, to have a wealth of information available to all of us all the time. But on the other hand, past a certain point more information causes as many problems as it helps solve.

For all of recorded human history (and likely even before that), human beings have stored knowledge and information in places where they do not have instant access. They have shared knowledge among themselves and come up with social systems for organizing and tracking who is in charge of knowing what and with regulatory systems to make sure that knowledge is shared (or withheld) in ways that somebody or other has deemed appropriate (Hirst et al. 2018). It's useful to examine those prior approaches to managing knowledge—how they have worked and what their limits are—as a way of thinking about new technologies, those that exist and those still in development, that help human beings store, retrieve, and analyze the things they have learned.

3.5.1 STORING KNOWLEDGE IN ANCIENT TIMES

How are advances in technological systems going to affect our brains and our ability to store, organize, and synthesize the things we know? As we noted above, modern technological systems represent a major shift in our ability to store knowledge, but it is not the first such shift. As we mentioned in section 3.1, Socrates was worried about these kinds of issues when the technology in question was written language.

Even before the development of writing, people have always needed techniques to manage knowledge; the human brain itself retains information and knowledge so that it is not always able to retrieve it without specific prompting (Tulving and Pearlstone 1966). Prehistoric cultures developed techniques for organizing their own mental storage and for requisitioning the information that is sequestered in their own brains. The "memory house," which was part of classical rhetorical training, was a device by which people mentally constructed and maintained a house of many rooms and "stored" different things that they had learned in different parts of the house. It has also been argued that the ancient practice of naming constellations did not represent actual belief that clusters of stars were connected to the associated myths, or even that they resembled the things they were named after, but rather that constellations served as memory devices intended to "store" culturally important stories and to impose an order on the map of the sky so that it, too, could be more easily remembered (Carruthers 1998, 25–27).

The technology of writing also has benefits for using and engaging with information that Socrates and Plato did not acknowledge. In fact, the shift to literate culture has in some respects helped human beings to think *more* deeply. As classicists and cognitive scientists have shown, abstract and conceptual thought are aided significantly by the existence of written texts; they are further aided by organizational tools, such as indexes and tables of contents, that can exist only in reference to written texts. These new tools are widely accepted as the reason for major advances in mathematics, philosophy, and many other branches of learning (Rubin 1997).

Yet it is important to understand *how*, exactly, these tools have helped to support human thinking. The advances of literate cultures have less to do with the expanding volume of available information made available by the written word and more to do with the increased range of ways that people can interface with their culture's texts. Written texts and images allow human beings to deploy visual and spatial skills, as well as aural ones, in remembering and processing a given text (Kuhn 2010). These advances, and others, have built on existing strengths of the human mind—strengths that had been untapped by the particular task of listening to and remembering oral texts. In this way, the technologies of writing changed the *conditions* of thinking and understanding, and "added to the collection of tools that can solve thought problems" (Rubin 1997); in so doing, they changed the kinds of thinking that could take place. Many premodern practices of memory were built on this principle: that visually interfacing with an idea could expand the ways that the mind could make use of it. In Europe, memory techniques such as Cicero's method of *loci* (the memory house), or medieval monastic *memoria* (which involved crafting mental images), aimed to reproduce a visual relationship between the rememberer and the thing being recalled. These techniques were aimed not at enabling regurgitation but rather at enabling the rememberer to draw on the remembered phenomenon in a flexible, generative manner. Building a multisensory relationship to the remembered item was central to the theory of these *memoria* techniques and crucial to their success (Carruthers, 1998, 10). Similarly, Australian aboriginal cultures use a variety of sensory and bodily practices to establish and communicate memories. Most notable of these is the songline tradition, in which a story—typically spoken but also possibly including song and/or dance elements—is held and maintained by an elder. This story typically incorporates aspects of the flora, fauna, and layout of the local area to encode detailed information including numerical, spatial, and temporal relationships of importance, such as tribal movements, hunting and farming information, and maps and directions. The key distinguishing feature in the traditions of Australian Aboriginals is that the story is closely tied to place and that the local references are deeply interwoven with the story. Interestingly, aboriginal techniques were taught to first-year medical students in 2021 as a technique for passing their exams, and the songline techniques worked better than memory palaces at helping students recall information (Reser et al. 2021).

However, new technologies of information may also expose weaknesses or shortcomings that were invisible or irrelevant under prior technological conditions. Technologies like hypertext have addressed historical concerns about the navigability of available information, but these same developments have also exposed the fact that navigability cannot overcome the essential limits on the human mind's capacity to absorb. Although it is not always possible to anticipate the possibilities and pitfalls of a new platform for information or knowledge, it is vital to be aware that platform shifts can impact the conditions under which we learn, think, and remember, and therefore very helpful to be aware of the history of some of these shifts.

3.5.2   HOW COMPUTERS CHANGE THINGS

If Socrates in *Phaedrus* worried that the convenience of written information would cause forgetfulness, he would certainly find it troubling that we can now browse through a zettabyte-sized trove of information on demand. And the internet has amplified not only the volume and accessibility of information in our daily lives but also the ease with which we can navigate it; beyond indices and tables of contents that help us find our way around single texts, we now have search engines and hyper-links that meld the individual sources into one interconnected body in which we can easily discover new relevant content. Modern thinkers who echo Socrates argue that the ready availability of information housed in computers instead of brains is weakening cognitive abilities like short-term memory, a phenomenon nicknamed "digital dementia" (Spitzer 2012). There is some empirical evidence supporting this claim; in an experiment in which subjects were asked to transcribe a provided list of trivia facts onto a computer and then to recollect those facts, people who were told the typed document would be saved remembered less than people who were told the document would be erased (Sparrow et al. 2011). Studies have shown that people engage differently with digital and print text. In particular, people are skimming more and doing less deep reading (Liu 2005).

3.5.3   THE VULNERABILITY OF STORED IDEAS

Externally stored information can always be lost or destroyed, no matter the medium. In some ways, information faces the same hazards residing in a digital medium as it does written on paper. Fires and floods can destroy a server room just the same as a library. Exposure to the elements and the hands of readers gradually put wear on books; cosmic radiation and repeated access gradually cause corruption in computer files. In fact, today's digital storage devices tend to expire much more quickly than the analog media of old (Bollacker 2010); contrast the millennia-long survival of many ancient Egyptian papyrus documents to the mere years-long life span of an optical disc. Also amplified with digital storage is the potential for loss due to human malice (e.g., a computer virus) or negligence (e.g., a careless press of the Delete key).

Fortunately, the ease with which digital files can be replicated mitigates these problems. Database servers like Google keep copies not only across many machines but also across many geographic locations, ensuring that nothing is lost to a localized catastrophe (Barroso et al. 2003). On the web, even if the original server for an important piece of content has gone down, there is a high chance that another entity (e.g., the Internet Archive) has mirrored it. There is some truth to the saying: "Once it's on the internet, it's there forever." It is tempting to believe that this ease of replication alone is enough to create permanence.

However, digital storage brings a newer, more insidious issue. Although classical media is directly human-readable, we depend on hardware and software to translate electronic bits into an intelligible form. These devices are not universal but rather rely on knowledge of the specific scheme used to encode the bits. The new threat is having the *data* but losing the means to extract the *information*. (A less catastrophic version of this can happen in written human languages, too, when the usage of particular words shifts and some shades of meaning are lost to later readers.) For instance, years after the Viking probes visited Mars, NASA scientists struggled to process images from the mission because the data format was obsolete and the original programmers were gone; months of effort reverse-engineering the probes' recording machines were needed before the processing could be done (Smit et al. 2011). Files that no longer have software that can read them, software that no longer has operating systems that can run it, and storage devices that no longer have hardware that interfaces with them—these are consequences of the rapidly changing digital landscape, and some media experts warn of a "digital dark age" (Bollacker 2010) in which we lose a great deal of information to technological obsolescence.

The vulnerability of digital archives raises the specter of lost data or information. Perhaps more troubling, however, is how technology creates the conditions for some kinds of *knowledge* to be lost. When the people who are entrusted with a certain kind of knowing become reliant on technology, there is the risk that they themselves will lose that knowledge. One high-stakes example is airline pilots. In an article entitled "The Great Forgetting," Nicholas Carr (2013) writes of several plane crashes—each one both fatal and preventable—that occurred when the jet's autopilot failed and the human pilots in the cockpit failed to steady the plane, even though their training should have prepared them to do so. For Carr, this is only one example of how, in relying on AI and automation to perform tasks that were once considered highly skilled, human beings have allowed their own knowledge of those tasks to atrophy, a process called *deskilling*, which we discuss in chapter 5.

The main issue here is not that the human mind has been so altered by technology that some kinds of knowledge are now out of reach. Rather, the argument is that technology can seem sufficient to *replace* human knowledge; as a result, individuals who have been entrusted with specialized knowledge will lose their easy command of that knowledge and, in the event of technical failure, fail to regain it in time to avert disaster or loss.

Story Point: "Lacuna Heights," by Theodore McCombs

> What if Andrew is not so tech illiterate as he claims? What if he's actually quite good at using the Aleph, at least when he's using it for Privacy Mode, to keep his affair secret even from himself? This is what scares him about neurotech: We've outsourced so much of our selves to an array of companies, so their algorithms shoulder the hard work of living—and now he's doing it himself, to outsource the planning, lies, and guilt of adultery to this other, shadow self. And if he doesn't figure out what's going on, soon, he'll lose Madeleine and this world he has with her.

At the center of "Lacuna Heights" is the Aleph, a neural implant that connects your mind directly to the internet. This story shows both the micro- and macro-level costs of this radical innovation in knowledge technology, which bundles individual users' experiences and memories and makes them available for public searches. Andrew is worried about something, but he doesn't know what: he has been using Aleph's privacy mode to store some of his experiences, which protects them from being bundled but also means that he himself can't remember them when he goes back into public mode. In that absence of information about his own life, his best guess is that he's having an affair, a theory that reflects the limits of what he can understand without that information. As we learn over the course of the story, Andrew has hidden large parts of the past from himself in an effort to block out the grief and guilt they cause him. Yet those feelings continue to seep into his public-mode life, even though he doesn't have access to the knowledge that would help him understand or respond to them. Andrew's individual act of forgetting is aided and abetted by a larger social forgetting, as the denizens of this future San Francisco strive to move on from a tragic and traumatic past. By giving us an up-close look at Andrew's struggles, "Lacuna Heights" illuminates the social, emotional, and ethical costs that can occur when we outsource the task of some kinds of remembering.

3.6 STORING KNOWLEDGE OUTSIDE OURSELVES: HOW DOES IT AFFECT OUR COMMUNITIES?

So far, we have discussed the issue of how and where information and knowledge are stored and used relative to the individual. But knowledge and wisdom have always been the product and possession of communities, as much as of individuals. In this section, we consider how knowledge has historically been stored, synthesized, and shared by communities, and how the advent of modern technological systems that process vast amounts of data builds on existing patterns as well as offering new possibilities. We further discuss how modern information and communication technologies have helped to create new communities of knowledge and to mediate existing ones.

3.6.1 COMMUNITY CONTEXT AND DIKW

Cultures of knowledge can take many forms: they can be bound together by professional, philosophical, political, or cultural concerns. Think about certain kinds of trade, crafts, sports training, or playing music. If a person is learning a new skill or discipline, it becomes necessary for them to "learn the ropes" from their more experienced counterparts. Even if they do not receive formal education or training in that field, they will still need to figure out how to do the work successfully and so will emulate more skilled and experienced practitioners.

A culture of knowledge supplies its participants with basic frameworks for understanding the world (or some limited part of it). It also offers participants a set of practices to help them get better at doing the tasks central to that knowledge culture, or at understanding the world from the perspective of that knowledge culture. For instance, neurochemists and psychotherapists will both see depression as a problem, but will describe the problem itself differently, and will of course see different solutions. Both cultures (or at least some of the people in each of them) know that their own solution is limited and will not fix the problem entirely. But one side effect of belonging to a knowledge culture that has control of partial solutions is that it can shape an individual participants' beliefs about which parts of the problem can be meaningfully addressed and which must be endured.

A given knowledge culture will often have a shared point of view on important questions. Even when there is disagreement, that disagreement typically occupies a limited range of points of view and relies on basic shared ideas about what the questions are and how to ask them. But within the broad consensus of a given culture of knowledge, the information and knowledge held by individual members is not necessarily consistent. In medicine, for example, a patient with a complex problem is usually treated by several different specialists working together in a team. Those specialists share the same broad knowledge base (the human body, or even a specific system within it) and a common normative commitment (to heal their patient), but they are in command of different bodies of specialist knowledge and have access to different kinds of information about that patient. Even when they share that information with one another, the things they learn may be different, depending on what their particular expertise enables them to notice or infer. This distributed responsibility for collecting and interpreting data and information, in light of different but complementary bodies of knowledge, allows a medical team with varied skills to treat a patient far more effectively than they could have if they all knew the same things or processed information in the same way.

A preference for tried-and-true methods, building on the insights of earlier generations of a given knowledge culture, is not always easy to distinguish from a reflexive reliance on what is familiar and comfortable. One concrete example of this pattern is that knowledge cultures tend to elevate and celebrate those who most resemble past leaders, which is a disadvantage for those who are different or who are not trained by previous elites to "fit the mold." On the other hand, discontent with this system can spur the elevation of individuals simply on the basis of their being different—for instance, an actor or athlete who is elected to political office as a protest against the usual politicians.

The effects of authority figures on community knowledge include the way teachers, religious leaders, politicians, celebrities, and journalists, editors, and bloggers frame material, as well as the choices that social networks, search engines, and news aggregators make about what we see. Imagine the following scenario. A celebrity doctor advocates a particular diet, based on a food grown only in Central America. Other celebrities praise the diet and publicly follow it. Bloggers write about it, and journalists publish stories about the popularity of the food and about the communities that grow it. Trendy restaurants feature it. The food focuses attention on one aspect of diets, such as protein, and suddenly, that aspect is part of standard discourse about diet. The Central American communities that grow it switch their economic model to export most of their crop, changing both the local economies and the local diet. This example shows how authority figures can affect what we know about something (in this case, diet), and how that change in the knowledge culture can have huge ramifications beyond the knowledge itself.

3.6.2 TRANSMITTING KNOWLEDGE ACROSS CULTURES

Information and knowledge do not circulate only within a given community. Any record of a knowledge culture can reach and have an impact on those outside the culture. People sometimes travel outside their own communities. They have new experiences that transform them, and they pick things up and bring them home to influence their home cultures.

This sort of boundary crossing is celebrated by the open information movement and others on the grounds that it can break us out of the traps of rigid knowledge cultures: narrow-mindedness, ignorance, homophily, and mob mentality. The historical record shows that this is true: with greater access to information, people can discover that their interests are legitimate and possible to pursue. They can be exposed to ideas that disrupt the explicit or implicit consensus of their communities.

On the other hand, although we can celebrate the idea of a person being exposed to new political ideas that resonate with their values or find others who share their specialized interest, that is not in itself a good. Consider the cases in which the new political ideas are grounded in a hate-based ideology, or the interests-and-skills community teaches people how to groom children for sexually abusive relationships. For specific examples of these kinds of new connections, see chapter 5, section 5.3.

Boundary crossing, learning about unfamiliar things, is challenging, but people don't always rise to a challenge. As with any information or knowledge, it's all too easy to get false or incomplete information. When we are working in a new domain or with new frameworks, we don't always have the resources or experience to correct misinformation or misinterpretations. A person's capacity to collect and synthesize ideas is limited and is shaped by what they already know. Given that reality, selective engagement with other cultures—which is much easier to do when that engagement is technologically enhanced and doesn't involve going to a place and meeting people—can lead one to misunderstand what those other people are doing. This can lead to a belief in their inherent ignorance or inferiority. It can also lead to cultural appropriation, the out-of-cultural-context bastardization of a culture's intellectual and creative property. Although transmission of knowledge can lead to solutions to problems both known and previously not observed, it is not a categorical solution. Because human beings are finite and also get tired sometimes, a larger and diverse sphere of available ideas does not, by itself, transform humans as knowers and actors and participants in culture. Rather, that expanded sphere creates a different set of conditions under which the same basic problems of human conduct can take root, and a different set of conditions for contending with those problems.

3.6.3 EVERYONE'S AN EXPERT: INFORMATION AND KNOWLEDGE IN THE AGE OF MASS PLATFORMING

The pre-internet technologies of writing and the printing press, as well as other information technologies not discussed here, such as the telegraph, have each, in their own way, changed the way that information and knowledge is circulated both inside and outside certain knowledge cultures. In this sense, the internet does have precedents, although these precedents were themselves quite dramatic in their effects. The effects were not always what their creators or advocates had envisioned—such as President James Buchanan's prediction that the telegraph "would prove to be a bond of perpetual peace and friendship between the kindred nations," a hope that was almost instantly disproven (Brooking and Singer 2016)—but they were inarguably transformative.

What *is* new about the internet, however, is the way it enables many-to-many exchanges of information and knowledge. In a high-literacy age when a significant portion of the global population has consistent (if not constant) internet access, the ability to share ideas is available on an unprecedented scale. Relatedly, the structure of many parts of the web allows for and even encourages undirected wandering so many users may stumble upon many more different sources of information than before. In contrast, some sites, particularly social media sites, offer the user a much more "tailored" experience that makes it far less likely that you will encounter something unfamiliar or surprising. We return to these ideas in chapter 5 when we investigate the effects of the scope and scale of many-to-many exchanges in society.

The internet makes it very easy to get outside of one's native cultures of knowledge and observe what other people are up to, benevolently or otherwise. It makes it possible for people to form new communities that develop their own cultures. It allows for crowdsourcing solutions to problems. It creates a context in which expertise is less powerful than it once was because people other than officially sanctioned experts now have the means to speak and be heard. Relatedly, the internet also makes it possible, at least in principle, for individuals or groups with limited resources and prestige to announce events, to launch campaigns of harassment including revenge porn and other forms of bullying, to start memes, to raise money, or even to produce a social good such as helping to track down criminals from open-sourced data (Gross 2021) or calling attention to police brutality and social injustice (Kornfield 2021). For some, this is all straight-up good news: it's the democratization of information. For others (Brooking and Singer 2016), the potential use is terrifying.

One concern that has information scientists worried (Bawden and Robinson 2009) is the legitimacy of information and knowledge. Even before the advent of the internet, public conceptions of knowledge and expertise were shifting as fracturing knowledge cultures made it harder to find consensus about what counts as truth or proof (Lepore 2016). Drawing on the Yorùbá approach to knowledge described in chapter 2 (Hallen and Sodipo 1986), and considering a source's personal character in epistemological terms, can help one make decisions about whose version of truth is likely to be trustworthy; however, it's challenging to bring this communitarian approach to bear on everyone you encounter on the internet. We have all seen self-professed "experts" spouting nonsense in the comments sections of websites. Those commenters are not always easy to distinguish from experts (self-taught or otherwise) who truly know what they are talking about, but comment sections at least make it relatively easy to see who is contributing which ideas. There are also, however, some types of sites that do not call attention to the fact that their information is sourced from all kinds of experts, trustworthy and otherwise. For instance, the decentralization and anonymity of Wikipedia allows for biased or incorrect articles to stand, sometimes for a long time, and for certain pages, such as on abortion or on major political figures, to be sites of ongoing battles for control of content and presentation.

This devaluation of expertise presents a challenge to the methods that academically trained experts have used to maintain the quality of scholarly work. Academic culture in the West has long depended on peer review to validate research. Venues such as professional or academic journals or conferences depend on experts in the field to validate the correctness, relevance, and novelty of published works. There are several trends that are challenging the expertise-based model of publication. One is the existence of online research repositories, such as arXiv, where one can post without peer review, although there is space for public in-site reviewing. Another is the rise of "predatory" conferences and journals, which claim to use peer review but in fact do not. Many academic experts maintain blogs or Twitter feeds to discuss new work and present informal thoughts related to their expertise, but there are also myriad blogs where posts are not supported by the evidence or theoretical knowledge that the academic community of knowledge expects.

Thus, one effect of the broad availability of information is to challenge, in both productive and unproductive ways, the notion of expertise and of control of production and distribution. On the one hand, anyone with an open internet connection can encounter a wide range of ideas, arguments, and viewpoints. On the other hand, people still sometimes respond to this overwhelming array by putting trust in a single authoritative voice, or by succumbing to mob mentality and letting themselves be steered by the emotions and values of the group in which they find themselves.

Although mob mentality rarely has positive results, it is especially dangerous when it is deliberately weaponized. There have been studies of how, for instance, Adolf Hitler's government used timing (dusk) and lighting, music, and other effects to amplify the emotional resonances of his rallies and thus of the pressure on individuals to collaborate with his regime (Rawson 2012). Although the structure of the internet has diffused some of the factors that can promote mob mentality, in other respects it offers an ideal staging ground for mobilizing large groups of people on the strength of shared emotion. Although the internet is highly decentralized, big corporations—search engines and social media, as well as news media—have gathered enormous control over what information we receive and how it is presented. However, the internet and often the sites of these large corporations also offer the opportunity for individuals and small groups to organize large events or movements. Using search engines, social media, and news media, as well as carefully placed bloggers and carefully staged and reported actual real-world events, groups with financial and social capital can manipulate the emotions, beliefs, and ultimately, the actions of a population. Doing so successfully is difficult, but with the growth in data mining and machine learning as well as the ability to outsource human expertise, it is increasingly possible.

Story Point: "Welcome to Your Authentic Indian Experience™," by Rebecca Roanhorse

"Nobody wants to buy a Vision Quest from a Jesse Turnblatt," you explain. "I need to sound more Indian."

"You are Indian," she says. "Turnblatt's Indian-sounding enough because you're already Indian."

"We're not the right kind of Indian," you counter. "I mean, we're Catholic, for Christ's sake."

What Theresa doesn't understand is that Tourists don't want a real Indian experience. They want what they see in the movies, and who can blame them?

Jesse Turnblatt is both a personal and a professional American Indian. He works for a virtual reality outfit called Sedona Sweats, guiding Tourists on a spiritual journey into a Hollywoodized version of Indigenous American culture. In the VR world, Jesse changes his name and appearance to match the Tourists' popular-culture-based expectations of what an authentic Indian experience (and an authentic Indian) will look like. Although Jesse's identity is his most important credential as a Spirit Guide, he prioritizes the Tourists' ideas about authentic Indian-ness above his own—after all, they are the ones paying for the Experience, and Jesse knows that money is what matters to his Boss. By taking us inside Jesse's life, this story illuminates not only the fraught dynamics around minority identity and authenticity, but also how the power of knowledge often depends less on the quality of the knowledge and more on the power of the knower.

3.7 CLOSING THOUGHTS: KNOWLEDGE AND SELFHOOD

The paradigm with which we opened this chapter, and that is widely assumed within STEM fields, characterizes data and information as objective. By extension, it assumes that the knowledge and even wisdom rooted therein will be broadly recognizable as correct in some way because it is synthesized from data and information that is essentially objective.

As we have seen in this chapter, however, that account of data, information, knowledge, and wisdom falls short in several ways. It fails to account for the way in which cultures of knowledge shape the kind of information that seems intelligible or instructive and the sorts of data that seem useful or available. These cultures of knowledge are durable: for better *and* for worse, a given knowledge culture is far more likely to find a way to assimilate information that might otherwise challenge the collective, shared knowledge and wisdom than it is to discard, or even adjust, its existing frameworks for making sense of the world. Moreover, the human mind can absorb a finite amount of data, information, and knowledge before its ability to manage those things begins to break down. Even before the advent of the information age, persons and cultures were forced to develop strategies to manage their interaction with the vast quantities of data, information and knowledge that confronted them.

The challenges that necessitated these strategies of managing knowledge have not been solved in the information age; they have, on the contrary, intensified. In order to understand how data, information, and knowledge circulate within the world, and to participate productively in this world, it is essential to understand the strategies—and the strategic weaknesses—of human individuals and cultures in coping with the overwhelming onslaught of all the things there are to know.

Once we reformulate our understanding of DIKW, we can gain a clearer picture of how these things circulate in the world and how technological systems shape that circulation and the lives of its users. As this revised paradigm makes apparent, DIKW are not just freestanding bits that circulate in the world, independent of the human beings who know them or resist them or are guided by them. Rather, the things a person knows inevitably become a part of who that person *is*. This is also true on a broad-strokes historical level, with regard to the traditions of knowledge that we inherit: as the philosopher Hans-Georg Gadamer notes, "we do not conceive of what tradition says as something other, something alien. It is always a part of us, a model or exemplar" (1992, 283).

Our choices about what kinds of knowledge to be guided by, what kind of wisdom to aim for, what kind of data to look for, and what kind of information to accept as real are ultimately choices about what kind of individuals and citizens we will be.

This interrelationship between knowledge and the self raises some key ethical questions that are ever more urgent in the information age. If we as individuals and citizens are to an extent constituted by our knowledge or wisdom or ability to collect information, does the existence of modern information and communication technologies render us, ourselves, less valuable? If the DIKW that constitutes our experience is more widely available to everyone, what makes us unique? To what extent does using modern information and communication technologies make us more autonomous and more free to choose our course—and to what extent does it limit our freedom?

Even asking these questions tips the deck in one direction: it implies that we as individuals can choose to what degree we interact with technological systems and let them shape our decisions. This isn't necessarily true. Even if we can make some choices about how we ourselves engage with the things that we can learn from or know, we can't control the choices of others in our community, and we can rarely choose what is known about us. To an ever-growing degree, individual people are being commodified as components of data sets, which are available to marketers and politicians alike. This reality poses a particular ethical challenge for technologists: how can one build systems that serve the people who use them and enhance their lives, instead of converting them into products for consumption by the powerful few?

The next chapter digs more deeply into these questions of privacy, and of what makes a person who they are and helps them to flourish. As will become clear, these two subjects—which may on their face appear to be unconnected—are in fact intimately related.

REFLECTION QUESTIONS

1. *Pick a decision-making process that incorporates technology, data, information, and knowledge in modern society, such as college admissions, targeted advertising, or bail and sentencing. Describe what needs to be collected or processed, such as what data, processed how, incorporating what background knowledge or judgments, at each stage of the DIKW paradigm to arrive at these decisions. What risks are involved with formalizing such a system in this way? How could errors at one level of the hierarchy impact the functioning at other levels?*

2. *Find an example of a data storage methodology that predates computing. What were the goals of this methodology? What had to be left out because it could not be recorded in this way? Find an example of a modern, digital age, technology that addresses this shortcoming. How have we, as a society, come to use this new technology? Has it helped or hurt our decision-making abilities?*

3. *When you are preparing to make even a small purchase decision—for example, what to eat for lunch—what data goes into your decision making? Do you consider price,*

nutrition, where it comes from, aesthetic qualities, reviews from friends, reviews from the internet, or other sources? Make a list of all the sources of data and/or information that you could consult when making this decision. Do you always use all these when making a decision? Why or why not?

4. *Dive deeper into one of the situations in which bias encoded in an algorithm has had impacts in our society—predictive policing; weblining and the digital divide; targeted advertising; bail and sentencing software; or hiring and admissions—and find a recent example in popular media. Take this article and analyze it through one of the ethical frameworks we discussed in chapter 2. For example, if we are looking at hiring software through a utilitarian lens, what is the objective of using the software? What values, such as time or money, are being traded off? Or if we look at sentencing software through a virtue ethics lens, what are the virtues, such as fairness and generosity, that the system design strives to emulate?*

5. *Think about attempting to verify something as complex as a medical research paper (e.g., more chocolate makes you less likely to have a heart attack) or a data-based claim in local media (e.g., crime is on the rise). Pick one of these and attempt to find the data necessary to verify the claim. Were you able to? Given the data that you have access to, are you able to verify the claims given the knowledge and information-processing abilities that you currently possess? Why or why not?*

6. *Consider the discussion in section 3.6.3 about the idea that social media and mass platforming will bring together communities and usher in a "bond of perpetual friendship between the kindred nations." In what ways have things gotten easier or better through the use of mass platforms such as YouTube, Facebook, TikTok, and other social media? In what ways has this changed how we communicate with each other and where we get our information? How has this changed how we come to know things?*

7. *In this chapter we have claimed that knowledge is a shared enterprise between ourselves and society. Justify this claim by breaking down the ways in which we decide that a particular piece of information is legitimate—for example, that a particular historical event really happened, that vaccines prevent illness, or another topic of your choice.*

REFERENCES CITED IN THIS CHAPTER

Abebe, Rediet, Solon Barocas, Jon Kleinberg, Karen Levy, Manish Raghavan, and David G. Robinson. 2020. Roles for computing in social change. In *FAT '20: Proceedings of the 2020 Conference on Fairness, Accountability, and Transparency*, 252–260. Association for Computing Machinery.

Ackoff, R. L. 1989. From data to wisdom. *Journal of Applied Systems Analysis* 16, no. 1: 3–9.

ACLU. 2019. In historic decision on digital bias, EEOC finds employers violated federal law when they excluded women and older workers from Facebook job ads. ACLU, September 25. https://www.aclu.org/press-releases/historic-decision-digital-bias-eeoc-finds-employers-violated-federal-law-when-they.

Angwin, J., J. Larson, S. Mattu, and L. Kirchner. 2016. Machine bias: There's software used across the country to predict future criminals. And it's biased against blacks. ProPublica, May 23. https://www.propublica.org/article/machine-bias-risk-assessments-in-criminal-sentencing.

Bakker, Egbert J. 2008. Epic remembering. In *Orality, Literacy, Memory and the Ancient Greek and Roman World*, edited by E. Anne Mackay, 65–78. Brill.

Barocas, S., and A. D. Selbst. 2016. Big data's disparate impact. *California Law Review* 104, no. 3: 671–732.

Barroso, Luiz André, Jeffrey Dean, and Urs Hölzle. 2003. Web search for a planet: The Google cluster architecture. *IEEE Micro* 23, no. 2: 22–28.

Bawden, David, and Lyn Robinson. 2009. The dark side of information: Overload, anxiety and other paradoxes and pathologies. *Journal of Information Science* 35, no. 2: 180–191.

Bellinger, Gene, Durval Castro, and Anthony Mills. 2004. Data, information, knowledge, and wisdom. http://www.systems-thinking.org/dikw/dikw.htm.

Bietti, Elettra. 2020. From ethics washing to ethics bashing: A view on tech ethics from within moral philosophy. In *FAT '20: Proceedings of the 2020 Conference on Fairness, Accountability, and Transparency*, 210–219. Association for Computing Machinery.

Bogen, Miranda, and Aaron Rieke. 2018. *Help Wanted: An Examination of Hiring Algorithms, Equity, and Bias*. Upturn. https://www.upturn.org/static/reports/2018/hiring-algorithms/files/Upturn%20--%20Help%20Wanted%20-%20An%20Exploration%20of%20Hiring%20Algorithms,%20Equity%20and%20Bias.pdf.

Bollacker, Kurt D. 2010. Avoiding a digital dark age. *American Scientist* 98, no. 2: 106–110.

Bonta, Mark, Robert Gosford, Dick Eussen, Nathan Ferguson, Erana Loveless, and Maxwell Witwer. 2017. Intentional fire-spreading by "Firehawk" raptors in Northern Australia. *Journal of Ethnobiology* 37, no. 4: 700–718.

Briggs, Asa, and Peter Burke. 2002. *A Social History of the Media: From Gutenberg to the Internet*. 3rd ed. Wiley.

Brooking, Emerson T., and Peter W Singer. 2016. War goes viral: How social media is being weaponized across the world. *The Atlantic*, November. https://www.theatlantic.com/magazine/archive/2016/11/war-goes-viral/501125/.

Burke, Lilah. 2020. The death and life of an admissions algorithm. *Inside Higher Ed*, December 14. https://www.insidehighered.com/admissions/article/2020/12/14/u-texas-will-stop-using-controversial-algorithm-evaluate-phd.

Calzon, S. 2020. 21 Examples of big data analytics in healthcare that can save people. *The Datapine Blog*. https://www.datapine.com/blog/big-data-examples-in-healthcare/.

Carr, Nicholas. 2013. All can be lost: We rely on computers to fly our planes, find our cancers, design our buildings, audit our businesses. *The Atlantic*, November. https://www.theatlantic.com/magazine/archive/2013/11/the-great-forgetting/309516/.

Carruthers, Mary. 1998. *The Craft of Thought.* Cambridge University Press.

Chafe, William Henry, Raymond Gavins, and Robert Korstad, eds. 2011. *Remembering Jim Crow: African Americans Tell about Life in the Segregated South.* New Press.

Corbett-Davies, Sam, and Sharad Goel. 2018. The measure and mismeasure of fairness: A critical review of fair machine learning. arXiv preprint arXiv:1808.00023.

Cowgill, Bo, Fabrizio Dell'Acqua, Sam Deng, Daniel Hsu, Nakul Verma, and Augustin Chaintreau, Augustin. 2020. Biased programmers? Or biased data? A field experiment in operationalizing AI ethics. In *Proceedings of the 21st ACM Conference on Economics and Computation,* 679–681. Columbia Business School research paper. https://papers.ssrn.com /sol3/papers.cfm?abstract_id=3615404.

Crawford, K. 2016. Artificial intelligence's white guy problem. *New York Times,* June 26. https://www.nytimes.com/2016/06/26/opinion/sunday/artificial-intelligences-white -guy-problem.html.

Danziger, Shai, Jonathan Levav, and Liora Avnaim-Pesso. 2011. Extraneous factors in judicial decisions. *Proceedings of the National Academy of Sciences* 108, no. 17: 6889–6892.

D'Ignazio, C., and L. F. Klein. 2020. *Data Feminism.* MIT Press.

Edmunds, Laura. 2020. Two teenage girls, aged 14 and 15, "abducted at gunpoint fought off their attacker by grabbing his steering wheel and causing a car crash in Louisiana." *Daily Mail,* November 2. https://www.dailymail.co.uk/news/article-8906989/Two-teenage -girls-abducted-gunpoint-fought-attacker-causing-car-crash.html.

Eubanks, V. 2018. *Automating Inequality: How High-Tech Tools Profile, Police, and Punish the Poor.* St. Martin's Press.

Everitt, Brian S., and Anders Skrondal. 2010. *The Cambridge Dictionary of Statistics.* 4th ed. Cambridge University Press.

Fjeld, Jessica, Nele Achten, Hannah Hilligoss, Adam Nagy, and Madhulika Srikumar. 2020. *Principled Artificial Intelligence: Mapping Consensus in Ethical and Rights-Based Approaches to Principles for AI.* Berkman Klein Center Research Publication no. 2020-1, January 15. https://papers.ssrn.com/sol3/papers.cfm?abstract_id=3518482.

Floridi, Luciano. 2010. *Information: A Very Short Introduction.* Oxford University Press.

Fraser, Alan G., and Frank D. Dunstan. 2010. On the impossibility of being expert. *British Medical Journal* 341, no. 7786: 1314–1315.

Gadamer, Hans-Georg. 1992. *Truth and Method.* 2nd ed. Translated by Joel Weinsheimer and Donald G. Marshall. Crossroad.

Gantz, J., and E. Reinsel. 2011. *Extracting Value from Chaos: IDC's Digital Universe Study.* Sponsored by EMC. International Data Corporation.

Gardner, Megan. 2012. Percy Bysshe Shelley frets about information overload . . . in 1821. *The Atlantic,* July 29. https://www.theatlantic.com/technology/archive/2012/07/per cy-bysshe-shelley-frets-about-information-overload-in-1821/260454/.

Gitelman, Lisa, and Virginia Jackson. 2013. Introduction. In *"Raw Data" Is an Oxymoron,* edited by Lisa Gitelman, 1–14. MIT Press.

Goodman, B., and S. Flaxman. 2016. EU regulations on algorithmic decision-making and a "right to explanation." arXiv preprint arXiv:1606.08813.

Greshko, Michael. 2018. Why these birds carry flames in their beaks. *National Geographic*, January 8. https://news.nationalgeographic.com/2018/01/wildfires-birds-animals-australia/.

Groesbeck, Amy S., Kirsten Rowell, Dana Lepofsky, and Anne K. Salomon. 2014. Ancient clam gardens increased shellfish production: Adaptive strategies from the past can inform food security today. *PLOS One* 9, no. 3: e91235. https://journals.plos.org/plosone/article?id=10.1371/journal.pone.0091235.

Gross, Terry. 2021. How a group of online sleuths are helping the FBI track down Jan. 6 rioters. NPR, *Fresh Air*, December 23. https://www.npr.org/2021/12/23/1066835433/how-a-group-of-online-sleuths-are-helping-the-fbi-track-down-jan-6-rioters.

Hallen, B., and J. O. Sodipo. 1986. *Knowledge, Belief and Witchcraft: Analytic Experiments in African Philosophy*. Ethnographica.

Himma, Kenneth Einar. 2007. The concept of information overload: A preliminary step in understanding the nature of harmful information-related condition. *Ethics and Information Technology* 9: 259–272.

Hirst, William, Jeremy K. Yamashiro, and Alin Coman. 2018. Collective memory from a psychological perspective. *Trends in Cognitive Sciences* 22, no. 8 (August): 737.

Hoffman, Frederick Ludwig. 1896. *Race Traits and Tendencies of the American Negro*. Vol. 11, nos. 1–3. Published for the American Economic Association by Macmillan.

Kahneman, Daniel. 2011. *Thinking, Fast and Slow*. Macmillan.

Kendi, Ibram X. 2016. *Stamped from the Beginning: The Definitive History of Racist Ideas in America*. Hachette UK.

Kendi, Ibram X. 2020. The American nightmare. *The Atlantic*, June 1. https://www.theatlantic.com/ideas/archive/2020/06/american-nightmare/612457/.

Klayman, Joshua. 1995. Varieties of confirmation bias. *Psychology of Learning and Motivation* 32: 385–418.

Kornfield, Meryl. 2021. How a shaky cellphone video changed the course of the Ahmaud Arbery murder case. *Washington Post*, November 24. https://www.washingtonpost.com/nation/2021/11/24/arbery-video-conviction/.

Kuhn, Annette. 2010. Memory texts and memory work: Performances of memory in and with visual media. *Memory Studies* 3, no. 4: 298–313.

Laney, Doug. 2001. 3D data management: Controlling data volume, velocity, and variety. Gartner. File no. 949. February 6. https://web.archive.org/web/20200208140930/http://blogs.gartner.com/doug-laney/files/2012/01/ad949-3D-Data-Management-Controlling-Data-Volume-Velocity-and-Variety.pdf.

Lepore, Jill. 2016. After the fact: In the history of truth, a new chapter begins. *New Yorker*, March 12. https://www.newyorker.com/magazine/2016/03/21/the-internet-of-us-and-the-end-of-facts.

Levav, Jonathan, Mark Heitmann, Andreas Herrmann, and Sheena S. Iyengar. 2010. Order in product customization decisions: Evidence from field experiments. *Journal of Political Economy* 118, no. 2: 274–299.

Levy, David. 2005. To grow in wisdom: Vannevar Bush, information overload, and the life of leisure. In *JCDL '05: Proceedings of the 5th ACM/IEEE-CS Joint Conference on Digital Libraries*, 281–286. Association for Computing Machinery. https://doi.org/10.1145/1065385.1065450.

Levy, Steven. 2004. All eyes on Google. *Newsweek*, March 28. https://www.newsweek.com/all-eyes-google-124041.

Lewandowsky, Stephan, Michael E. Mann, Nicholas J. L. Brown, and Harris Friedman. 2016. Science and the public: Debate, denial, and skepticism. *Journal of Social and Political Psychology* 4, no. 2: 537–553.

Liu, Ziming. 2005. Reading behavior in the digital environment: Changes in reading behavior over the past ten years. *Journal of Documentation* 61, no. 6 (December 1). https://www.emerald.com/insight/content/doi/10.1108/00220410510632040/full/html?skipTracking=true.

McDermott, Richard. 1999. Why information technology inspired but cannot deliver knowledge management. *California Management Review* 41, no. 4: 103–117.

Minchin, Elizabeth. 2008. Spatial memory and the composition of the *Iliad*. In *Orality, Literacy, Memory and the Ancient Greek and Roman World*, edited by E. Anne Mackay, 9–34. Brill.

National Commission on Terrorist Attacks upon the United States. 2004. *The 9/11 Commission Report: Final Report of the National Commission on Terrorist Attacks upon the United States*. US Government Printing Office.

Netflix. 2021. How Netflix's recommendations system works. Netflix Help Center. https://help.netflix.com/en/node/100639.

Nicholas, George. 2018. When scientists "discover" what Indigenous people have known for centuries. *Smithsonian Magazine*, February 21. https://www.smithsonianmag.com/science-nature/why-science-takes-so-long-catch-up-traditional-knowledge-180968216/.

Noble, S. U. 2018. *Algorithms of Oppression: How Search Engines Reinforce Racism*. New York University Press.

O'Neil, C. 2016. *Weapons of Math Destruction: How Big Data Increases Inequality and Threatens Democracy*. Crown.

Oyěwùmí, Oyèrónkẹ́. 1997. *The Invention of Women: Making an African Sense of Western Gender Discourses*. University of Minnesota Press.

Perez, C. C. 2019. *Invisible Women: Exposing Data Bias in a World Designed for Men*. Random House.

Pinker, Steven. 2018. *Enlightenment Now: The Case for Reason, Science, Humanism, and Progress*. Penguin.

Plato. 1995. *Phaedrus*. Translated by A. Nehamas and P. Woodruff. Hackett.

Rawson, Andrew. 2012. *Showcasing the Third Reich: The Nuremberg Rallies*. History Press.

Reser, David, Margaret Simmons, Esther Johns, Andrew Ghaly, Michelle Quayle, Aimee L. Dordevic, Marianne Tare, Adelle McArdle, Julie Willems, and Tyson Yunkaporta. 2021. Australian Aboriginal techniques for memorization: Translation into a medical and allied health education setting. *PLOS One* 16, no. 5: e0251710. https://journals.plos.org/plosone/article?id=10.1371/journal.pone.0251710.

Rowley, Jennifer. 2007. The wisdom hierarchy: Representations of the DIKW hierarchy. *Journal of Information Science* 33, no. 2: 163–180.

Rubin, David C. 1997. Introduction. In *Memory in Oral Traditions: The Cognitive Psychology of Epics, Ballads, and Counting-Out Rhymes*, 1–4. Oxford University Press.

Schumann, Candice, Jeffrey S. Foster, Nicholas Mattei, and John P. Dickerson. 2020. We need fairness and explainability in algorithmic hiring. In *AAMAS '20: Proceedings of the 19th International Conference on Autonomous Agents and MultiAgent Systems*, 1716–1720. Association for Computing Machinery.

Shaw, Al, and Lylla Younes. 2021. The most detailed map of cancer-causing industrial air pollution in the U.S. ProPublica. November 2. Updated March 15, 2022. https://projects.propublica.org/toxmap/.

Smit, Eefke, Jeffrey Van der Hoeven, and David Giaretta. 2011. Avoiding a digital dark age for data: Why publishers should care about digital preservation. *Learned Publishing* 24, no. 1: 35–49.

Sohn, Pow-Key. 1959. Early Korean printing. *Journal of the American Oriental Society* 79, no. 2 (April–June): 96–103.

Sparrow, Betsy, Jenny Liu, and Daniel M. Wegner. 2011. Google effects on memory: Cognitive consequences of having information at our fingertips. *Science* 333, no. 6043: 776–778.

Spitzer, Manfred. 2012. *"Digitale demenz": Wie wir uns und unsere Kinder um den Verstand bringen*. Droemer.

Staab, S., S. Stalla-Bourdillon, and L. Carmichael. 2016. Observing and recommending from a social web with biases. https://arxiv.org/abs/1604.07180.

Tierney, John. 2011. Do you suffer from decision fatigue? *New York Times Magazine*, August 17. https://www.nytimes.com/2011/08/21/magazine/do-you-suffer-from-decision-fatigue.html.

Tuerkheimer, Deborah. 2021. *Credible: Why We Doubt Accusers and Protect Abusers*. Harper Wave.

Tulving, Endel, and Zena Pearlstone. 1966. Availability versus accessibility of information in memory for words. *Journal of Verbal Learning and Verbal Behavior* 5, no. 4: 381–391.

Vila-Rodriguez, Fidel, and Bill G. Macewan. 2008. Delusional parasitosis facilitated by web-based dissemination. *American Journal of Psychiatry* 165, no. 12: 1612.

Vopson, Melvin. 2021. The world's data explained: How much we're producing and where it's all stored. The Conversation, May 4. https://theconversation.com/the-worlds-data-explained-how-much-were-producing-and-where-its-all-stored-159964.

Weinberger, David. 2010. The problem with the data-information-knowledge-wisdom hierarchy. *Harvard Business Review*, February 2. https://hbr.org/2010/02/data-is-to-info-as-info-is-not.

CHAPTER 4

PERSONHOOD AND PRIVACY

Learning Objectives

At the end of this chapter you will be able to:

1. *Explain the difference between metaphysical, epistemological, and axiological questions and then build on this understanding to identify and evaluate the conceptual underpinnings of common questions about personhood.*

2. *Summarize different perspectives and definitions of the concept of personhood and then compare and contrast how the fundamental notions of self, identity, and responsibility are understood from these perspectives.*

3. *Discuss concepts that are commonly used to draw boundaries between persons and technologies, such as identity, responsibility, memory, and bodies as well as the limitations of each of these boundary drawing concepts.*

4. *Outline the history of the concept of privacy in society and identify major points of contention including privacy as a right, ownership, and a value.*

5. *Justify privacy as a necessary condition for personhood and defend this condition in the presence of ever-changing technology.*

4.1 INTRODUCTION

You probably have some basic intuitions about the boundaries of personhood. You are a person, but your desk lamp is not. A man who relies on medical technology to stay alive is a person, but your final project for your data management course is not. These might seem like trivial examples, but when it comes to explaining how and why we make such distinctions, things can get quite complicated.

Consider a more challenging question: are all human beings persons? What about nonhuman animals like dolphins and chimpanzees? What about artificially created life?

Some of the most important ethical questions surrounding technology today are based on presuppositions about who or what counts as a person and who or what does not. These presuppositions often influence design. For example, many automobiles today are designed to crumple in a crash, destroying the car but protecting the person (or people) inside of it, on the grounds that the person inside is what matters and needs to be protected. But if the car in question is a highly advanced autonomous vehicle, capable of some measure of independent decision making, our assumption that the car should be hurt instead of the human might begin to seem less obvious. For instance, imagine a car that is equipped with facial recognition, customized to fit your unique preference and even to ask you about your day. To what degree would the car need to possess these typically "person-like" qualities before it begins to matter, in an ethical sense? If we are accustomed to seeing these qualities primarily if not exclusively in human beings, how do we know that we would recognize these qualities in a different kind of entity, if they should develop?

The underlying question in this situation, and many others like it in ethics, is: What counts as a person? How do you distinguish a person from other kinds of entities? Does this distinction tell us anything about how we should act toward such entities? Why or why not?

The dictionary can't answer such questions for us, nor can this textbook. In order to gain clarity about our own intuitions about personhood as well as broaden our understanding about what personhood might be, this chapter lays out several conceptions of personhood and how they have changed over time. Perhaps some of these definitions will be familiar to you; others might be strange or seem wrong to you. Whatever the case, the purpose of this chapter is to help you gain clarity about how your own ideas about personhood influence the way you approach ethical problems and decision making.

Rather than providing a list of definitions or an argument for how to draw distinct boundaries around personhood in such a way that it includes some and excludes others, we invite you to use this chapter as a guide or conceptual roadmap. It can be used not only to gain clarity about your own conception of personhood but also to help you identify some of the ways ideas about personhood inform public debate and impact your life as a professional.

In reflecting on our own ideas and thoughts about personhood, we may find that our ideas and thoughts differ from those of our peers, our classmates, and even our friends and family members. These differences do not imply that one definition is better than others. However, the fact that these differences exist may indicate that we may need to evaluate our own intuitive criteria for determining what makes a person a person and to consider whether there is room for clarification of the definition or space to fill in some gaps within our working definition.

Sections 4.2 and 4.3 of this chapter use philosophical thought and methods to describe some of the ways in which personhood has been defined over time. After clarifying some of the common assumptions that underlie different accounts of personhood, we look at some of the ways in which contemporary technologies complicate and challenge these definitions. Sections 4.5 and 4.6 bring these ideas and questions into conversation with the topic of privacy.

Linking privacy and personhood together in this way is important for a variety of reasons, many of which we discuss in this chapter. In brief, although privacy is widely recognized as a pressing subject in technology ethics, many discussions and debates ignore the ways in which our ideas about privacy, and our implementation of those ideas, depend on unstated views, opinions, and beliefs about what counts as a person. In discussing privacy and personhood together, this chapter showcases how a deeper reflection on varying conceptions of personhood can help us identify and speak more directly to pressing ethical issues surrounding technology today.

4.1.1 WHY PRIVACY?

Many ethical concerns related to computing and privacy are rooted in unstated beliefs about what human personhood is, in what context it applies, and what it entitles one to do. Why is it sometimes hard to talk in class, to state your ideas publicly, or to admit you don't know something in front of your peers? Why is this easier (or harder) to do when it can be done anonymously? Why do social media platforms, whose stated main product is interconnection and whose source of revenue comes from that interconnection, give users the option to make their profiles private? Going further, why, in some cases, do these same sites give users the option to remove content including images or quotations that others have posted about them?

Most people in the West agree that a "right to privacy" exists, even if they don't agree about what this right includes or how much it matters in practice. However, few people ask themselves *why* this right exists.

In fact, few people realize that the right to privacy was not conceptualized as something protectable by law until the end of the nineteenth century. Long before we ever posted selfies online, Samuel Warren and Louis Brandeis (Warren and Brandeis 1890) authored the first legal publication in the West to call for a "right to privacy." This article has had an enduring influence on how privacy is imagined and understood. It was written primarily in response to the recent invention and proliferation of radical new technologies, including the simple, affordable box camera, the kinetoscope (motion-picture recording), and the phonograph (audio recording), which made it possible to document someone or something and to subsequently transmit, reproduce, and republish the documentation with or without the knowledge of the person being recorded.

Despite its foundational status in privacy debates, this legal opinion is a product of its era, written in response to the technological changes that were happening at a specific moment in time. Therefore, it does not always line up clearly, or indeed at all, with the potential incursions on privacy that more recent technologies have made possible. Similarly, Warren and Brandeis's underlying assumptions about personhood are a product of that time. But because personhood is so rarely recognized as a core element of how we understand privacy, it can be harder for us to notice how these authors' presuppositions surrounding personhood have carried over into contemporary debates. This inherited concept requires some re-examination, just as privacy does.

However privacy is defined, the implicit role of personhood is visible once you think to look for it. Generally speaking, the language of "privacy" is used to refer to an expectation that a person's actions, words, and information about their identity will not be shared in certain contexts or with certain people or institutions without that person's consent. Indeed, as technology and society scholar Shoshana Zuboff notes, Justice William Douglas stated in 1967 that "the individual should have the freedom to select for himself the time and circumstances when he will share his secrets with others and decide the extent of that sharing." Zuboff sees this "freedom to select" as central to the epistemic right to know ourselves and how we are shared (Zuboff 2021). Even this very basic definition of privacy links it to *self-determination*: that is, a person's ability to exercise some level of control over themselves, how they represent themselves to others, and how they choose to live out their life. Without an operational understanding of personhood, arguments for privacy would make little sense. But only rarely are these presuppositions stated and defended.

By grouping privacy and personhood together, this chapter underscores ways in which developments in technology change not only our understanding of what privacy is but also the ways in which technological developments more broadly speaking can alter our sense of what should be permissible for a person. But first let's tackle the concept of personhood.

4.2   WHAT IS PERSONHOOD? DEFINING THE QUESTION

According to the *Oxford English Dictionary*, personhood is "the quality or condition of a person." But what is a person?

The concept of personhood has a long philosophical history, where it is closely associated with a number of other complex topics. These topics include but are not limited to self-consciousness, agency, identity, selfhood, and human subjectivity and conceptions of the soul, both material and immaterial (Thiel 2011). The history of philosophy has proven that defining any one of these terms is not easy. Rather than trying to offer one clear-cut definition that will satisfy everyone, most philosophers think that the ongoing discussion about what these terms mean is part of what makes the terms valuable in the first place.

Still, though absolute precision and accuracy may not be our objectives, we can nevertheless get *some* clarity if we first consider what kind of question we are asking when we ask the question "what is a person"?

One of the reasons this question is difficult to answer is that personhood is a normative concept. Norms, as we noted in chapter 1, are rules or standards for action. Normative concepts are not real in the same way that physical objects, like a piece of wood or the chair you are sitting in, are real. Instead, they have reality insofar as they regulate and motivate action. Therefore, when we say that personhood is a normative concept, we mean that it is a concept that comes with certain value judgments and ideas about how we ought to live, and that it has implications for how we ought to treat others.

These kinds of value judgments do not, typically speaking, emerge out of thin air. Whether or not we are conscious of it, the things we value, and our understanding of the world and our place in it, are all factors that help us decide what matters ethically. Although these things are often bound up together, conceptual distinctions can help us zero in on what we are talking about when we refer to "persons" or personhood.

Philosophy supplies us with several different branches, or subfields, from which to approach the questions of personhood. Classically, the three main branches are metaphysics, epistemology, and axiology (Rescher 2006; Samuel 2011). Although these different branches and their subdivisions are interconnected, they ask different kinds of questions and therefore expect different kinds of answers to those questions. We can use these categories to help us decide what kind of question, and therefore what kind of response, we are looking for when it comes to defining personhood.

- **Metaphysical questions**: Metaphysics is the study of how the world works, the nature of reality, and the place of human life in the grand scheme of things. It is a way of talking about the main structures and elements of reality and explaining how those things hang together in a system. Such lofty reflections might seem far removed from our own ideas about personhood, but it is often the case that ideas about personhood are tied to our beliefs about the world and what we believe we owe to others and to ourselves. Some examples of metaphysical questions include: What is the true essence or nature of human beings? How did the universe come into existence?

- **Epistemological questions**: Epistemology is a branch of philosophy that studies how human beings come to gain knowledge, whether that knowledge is about themselves, their immediate context or environment, or other living and nonliving objects and entities. Some epistemological theories draw from evidence in the human and natural sciences (sociology, cognitive sciences, and theories of perception and embodiment), whereas others are strictly conceptual theories that explore the limits and scope of human knowledge. Examples of epistemological questions include: What are the cognitive capacities that allow us to arrive at an objective fact? How is it that we can know that other entities have self-consciousness? How do we come to *know* what counts as right and wrong?

- **Axiological questions**: Axiology is a branch of philosophy that studies values and principles. It considers questions about what is ultimately good and how and why people come to value what they do. Some examples of axiological questions include: Why ought we to value human persons over material goods? What activities should occupy most of our time? Why should we create software that helps people live happier lives?

With these different types of questions in mind, let's return to the question at hand: *what is a person?*

When someone asks this question, they may be asking a question about what kinds of essences, features, traits, or properties a specific entity has that awards them the title of "person." Or, similarly, they may be asking about the true nature of such entities and how those entities fit in the grand scheme of things. Answers to such questions will likely include one's beliefs about how the world works or ideas about ultimate reality. We can recognize these as metaphysical questions.

Alternatively, when a person asks the question "What is a person?" they could be more specifically asking about the perceptive and cognitive capacities that enable us as perceivers to recognize a person as a person. To ask about how it is that we can know something is to ask a kind of epistemological question.

And still, when someone asks the question "What is a person?" they might be asking about which characteristics or traits are *most important* or most essential when it comes to defining human life. This is a question of value and prioritization—in other words, it is an axiological question. On the other hand, if a set of traits or properties is provided, the inquirer might wish to know why we have selected these traits and properties as the most important as opposed to all others. Arguments in defense of such positions could again branch off into areas of epistemology or metaphysics, or develop into more specific questions about value.

Consider: when someone asks you the question "What is a person?" What do you typically understand them to be asking about? When you ask the question, what are you trying to understand?

4.2.1 PERSONHOOD AND HUMAN IDENTITY

In our analysis in the previous section, it was suggested that when someone asks the question "What is a person?" they may be asking about what counts as *human* life. The simple explanation for this is that for various cultures, past and present, "personhood" is a term that is used synonymous with "human being." But this explanation may not please everyone—and there are good reasons to be cautious here.

If we automatically assume that the concept "person" can be used interchangeably with "human being," then when we ask the question "What is a person?" we may be asking about what is the most fundamental description of human life that we can think of. A common and indeed ancient way of answering this question has been to find a trait or property that humans are said to possess but that other living or nonliving entities do not. For instance, one might suggest that human beings have self-consciousness or that they use language, and that other living beings do not have these things; for this reason, what it means to be a person (i.e., a human) is to be a being that uses language or possesses the capacity for self-consciousness—that is, there is a quality intrinsic to all persons that all other beings lack. Although this might sound familiar or seem like a straightforward way to proceed, it tends to blur together metaphysical beliefs, epistemological presuppositions, and axiological commitments. This means it is hard to gain insight into what kind of answer we are expecting. Complicating matters further is the fact that not everyone agrees about what set of properties, characteristics, or traits are the most essential or fundamental to human life (this is an axiological question).

In the context of many Western societies, personhood is said to be what accords human individuals universal worth and inalienable rights. The notion that human persons have intrinsic rights historically has been used to define the rights of citizens against others and in relation to the state (Mill [1869] 1991; Rawls 1972; Rousseau [1762] 1997). Within this intellectual tradition, rights and duties are understood to be rooted in the human being's status as, for instance, a self-conscious and rational agent. But historically speaking, not everyone was said to fully possess these capacities and traits. In some cases, these capacities were presumed absent from those of a particular gender, race, or class. For example, Aristotle, who is well known for his arguments regarding the rational status of human beings, denied rationality to women and claimed that Africans were naturally slavish due to living in warmer climates. This form of argumentation, known as "climate theory," was used to justify and naturalize slavery for centuries (Kendi 2016).

These same biases shaped how personhood was accorded in the United States as well. The politicians and philosophers who molded the United States in its early decades were almost all of European descent. These men, who framed both the laws and norms of this new society, denied legal standing and hence personhood to Native Americans on the grounds that they were of a different nature or had not yet progressed enough developmentally to be fully human. Black Americans were denied equal standing because they had been declared property rather than persons (Kendi 2016). Hence, what appears to be a universal claim about all human beings has, historically, worked in ways that exclude certain people and include only a select few, and the practical implications for those excluded are broadly, profoundly, and enduringly harmful.

Despite this unsettling legacy, intrinsic notions of personhood are quite prevalent even today and are frequently called upon to support human dignity and to argue for the value of human life in all its forms. Indeed, Article 1 of the Universal Declaration of Independence, for instance, states that "All human beings are born free and equal in dignity and rights. They are endowed with reason and conscience and should act towards one another in a spirit of brotherhood" (United Nations 1948). Article 6 further specifies that "Everyone has the right to life, liberty and security of person" (United Nations 1948).

Intrinsic notions of personhood are not the only way of conceiving of personhood, however. For instance, political philosopher Martha Nussbaum suggests that human dignity and the respect that we owe to others, at least from a political perspective, need not be based on controversial epistemological or metaphysical assumptions but rather on political commitments and, with time, overlapping consensus (Nussbaum 2011; Formosa and Mackenzie 2014).

Furthermore, as we saw in chapter 2, sub-Saharan African communities largely understand personhood as a community matter: personhood is not intrinsic but is rather something achieved in and by means of a given community (Menkiti 1984). Furthermore, these traditions generally do not ascribe major metaphysical significance to humanity, as distinct from plants, animals, natural phenomena or spirits. In the context of this "cosmic humility" (Tangwa 2005), personhood is a social concern that is of practical importance to other persons who live in the same community but is not necessarily a matter of unique philosophical significance.

4.2.2 PERSONHOOD AND PERSONAL IDENTITY

Leaving aside the presupposition that personhood is fundamentally a question about what it is that defines human identity or a question about what qualities or characteristics are most fundamental for human life, we can approach the topic of personhood from a different angle.

You might feel that the best way to attack a big, nebulous concept like personhood is to "solve" for a smaller part of it; in other words, figure out what makes an individual person distinctive and then generalize from that to persons in general. But of course, we aren't just smaller units of personhood; each of us is made distinctive by our *personal identity*.

Your "identity" can be broadly defined as certain demographic facts about you that you may have little to no control over—for instance, that you are a human being, the location of your birth, your ethnicity, and the gender that you were assigned at birth. It may also refer to things over which you (in most cases) have some control: your credit score, your marital status, and your religious affiliation.

Personal identity, however, points to something more specific—something, well, more "personal." Personal identity is typically understood as those aspects, characteristics, and features that define a person's existence in relation to all others—the things that make you who you are. What makes you who you are can involve things that you have chosen for yourself, such as your actions and beliefs, as well as other things that you may not have chosen that influence you.

For instance, we may consider our relationships with others, our place within society, and/or the communities in which we live to be part of our identity. It is also common to think of personal identity as having something to do with the choices that we have made, the way we live our lives, and the way we define ourselves in relation to others.

Your identity may also include certain details about your temperament, your aesthetic preferences and other psychological traits. As we saw in our discussions of both virtue ethics and communitarianism, a person's present self is largely a product of influences that they have not necessarily chosen or that affect them in ways beyond their control: where they grew up, the experiences they have had, and the people with whom they spend their time. Aesthetic preferences—that is, questions of personal taste— are an interesting example of how each of us is shaped by our time and place. It would be ridiculous to claim that our tastes are only a matter of when and where we come from: after all, it's unlikely that you and your friends all have identical taste in music or movies or food, among the various options on offer in the present. You may even know some-one (or be someone) who particularly enjoys mid-twentieth-century noir films, or who refuses to listen to music made after the early 1990s. All the same, our preferences are shaped, at least in part, by what we are used to and what is available to us.

But if we focus only on these things—the things that make you who you are or make me who I am—the question of personhood becomes primarily about self-knowledge (who am I?). But surely this cannot be the only way to think about person-hood because, as stated above, personhood is, in most cases, a *normative concept*.

If personhood is the "quality or condition of being an individual person," clearly what it means to be a person or the quality or condition of being a person is not rooted only in the way that you define yourself but is rather a larger concept that we can apply to others. In other words, we cannot use the same things that you uniquely attri-bute to yourself to define what personhood means more generally. Personhood must encompass both the possibility of those individual differences and some underlying commonality.

It was noted above that personhood is sometimes understood to be an intrin-sic property—a property that all human beings share. Let's assume for a moment that each of us has a unique identity or essence that makes us who we are. Is this unique identity something that we achieve over time? Is it something that can be lost? Is it something that can be duplicated or repeated? Or is it possible only for one individual to have one unique identity? In the Western philosophical tradition, these questions take us to the very heart of the matter of personal identity.

Historically, the debates about personal identity in this Western context are cen-tered around the following two main themes:

- A search for a principle of *individuation*, that is, a principle or rule by which some-thing can be distinguished from other things or from others within the same group or classification.

- A way to account for the *continuity* of identity, that is, what allows something to acquire and maintain an identity over time.

Both of these themes might seem far removed from ideas about personhood, but not if we consider some of the questions that come up in popular discourse about technology. For instance, is it possible to upload our minds into machines? Do clones have the same identity as their originals, and if not, how do they differ and why? Can we replace most of our organs and limbs with technological prosthetics and still have the same identity?

Although these specific questions may reflect the contemporary popular imagination, philosophers have long wondered about what constitutes personal identity and how it can be lost, achieved, or maintained. We will touch on some of these ideas so that in the next section we can see how ideas about human identity and personal identity shape the ways we think about technological change and mediation.

4.2.3   INDIVIDUATION AND CONTINUITY OF IDENTITY

Theories of individuation look for the cause or principle that can account for the distinct nature of an individual entity. A principle of individuation can be used to explain how humans can be distinguished from other things (chairs, dinosaurs, rainforests, microwaves); alternatively, it can be used to explain how persons are distinguished from other persons. The principle in question is not something that pertains to specific traits like the type of haircut you have, or the kind of music you like, or where you grew up. It is said to be a fundamental or essential principle, cause, or mechanism that *makes it possible* for there to be a fundamental distinction between individuals. It amounts to a kind of metaphysical claim about what something is in its essence and how that essence can be distinguished after it has been assigned a category, such as human beings or rational agents (Gracia 1984).

For many ancient thinkers, and still for many people today, what makes a person a person hinges on a metaphysical belief about an intrinsic property that occurs only within human persons, for instance, a soul or an essence.

In the context of the Western intellectual tradition, historically speaking, the principle of individuation was about a substance of some kind, something either material (for instance, the body) or immaterial (for instance, the soul or intelligence) that could endure through time. This enduring substance accounts for individual personal identity, because the individual is one and the same substance over time. Whether that consistent substance is material (like a body) or immaterial (like a soul), the continuity of the substance helps to explain the continuity of the distinct individual composed of it.

When this view is maintained, there is no real need to determine what remains continuous over time since what remains continuous, as well as what differentiates one being from other beings, comes down to an essential substance. But not all theories of personal identity look for a principle of individuation to determine what constitutes personal identity. And not all theories of personal identity rely on metaphysical claims about an underlying substance or essence.

In fact, there have been various shifts in the way personal identity has been understood within the context of philosophical, legal, and political thought. In sixteenth-century Europe, the term "person" was frequently used to talk about participating in a community and being subject to that community's laws. (You may notice a resemblance here to how "being a person" is understood in sub-Saharan communitarian traditions.) Thomas Hobbes, for instance, tied the concept of person to *representation* and how we *appear* to others. He appealed to the Latin roots of the term: "*persona*" in Latin originally referred to the theatrical mask used in both Greek and Roman theater. Accordingly, Hobbes was able to describe two different kinds of persons: *natural persons* who represent themselves and *artificial persons* who represent other persons or things (Hobbes [1668] 1992).

David Hume, notably, argued that personal identity doesn't have any enduring status at all (Hume 1978). For him, if we think there is an underlying substance or unity that supports the continuity of the self, we are mistaken. The mind, he insisted, is nothing more than a bundle or collection of different perceptions. Individuals are differentiated by their perceptions and experiences but there is no essential self onto which these perceptions and experiences adhere. According to Hume, although we have a tendency to think of ourselves as the same person that more or less remains the same over time, this is nothing more than an illusion—a trick of the mind.

John Locke (whom you may remember from our discussion of utilitarianism) had similar ideas about personal identity. However, his account is particularly noteworthy because he is often cited as the one who cut the knot that for centuries tied personal identity to metaphysical ideas about substance. For him, the question of personal identity is not a matter of determining "what the person is made of" but rather "what it is that allows a person to maintain their identity over time." He insisted that the identity of a person is not to be based on a material substance such as the human body nor does it rely on an immaterial substance such as a soul or spirit. Instead, the identity of persons rests solely on consciousness. A person, Locke maintained, is best defined as "a thinking intelligent being, that has reason and reflection, and can consider itself as itself, the same thinking thing in different times and places; which it does only by that consciousness" (Locke [1690] 1990, II, §26).

Locke also thought it was important not to conflate our understanding of persons with our formal definitions of humanity (Thiel 2011). To refer to a person is to refer to an intelligent agent that is able to extend "itself beyond present existence to its past, whereby it becomes concerned and accountable" for past actions (Locke [1690] 1990, II, §26). The unity and identity of the person is a continuous process through which the past is brought into the present.

Locke understood the idea of persons to be a "forensic" notion, which is to say that he thought that our ideas about personhood were crucial for understanding how we ascribe moral and legal responsibility. In order to ascribe actions to an individual, we have to be able to see them as the cause of actions. In order to ascribe moral and legal responsibility to a person, that person needs to be seen as a coherent and consistent self. In Locke's judgment, without memory and without the capacity to reflect on our own actions, this type of consistency and coherency is impossible. In other words,

Locke's approach to personal identity was unique in part because of the emphasis it placed on *self*-awareness or *self*-consciousness.

It is worth noting that self-consciousness is another of those traits or capacities that are often used as a way to distinguish persons from other entities or things. Not surprisingly, this also means that people continue to debate its exact meaning (Crone et al. 2012).

Generally speaking, self-consciousness refers to a capacity that makes it possible for individuals to distinguish themselves from the world in which they are embedded. Self-consciousness is also what makes it possible for individuals to recognize feelings, thoughts, and emotions as their own. Some philosophers have argued that self-knowledge—knowledge that we gain about ourselves—presupposes self-consciousness and, therefore, it is an essential aspect of what it means to be a person. Others argue that self-consciousness is necessary for us to deliberate about our actions and intentions and, therefore, it plays a fundamental role in ethical thinking (Korsgaard 1996).

4.2.4 THE SELF, IDENTITY, AND NARRATIVIZATION

As noted above, Locke's conception of the human person is forensic; that is, it is concerned with how it is that we attribute responsibility. This is one of the reasons that self-consciousness is so important for Locke's conception of persons. Without it, a person could not be understood to be the author or agent of their actions and therefore could not be held accountable for their actions. As influential as this approach to personal identity may be, it prioritizes the human's capacity to know and do.

A different perspective on personal identity can be found in what is sometimes described as the narrative approach to personal identity (McLean 2008). Narrative identity begins with the basic claim that human beings are not simply knowers and doers—they are also storytellers. As both highly intelligent and imaginative creatures, humans use narrative to explain where they come from, where they are, and who they wish to become. These narratives reflect aspects of one's life that they find meaning and valuable. Because narrative plays such an important role in human life, theorists across various fields have used the term "narrative identity" to capture something about what it means to be a person that other accounts leave out. As DeGrazia explains,

> Your narrative identity involves your self-conception, your self-told story about your own life and what's most important to you. Rather than merely listing events involving you or facts about you, it orders these events and facts, highlights certain features and people, and organizes what's highlighted into a more or less coherent story of your life and character. (DeGrazia 2005, 179)

The experiences, events, and happenstances of our existence can be construed as information about who we are as people. But they do not seem to present us with a complete picture of who we understand ourselves to be. Narrative identity helps capture the fact that our identities are not something that happen to us: they are something that we creatively express as we talk about ourselves and share the events of our lives with others. Self-knowledge, in other words, is not something we can gain through immediate and direct analysis. Instead, narrative identity theorists argue that self-knowledge is something we gain through mediation. In particular, narrative identity theorists argue that we gain self-knowledge through the stories that we have been told, or through the symbols and myths that we have inherited through different social settings (Ricoeur 1991). Therefore, it is often the case that our feelings and judgments about how best to represent ourselves reflect the values and prevailing story lines of our own sociocultural context.

Story Point: "Welcome to Your Authentic Indian Experience™," by Rebecca Roanhorse

> The first sensation is always smell. Sweetgrass and wood smoke and the rich loam of the northern plains. Even though it's fake, receptors firing under the coaxing of a machine, you relax into the scents. You grew up in the desert, among people who appreciate cedar and pinon and red earth, but there's still something home-like about this prairie place.
>
> Or maybe you watch too much TV. You really aren't sure anymore.

Jesse works as an Experience Guide for a local VR company. Both a personal and a professional American Indian, Jesse offers Vision Quests to Tourists in search of a spiritual experience. Jesse's skill as a storyteller and guide helps Tourists feel like they've had a deeply authentic experience, even though—or perhaps because—Jesse bases his Vision Quests in familiar movie stereotypes instead of his own history and life experiences. Throughout the second half of the story, we see the slow theft of Jesse's life, and arguably his personal identity, by a White Tourist who believes he understands and deserves Indian identity more than Jesse does. But even before this particular act of individual theft we can observe a subtler, broader theft at work by noticing how the disjunction between pop-culture Indian stereotypes and everyday reality has made it difficult for Jesse to feel like he really counts as Indian. This story raises important questions about who gets to claim what identities, how shared identities get constructed, and who decides what counts as "authentic"—a word that, much like personhood, is very powerful but also difficult to define.

4.2.5 MORAL AND LEGAL RESPONSIBILITY

In most classically liberal and democratic nations, individual freedom and individual control over one's actions, body, and choices are considered valuable and worth protecting. Accordingly, these societies typically recognize certain rights in order to protect them. With this individual freedom and control comes responsibility, both legal and moral. Individuals are understood to be the authors of their own actions and therefore are responsible for those actions.

What does responsibility entail? Although moral responsibility and legal responsibility are closely related, they do not always carry the same meaning. In this section we will identify how they overlap and why in some instances they differ.

Legal responsibility is assessed (and intentionally so) according to criteria that closely align with moral responsibility. One such criterion is that the individual in question, in order to be legally culpable, must have *moral agency* in order to be responsible. Generally speaking, moral agency is comprised of the following five aspects:

- Agency: Defined as the capacity to act and effect change on our environment and to affect other living and nonliving entities.

- Self-determination (autonomy): The idea that we can act in ways that are not fully determined by the environment in which we find ourselves nor by our historical or social context.

- Self-awareness: An awareness of one's self in distinction from objects and from other agents acting in the world.

- Rationality: Narrowly defined as the cognitive process through which we determine that for every action or event there is a cause.

- Responsibility: The capacity to identify our actions as our own and take responsibility for those actions, and/or the capacity to accept that we are part of relations through which we come to determine what we are responsible for and what we are not responsible for.

Another important criterion for legal responsibility, which likewise overlaps with many ethical frameworks, rests on the distinction between voluntary and involuntary actions. As we discussed in chapter 2, in the context of most traditional Western ethical frameworks such as deontology, virtue ethics, and utilitarianism, moral responsibility and culpability apply only in those cases in which voluntary action is performed. Involuntary actions (understood in this context as actions performed out of ignorance or action compelled by force or coercion [Aristotle 1999, III.1]) may be considered good or bad with respect to their consequences, but the moral agent is responsible only indirectly, if at all, for involuntary actions (Sidgwick [1874] 1981). Similarly, a moral agent is less responsible—if at all—for involuntary action (which typically means action performed out of ignorance or action compelled by force or coercion [Aristotle 1999, III.1]) than for voluntary ones. Similarly, most criminal offenses require proof not only of an *actus reus* (a "guilty act") but also of a *mens rea* ("guilty mind"), a specific state of mind on the part of the accused. Furthermore, an *actus reus* must include a voluntary act for someone to be charged with a crime. As the legal theorist and philosopher Anthony Duff explains: "In the absence of a voluntary act, any movements of my body are involuntary; and whilst those movements might cause harm, thus making me (or them) causally responsible for the occurrence of that harm, I lack control over those movements and cannot justly be held either morally or legally responsible for their effects" (Duff 2009).

Yet although the criteria overlap significantly, it's not hard to find situations in which legal and moral responsibility do not align. For example, it is not illegal to entice children to spend money on virtual goods or hook people into pay-for-play platforms, but there may still be moral reasons for finding ways to prevent these practices (Hohn 2018).

On a more personal level, most of us can recall times when we felt morally guilty or responsible for something we did even though we did not break the law. Even if we cannot recall a specific time we felt this way, we can likely think of examples in which people explained their actions in terms of guilt and remorse. Similarly, most people would agree that while, for example, jaywalking and stealing money from your neighbors are both illegal actions, they do not necessarily come with the same moral weight. The point is that even if we believe in some cases that the parameters of legal responsibility could be fine tuned or updated to match our own moral judgments, the existence of these gaps helps to underscore that they are not the same thing. Legal theorists, philosophers, sociologists, and psychologists have offered various explanations for how and why they are different. One fairly common suggestion is that moral responsibility can be distinguished from legal responsibility because it is more deeply concerned with an agent's attitudes and responses. When we hold a person (whether ourselves or another) to be morally responsible, we evaluate them not only on the basis of the things that they do or don't do, but also on the basis of the underlying dispositions that shape the actions and attitudes that they express about others' actions (Strawson 1962; Watson 2004). Blaming somebody in a moral sense (as opposed to a legal one) involves passing judgment on their character. It involves finding fault with their ideas about good or bad ways to act, or their responses to external actions or events (Nozick 1981). These shortcomings or defects do not imply illegal or criminal behavior, but they may still be considered failures of moral responsibility. Imagine, for example, that you learn that a friend of yours has died in a horrible accident. When you call to inform one of your mutual friends and share the date of the funeral, he acts indifferent to the news and says he can't attend the funeral because he can't miss soccer practice. Although this might be a simple matter of bad judgment, it is an example of a situation in which we expect certain responses from people, not only with respect to how they act but also how they feel about certain things. We can be let down or change our opinion about their character when they do not respond in the ways that we have come to expect.

Another difference is that the consequences of legal responsibility are also more concrete: the legal system holds you accountable, or doesn't, as the case may be. The consequences of moral responsibility are far less straightforward. For instance, it may briefly feel rewarding to tweet "history will judge you harshly" at somebody whose actions you condemn, it's difficult—perhaps impossible—to enforce a felt sense of moral responsibility in others. The simple reason for this is that you cannot control how people will respond to certain situations, let alone control how they feel, at least as an individual. This sort of enforcement does often take place at a broader social level: communities or societies always express expectations about how people should think and act. These expressed expectations can be very powerful, but there are no clear ways to measure or fully predict their impact on a particular human individual.

4.3 PERSONHOOD AND TECHNOLOGY

Thus far we have considered ideas about personhood by first considering what kind of question we are asking when we ask, "What is personhood?" We then considered the connection between personhood and human identity as well as between personhood and personal identity. The discussion of personal identity included a conversation about how Western philosophers throughout the centuries have approached the topic of personal identity through individuation and continuity. After discussing some metaphysical arguments regarding personal identity, we noted the shift in the sixteenth and seventeenth centuries away from metaphysical beliefs about substance to ideas about the individual as constituted by consciousness and memory. We also discussed how personal identity is shaped by the narratives that we creatively piece together as we live our lives. The sections on personal identity required us to consider the role of self-consciousness and self-knowledge and how ideas about both are featured in various accounts of what it means to be a person. This final point led us to the topic of responsibility, and we made a point to distinguish between moral and legal responsibility.

Now that we have an overview of varying conceptions of personhood in the history of Western thought including legal, political, and philosophical viewpoints, in this section we consider how these conceptions are at play in some of the contemporary debates surrounding technology.

The topics that we explore are organized around four main themes:

1. **Technology and human identity:** As we saw above, sometimes what it means to be a person coincides with definitions of what it means to be human (human identity). Does technology change our conception of what it means to be human? What is at stake in defining other entities, such as artificial life forms, as "persons"?

2. **Technology and responsibility:** We have seen that understanding personal identity is often a question about how it is that we remain the same person over time and how it is that we can attribute actions, behaviors, creations, and thoughts—in short, agency—to individuals. How do advanced technologies interrupt common conceptions of what it means to be an agent and an agent responsible for one's actions?

3. **Technology and memory:** It was noted that memory seems to be an important ingredient in understanding personhood. If memory is important, does it matter that we rely on technology to remember things about ourselves and others? Do we give up anything when we outsource memory? Is it possible to live a human life without relying on external memory supports?

4. **Technology and narrative identity:** We also noted that personal identity has something to do with the way we understand ourselves and the narratives that we use to describe and define ourselves. How does our conception of ourselves and others change with online interactions?

These days, it's obvious how much technology is part of human life. Fitness trackers monitor our blood pressure, heart rate, and activity levels, while possibly influencing us to adopt practices that result in better metrics. Smart speakers give us voice control over lights, temperature, music, and many other aspects of our house, all by talking to a friendly disembodied voice. There are so many new technologies of this kind that the concepts of both "cutting-edge" and "normal" are constantly shifting.

Though the pace at which the "normal" changes may be new, technological mediation between ourselves and our surroundings is not. Anthropologists and paleontologists have long noted that the earliest humans used tools, created artifacts, and thought creatively about how to use the objects and space around them. Constructed shelters shape the world that we inhabit as well as our experience of it. Eyeglasses and canes are not as invasive as manipulating stem cells or altering DNA, but technology broadly understood has been with humanity since the beginning and has always influenced the way individuals and societies experience the world.

This is not the place to review the history of technological developments over the course of human evolution (although if you're interested, you should check out classes on the history of science). But it is important to remember that although technology is often associated with the new, what we conceive of as technology is often shaped by what we consider to be normal. In other words, our baseline or default sense of how humans interact with the world is a product of our particular time and place rather than a meaningful universal baseline. Once new technologies are adopted and become part of our lives they create a new normal, and it is not always easy to imagine life without them. Some would argue that it is impossible.

4.3.1 TECHNOLOGY AND HUMAN IDENTITY

Various works of science fiction invite us to consider the personhood of androids so advanced and human-like that they can pass unmarked in human society; these stories seem to urge us, by and large, to expand our definition of personhood to include these not-quite-humans or warn us of the dangers of technology so advanced as to be worthy of personhood. While it may be true that such advanced technology is many horizons ahead of us, these stories also push us to consider the scope and boundaries of personhood.

Humans have been grappling with questions concerning the personhood of non-human entities, such as advanced artificial intelligence (AI), for some time now. In fact, societies have considered the personhood of animals and other nonhuman entities for centuries! The notion of building an autonomous entity goes back very far: many ancient Greek myths imagine the construction of autonomous beings, including but not limited to humans themselves. The Jewish myth of the Golem, an anthropomorphic creature built out of mud, goes back at least to antiquity. After the fall of Rome, widely circulated tales reported that the Empire had left behind vast underground troves of wealth, guarded by golden automata. But only in the nineteenth century did the stories of such built entities start to consider the potential personhood of these entities and the ethical problems that their personhood would entail. Perhaps one of the best-known stories is Mary Shelley's 1818 novel *Frankenstein; or, The Modern Prometheus*, which is really a story about the creature (the Monster) as a person rather than the scary-monster story of later film adaptations. The word "robot" was introduced by Czech writer Karel Čapek in a 1920 play, *R.U.R.* The play includes notions of the "mechanization" of labor (though Čapek's "robots" are more biological than mechanical) and includes ideas that are now referred to as the "robopocalypse," wherein robots destroy humanity.

It can be very tempting to jump right into debating whether robots should count as people. Many mainstream news stories and think pieces about advanced AI systems are organized around this exact question. Before we start down that road, consider other, more familiar nonhuman entities that could (and perhaps should) provoke the same question. Crows, for instance, make tools out of sticks and their own feathers and drop stones into containers of water to raise the water level. Elephants take their young to visit the bones of their ancestors. Elephants and crows are not human beings, but they use tools and engage in burial practices, both of which have been suggested as markers of a distinctly human life (Jonas 1996).

One of the reasons it is both strange and familiar to think of nonhuman entities as having personhood is that there is a long tradition in which nonhuman entities are said to have some kind of legal or political standing. In ancient cultures animals were charged with and punished for crimes. In the United States, corporations are considered to be legal persons. Legal personhood often determines what sort of entities are deserving of certain rights and protections. Should certain forms of artificial life be granted with legal personhood? In 2017 the robot Sophia was granted citizenship in Saudi Arabia (Reynolds 2018). Earlier that year, the European Parliament adopted a proposal to explore the implications of how robots are classified legally and the possibility of creating a special legal status for them. The possible consequences of applying an existing framework like corporate personhood to robots are complicated: for example, the possibilities that such a law could allow humans to shield themselves from liability for robots' violations of human legal rights (Pagallo 2018). (See the sidebar "Metaphors, Personhood, Technology, and Autonomous Cars" in chapter 6 for some of the complications with defining what should even be regulated as a "robot" to begin with.)

Story Point: "Message in a Bottle," by Nalo Hopkinson

"Human beings aren't the only ones who make art," she says.

Greg's friends all want to be parents, but Greg himself has different ideas about how to create a legacy for himself. Greg is an artist, and part of why he values art is because it can help people connect with each other even across differences of culture and generation. Greg's art is motivated by the idea of giving a voice to the unheard and the unseen, using everyday objects to bring overlooked people back into our shared story. But although Greg wants to widen the circle of personhood to include many who are often forgotten, he balks at widening the circle further—especially when it seems to come at the expense of his own artistic legacy. Both Greg's own art and the challenge he faces can help us think about how our tools and our creations are a part of our selves, and how they shape the way others understand us.

4.3.2 ANTHROPOMORPHISM: PERSONIFYING TECHNOLOGICAL ARTIFACTS

Even if we don't think that technological objects of any kind can have personhood, there is still the matter of anthropomorphism—of attributing what we identify as human-like qualities to nonhuman entities such as a boat or a robot—and the effects of this practice. Darling (2017) details an experiment with two groups: one group that interacted with a robot who was given a human name and a backstory, and another group that was given just the robot with no backstory. The group given the backstory showed more standard markers for empathy when interacting with their robot, including asking if it was okay when struck with a falling object. This study suggests that the framing of the robot relationship can influence how people perceive, and subsequently interact with, a given piece of technology.

There are arguments against encouraging anthropomorphizing technology, especially forms of social robots that are specifically designed to interact with humans such as caregivers or teachers. The predominant argument is that these technologies may "replace" relationships with real humans because interacting with robots may be "easier" in some cases (Turkle 2010). For instance, roboticist Carme Torras, in her novel *Vestigial Heart* (Torras 2018), imagines a world in which each person has a helper robot rather than meaningful personal relationships. However, as Darling (2017) points out, there are many examples in which instead of replacing a relationship with a human, a robot can augment and enhance a human relationship. Consider an instance of an authority figure and a robot: Darling details specific cases including using both a Paro, baby seal robot, and a social worker in which interviews with young children were more productive with both the human and the robot present. Hence, social robotics can sometimes encourage and facilitate communication.

Story Point: "Dolly," by Elizabeth Bear

> Sven was cooking shirtless, and she could see the repaired patches along his spine where his skin had grown brittle and cracked with age. He turned and greeted her with a smile. "Bad day?"
>
> "Somebody's dead again," she said.
>
> He put the wooden spoon down on the rest. "How does that make you feel, that somebody's dead?"
>
> He didn't have a lot of emotional range, but that was okay. She needed something steadying in her life. She came to him and rested her head against his warm chest.

"Dolly" is an old-fashioned detective story set in a technologically new-fashioned world. In this version of the future, many people have home companion robots that they use for housekeeping and for intimacy. One such person is the billionaire Clive Steele, who is killed by his advanced prototype Dolly. Another is Roz Kirkbride, the detective tasked with investigating Steele's murder. Roz longs for a companion who can truly love her, but in the absence of such a partner, her home companion doll Sven fills some of the gaps. As the Dolly case unfolds, it challenges many of Roz's most basic assumptions about what is possible and what she must do—and by the end, even her uncomplicated use of Sven has come to feel more complicated. As readers in a world in which companion AIs are less advanced than Sven but nonetheless manage to inspire our attachment, "Dolly" helps us reflect on the costs and benefits of relying on machines for emotional needs.

4.3.3 AI AND RESPONSIBILITY

Personal assistants have become more commonplace in the home. Many people have started to think of the disembodied voice of Alexa or Siri as part of their households, another person with whom to discuss, interact, and ask to play music. These digital assistants are oftentimes deployed within a home with multiple users, be they parents and children or a set of roommates. This situation of an entity within a shared space raises a number of curious questions related to the personhood, and more specifically the responsibility, of digital assistants. Consider a case in which two younger children are home alone and a friend brings beer to the house for all of them to drink: should the digital assistant immediately call the parents? Does the assistant have a responsibility to ensure the safety of the children? Or is it subservient to their desire to drink beer and have a good time breaking the rules? Or consider the simple question of a shared digital assistant between two partners: if the agent is instructed by one partner to lie to the other about a date or a location, should it? Can an assistant read your partner's email to you? These questions are explored in detail by Luria et al. (2020), who find that many of these questions largely come down to how we perceive the social role of the technological artifact itself. Do we see the assistant as a full person, a member of a family? Is the assistant "owned" by a particular person? Or does the assistant have responsibility to everyone within the home? Does the way in which you define the personhood of the assistant make a difference in deciding to whom or to what the assistant owes responsibility?

These examples show us that it is important to be aware of how the language we use, or are encouraged to use, to refer to technological artifacts may influence both our interactions with and our thinking about these artifacts and their personhood.

Story Point: "Asleep at the Wheel," by T. Coraghessan Boyle

> Warren's grinning, so Jackie starts grinning, too. "What?" he says.
>
> "Let's us play chicken. Re-enact the scene, I mean. For real."
>
> He just laughs. Because it's a joke. Real cars, cars that do what you want, cars you can race, are pretty much extinct at this point, except for on racetracks and plots of private land in the desert, where holdovers and old people can pay to have their manual cars stored and go out and race around in them on weekends, though he's never seen any of that, except online, and it might just be a fantasy, for all he knows.

Cindy is a single parent with a high-stress job, and so she's happy to outsource logistics and scheduling to the operating system of her self-driving car, "Carly." Carly keeps track of Cindy's schedule, the traffic conditions, and her recent shopping interests; it also monitors the movements of Cindy's teenage son Jackie more closely than Cindy does. Cindy harbors some deep frustrations with her car, especially when its safety protocols lead it to operate in ways Cindy dislikes. But these frustrations do not reflect individual personality quirks of the car (which is ultimately an AI with recognizable decision trees, even if it's been programmed to speak in a chatty voice). Rather, they reflect a larger technological social apparatus that has been designed to smooth away difficult decisions and dangerous edges for users like Cindy—both when those users like it and when they don't.

4.3.4   DO WE NEED BODIES?

As we have seen, some definitions of personhood rely on metaphysical beliefs about an underlying substance that is said to constitute the unity and continuity to individual entities. We have also discussed how memory and self-consciousness can function to support notions of personal identity. Perhaps you have heard the hypothetical situation in which humans will eventually be able to upload their consciousness or their memories to a computer and "live" forever. How do the definitions of personhood provided above operate in ways that support these kinds of claims? How do they limit them?

Consider how our definitions play out in an example in which one consciousness, or at least appearance of consciousness, has the ability to "jump" between bodies. You may have had an experience close to this already if you've interacted with

Alexa, embodied in multiple different speakers throughout your home, or Siri on your iPhone versus your laptop. Consider now a computer assistant/persona that moves between locations and even embodiments while interacting with you. There are multiple ways that this assistant can inhabit various embodied forms, which is totally different from what we as humans can do. Consider the following options for the embodiment of an assistant:

- One-for-one: Each assistant has a single physical embodiment. This is closest to our experience as humans.

- One-for-all: One assistant is embodied in several forms at the same time. This would be like talking to Siri on your smartphone and then interacting with Siri on your laptop. The physical embodiment and modes of interaction may change but the "consciousness" inhabits all of these embodiments simultaneously.

- Re-embodiment: One assistant is embodied in several forms in sequence, but only one form at a time. This is similar to the Siri example but instead of being able to control multiple embodiments at the same time, Siri would have to "move" between them.

- Co-embodiment: Multiple assistants can be embodied in the same physical form. This would be akin to interacting with both Siri and Alexa on your smartphone, at the same time. Both consciousnesses inhabit the same physical embodiment.

How do these options change your perception about the personhood of the digital assistant? Would you treat all these the same? Would the embodiment change how you perceived the assistant or what it could do? One such experiment was carried out first with a robotic butler that helps to check you into your room at a hotel reception, in the form of a humanoid robot with a name and identity that connects to you. The persona then could jump to different artifacts later, a TV screen in another room or a speaker overhead, to help you as you moved through a new space. How would you feel about interacting with such an entitity? It might seem nice to have one "personality" follow us between locations, but it also seems that it could be unsettling in some contexts (Luria et al. 2013; Reig et al. 2020).

This technology in particular presents a challenge for some views of personhood. Specifically, it directly puts in tension the distinction between the material substance-based metaphysics (i.e., that people have the same body over time), and the nonsubstance-based view (i.e., that people have memory and self-awareness). In the case of the body-hopping assistant, it turns out that individuals' reactions depended heavily on the context. In the example noted above involving a robotic butler, people reported enjoying the fact that the personal assistant could follow them around and look after them in such a personalized way. However, when the same experiment was performed at a hospital, where the robotic concierge checked them in and then jumped to a smart TV in the room to offer assistance on dosing for medications, many participants thought that the assistant was not "smart enough" to give advice on dosing because it was working the front desk/check-in, saying "when you go to the X-ray room, [you want to] have a professional person [social presence] there, not [the one from] the reception area" (Luria et al. 2013).

4.3.5 TECHNOLOGY AND RESPONSIBILITY

The complications that technology brings to the concept of personhood doesn't need to involve a robot. Instead, consider the ease and ability that many of us have to act at a distance from our physical selves and to act as a part of a larger collective. With the internet we can turn on cameras in remote homes, instruct Alexa to turn up the bedroom thermostat while we're on an airplane, complete a digital art project online, or even commit code to a large open-source project used by millions of people (see the thousands of individuals who work on Linux in a completely distributed manner, listed under "Contributors," at https://github.com/torvalds/linux).

The ability to act and create at a distance, or even as a part of a collective, complicates the notion of responsibility that we talked about in the definition for moral agency. The very notion of responsibility implies that we can identify our actions as our own. At a very basic level this can be understood with reference to our own bodies: I push a glass off the table, and it falls and breaks. I am the cause of that action and thus I am responsible for the result. However, various forms of technological mediation seemingly remove our actions from the immediate effects. Both of these features cause problems for many of the traditional Western theories of agency.

Consider also: is it possible for collectives, organizations, or simply groups of individuals to share responsibility? It could be argued that the internet and other networking and communication technologies create new forms of agency and action in which it becomes hard to find the locus of action. Do individuals invent new technologies? Or are new inventions and the production of technological objects the results of collective action? In chapter 5 we discuss the field of science and technology studies and stress the way society and tech operate together in an inseparable way. But here we show how our ideas about personhood are adjusted when we grant personhood to collectives and organizations.

Some collectives are made possible by technology, and in some cases such as Reddit, they are technological in how they function (i.e., as a nonphysical person), but the individuals act together as if they were a person in terms of collective action.

One metaphor we can look to is past complications brought about by a corporation or a state. In these examples, as with modern internet groups, collectives are enabled to create implicit persons in ways not before seen. In the United States, the legal workaround for this has been through the use of courts and recognition of these "implicit persons." In the United States, public institutions and corporations are considered to be "juristic persons." Juristic persons, much like "natural persons," most notably human beings, are considered to have certain rights and duties (*Oxford Reference*, n.d.).

Consider the development of a software product via a distributed platform, where none of the team members ever meet. It becomes unclear how rationality, the process of determining cause and effect and self-awareness, and the process of understanding ourselves as distinct from other agents are to be maintained under such circumstances. For instance, in communal development it may become unclear whose words or code end up in a product. There are ways that we have attempted to maintain this connection through our tools; for instance, the distributed code management tool Git has full history, carefully logging each persons' contributions, and it even includes a BLAME function to tie code back to individuals.

This last point is important not only for thinking about responsibility in a digital age but also specifically in the computer professions, where responsibility is dispersed among many individual agents. It is highly relevant to other engineering fields as well, in which a single project often involves hundreds or even thousands of people who work on the design, engineering, and building of an object.

Some theorists argue that responsibility can apply only to individuals (Nissenbaum 1996; Tavani 2012). Technology scholar Helen Nissenbaum, for instance, thinks it is important to use the term "accountability" rather than "responsibility" in contexts in which there is no single actor but rather a collection of actors. Accountability, Nissenbaum argues, is a broader concept than responsibility. Individuals, groups, and organizations can all be answerable. Only individuals are responsible. In a computing context, she notes that "accountability means there will be someone, or several people to answer not only for malfunctions in life-critical systems that cause or risk grave injuries and cause infrastructure and large monetary losses, but even for the malfunctions that cause individual losses of time, convenience, and contentment" (Nissenbaum 1996, 26–27).

Nissenbaum's focus on accountability rather than responsibility raises some important ethical questions. Is it fair to say that only people can be morally responsible for an action? What counts as action? We might note, for instance, that sometimes responsibility is removed from the effects of our own individual actions and applies to our compliance or failure to act. For example, though it cannot be said that any one human being is responsible for climate change, there are plenty of moral arguments that call for all of us to take responsibility for the planet that we inhabit and for future generations, such as decreasing our carbon footprint. From this perspective, there is no reason to think responsibility cannot be collectively shared.

These debates are ongoing, and that is in part because the difference between responsibility and answerability also presupposes a conception of persons and how those persons can be distinguished from a group, organization, and community, or from a development team, project team, or company. Under further reflection, do these distinctions make sense? Do they help us think about responsibility and accountability? Do they help us to identify our values and promote those values? Or do we need to change our conceptions of personhood to better align a conception of human action that is embedded in technosocial systems?

4.3.6 TECHNOLOGY AND MEMORY

As we have noted above, one's conception of one's self is shaped by our life experiences and the particular ways in which we have internalized these experiences. This requires consciousness, certainly, but also the capacity to retain and remember past events. Insofar as your memories play a crucial role in shaping your goals and your values, they are an important part of your ability to think and act in a way that feels like it is "really" you.

It can be tempting to think that this "real" you, the one that is informed by your memories, somehow exists apart from the material world, including technology. But as we have noted, the boundary between the self and the world is not so easy to pin down in practice. Vision-correction technology or hearing aids can seem external to the self, but they can have a huge impact on how a person perceives and engages with the world. Similarly, technologies designed to collect, store, and transmit memory in many ways determine the means and conditions under which a person interacts with their own memories, either by storing things externally or by "remembering" our passwords and preferences so that we ourselves do not have to reproduce them regularly.

The outsourcing of "remembering" to smart technologies might seem like a threat to "real" you, but trying to remember a time when human beings did not rely on external memory supports seems impossible (as we discussed in chapter 3). In many ways, memory is experienced and realized differently depending on the technologies that are available and considered "normal."

To put this into perspective, consider Locke's notion of personal identity above, which emphasizes the importance of continuity of the self by reference to consciousness and memory. If memory is taken to be one of the fundamental aspects of our definition of personhood, is there more at stake when we rely on external memory supports? If, for instance, one's personal memories and experiences could be transferred into a different entity, does that person exist in two places? Do they really die when their body dies? As more and more of our memories are stored in images, text, and other digital forms, some people have wondered: do multiple copies of ourselves exist? Answering these questions requires us to gain clarity about our own ideas about what constitutes personhood.

4.3.7 TECHNOLOGY AND NARRATIVE IDENTITY

"On the Internet, nobody knows you're a dog" (see figure 4.1). While tongue-in-cheek, this saying illustrates an important facet of the online world. Since people are not physically observable when they interact as end users on a particular online platform, these interactions can give an impression of privacy. For example, when posting on Reddit or playing Fortnite there is a reasonable expectation that your personal information is private unless you choose to reveal it on the platform. Hence, when interacting online as end users, interactions require deliberate button presses, so we start with a "blank slate" of anonymity granted by this privacy. This gives us a freedom, to craft the personas through which others on the internet interact, far beyond what we typically have in the

FIGURE 4.1
A dog enjoying online pseudonymity. "On the internet, nobody knows you're a dog" is an adage and internet meme about anonymity that began as a caption to a cartoon drawn by Peter Steiner, published in the *New Yorker* on July 5, 1993.

offline world (sometimes called "meatspace," to designate it from online spaces (Glabau 2021; Sullins 2000), where we constantly send out signals about ourselves to those around us even when we don't intend to. In this section we focus on what this combination of a blank slate and intentionality means for the formation of identity.

One of the classic draws of privacy is that it allows a person to maintain an acceptable "public" face while still letting the person "be themself" in more personal settings. This conception of different spaces of privacy or "spheres" may be familiar to you, and we return to it when we look at conceptions of privacy. This partitioning of space into public and private is sometimes used to strengthen intimate bonds between persons with less fear of harming their overall image. The internet adds a new dimension to this public-private distinction: people can put out content that is thoroughly public in terms of visibility, yet through their online personas they can separate this content from their offline public selves, which lets them share things about themselves that would normally be reserved for the private/personal/intimate sphere.

On the other hand, the internet can also let people present an even more carefully cultivated version of their real-world public selves. Consider how often social-media photos feature vacations, parties, and the other "glossy" moments of the subjects' lives. Traditional media has long glamorized the lives and appearances of celebrities. This practice has carried on to social media, except with ordinary folks as the producers and subjects. This offers producers of such content avenues to fame and occasionally fortune through self-promotion. In this way, social media democratizes industries that used to be controlled by a small number of people. However, a downside of increased accessibility is increased pressure to compete with the high standard set by the curated lives of our friends and neighbors (Thal 2020).

A person's public persona has never been entirely within their own individual control: even in a society that strongly enforces slander and libel laws, no individual can exert complete control over how other people interpret the things one says and does or the opinions that others form. Online, information about a person can disseminate faster and become even harder to erase, leading to reputational changes overnight. This process can be an instrument of aggression or of accountability; compare a situation in which a person's photo, originally shared in a private and consensual setting, gets used against them as "revenge porn" versus a situation in which a person's abusive communications, which they may have been able to cover up or deny in an offline setting, are used to warn other potential victims.

There are, broadly speaking, three ways to situate a given online persona relative to one's offline identity. One of these is for an online user to lay claim to their offline identity, using either their legal name or other key identifying details of their offline self. (Legal names aren't the only way to do this, but they are by far the most common, so we will refer to these as "real name IDs.") A second approach is to use a pseudonym, and to consistently use the same pseudonym (or set or series of pseudonyms) when participating in an online community or group of communities that can be recognized by other participants as attributable to the same user. The third approach is anonymity, either by avoiding any identifying details or by disposing of user IDs frequently enough to prevent others from connecting one ID to another.

It can be tempting to see pseudonymous IDs as a middle ground between a real name ID and pure anonymity. On a very surface level, there is some truth to this: participating pseudonymously in a community over a period of time offers many of the same experiences of connectedness as offline participation in a meatspace community. These similarities include forming relationships and developing a publicly recognizable history and body of contributions. However, the online space remains distinct from the connectedness and selfhood of the offline person who uses that pseudonymous ID. But this simplistic spectrum makes sense only if we assume that the fullest and most authentic disclosure takes place under the real name ID, or that using one's real name is equivalent to a full and authentic disclosure—and that is an assumption that does not hold up to scrutiny. Almost anyone who has had pseudonymous experiences in online communities is likely to have formed intimate connections through that medium, or at least to have disclosed things about themselves that they would not be comfortable sharing with many people in their offline lives. And any person with any experiences in offline communities will be familiar with the experience of editing oneself in front of others.

Story Point: "Lacuna Heights," by Theodore McCombs

> How desperately unhappy do you have to be, Andrew thinks, to pothole your own mind like that?

Andrew, like many residents of future San Francisco's upper crust, has an Aleph—a neural implant that connects directly to the internet. Unlike most of Aleph's client base, though, Andrew makes use of the device's privacy mode, a function that prevents privacy-locked memories and experiences from being bundled for public consumption but that also necessarily locks out the user themselves once they exit privacy mode. While in public mode, Andrew's experience of his own life is fractured and disjointed. He is troubled by the knowledge that he is hiding something important from himself and also haunted by grief and anxiety about this important thing he cannot quite remember. The reader's journey alongside public-mode Andrew—an unreliable narrator if ever there was one—conveys both the impossibility of living life well when one's life story is radically incomplete, and the emotional and ethical toll it takes on Andrew himself that he is unable to fashion a coherent story about his own life.

4.3.8 AVATARS AND SELF-PRESENTATION

Up to this point, we have assumed a "disembodied" online experience, in which other people perceive a user only through content that user has written or uploaded. But since physical appearance is such a powerful component of personal identity for many people, it is not surprising that developers have tried to recreate the experience of embodiment virtually as well.

Consider a massively multiplayer game like *World of Warcraft* (figure 4.2) or a social simulator like *Animal Crossing*, which allows players to interact with each other through customizable avatars. In some cases, players create their avatars to be virtual doppelgangers of their physical selves. It's possible for players to base their avatar choice on elements that are not related to personal identity, at least consciously ("I play as a Night Elf because they have useful stat bonuses"). But avatar creation can also be a unique means of self-exploration or self-expression.

For instance, many players create avatars with gender presentation different from the players' offline gender presentation. Some of these players are transgender people missing either the material resources or the supportive social environment they need to transition in the offline world; avatars can offer these players an opportunity to present themselves as they wish (Jamison 2017). Other players may choose their avatar's gender expression to challenge or play with social norms, to experience life from new perspectives, to alter others' behavior toward them, or to tell stories through role playing (Osborne 2012). While the opportunity for such elastic

FIGURE 4.2
In a 2007 *World of Warcraft* commercial, Mr. T shows off his Night Elf likeness.

self-determined presentation has been available for a long time through role-playing games (decades if one considers only tabletop role-playing games, millennia if one considers the more basic act of collaborative imaginative storytelling), such games have relied on the player's own narration or description to create and sustain their fictionally crafted appearance. In online games, a player's chosen appearance is visible to others without that player having to mediate or explain, making it more like meatspace life in that one way.

Story Point: "Not Smart, Not Clever," by E. Saxey

> The polite term for what I do is ghost-writing. Sometimes I'm ghostly. When I creep into lecture theatres. When I need someone to swipe me in and out of buildings, to pick up their device, to type for me, as though I can't touch objects. When the lecturer says, "Any questions?" and I'm bursting with questions, but I can't have a voice. When I see my face reflected in the screen of someone else's device and I pull away before the face-rec can catch me.

Lin's new friends at University are in the same boat as undergraduates everywhere—they're struggling to keep up with their schoolwork, terrified of flunking out, and on the hunt for a form of plagiarism that can outwit the face-rec and brain-rec software their University is using. Lin is a talented researcher and writer with a knack for copying other people's writing style, so she helps her peers out—for a price. Over the course of the story, it becomes clear that Lin herself pays a different kind of price for submerging herself, over and over again, into other people. "Not Smart, Not

Clever" offers a powerful reflection on how transformative it can be for the self—for good or for ill—to take on other identities.

4.4   THE POWERS AND LIMITS OF DEFINITIONS

In sections 4.2 and 4.3 we explored multiple ways of defining personhood. These definitions have helped us explore the ways in which technology, as something fundamental to human life, influences the way we think about personhood.

But although definitions can help us think more clearly and guide our reasoning, they do not by themselves solve the problems they help us recognize and describe. After all, acknowledging the personhood of someone or something doesn't automatically mean we will know how to treat that person or thing, nor does it mean that we will always do what is right by that person or thing.

Definitions can also give us the impression that categories and concepts we use are fixed and permanent. But as we have seen so far in this chapter, technological, political, and societal changes have continually challenged us to rethink our definitions of human identity and personal identity. There are no final definitions of what does and does not count as personhood, even if we may need to rely on operational definitions.

At the same time, personhood is, as we suggested, a normative concept. This means that it is a concept that has implications for how we are to treat others. For this reason, it remains important to pay attention to situations in which it is applied as well as to situations in which it is being denied or retracted.

It is an unfortunate fact that sometimes human beings are put into situations where their personhood is severely constrained or limited. Over time, these constraints can lead to an insidious feedback loop, where the normative claim to personhood grows weaker and weaker in the eyes of those who do not suffer from the same constraints. One front where this dynamic plays out most visibly is with respect to privacy.

4.5   PRIVACY AND PERSONHOOD

In the early months of the COVID-19 pandemic, many people's lives were transformed to revolve around online interactions for nearly everything: socializing, work, school, shopping, and medical care. Moving so many of our interactions online allowed data about these interactions to be recorded and tracked in an unprecedented manner. Even before the pandemic, a Pew Research Poll found that 81% of Americans felt they had very little or no control over the data collected about them (Auxier et al. 2019). With so much of our activities moving online during the pandemic, these concerns have only grown as we have generated more and more data about how we work, play, and live; in a recent survey, 75% of adults agreed that there should be more governmental regulation on how companies can use personal data (Brennan-Marquez 2021; Kite-Powell 2020; Véliz 2021). When the 2019 Pew survey asked participants to describe privacy in their own words, 28% of respondents mentioned other people or organizations, 26% mentioned the ability to control or decide accessibility to parts of their lives, and 15% mentioned control of possessions or information without reference to others. However, 9% of respondents responded that privacy, and especially digital privacy, is a "myth" and does not exist.

The developments in information and communication technology over the last several centuries have made it easier than ever to collect data about people's movements, spending, and interactions with others. Modern information and communication processing systems have also made it much more efficient to analyze that data, and much easier to bundle the data into a form that is usable by a marketing department or a political organization. As physical devices like smartphones and platforms like Facebook or LinkedIn have become de facto requirements for maintaining one's social or professional connections, users have submitted to the loss of different kinds of privacy with varying degrees of consternation.

But simply enumerating these encroachments on our privacy does not help us understand what they mean in practice, or how much they ultimately matter, or why. Connecting privacy to personhood will help us gain a better understanding of why these things matter. In particular, we will see that several of the core elements of personhood—memory, identity, decision making, self-consciousness, self-understanding, and responsibility—rely on privacy. These parts of our existence can operate fully only when some measure of privacy exists in order to protect them.

Although this chapter offers you some fairly sophisticated philosophical tools to understand privacy and personhood and how they are connected, it is also possible to appreciate this connection at a more intuitive level when we immerse ourselves imaginatively into specific settings where they are in play. The following three examples offer some situations for feeling your way through how deeply connected personhood and privacy are. As you will see, each of these three examples becomes easier to understand fully, and to respond to, when we bring to bear well-developed ideas about personhood. But you can probably also see how, even without those well-developed ideas, there is in each case an intuitive connection between the threats to privacy (potential or actual) and the ways in which those threats diminish one's own capacity to regard oneself as a person and are thus worthy of respect and fair treatment.

4.5.1 TYING PERSONHOOD AND PRIVACY TOGETHER: THREE EXAMPLES

Example 1: Open Offices

One recent case study can be found in the move toward open-plan offices in many white-collar jobs. Open office plans represent a shift away from a more "traditional" plan of white-collar offices, in which most employees work in private or semiprivate spaces. Because many other aspects of these formerly individualized office workplaces have remained stable during this shift from private or semiprivate workspaces to open-office plans, comparing working conditions and practices under these two layouts can help illuminate how the shift in layout affects other aspects of workers' lives.

The private office, as we know it, came to exist during the eighteenth century in the major cities of Europe, when lawyers, civil servants, accountants, and other professionals who had previously worked from home began renting separate rooms for work; in response, the companies that employed them began to build buildings specifically to provide them such space (Dishman 2018). This model remained standard until the middle of the twentieth century for some classes of professionals. (It is worth noting that individual office space was never representative or standard for most workplace conditions; most types of physical labor such as construction, agricultural work, and factory jobs take place in open spaces, and many types of office workers have shared space with other workers doing the same job.)

Since the 1960s, there have been two waves of effort to shift away from private workspaces and toward open offices. In the 1960s, German and American companies began to experiment with open-plan offices in an effort to democratize the workplace and promote greater collaboration. There was significant backlash to this move among workers, and most of the companies involved in this shift converted their open workspaces into semiprivate cubicles in the 1970s and 1980s. But the open-office idea rose to popularity again in the mid-2000s, especially as a workplace design for tech startups. Many of these startups relied on open plans because they were cheaper to establish and maintain. But even as these companies became more profitable and were able to design workspaces of their choosing, an increasing number of companies—in the tech industry and elsewhere—began to choose open-plan offices, again in an attempt to decrease hierarchy and to promote collaborative innovation (Bernstein and Turban 2018; Oldham 1988) as well as to capitalize on the symbolic power of the open-office plan, which had quickly become recognizable as characteristic of the exciting and innovative (and profitable) Silicon Valley startup.

Yet studies have repeatedly found that open-plan offices rarely succeed in promoting the work culture for which they aim. According to a 1988 study, workers who had switched to private workspaces reported feeling less crowded and better able to accomplish tasks (Oldham 1988). A 2018 study of workplaces that transitioned to open-plan offices, which tracked actual behaviors rather than relying on self-reporting, found that face-to-face communication among workers dropped by 70% and that these workers took almost two-thirds more sick days (Bernstein and Turban 2018). Workers interviewed for the study reported feeling unable to do their work amid the distraction of the open office plan, and significantly lower levels of overall health and happiness.

The reason worker productivity suffered in open-plan offices is not that the architects' vision was entirely mistaken; in fact, researchers found that there was indeed more communication and collaboration between teams, whose members would be less likely to interact in a traditional cubicle-based office space. But when we consider the larger work patterns in this study, specifically the amount of face-to-face interaction between coworkers, it becomes clear that open-plan offices are quite damaging both to office productivity and to the workers' self-reported well-being, and that these forms of damage are related.

In both the 1960s open-plan offices and those of the new millennium, workers felt that their ability to do certain kinds of work was undermined by the lack of privacy. Additionally, workers in several studies reported feeling pressure to look busy, and one study in particular reports that women feel an additional pressure to dress nicely (Hirst and Schwabenland, 2017; Morrison and Smollan 2020). In other words, workers in open-plan offices find themselves stripped of many different kinds of privacy: the ability to control the terms of their interactions with others, the ability to think and work without distraction or interruption, and the ability to make choices about how to spend any given moment of their time without feeling evaluated or scrutinized. The need for this personal sense of control also explains why, after the proliferation of remote work in the beginning of the COVID-19 pandemic, which we discuss further in section 5.5.6, many workers were hesitant to return to in-person workplaces even when they were not specifically concerned about COVID exposure.

As these studies found, the ideal of the nimble, highly collaborative open-plan office is undermined by the fact that many workers find it necessary to manage their experience of being there, and the concept of personhood equips us to understand how and why. If we imagine workers only as entities that aim for maximal productivity, these workers' choices—taking lots of sick days or communicating primarily by email—seem counterproductive. But if we take seriously the idea of workers as persons—that is, not only as entities capable of reflecting and making choices but as entities who need to be able to reflect and make choices for their own (and others') basic well-being—then we can better understand how the constraints imposed by open-plan offices might seriously impede a worker's ability to accomplish tasks.

Most office jobs that have not yet been automated require distinctly human capacities from their workers, including creativity, sustained attention, and flexibility—the same qualities that are most at risk from the lack of privacy in an open-plan office. Without some measure of control over her interactions with others and the pace and method of doing her work—that is, *without any privacy*—a worker's personhood is put under pressure.

Example 2: Re-evaluating Gender Identity

Many, if not most, people would say that their gender is an important part of who they are and how they experience the world. Many of our expectations about what other people will be like—and about what we ourselves will be like—are rooted in deeply embedded cultural norms about how men and how women are each supposed to behave. Even if you don't notice it in your own social interactions with others, reading or watching stories in popular media will reveal some deeply embedded assumptions about how being a man, or being a woman, is related to how a person thinks, acts, and feels.

Sidebar: Sex and Gender

The concepts of sex and gender are often conflated. When a doctor pronounces "It's a boy" or "It's a girl" for a newborn, this pronouncement is based on the child's apparent *sex*, but it prescribes a *gender* in which the child will likely be raised and that the child ultimately may or may not be satisfied with.

Sex describes a set of physical traits. While chromosomes are influential, exposure to hormones will also shape the sex traits a person develops—primary sex traits (i.e., genitalia) in the womb, and secondary sex traits (such as facial hair or breast growth) later in life.

Gender is a social rather than physical category. It is difficult to define precisely and has many aspects. For instance, *gender identity* is a person's internal sense of being a man, woman, or something else. *Gender presentation* is how a person conveys gender to others through their appearance and mannerisms. *Gender roles* are the templates for how a culture expects people of particular genders to behave.

Even people who are cisgender (i.e., people who identify with the gender assigned to them at birth) often struggle with how much, and in what way, their gender identity informs the way they perceive themselves, others' gender-related expectations of them, and their social capabilities. These struggles are far more complicated for people who are transgender (trans, i.e., people whose gender identity differs from the gender assigned to them at birth). In order to claim publicly the gender identity that feels true to them, trans people must argue for, or at least assert, an identity that does not match what others have previously expected them to be. Because many people are unfamiliar with the concept of gender nonconformity, or even hostile to it, any trans person who wishes to change their public identification will often be put in the position of explaining this basic aspect of their existence over and over again, and in many cases will have to contend with others who resist or deny their self-identification (Austin 2016).

When a trans person publicly claims a gender identity that feels comfortable to them, or rejects gender identity altogether, they are making something personal about themself part of their public life. That public can include friends, family, coworkers, or the Department of Motor Vehicles (DMV). Before reaching that point, almost all trans people go through a period of reflection or introspection in which they reconsider their gender identity (Planned Parenthood, n.d.). In materials designed to help those who are transitioning, it is often advised to find a small circle of trusted friends to tell first. One trans woman reports that "I tell myself first, repeatedly. I keep it up until . . . my mind has completely come to terms with what I'm telling it. Then I'll be ready to tell others." (Advocates for Youth 2020). This introspection is often quite challenging. Even when they are not facing prejudice and threats of violence—as is frequently the case (Wirtz et al. 2020)—trans people often report having had difficult periods of time when they found it hard to make sense of their own feelings and experiences, because they were trying to interpret themselves by means of categories that didn't fit; even those trans people who report a clear sense of their own gender identity early in life almost uniformly face significant pressure to conform to their birth-assigned identity.

The challenge that trans people face—whether they identify clearly with a different gender or understand themselves to be fluid or nonbinary—helps make clear the complex nature of gender identity: it is both internal to the individual and part of the public sphere. It is internal in the sense that individuals identify their own gender identity; but it is also public in the sense that a person's gender identity is part of how they interact with the world and with other people. Because of this complex intersection, a trans person's decision to change their social and/or physical presentation to match their gender identity is a difficult one, because it requires them to demand changes in how other people (such as family and friends) and institutions (such as the DMV) recognize and categorize them.

In addition to ascertaining their gender identity, trans people also have to, or get to, decide how much of this personal history to make public. A person who has undergone some form of "transitioning"—the process of integrating their gender into how they present themselves to others—may wish to keep their *trans status* private even if they publicize their *gender*—for instance, changing the sex marker on their driver's license to avoid harassment at traffic stops and ID checks.

A trans person's control over information about their assigned-at-birth gender, their true gender, and the relationship between them is important in part because it affords them the physical and legal safety to live as they wish. Individuals report being harassed or harmed or fired when their gender identity was revealed (Thompson 2016). In the United States, this disclosure can happen through electronic medical records, through interactions with employer-provided insurance (for instance, if the person wants the insurance to pay for trans-related medical expenses such as hormones) (Thompson 2016), through legal records such as birth certificates, through drivers' licenses, or through old online presence under their "deadname" (the one that they no longer use). Having privacy about their body and their gender expression allows them to live more securely.

Just as importantly, privacy is directly essential for this process of transformative self-knowledge, because a trans person's process of coming to understand their own gender identity necessarily involves breaking away from other people's expectations and ideas. Even the most sympathetic and receptive friends are likely to think of a trans person in terms of their prior gender identity before the newly claimed identity has been disclosed to them. It is necessary for a trans person to feel entitled to know themselves in a way that is not available to anyone else and to try on new identities without scrutiny or judgment from others. Without that private space, they will not have the opportunity to test new possibilities against their own ideas of their own unique personhood.

Example 3: Prisons

Prisons are a more deliberately constructed case study on the intersection of privacy and personhood because prisons are institutions in which persons—those convicted of crimes or accused of crimes—are systematically denied privacy in order for that institution to function as intended.

Prisons serve two major functions. The first function, to detain people convicted of (or accused of) crimes, answers the practical and logistical needs of the state in maintaining civic order; it is much easier to oversee and contain a group of people if they are all concentrated in the same place rather than scattered. The second function, to punish or reform the prisoner by subjecting them to the experience of prison, is not about logistics; this function is fulfilled only when the imprisoned person suffers from the experience of being imprisoned, so that they will be either deterred from or trained not to pursue future crime. The psychological and social theories behind prison-as-punishment and prison-as-reform are quite different in several important ways, but these approaches share the assumption that the prisoner needs to experience suffering at a deep level in order for prison to "work."

This second reason for using prisons dates back to the Roman Empire and was used intermittently throughout the Middle Ages (although slavery was a common alternative). In the modern age, as prisons have become the primary form of punishment, this second function has become central to how prisons function, both in practice and as a part of how civil society is conceptualized (Grassian 2006).

With rare exceptions, incarcerated people do not have access to any kind of physical space where they can be alone and/or not surveilled; in high-security facilities, even restrooms are typically subject to oversight (Feeney 2016; Foster 2017). They also lack most of the legal protections they had or would have outside the prison, and what protections they have are often fragile. In the United States, for example, the Supreme Court determined in the middle of the twentieth century that imprisoned people cannot claim the protection of the Fourth Amendment (which protects against unreasonable search and seizure of one's property), because this protection depends on a "reasonable expectation of privacy," which prisoners cannot claim to have (*Bell v. Wolfish* 1979, 559). Imprisoned people also have less control over their own bodies than nonprisoners; for example, a person incarcerated in the United States who fasts or goes on a hunger strike can be force-fed, on the grounds that the state's interest in keeping the prisoner alive outweighs their right to privacy (Stahl 2019; West and Noffsinger 2019). As British ex-inmate Carl Cattermole wrote in *Prison: A Survival Guide*, "Prison and privacy are two words that share the first three letters, but they have no overlap whatsoever" (Cattermole 2019).

Prison sociologists have argued that this lack of privacy for inmates is not just a by-product of the prison's physical needs (the first, organizational function) but is also a goal in itself, intended to degrade the prisoners in front of both the guards and themselves (the second, experiential function) (Schwartz 1972). A great many firsthand accounts from current and former incarcerated people describe feeling degraded by the lack of privacy. (*Bell v. Wolfish* 1979; Brown 2020; Cattermole 2019; Stahl 2019). When prisoners are forced to be naked or subjected to strip searches in front of others, they are unable to exercise control over how others see them. This involuntary nudity can also be degrading for others who are nearby, because they do not have the choice to avoid seeing one another naked and without control over themselves (Schwartz 1972). An incarcerated person's contact with nonimprisoned friends and family is also heavily surveilled, whether it takes place in person or by remote correspondence (Schwartz 1972; Cattermole 2019). At best, this surveillance limits some aspects of meaningful exchange; some argue that the heavily constrained nature of this contact makes it potentially more damaging than not being in contact at all (Schwartz 1972).

But prisons also help us understand that privacy is not the same thing as simply having time alone. One of the most serious punishments to which incarcerated people are subject is solitary confinement. In solitary confinement, each inmate has their own small space, where (unlike in other prison settings) they do not have to share space with others, and in many cases cannot hear them. Yet study after study shows that solitary confinement is highly psychologically destructive (Grassian 2006; Haney 2018; Shalev 2017; Smith 2006). Inmates in solitary very often suffer from headaches, abdominal pains, decreased alertness, hallucinations, and delirium, frequently within a matter of hours of being moved into isolation (Grassian 2006). After a 2011 investigation, the United Nations recommended classifying any solitary confinement stint over 15 days as torture (Smith 2018). Even though most incarcerated people desperately crave time alone, the enforced isolation of solitary confinement, without any ability to determine the terms for themselves, does not afford them relief.

In addition to affecting the personhood of the imprisoned, the existence of prisons also indirectly shapes the experience of personhood of those on the outside. More recent studies have emphasized that although the wider public does not witness these degradations directly, the full social function of the prison requires that the public be at least indirectly aware of the degrading nature of imprisonment (Novek 2009). The first reason is so that the threat of prison can serve as a deterrent to those who are not imprisoned, and so elicit socially compliant behavior from them. The second, more subtle reason is that nonimprisoned persons can better understand their status as free citizens when they are able to compare themselves to the imprisoned (Foucault 2007). When a society frequently exercises its power to imprison its citizens, on the grounds that they deserve it—as the United States, China, and many others do—that reality becomes part of the backdrop for everyone else. Those who are not in prison can tell themselves stories about how they deserve their freedom, unlike those who have been incarcerated; but their freedom remains contingent, because the government has the power to take it away.

The existence and operation of prisons thus illustrate that denying a person the opportunity for privacy degrades them in their own eyes and in the eyes of others; often, the long-term consequence for the person who has suffered this degradation is that they are seen as less valuable, or even as less deserving of the rights and protections extended to those who have not suffered in this way.

As these three examples show, privacy is crucial for some of the basic ideas and activities that help us understand ourselves as persons, both in the sense of being uniquely ourselves and in the sense of being worthy of consideration. Each of these examples centers on personhood in some way: as a matter of reaching clarity about one's own identity and values, or of remaining recognizable to oneself and others as worthy or capable of these things. And in each case, in different ways, some kind of privacy is a necessary ingredient. The implication of these combined examples is clear: when privacy is eroded, then the conditions for personhood are changed, even diminished.

But protecting privacy requires more than a commitment: it also requires a mutual understanding of what it is and how it intersects with personhood. The problem is that debates about privacy today seem to incorporate various ideas about personhood without ever explicitly naming them. The remainder of this chapter clarifies how our ideas about privacy have changed and how they intersect with personhood.

4.5.2   What Is Privacy?

In the three examples provided in the previous section, privacy is understood as a separate sphere, space, or place for reflection. In the first example, workers who are continually exposed to others in an open environment are unable to perform their work and maintain a healthy sense of well-being. In the second example, gender identity moves from the private or personal sphere to the public sphere when someone rejects or claims a particular gender identity. In the final example, incarcerated people are denied the right to a physical space where they are free from surveillance and coercion. Each of these examples highlights how removing access to privacy as a discrete space fundamentally alters someone's sense of personhood, in their own eyes and in others.'

For most of recorded history, privacy has been conceptualized in terms of physical space. Aristotle, for instance, made a distinction between public and private spheres (Shields 2022; DeCew 2018). He made it clear that there are some things that are meant to be shared broadly, and some things that are not. This distinction between private and public was also important during the seventeenth and eighteenth centuries in the context of the European Enlightenment. In the Victorian era (a designation for British history and culture during the rule of Queen Victoria, 1837–1901) the financially well-off classes, professional or rich, distinguished between the public and private spheres (Digby 1993). Much as for the ancient Greeks, for the Victorians the public sphere encompassed commercial, political, and nonfamily social interactions (Arendt 1958). According to this classification of space and social function, the public sphere was exclusive to adult males (Digby 1993). The private sphere was physically located within the house and included the management of the household and of relationships within the household. This of course implies that anything that happened within the context of the household was "private." This did not mean that servants or family members were confined to themselves. It meant that the spaces were "private" because certain people were excluded from that private space and those "private" interactions. Likewise, those people who were seen as primarily belonging in the private sphere—women, children, servants, and slaves—were rarely, if ever, able to participate in public life. Indeed, in Greek culture, nothing was more embarrassing to a public figure than to have their private life, such as their wife or their family, discussed in public spaces (Garland 2009).

The literal division of spaces helps us separate some aspects of our physical, daily lives: the home, the office, or public institutions. And as noted in sections 4.3.7 and 4.3.8, because we are not physically present when we interact with others online, we can conceal parts of our identity that would be difficult to conceal in other types of interaction. But the organization of the world into spheres or spaces was not and is not only a literal configuration: it also functions conceptually. Although this distinction doesn't specify the legality of what information can be obtained, it does enable us to think about what it means to be private by helping us think about whom and what "privacy" excludes.

When we talk about the division of online "spaces," we are talking about a conceptual division, using physical space as an analogy. It can help us think about the way we perceive the space of our interaction, who is included, and who is not. You can see the space analogy at work even in dealing with the operating system of your computer or the design of networks. Access is partitioned into "zones"; networks exposed to the outside are called "demilitarized zones," or DMZs, to denote when the network has gone from a private local intranet to the public internet (Kurose and Ross 2016; Silberschatz et al. 2008).

However, this analogy of space can break down in important ways that we often do not realize, and this breakdown can cause us to misinterpret the degree and kinds of privacy that are actually available in the places we access virtually (Betz and Stevens 2013). For instance, email servers, both online like Gmail or available through your school/workplace, may seem like private spaces: they have all the trappings of a private space, including locks, such as login screens, to restrict access. But in general, these spaces are not exclusively yours. In the United States or the European Union (EU), if you work for a government or a private company, you have no claim to privacy over the email that you write or send through the employer's servers (Privacy Rights Clearinghouse 2019). Similarly, for most online providers, algorithms read your email to place ads and direct you to relevant content on the internet. Likewise, with the proliferation of digital assistants in the late 2010s, such as Amazon Alexa and Google Assistant, the words spoken by you in a number of spaces are recorded and transmitted to servers for processing.

4.5.3   PRIVACY, OWNERSHIP, AND SELF-POSSESSION

As we have seen, even in ancient times there was an idea of a distinction between our bodies, our homes, and property versus things that belong to the social, the public, or the state. But over time privacy has also expanded to include a sense of control over information—information about oneself. It is not all that surprising that privacy is associated with control over information. Notice that in all three of the examples given in section 4.5.1, control of various kinds is essential for privacy: control over how one presents oneself socially, control over one's interactions with others, and control over how one organizes one's time. This focus on control in many ways echoes the division of public and private: things that are private are the ones that we "own" or "are master of," in contrast to those that are public.

This sense (real or perceived) of ownership does not reflect only our practical control; it is quite common for people to see one's possessions as either part of or reflective of one's sense of identity. The car we drive, the clothes we wear, the kinds of books and art objects we own, the equipment that we use to play certain sports, or the specific tools for the hobbies we have; these things we own tell others something about ourselves. In fact, there is even a tradition that suggests that the ownership of property is necessary for the development of personhood. This tradition suggests that "Personality develops itself in its interaction with the world; without a sphere of property over which we exercise control, for example, moral responsibility is unlikely to develop" (Palmer 1990, 819).

Yet in practice, the notion of ownership does not translate quite so easily when we think about privacy in terms of access to information. Information, unlike a house or a car, can be held by multiple parties. In fact, this shared possession of information does not, by itself, mean that either party impacts (or even knows about) the other's possession of it. A given set of information can still confer a lot of power and control upon those who possess it, and yet that possession can happen silently or invisibly. For these reasons, questions of intellectual property—that is, questions about who owns an intangible creation—are not only legally complicated but also have significant implications for questions of personhood.

Intellectual property law exists to protect a creator's right to profit from the things that they have created in cases when that thing can be cheaply replicated, such as a poem, or a game, or a vaccine (Moore and Himma 2018). The existence of this area of law is, itself, evidence of the widely held belief that people own what they create and cultivate. But the ownership of something like this is not the same as what is considered to be personal information about us.

When we start to think about *privacy in terms of property*, some very interesting questions arise. For instance, do we consider our bodies to be our property? What about our talents, feelings, and experiences? If so, is information about our body also our property? Why or why not? There seems to be some basic consensus (e.g., Health Insurance Portability and Accountability Act [HIPAA] law) that our medical history is private information that an individual owns. But we also understand that this kind of information is necessary to share with medical professionals whom we rely on and trust.

If we own our bodies and our actions, then certainly we own these things in a different way from the objects we have in our home. But how is the information about our bodies something worth protecting, whereas information about our actions, behaviors and choices are open for anyone to collect and view? What about the things that we post about ourselves? What about the websites and other digital objects that we create?

These questions are becoming more and more important today when we have various forms of technologies that can be used to obtain, store, reproduce, and share information and data about ourselves and others. The traces that we leave are more abundant, and the transmission and sharing information about ourselves has a wider scope.

4.5.4 PRIVACY AS A LEGAL RIGHT

Warren and Brandeis defined privacy primarily as the "right to be let alone." Their article, like many legal arguments today, suggested that the law needed to change to keep up with technological change. They noted that such constant updating was necessary, given the "expanding communication technology such as the development of widely distributed newspapers and multiply printed reproductions of photographs" (DeCew 2018). In order to understand privacy's relationship to technology, it is important to keep in mind the legacy of Waren and Brandeis's article and to think about other framings as well.

Over the years, various legal versions of privacy, typically tort or civil laws, have expanded what is covered and what is not covered by the interpretation of privacy as a protection over information. Prosser (1960, 389, qtd. in DeCew 2018) broadly outlines the four types of privacy that are covered under tort law in terms of *informational privacy*:

1. Intrusion upon a person and/or their private affairs.

2. Public disclosure of facts/information about an individual.

3. Placing someone in a false light publicly.

4. Appropriating one's likeness for gain.

Rulings in recent years in the United States have also expanded the scope, concept, and reasoning behind informational privacy. Indeed, the courts have repeatedly stressed that the legal protection for privacy finds its source in the Fourth Amendment to the United States Constitution, which prohibits "unreasonable searches and seizures" (DeCew 2018). The reasoning following from the Fourth Amendment is again about control over information: you may have some information on your person that is necessarily protected from the government when the government has no reason to examine you.

Another operational definition of privacy comes from the training programs used for academic and professional researchers, for use in interacting with human subjects. These programs were established in the twentieth century as a direct response to research abuses that took place in the United States and Europe before and during World War II. These definitions were encoded into law with the National Research Act of 1974. The initial set of codes was adapted from the Nuremberg Codes at the close of

World War II. From this training program we are given the following definition of both privacy and confidentiality (Hicks, n.d.).

- Privacy can be defined in terms of having control over the extent, timing, and circumstances of sharing oneself (physically, behaviorally, or intellectually) with others.

- Confidentiality pertains to the treatment of information that an individual has disclosed in a relationship of trust and with the expectation that it will not be divulged to others in ways that are inconsistent with the understanding of the original disclosure without permission.

These definitions for research practice have subsequently been used as the basis for definitions of privacy in a number of US regulatory requirements including HIPAA (Annas 2003).

In the second half of the twentieth century, a different legal interpretation of privacy emerged: *decisional privacy*, sometimes called the constitutional right to privacy (DeCew 2018). The notion of decisional privacy in US law extends from a reading of the Fourth Amendment against unreasonable search and seizure that differs from the one about information. In 1965, the Supreme Court heard the case of *Griswold v. Connecticut* (1965), which questioned whether a Connecticut law, restricting access to contraception, was legal under the Fourth Amendment. In the ruling, Justice William O. Douglas stated that the law violated the "right to marital privacy," which laid the foundation for later rulings that protect a "zone of privacy" around marital relationships. The most controversial and most discussed implication of this decision was *Roe v. Wade* (1973) which, temporarily, made abortion legal in the United States (DeCew 2018). The fundamental idea is that there are certain decisions and spaces within which we are entitled to privacy and that the government should not be allowed to deny us this space. There is also an implicit idea here of the body as property: that we own ourselves and that what we do with ourselves is private.

This new reading of the Fourth Amendment in *Griswold* opened the door for many current discussions of privacy that include both control over personal information and control over our bodies and personal choices that are necessary for our concept of self (Kupfer 1987). The idea, generally, is that one can read the Fourth Amendment, and especially the ideas behind search and seizure, as protecting both information about oneself and the ability to make decisions, and that these rights should not be violated without some sort of due process. In a more modern reading, one can expand the definition of privacy to more abstract concepts of information—that is, all information about ourselves and what we do, wherever it may reside, is worthy of protection.

Perhaps the most comprehensive law passed with respect to informational privacy in recent years was the 2016 passage of the European Union's General Data Protection Regulation (GDPR) (Voigt and Von dem Bussche 2017). This law defines a number of protections for EU citizens with respect to their data, independent of where that citizen is located or where the data is being processed. For example, the law includes the "right to be forgotten," which means that any citizen may request that any company handling their data delete all data collected about them. The law also places strict requirements and penalties on companies that handle personal data of EU citizens and do not take proper care in safeguarding the data and/or do not share it in narrow and explicitly allowed ways.

The passage of GDPR has influenced and continues to influence privacy legislation around the globe. Understanding the complex legal and social history of the concept of privacy enables us to see that definitions about privacy are neither universal nor static. The United States, the European Union, and Asian countries have approached the legal protection of privacy differently. Within the United States there is a patchwork of laws, but generally there is only a limited system of formal protection (e.g., HIPAA), and personal data is readily available and easily collected for business or governmental use. This stands in stark contrast to the direction taken by the EU, especially with respect to the provisions in the GDPR which firmly place control of personal information in the hands of individuals, with strong protections accorded to individuals to control the flow and storage of information about themselves.

As we have seen, ideas of privacy have their roots in something fundamental that philosophers have struggled with for millennia. The modern definitions from which we sometimes (unknowingly) operate typically emanate from the developments of the legal tradition. In the United States especially, privacy has expanded beyond control over information and the right to be let alone, and now includes certain protections that protect the individual's right to make choices about their lives and lifestyle (DeCew 2018). However, it's important to realize that the law often lags broader social or cultural movements, the "policy vacuum" (Moor 1985), and this can also be observed in the legal history of privacy in the United States.

We have been revising and changing our definition of privacy for millennia, which reflects an ongoing commitment to the idea that there is something inherently valuable about privacy that is worth protecting. It has also changed with our conception of personhood. In the next section we discuss the value of privacy, and why privacy is necessary for one to develop a concept of self as a purposeful and self-determining agent.

4.5.5 PRIVACY AS A VALUE

So far we have discussed privacy primarily as a right that is typically associated with the ideas of property, bodily security, decision making, and especially personal information. This is arguably a legacy of the work of Warren and Brandeis, as most privacy discourse borrows their framing to describe *privacy as a right*—that is, a thing to which persons are entitled—rather than *privacy as an interest*—that is, a thing that people want and that benefits them in a morally significant way. Or, to use the language that we introduced earlier in the chapter, privacy might be construed as something that we have an interest in protecting because we value it (an axiological claim). We may value privacy because it is intrinsically valuable, or we may value privacy because it allows us to protect something we find especially valuable; for instance, personhood itself.

Matters of value, interest, and rights are deeply connected both conceptually and in everyday practice. But in this section, we consider more directly some reasons we might have an interest in protecting privacy for its own sake. In other words, we provide some answers to the question, why is privacy good? Why is it worth protecting?

Social and Political Values

In liberal democracies and societies, many people believe—as the laws often indicate—that privacy is necessary for freedom: freedom to form opinions and to make decisions about our own lives without interference from others. (One instance of this can be seen in section 4.5.1 in our second example, which concerns transgender persons.) Historically, this kind of freedom was associated with the protections granted to citizens against the power of the state. Within the context of US law, these rights are understood to be protected by the Constitution and specifically by the Fourth Amendment.

The public sphere is the realm in which other people are viewed and the government can interfere to a certain extent, whereas the private sphere is the realm for self-regulation and self-determination. Instead of allowing the government to make all our decisions for us, it is assumed, as John Stuart Mill believed, that we are experts in our own best interests: "The only freedom which deserves the name, is that of pursuing our own good in our own way, so long as we do not attempt to deprive others of theirs, or impede their efforts to obtain it" (Mill 2002, 14).

Thus, one way to defend our interest in privacy is to make the claim that privacy is instrumentally necessary for something else we value, such as liberty or freedom. Warren and Brandeis grounded their arguments in a claim about social value and "inviolate personality." Inviolate personality, as it was later defined, is what "defines one's essence as a human being and it includes individual dignity and integrity, personal autonomy and independence" (DeCew 2018). Inviolate personality both unifies and grounds the concept of privacy and lends support to the claim that invasions of privacy are affronts to human dignity (Bloustein 1964; DeCew 2018; Warren and Brandeis 1890). For example, as applied to our gender and gender identity example from section 4.5.1, the concept of inviolate personality offers specific conceptual grounding for the idea that an individual gets to determine their own gender identity.

Rather than seeing privacy as instrumental, it can also be argued that privacy is an intrinsic good that is required in order for human beings to understand themselves as purposeful, self-determining agents, which some people argue is presupposed in the very idea of citizenship and is the basis for democracy (Allen 1988). Bringing in our example of prisoners from section 4.5.1, the lack of privacy in prisons, indeed the act of denying this privacy to individuals, serves as a punishment in itself. It is specifically the lack of privacy that can strip away one's feeling of being a purposeful, self-determining agent. Hence one can argue that privacy is in fact necessary for a functioning society and thus is good for this reason alone.

> **Sidebar: The Privacy vs. Security Fallacy**
>
> Privacy is often framed as an individual concern, as something that benefits particular individuals instead of, or even at the expense of, society as a whole. This framing is particularly common when it comes to matters of national security. Those who advocate for limiting government surveillance—even when it is undertaken in the name of national security—are often criticized by their opponents for prioritizing individual liberties over the public good. But many kinds of public good require privacy in its various forms. Some of this is simply because society is made up of individuals, who are more likely to be happy if they have some privacy. But there are many kinds of prosocial actions that likewise require privacy, such as whistleblowing (that is, calling public attention to corruption or abuses), political organizing on behalf of the public that challenges powerful interests, and the formation of personal relationships that are the basis of any community.

Character Development

It has also been suggested that a certain amount of privacy is necessary for how we develop our own character and view ourselves as persons (Kupfer 1987). It may also be understood as a way for us to develop personal expression and choice (DeCew 2018; Schoeman 1992).

We can think about this through the notion of character development and virtue ethics. From within the framework of virtue ethics, our choices and opinions are not based on gut level reactions or automatic behavior. Instead, they are accumulated over time and require careful deliberation. In order to have the time and emotional space to form reasoned judgments, we may need to be cut off from the steady stream of others' opinions. Privacy affords the opportunity to test new ideas before exposing them to the flood of reactions and counterarguments. From our examples in section 4.5.1, we can see that this need of space for reflection is one thing that is being removed in open office plans. Workers in these open offices may need space to make judgments, both the big ones about their work and also the many tiny judgments that we make every day in the course of our labor. Here we can see the interplay of the various definitions of privacy discussed above: we may need to be able to keep our opinions to ourselves for a period of time (i.e., to be left alone), *and* we may need to be able to make decisions on our own, without fear of interference or judgment. Privacy is the concept that both articulates and defines the space necessary to fulfill these needs.

More broadly speaking, privacy does not just make provision for individuals' ability to know themselves; it is also understood as a justification for why people should be allowed to act in accord with their own personal beliefs and convictions, as long as their actions are not harming others. This form of freedom is sometimes described as concerning "matters of inward conviction" (Strohm 2011, 88). A similar principle is behind arguments regarding the "freedom of conscience." Article 18 of the Declaration of Human Rights states that "Everyone has the right to freedom of thought, conscience and religion" (United Nations 1948).

Interpersonal Relations

Privacy is also a factor in developing and maintaining meaningful relationships with others. Many social media sites classify all the people to whom we are connected as "friends," but in reality our relationships with others and different relationships come with different expectations, including different expectations about what we disclose to others. Choosing to share details about our lives, our feelings, beliefs, values, desires, and opinions can be an important part of how we establish and maintain social bonds. This is something we recognize when we compare how we interact with strangers with how we behave around our friends. In the examples from section 4.5.1, we can see how both open offices and prisons, to varying degrees of severity, hamper our ability to exercise this control.

When we have control over our personal information, we are able to reveal information about ourselves to others on our own terms. Sharing information with others may put us in vulnerable situations, but it is also a way to show trust and strengthen and affirm our relations with others. We can see that all three of our examples in section 4.5.1 touch on how privacy (or lack of it) inflects our interpersonal relationships. From the first example on open offices and the third example on prisons, it's clear that there may be aspects of our lives or our work that we want to choose to disclose at particular times. This option is stripped away from us in environments like open offices and prisons, inhibiting our capacity for relationship building. Likewise, our second example on gender identity makes clear the need for control over disclosure, both to whom and at what time, in order to build trust and intimacy. It seems that it is necessary to have some form of privacy *from others* in order to both build and maintain these relationships.

Even though our own individual ideas about privacy have likely been shaped by the social perspectives and legal history discussed above, it is also the case that our reasons for valuing privacy reflect our own personal values and are based on our own experiences and beliefs. Of course, this doesn't mean that we will all agree on why privacy is valuable. But if you are willing to grant that our own individual experiences and beliefs are relevant in this discussion, then it is likely that you already support some version of one of the arguments provided above.

4.5.6 IS PRIVACY ALWAYS GOOD?

Thus far, in our discussion of privacy, we have been emphasizing how valuable privacy is for a variety of reasons. But we should also consider how privacy is often in tension with other crucial goods and how it can create conditions for harm, abuse, and inequality.

Freedom from interference, whether from the government or the scrutinizing eyes of the public, seems like a good idea in principle, but it is an unfortunate fact that people sometimes use privacy to hide malicious behavior. This is especially the case in situations where people are underrepresented and/or are in vulnerable positions. For instance, from the perspective of feminist ethics, privacy is often questioned due to its historic use to hide physical, emotional, and financial harm (DeCew 2019).

Over the years, and especially in more modern settings, the value of privacy has been questioned, because privacy itself has been used as a shield to hide damaging actions (including domestic violence) and to preserve relationships of inequality.

Even if we grant that privacy is a right and something that is either good in itself or is a means to some other good end, this does not mean that it is something that will benefit all equally. In fact, it is quite easy to think of plenty of examples in which privacy is granted to some and not others. Both the parent-child relationship and the patient-client relationship contain not only an imbalance of power but also an imbalance of privacy.

Consider also situations in which someone we know is in need of care but refuses or is unable to reach out for help. For example, consider a scenario in which someone we know expresses suicidal thoughts, we may feel torn between sharing that knowledge with a licensed professional or someone else who can help and keeping that knowledge private. There are also times when people are called to make decisions for others for the sake of that person's well-being, such as when a loved one begins showing signs of dementia.

These examples help us see that privacy is not a simple concept that can be applied to everyone equally, rather it is important to contextualize specific concerns about privacy. Returning to the questions we outlined in our discussion of utilitarianism in section 2.3.5 in chapter 2 might help to illustrate this point.

When we are considering the right to privacy, whose privacy rights are we protecting? Is anyone being left out, and if so, why? What is the value or the cluster of values that we are associating with privacy protection? Who has the most to lose if privacy is not protected? Who has the most to lose if privacy is protected? Should information about someone always be protected? Or is it OK to publish someone's diary, for instance, after they die? If so, why is this OK?

Asking these questions and pushing ourselves to answer them helps us to sharpen our understanding about what privacy is and why it matters.

Story Point: "Here-and-Now," by Ken Liu

"I got something cool out of Tilly." Lucas had an evil grin.

Even knowing he was being baited, Aaron couldn't help asking, "What?"

Aaron is an avid user of Here-and-Now, the task app designed by Centillion (evocative of a certain real-world search engine company also named after a large number). The app allows users to place anonymous requests and to fulfill those requests in exchange for "bounties." Aaron loves knowing that he is helping others and enjoys feeling more connected to his community. But over the course of the story, Aaron begins to notice how the app makes it easy for users to invade others' privacy, including his own. Whereas technologies like Here-and-Now can easily be weaponized by malicious actors, "Here-and-Now" also offers a chilling account of how shifting norms around privacy can make it difficult even for well-intentioned people to honor the privacy of others.

4.6   BRINGING IT ALL TOGETHER: PRIVACY, INFORMATION TECHNOLOGY, AND PERSONHOOD

In this section we provide a concrete discussion of how our ideas about personhood and privacy have shifted over the last century toward being significantly more data-oriented. As we have seen, conceptions of privacy have changed a number of times over the years, and they have never been applied uniformly to everyone and everything. Additionally, all of the concepts of privacy discussed above seem to underscore the idea that, for a variety of reasons, people want space, both literally and conceptually, to live their lives. Let us discuss specifically how developments in internet and communication technologies (ICT) that have enabled data to be collected at a scope and scale never before imagined, have profound consequences for our conceptions of privacy and personhood.

4.6.1   THE EVOLUTION OF DATA AND DATA COLLECTION

Before the beginning of the twentieth century, data about individuals was typically only collected through laborious, large-scale censuses (Wikipedia 2022a). As we have seen, beginning in the early 1900s and accelerating rapidly ever since, the ability to collect, store, transmit, and process data has rapidly increased. In addition to the ability to store it, we have also expanded what we consider as *data*. As discussed in chapter 3, what types of data are collected and how they are interpreted are neither natural nor obvious; rather, they result from viewpoints and framing, conscious or otherwise. Before the 1940s, the idea of tracking our shopping or TV watching habits was not considered *data* or even important (Wikipedia 2022b). As the twentieth century progressed, the use of behavioral data and simulation to understand populations, and the decisions they make, has continued to increase (Lepore 2020).

It used to require efforts on the level of governments to gather even basic information, such as the number of people living in a city. Now, tremendous volumes of data are available, and the collection and processing of this data is nearly real-time. It is now possible to track where a person was (e.g., through webcams, GPS data, and communications traces), what they did (e.g., through purchase records, check-ins, and Instagram posts), and how it affected them (e.g., through social media posts, biometric data from smartphones and activity trackers, and reviews). That data is often available to a number of different entities—human, corporate, enforcement-related, or otherwise—anywhere there is internet access. Much of this data is largely personal, extending beyond basic information to reveal a part of our personal identity, demographics, aesthetic preferences, and history of decisions made.

As we discuss in chapter 5, much of what has enabled this data collection is the rapid pace at which technology has intermediated many of our relationships: from the friends that we talk to on our smartphones to the movies that we watch on Netflix. This data has been called "the new oil," and we are only starting to understand the implications of this data for business, the decisions we make, and what it means to us personally (*Economist* 2017). The continued increase in the availability of this data has fueled the modern revolution in AI and machine learning, which requires tremendous volumes of data to analyze, understand, and find patterns in our data (Kambhampati 2019).

Our own creation of data may be active, as when we post pictures online or fill out webforms, or it may be passive, happening without needing our awareness or express consent. Advertisers see which links produce incoming traffic and Netflix knows down to the second when you stopped watching that show. The data that we generate may be collected, accessed, used, or resold in ways that we did not fully understand (Madden 2014). According to Pew Research, most Americans feel that they have lost control of their personal data and what it is being used for. In this 2014 study, 91% of adults agreed that they had lost control of their data and more than 50% agreed that they did not understand how that data was going to be used by various companies (Madden 2014). In a 2018 study of users on Twitter, many were found to be unaware that their public posts could be collected and used by academic researchers, media companies, and many other entities. The majority of those surveyed felt that even researchers should get consent from users before collecting data, but that this feeling was highly contextual and depended on the type and scope of the research (Fiesler and Proferes 2018). However, in an updated Pew report, many were conflicted about the benefits of data collection and sharing (Auxier et al. 2019). For example, although many expressed concern about their map applications tracking their locations and saving it, many realized that the real-time traffic information, which is generated from the aggregated locations of everyone, presented a real value to them. Hence, in some instances users were happy to share personal data in exchange for the value received. However, in total, 81% of Americans thought that the potential risks of data collection by companies about them outweighs the benefits (Auxier et al. 2019).

How this data is collected and shared has been the subject of many studies, congressional oversight panels, and investigative journalism pieces. Dance et al. (2018) detailed a number of problems in the context of Facebook that allowed other companies to access more data than the company's own privacy rules allowed. For instance, things such as private messages, photographs, and demographic information marked as for friends and family were used by Facebook to sell ads and were also provided to third-party companies including Netflix, Spotify, Yahoo, Amazon, and Bing for product targeting. Whereas this is ostensibly to allow for "instant personalization" across a number of sites, with the stated goal of providing more value to users, there have been a number of problems with these schemes. The Federal Trade Commission (FTC) investigated these data-sharing practices in 2011 and some were shuttered in 2014, but there are still many programs that share data that users have flagged as private (Dance et al. 2018).

The issue is bigger than any individual company that possesses our data. The collection processes and exchanges that they use allow them to build up fine-grained and detailed views of users, and their history of decisions, that is then used to fuel prediction models and advertisers (*Economist* 2017; Kambhampati 2019). Data is used not only by the companies whose products and services you use directly but also passed between numerous data brokers and aggregators, combined across websites, survey firms, political organizations, public records, and private companies; this data exchange creates profiles of individual users and their desires. A Vermont law requires all companies that buy and sell third-party data to register, providing a window into the process: customer management services such as Mailchimp, advertisers like Nielsen, and consultancies and businesses like ACXIOM, Oracle, and McKinsey, as well as financial data markets like Experian and other credit firms all buy, sell, and swap your data between themselves and other companies for uses including targeted marketing and parole risk prediction. These companies are third-party, meaning that they don't have relationships with individuals and hence were required to register in Vermont, whereas Google, Facebook, and others would not need to register to pass on data (Christl 2017; Melendez and Pasternack 2019; Ram and Madhumita 2019). All this buying and selling, recirculating, and reconceptualizing of our data happens in ways that are not visible to us or might not make sense to us, even if we could know about it.

These collection practices raise concerns about the ownership of and access to our data, and how it translates to "public" and "private" space online isn't at all clear, but, as Nissenbaum persuasively argues, this doesn't mean that the online "world" is somehow a separate space, divorced from real life (Nissenbaum 2011). Online activity is deeply integrated into our social lives. "Answering questions about privacy online, like those about privacy in general, requires us to prescribe suitable, or appropriate, constraints on the flow of personal information" (Nissenbaum 2011, 14). Like offline activity, interactions online happen in a variety of social contexts. This is why it is a mistake to think about the "web" only in terms of a "commercial context where protecting privacy amounts to protecting consumer privacy and commercial information" (Nissenbaum 2011).

4.6.2 Why Privacy Matters for Personhood

Reflecting on our discussion of personhood, we can start to see why losing control of our data feels so *personal*.

Are You Your Data?

When we lose control of our data, it can feel like something has been taken from us. When digital images, records of our purchase or watch history, what music we've listened to, and many more details are collected and packaged, that a piece of ourselves is out there in the world, being bought, sold, and traded, can make us feel violated. There have been many accounts of this, from photos stolen out of cloud services to outdated digital records following your new housing application, and many people who have had these issues describe the feeling as being one of losing part of themselves, literally needing to "buy themselves back" (Ratajkowski 2020) to recover photos or correct erroneous data profiles on websites or government service accounts (Eubanks 2018).

As advances in technology have allowed for the production and collection of more and more data, we have—perhaps unconsciously—given more power to those that collect and interpret that data. The sheer prevalence of data as a framework for operating in the world has led to what Cheney-Lippold (Saulnier 2017) calls "measurable type": that is, the limited-option templates that are constructed out of collected data, which require us to be identifiable within that data framework. For example, when we have a Netflix account, we have a binary gender, or we have an address and zip code that must fit into a web form even if we rent an unregistered basement apartment. As Cheney-Lippold (2017) argues, the data that we generate is not always unified across all sources or with our lived experience, but is instead bought, sold, traded, interpreted, and reinterpreted. This data may be old, out of date, or different between companies, leading to competing versions of our identity—some up to date and some out of date or inaccurate—in the digital world. In chapter 3 we saw that both the types of data collected and how they are interpreted are neither natural nor obvious, and as more data is collected and interpreted by others, the more power we give away in determining who we are in the eyes of both the data and those entities that collect and process that data. If we accept John Locke's approach to personal identity (section 4.2.3), for instance, we can see that our data, and the sequence of decisions we made to generate that data, can be interpreted as a non-substance-based view of personal identity. This leads us to the question: how do we distinguish ourselves, our personhood, from data *about* us? And when we lose control over that data, do we lose control of ourselves?

Expressed concerns regarding the loss of control of our data are not new in the internet age. One of the most commonly shared medical data sets, the Pima Indian diabetes data set, was started in 1965 as a longitudinal study of diabetes in the US Indigenous community (Radin 2017). This data is now freely available online and shared by thousands of researchers, a reality that the Pima Indians of the 1960s could never have imagined when agreeing to take part in the initial collection of their data. This loss of control over the information about themselves not only can be seen as a loss of agency and identity but also reveals the way in which these people were denied the full status of personhood. A similar example can also be found in the case of Henrietta Lacks, a Black woman whose cells were collected as part of a cancer biopsy in the 1950s. Though she died shortly after the collection, her cells live on, immortalized, and are still used for worldwide medical research to this day, despite the fact that neither she, nor her family, were aware of or consented to the initial collection and proliferation (Skloot 2017).

Data Can Change Who We Are

Another common fear is that many of these systems are using data about us to affect our agency, that is, our capacity to act and effect change on our environment, in both subtle and not-so-subtle ways. This data usage could also potentially impact our self-determination, that is, our capacity to act in ways that are not fully determined by the environment in which we find ourselves nor our historical or social context.

As we have discussed, the different digital versions of our persons are distributed across numerous websites, government databases, and online forms. These digital representations of us are then bought and sold between advertisers, political parties, financial institutions, and other entities for advertising and direct product marketing for things like insurance or movie tickets. This type of advertising can go so far as to predict that you are more likely to drink water on a hot day due to assumed medical conditions based on search and traffic history, highlighting the integrations between shopping, your current location and weather, and many other data points (Christl 2017; Ram and Madhumita 2019).

When advertisements appear on a platform, be it Google Search or Facebook, a complex system of algorithms and data comes together to serve up a particular ad. This overarching system, typically called online behavioral advertising (Dwork and Mulligan 2013), aims, like all advertising, to subtly affect our decision about what to purchase or consume. Whereas the jury is still out on whether or not internet ads actually affect what you buy, there is strong evidence that brand-based advertising, or even movies aggressively suggested by online platforms like Netflix, are more likely to be clicked on, purchased, or viewed (Matsakis 2018). Online behavioral advertising may subtly affect our personhood as well: decreased exposure to varying perspectives in what we are offered from online platforms may limit users' individual autonomy by linking users to profiles that frame their relevant experience in ways that may not be relevant anymore. This lack of control over the collection and use of personal data is often a focus of discussion on these topics, specifically that these classification systems may be tailored to a representation of our personal identity filtered through the data to which the system has access and not our entire lived experience.

Some of the high-profile disclosures of data being used in inappropriate ways have created efforts to increase end-user privacy by interest groups, researchers, and governments such as the Electronic Frontier Foundation's Do Not Track effort (Electronic Frontier Foundation, n.d.); the Princeton Web Transparency and Accountability Project (Princeton Web Transparency, n.d.); the European Union's GDPR (Voigt and Von dem Bussche 2017), which includes the right to be forgotten, to edit one's own data; and other numerous other data regulations. In a recent Pew poll, many users reported tension about wanting to protect their personal information and data, but in some cases they also felt that the benefits they receive from the services were worth the cost of sharing data on their preferences or interests (Auxier et al. 2019).

However, as Dwork and Mulligan (2013) argue, this is not inherently a privacy problem and changing privacy rules are not enough in this context. Privacy-based solutions focus largely on giving users access to the data to, optimistically, pierce "filter bubbles" (Pariser 2011). Although these filter bubbles may indeed create feedback loops and limit the worldview of an individual, simply giving someone control over this does not alleviate the concern that these bubbles may be freely chosen rather than imposed through classification. In fact, the reality is often a mix of the two (Dwork and Mulligan 2013). There is research to suggest that the design of online platforms can both help to expand our horizons and allow us to focus on and amplify our personal bias in areas as diverse as music selection (Mehrotra et al. 2020), news consumption (Collins and Dance 2018), and online dating (Hutson et al. 2018). We might not feel in control of the experience that we receive on these platforms. Equipped as we are with the conceptual framework of this chapter, we can now articulate this unease as a loss of agency and an infringement on our personhood.

The way that we understand data, both what it is and how data about ourselves is collected and used, has changed step for step with the technology of the times. Surely Warren and Brandeis could not have anticipated how much things would change in just 100 short years, neither could the Pima Indians who have been turned into grist for the data mill. The takeaway from this section is that it would be a mistake for us to assume that we can anticipate the nexus of personhood and privacy issues that will be created by future developments in internet and communication technologies. At the time of writing this book, a new company, Clearview.ai, was receiving massive attention for downloading and indexing billions of photos from the internet for use (Hill 2020). Hence, rather than assume one hard-and-fast rule will save us, we must maintain vigilance and be able to understand and articulate why it matters and what's at stake.

4.7 CONCLUDING REMARKS

Throughout this chapter we have seen that both privacy and personhood are complex concepts whose definitions are not as clear as they might seem, and whose many meanings have shifted over time. By clarifying the concept of personhood, we have seen both that privacy is a fundamental aspect of personhood and that the language of personhood can help us to articulate the importance of privacy. Learning more about how these concepts have functioned in the past and comparing that history with how those concepts are used today can help us dislodge our sense that our present ways of thinking and seeing with them are the only ways possible, and offer us alternate ways of thinking that challenge the assumptions of the present. While it can be tempting to try to distill universal ideas about what each of them means, there is always the risk of getting locked into a definition that limits our ability to understand our contemporary context.

An important theme throughout this book, and especially in this chapter, is that clarifying our concepts, both to ourselves and to others, can help us understand and analyze perennial issues in the design, implementation, and use of computing technology. In chapter 5 we explore how the ideas of society and context come to shape and are shaped by our understanding of what technology is or can be.

REFLECTION QUESTIONS

1. *When you ask yourself the question, "What is a person?" what do you typically mean? Describe the questions that you ask yourself in terms of metaphysical, epistemological, and axiological questions. Which of these questions do you consider most important? Which is the least?*

2. *We have discussed that privacy can be conceived as a right or an interest. Which of these arguments do you find more persuasive and compelling, and why? Which do you think is more applicable in considering the role of technology in society?*

3. *Suppose that someone uses demographic information associated with your online shopping or entertainment preferences to draw conclusions about how you will behave in the future. When data about you is used in this way, what is at stake in terms of how others might perceive you? What is at stake in how you perceive yourself?*

4. *In the following situations, consider how one might use the conceptual resources of this chapter, including notions of privacy and personhood, to articulate why these situations constitute invasions of privacy, or why they are acceptable.*

   a. *Doxxing a public figure.*

   b. *Publishing voting records at the aggregate level for a county, or at an individual level.*

   c. *An insurance company requiring you to install a tracking app on your car in order to be insured.*

5. *The time gap between when we create data and when someone accesses it does seem to matter, but it is not clear how and why it matters. Think about someone posting old photos of you online even if they were always available. Would it feel different from someone posting a photo from last week? Why or why not? What ideas from your understanding of personhood help you explain this?*

6. *Data about individuals is collected in the United States well before the individual is born, as electronic medical records of their time in utero, not to mention media posts and writings by parents and other connected individuals. This data could be used, in the aggregate, for demographic or epidemiological modeling. Can you imagine reasons that a person would want to control or delete such data? Who do you think should have the right to choose what happens to the data? Shared data about a pregnancy can have ramifications for multiple people. Which individuals have a stake in the data?*

7. *Choose a particular type of technology that has changed substantially over the years, such as social media. Justify privacy as a necessary condition for personhood, and defend this condition through iterations of the particular piece of technology that you have chosen.*

BACKGROUND REFERENCES AND ADDITIONAL READING

For a more comprehensive background on defining and categorizing questions in ethics, see the following:

Blackburn, Simon. 2008. *The Oxford Dictionary of Philosophy*. 2nd ed. Oxford University Press.

Craig, Edward. 2000. Metaphysics. In *Concise Routledge Encyclopedia of Philosophy*, 567–570. Routledge.

Fast, Nathanael J., and Arthur S. Jago. 2020. Privacy matters . . . Or does it? Algorithms, rationalization, and the erosion of concern for privacy. *Current Opinion in Psychology* 31: 44–48.

Klein, Peter D. 2000. Epistemology. In *Concise Routledge Encyclopedia of Philosophy*, 246–249. Routledge.

McIntyre, Jane L. 2008. Hume and the problem of personal identity. In *The Cambridge Companion to Hume*, edited by David Fate Norton and Jacqueline Taylor, 177–208. 2nd ed. Cambridge University Press. https://doi.org/10.1017/CCOL9780521859868.006.

Nissenbaum, Helen. 2009. *Privacy in Context: Technology, Policy, and the Integrity of Social Life*. Stanford University Press.

REFERENCES CITED IN THIS CHAPTER

Advocates for Youth. 2020. *I Think I Might be Transgender, Now What?* A brochure by and for transgender youth. http://www.ct.gov/shp/lib/shp/pdf/i_think_i_might_be_transgender.pdf.

Allen, Anita. 1988. *Uneasy Access: Privacy for Women in a Free Society*. Rowman and Littlefield.

Annas, George J. 2003. HIPAA regulations—A new era of medical-record privacy? *New England Journal of Medicine* 348, no. 15: 1486–1490.

Arendt, Hannah. 1958. *The Human Condition*. University of Chicago Press.

Aristotle. 1999. *Nicomachean Ethics*. 2nd ed. Translated by Terence Irwin. Hackett.

Austin, Ashley. 2016. "There I am": A grounded theory study of young adults navigating a transgender or gender nonconforming identity within a context of oppression and invisibility. *Sex Roles* 75, no. 5–6: 215–230.

Auxier, Brooke, Lee Rainie, Monica Anderson, Andrew Perrin, Madhu Kumar, and Erica Turner. 2019. Americans and privacy: Concerned, confused and feeling lack of control over their personal information. Pew Research Center, November 15. https://www.pewresearch.org/internet/2019/11/15/americans-and-privacy-concerned-confused-and-feeling-lack-of-control-over-their-personal-information/.

Bell v. Wolfish. 1979. 441 U.S. 520, 99 S. Ct. 1861. https://caselaw.findlaw.com/us-supreme-court/441/520.html.

Bernstein, Ethan S., and Stephen Turban. 2018. The impact of the "open" workspace on human collaboration. *Philosophical Transactions of the Royal Society B* 373: 20170239. https://doi.org/10.1098/rstb.2017.0239.

Betz, David J., and Tim Stevens. 2013. Analogical reasoning and cyber security. *Security Dialogue* 44, no. 2: 147–164. https://doi.org/10.1177/0967010613478323.

Bloustein, Edward J. 1964. Privacy as an aspect of human dignity: An answer to Dean Prosser. *New York University Law Review* 39: 962–1007.

Brennan-Marquez, Kiel. 2021. Beware of giant tech companies bearing jurisprudential gifts. *Harvard Law Review Forum*, June 20.

Brown, Melissa 2020. "American horror story": The prison voices you don't hear from have the most to tell us. *Montgomery Advertiser*, January 7. https://www.montgomeryadvertiser.com/in-depth/news/2019/11/13/alabama-department-corrections-prison-inmates-describe-horrid-conditions/2234480001/.

Cattermole, Carl. 2019. *Prison: A Survival Guide*. Ebury Press.

Cheney-Lippold, John. 2017. *We Are Data: Algorithms and The Making of Our Digital Selves*. New York University Press.

Christl, Wolfie. 2017. *Corporate Surveillance in Everyday Life: How Companies Collect, Combine, Analyze, Trade, and Use Personal Data on Billions*. Cracked Labs Institute for Digital Culture, June. https://crackedlabs.org/en/corporate-surveillance/.

Collins, Keith, and Gabriel J. X. Dance. 2018. How researchers learned to use Facebook "likes" to sway your thinking. *New York Times*, March 20. https://www.nytimes.com/2018/03/20/technology/facebook-cambridge-behavior-model.html.

Crone, Katja et al. 2012. Introduction: Towards an Integrated Theory of Self-Consciousness. In *Facets of Self-Consciousness*, edited by Katja Crone, Kristina Musholt, and Anna Strasser, v–xvi. Brill.

Dance, Gabriel J. X., Michael LaForgia, and Nicholas Confessore. 2018. As Facebook raised a privacy wall, it carved an opening for tech giants. *New York Times*, December 18. https://www.nytimes.com/2018/12/18/technology/facebook-privacy.html.

Darling, Kate. 2017. "Who's Johnny?" Anthropomorphic framing in human-robot interaction, integration, and policy. In *ROBOT ETHICS 2.0*, edited by P. Lin, G. Bekey, K. Abney, and R. Jenkins. Oxford University Press. https://doi.org/10.2139/ssrn.2588669.

DeCew, Judith. 2018. Privacy. In *The Stanford Encyclopedia of Philosophy*, edited by Edward N. Zalta. https://plato.stanford.edu/entries/privacy/.

DeGrazia, David. 2005. *Human Identity and Bioethics*. Cambridge University Press.

Digby, Anne. 1992. Victorian values and women in public and private. *Proceedings of the British Academy* 78: 195–215.

Dishman, Lydia. 2018. Hate your cubicle? Thank medieval monks. *Fast Company*, September 21. https://www.fastcompany.com/90236769/hate-your-cubicle-thank-medieval-monks.

Duff, Antony. 2009. Legal and moral responsibility. *Philosophy Compass* 4: 978–986.

Dwork, Cynthia, and Deirdre K. Mulligan. 2013. It's not privacy and It's not fair. *Stanford Law Review* 66 (September). https://www.stanfordlawreview.org/online/privacy-and-big-data-its-not-privacy-and-its-not-fair/.

Economist. 2017. The world's most valuable resource is no longer oil, but data. *Economist*, May 6. https://www.economist.com/leaders/2017/05/06/the-worlds-most-valuable-resource-is-no-longer-oil-but-data.

Electronic Frontier Foundation. n.d. Do not track. Accessed May 20, 2021. https://www.eff.org/issues/do-not-track.

Eubanks, Virginia. 2018. *Automating Inequality: How High-Tech Tools Profile, Police, and Punish the Poor*. St. Martin's Press.

Feeney, Kerry. 2016. The throne is king: Impact of the prison toilet. *American Institute of Architects*, April 6. https://www.aia.org/articles/5326-the-throne-is-king-impact-of-the-prison-toile.

Fiesler, Casey, and Nicholas Proferes. 2018. "Participant" perceptions of Twitter research ethics. *Social Media and Society*, March 10. https://doi.org/10.1177/2056305118763366.

Foster, Ryan. 2017. Sharing cells, open showers and masturbation: What privacy in prison is really like, *Metro,* June 19. https://metro.co.uk/2017/06/19/sharing-cells-open-showers-and-masturbation-what-privacy-in-prison-is-really-like-6647837/.

Formosa, Paul, and Catriona Mackenzie. 2014. Nussbaum, Kant, and the capabilities approach to dignity. *Ethical Theory and Moral Practice* 15, no. 5: 875–892.

Foucault, Michel. 2007 *Discipline and Punish: The Birth of the Prison*. Duke University Press.

Garland, Robert. 2009. *Daily Life of the Ancient Greeks*. 2nd ed. Greenwood Press.

Glabau, Danya. 2021. Imagination, whiteness, and the future of virtual reality. Under review.

Gracia, Jorge J. E. 1984. *Introduction to the Problem of Individuation in the Early Middle Ages*. Catholic University of America Press.

Grassian, Stuart. 2006. Psychiatric effects of solitary confinement. *Washington University Journal of Law and Policy* 22, no. 1: 325–383. https://openscholarship.wustl.edu/law_journal_law_policy/vol22/iss1/24.

Griswold v. Connecticut. 1965. 381 U.S. 479.

Haney, Craig. 2018. Restricting the use of solitary confinement. *Annual Review of Criminology* 1: 285–310.

Hicks, Lorna. n.d. Privacy and confidentiality module. CITI Program. Accessed October 2019. https://citiprogram.org.

Hobbes, Thomas. [1668] 1992. *Leviathan.* Edited by Edwin Curley. Hackett.

Hohn, Kaylin. 2018. The controversy with "loot boxes": How children become addicted to microtransactions. *Game Review*, December 4. https://gamervw.com/2018/12/04/the-controversy-with-loot-boxes-how-children-become-addicted-to-microtransactions/.

Hill, Kashmir. 2020. The secretive company that might end privacy as we know it. *New York Times*, January 18. https://www.nytimes.com/2020/01/18/technology/clearview-privacy-facial-recognition.html.

Hirst, Alison, and Christina Schwabenland. 2017. Doing gender in the "new office." *Gender, Work and Organisation* 25, no. 2: 159–176. http://uobrep.openrepository.com/uobrep/bitstream/10547/622118/2/c+schwabenland+Doing+gender+in+the+%27new+office%27+Hirst+and+Schwabenland+GWO.pdf.

Hume, David. 1978. *A Treatise of Human Nature.* Edited by L. A. Selby-Bigge. Revised by P. H. Nidditch. Oxford University Press.

Hutson, Jevan A., Jessie G. Taft, Solon Barocas, and Karen Levy. 2018. Debiasing desire: Addressing bias and discrimination on intimate platforms. *Proceedings of the ACM on Human-Computer Interaction* 2: 1–18.

Jamison, Lee. 2017. The digital ruins of a forgotten future. *The Atlantic*, December 15. https://www.theatlantic.com/magazine/archive/2017/12/second-life-leslie-jamison/544149/.

Jonas, Hans. 1996. Tool, image and grave: On what is beyond the animal in man. In Jonas, *Mortality and Morality: The Search for the Good after Auschwitz*, edited by Lawrence Vogel, 75–86. Northwestern University Press.

Kambhampati, Subbarao. 2019. What just happened? The rise of interest in artificial intelligence. *The Hill*, August 11. https://thehill.com/opinion/technology/457008-what-just-happened-the-rise-of-interest-in-artificial-intelligence.

Kendi, Ibram X. 2016. *Stamped from the Beginning: The Definitive History of Racist Ideas in America.* Hachette UK.

Kite-Powell, Jennifer. 2020. Here's how 2020 created a tipping point in trust and digital privacy. *Forbes*, October 27. https://www.forbes.com/sites/jenniferhicks/2020/10/27/heres-how-2020-created-a-tipping-point-in-trust-and-digital-privacy/?sh=1fa4b4984fc5.

Korsgaard, Christine. 1996. *The Sources of Normativity.* Edited by Onora O'Neill. Cambridge University Press.

Kupfer, Joseph. 1987. Privacy, autonomy and self-concept. *American Philosophical Quarterly* 24, no. 1: 81–89.

Kurose, James, and Keith Ross. 2016. *Computer Networking: A Top-Down Approach.* 7th ed. Pearson.

Lepore, Jill. 2020. How the Simulmatics Corporation invented the future. *New Yorker*, July 27. https://www.newyorker.com/magazine/2020/08/03/how-the-simulmatics-corporation -invented-the-future.

Locke, John. [1690] 1990. *An Essay Concerning Human Understanding.* Edited by Peter H. Nidditch. Clarendon Press.

Luria, Michal, Samantha Reig, Xiang Zhi Tan, Aaron Steinfeld, Jodi Forlizzi, and John Zimmerman. 2013. Re-embodiment and co-embodiment: Exploration of social presence for robots and conversational agents. In *Proceedings of the 2019 Conference on Designing Interactive Systems Conference*, 633–644. Association for Computing Machinery.

Luria, Michal, Rebecca Zheng, Bennett Huffman, Shuangni Huang, John Zimmerman, and Jodi Forlizzi. 2020. Social boundaries for personal agents in the interpersonal space of the home. In *Proceedings of the 2020 CHI Conference on Human Factors in Computing Systems*, 1–12. Association for Computing Machinery.

Madden, Mary. 2014. Public perceptions of privacy and security in the post-Snowden era. Pew Research Center, November 12. https://www.pewresearch.org/internet/2014/11 /12/public-privacy-perceptions/.

Matsakis, Louise. 2018. Online ad targeting does work—As long as it's not creepy. *Wired*, May 11. https://www.wired.com/story/online-ad-targeting-does-work-as-long-as-its-not -creepy/.

McLean, Kate C. 2008. The emergence of narrative identity. *Social and Personality Psychology Compass* 2: 1685–1702.

Mehrotra, Rishabh, Niannan Xue, and Mounia Lalmas. 2020. Bandit based optimization of multiple objectives on a music streaming platform. In *Proceedings of the 26th ACM SIGKDD International Conference on Knowledge Discovery & Data Mining*, 3224–3233. Association for Computing Machinery.

Melendez, Steven, and Alex Pasternack. 2019. Here are the data brokers quietly buying and selling your personal information. *Fast Company*, March 3. https://www.fastcompany .com/90310803/here-are-the-data-brokers-quietly-buying-and-selling-your-personal -information.

Menkiti, Ifeanyi A. 1984. Person and community in African traditional thought. *African Philosophy: An Introduction* 3 (1984): 171–182.

Mill, John Stuart. 2002. *The Basic Writings of John Stuart Mill: On Liberty, The Subjection of Women and Utilitarianism.* Modern Library.

Moore, Adam, and Ken Himma. 2018. Intellectual property. In *The Stanford Encyclopedia of Philosophy*, edited by Edward N. Zalta. https://plato.stanford.edu/archives/win 2018/entries/intellectual-property/.

Morrison, Rachel L., and Roy K. Smollan. 2020. Open plan office space? If you're going to do it, do it right: A fourteen-month longitudinal case study. *Applied Ergonomics* 82: 102933. https://doi.org/10.1016/j.apergo.2019.102933.

Nissenbaum, Helen. 1996. Accountability in a computerized society. *Science and Engineering Ethics* 2: 25–42.

Nissenbaum, Helen. 2007. Computing and accountability. In *Computer Ethics*, edited by J. Weckert, 273–280. Ashgate. Reprinted from *Communications of the ACM* 37 (1994): 37–40.

Nissenbaum, Helen. 2011. A contextual approach to privacy online. *Daedalus* 140, no. 4: 32–48. https://www.amacad.org/sites/default/files/academy/multimedia/pdfs/publications /daedalus/11_fall_nissenbaum.pdf.

Novek, Eleanor. 2009. Mass culture and the American taste for prisons. *Peace Review* 21, no. 3: 376–384.

Nozick, Robert. 1981. *Philosophical Explanations*. Harvard University Press.

Nussbaum, Martha C. 2011. *Creating Capabilities*. Harvard University Press.

Oldham, Greg R. 1988. Effects of changes in workspace partitions and spatial density on employee reactions: A quasi-experiment. *Journal of Applied Psychology* 73, no. 2: 253– 258. https://doi.org/10.1037/0021-9010.73.2.253.

Osborne, Heather. 2012. Performing self, performing character: Exploring gender performativity in online role-playing games. *Transformative Works and Cultures* 11. https://doi .org/10.3983/twc.2012.0411.

Oxford Reference. n.d. Juristic person. Acessed May 28, 2021. https://www.oxfordreference .com/view/10.1093/oi/authority.20110803100027393.

Pagallo, Ugo. 2018. Vital, Sophia, and Co.—The quest for the legal personhood of robots. *Information* 9, no. 9: 230.

Palmer, Tom G. 1990. Are patents and copyrights morally justified? The philosophy of property rights and ideal objects. *Harvard Journal of Law and Public Policy* 13, no. 3: 817–866.

Pariser, Eli. 2011. *The Filter Bubble: How the New Personalized Web Is Changing What We Read and How We Think*. Penguin.

Planned Parenthood. n.d. Coming out as trans. Accessed March 23, 2020. https://www .plannedparenthood.org/learn/gender-identity/transgender/coming-out-trans.

Princeton Web Transparency and Accountability Project. n.d. Accessed May 20, 2021. https://webtap.princeton.edu/.

Privacy Rights Clearinghouse. 2019. Workplace privacy and employee monitoring. March 25. https://privacyrights.org/consumer-guides/workplace-privacy-and-employee -monitoring.

Prosser, Willliam L. 1960. Privacy. *California Law Review* 48: 383–423.

Radin, Joanna. 2017. "Digital natives": How medical and Indigenous histories matter for big data. *Osiris: Data Histories* 32, no. 1. https://www.journals.uchicago.edu/doi/full/10 .1086/693853.

Ram, Aliya, and Madhumita Murgia. 2019. Data brokers: Regulators try to rein in the "privacy deathstar." *Financial Times*, January 7. https://www.ft.com/content/f1590694-fe 68-11e8-aebf-99e208d3e521.

Ratajkowski, Emily. 2020. Buying myself back. *New York Magazine*, September 15. https://www.thecut.com/article/emily-ratajkowski-owning-my-image-essay.html.

Rawls, John. 1972. *A Theory of Justice*. Oxford University Press.

Reig, Samantha, Michal Luria, Janet Z. Wang, Danielle Oltman, Elizabeth Jeanne Carter, Aaron Steinfeld, Jodi Forlizzi, and John Zimmerman. 2020. Not some random agent: Multi-person interaction with a personalizing service robot. In *Proceedings of the 2020 ACM/IEEE International Conference on Human-Robot Interaction*, 289–297. Association for Computing Machinery. http://samreig.com/files/reighri20.pdf.

Rescher, Nicholas. 2006. *Philosophical Dialectics: An Essay on Metaphilosophy*. State University of New York Press.

Reynolds, Emily. 2018. The agony of Sophia, the world's first robot citizen condemned to a lifeless career in marketing. *Wired*, January 6. https://www.wired.co.uk/article/sophia -robot-citizen-womens-rights-detriot-become-human-hanson-robotics.

Ricoeur, Paul. 1991. Narrative identity. *Philosophy Today* 35, no. 1: 73–81.

Roe v. Wade. 1973. 410 U.S. 113.

Rousseau, Jean-Jacques. [1762] 1997. *The Social Contract and Other Later Political Writings*. Edited and translated by Victor Gourevitch. Cambridge University Press.

Samuel, Ravi S. 2011. *A Comprehensive Study of Education*. PHI Learning.

Saulnier, Alana. 2017. Review of Cheney-Lippold's *We Are Data: Algorithms and the Making of Our Digital Selves*. *Surveillance & Society* 15, no. 5: 698–700.

Schwartz, Barry. 1972. Deprivation of privacy as a functional prerequisite: The case of the prison. *Journal of Criminal Law, Criminology, and Police Science* 63, no. 2: 229–239.

Shalev, Sharon. 2017. Solitary confinement as a prison health issue. In *Prisons and Health*, edited by Stefan Enggist, Lars Møller, Gauden Galea, and Caroline Udesen, 27–35. World Health Organization. https://www.euro.who.int/__data/assets/pdf_file/0011 /249194/Prisons-and-Health%2C-5-Solitary-confinement-as-a-prison-health-issue.pdf.

Shields, Christopher. 2022. Aristotle. In *The Stanford Encyclopedia of Philosophy*, edited by Edward N. Zalta. https://plato.stanford.edu/cgi-bin/encyclopedia/archinfo .cgi?entry=aristotle.

Sidgwick, Henry. [1874] 1981. *The Methods of Ethics*. Hackett.

Silberschatz, Abraham, Peter B. Galvin, and Greg Gagne. 2008. *Operating System Concepts*. 8th ed. Wiley.

Skloot, Rebecca. 2017. *The Immortal Life of Henrietta Lacks*. Broadway Paperbacks.

Smith, Dana. 2018. Neuroscientists make a case against solitary confinement. *Scientific American*, November 9. https://www.scientificamerican.com/article/neuroscientists-make -a-case-against-solitary-confinement/.

Smith, Peter Scharff. 2006. The effects of solitary confinement on prison inmates: A brief history and review of the literature. *Crime and Justice* 34, no. 1: 441–528.

Stahl, Aviva. 2019. Force-feeding is cruel, painful and degrading—and American prisons won't stop. *The Nation*, June 4. https://www.thenation.com/article/archive/force -feeding-prison-supermax-torture/.

Strawson, Peter F. 1962. Freedom and resentment. *Proceedings of the British Academy* 48: 1–25.

Sullins, J. 2000. Transcending the meat: Immersive technologies and computer mediated bodies. *Journal of Experimental & Theoretical Artificial Intelligence* 12, no. 1: 13–22.

Tangwa, Godfrey B. 2005. Some African reflections on biomedical and environmental ethics. In *A Companion to African Philosophy*, edited by Kwasi Wiredu, 387–395. Blackwell.

Tavani, Herman T. 2012. *Ethics and Technology: Controversies, Questions, and Strategies for Ethical Computing*. 4th ed. Wiley.

Thal, Adam. 2020. The desire for social status and economic conservatism among affluent Americans. *American Political Science Review* 114, no. 2: 426–442.

Thiel, Udo. 2011. *The Early Modern Subject: Self-Consciousness and Personal Identity from Descartes to Hume*. Oxford University Press.

Thompson, Hale M. 2016. Patient perspectives on gender identity data collection in electronic health records: An analysis of disclosure, privacy, and access to care. *Transgender Health* 1, no. 1: 205–215.

Torras, Carme. 2018. *The Vestigial Heart: A Novel of the Robot Age*. MIT Press.

Trimier, Jacqueline. 2003. The myth of authenticity: Personhood, traditional culture, and African philosophy. In *From Africa to Zen: An Invitation to World Philosophy*, edited by Robert C. Solomon and Kathleen M. Higgins, 173–199. 2nd ed. Rowman & Littlefield.

Turkle, Sherry. 2010. In good company? On the threshold of robotic companions. In *Close Engagements with Artificial Companions: Key Social, Psychological, Ethical and Design Issues*, edited by Yorick Wilks, 3–10. John Benjamins.

United Nations. 1948. *Universal Declaration of Human Rights*. https://www.un.org/en/about-us/universal-declaration-of-human-rights.

Véliz, Carissa. 2021. Privacy and digital ethics after the pandemic. *Nature Electronics* 4, no. 1: 10–11.

Voigt, Paul, and Axel Von dem Bussche. 2017. *The EU General Data Protection Regulation (GDPR): A Practical Guide*. Springer.

Warren, Samuel D., and Louis D. Brandeis. 1890. The right to privacy. *Harvard Law Review* 4, no. 5: 193–220.

Watson, Gary. 2004. *Agency and Answerability*. Oxford University Press.

West, Sara G., and Stephen Noffsinger. 2019. Absolute right to privacy for prison inmates. *Journal of the American Academy of Psychiatry and the Law* 37, no. 4 (2009): 563–565. https://www.aia.org/articles/5326-the-throne-is-king-impact-of-the-prison-toile.

Wikipedia. 2022a. s.v., "Census." Edited May 13. https://en.wikipedia.org/wiki/Census.

Wikipedia. 2022b. s.v., "Nielsen ratings." Edited May 13. https://en.wikipedia.org/wiki/Nielsen_ratings.

Wirtz, Andrea L., Tonia C. Poteat, Mannit Malik, and Nancy Glass. 2020. Gender-based violence against transgender people in the United States: A call for research and programming. *Trauma, Violence, & Abuse* 21, no. 2: 227–241. https://doi.org/10.1177/1524838018757749.

Zuboff, Shoshana. 2021. The coup we are not talking about. *New York Times*, January 29. https://www.nytimes.com/2021/01/29/opinion/sunday/facebook-surveillance-society-technology.html.

CHAPTER 5

TECHNOLOGY AND SOCIETY

Learning Objectives

At the end of this chapter you will be able to:

1. *Summarize some of the ways technology has been historically defined, consider how definitions limit or enhance our understanding of the complex relationship between technology and society, and evaluate the shortcomings of reductionist views of technology.*

2. *Identify tools and articulate strategies for evaluating technologies in context.*

3. *Name and provide examples of common traps in thinking when one is participating in technology development, and evaluate technological systems in light of these traps.*

4. *Apply tools and concepts for the analysis of sociotechnical systems to spheres of society, including structures of care; public discourse; friendship and communities; ecology and the environment; state power and force; and work and labor.*

5.1 INTRODUCTION

When we think about technology, our minds may jump to specific objects or artifacts: a camera, a traffic light, a gun, a smartphone, or even an app for a smartphone. If we use this as our starting point, the question of *how this technology affects society* may seem a narrow and obvious question of cause and effect: does a gun do bad things?

Are smartphones ruining human interaction? Will this particular app harm its users, and how? Although these questions are not bad questions, they stage the relationship between human beings and technology in a limited way and distract from some of the broader questions that address the myriad ways that *technology* and *society* interact and affect one another. Indeed, technology and society are bound together in an inseparable way: our choices about which technologies to develop, or not to develop, often grow out of the society in which we live. And those technologies, in turn, change our social settings and the subsequent technologies that are produced in the context in which we live.

Consider something as commonplace as a traffic light at an intersection (Jasanoff 2016). On its own, a red traffic light means nothing; it is just a colored light on a pole. Indeed, it may even be tempting not to think of a traffic light as technology at all, because for most of us they have always been around and do not seem as complicated or advanced as our smartphones. But if we let our description stop with "a colored light on a pole," it's easy to overlook how much there is to analyze about a traffic light, and how much we can learn about both the past and present by doing so. If we consider the traffic light in terms of its history, we learn that traffic lights, both electric and nonelectric versions, were invented and patented over a long development history spanning the 1860s to 1910s, as they increasingly became popular as a way to deal with the influx of cars on the road after the invention of the Model T (McShane 1999). The development of automated traffic lights even eliminated an entire class of jobs, because directing traffic used to be done exclusively by humans—a concern that is also often raised about today's automated technology. Just as in the not-so-distant past the traffic light fell under scrutiny for taking away jobs, today we worry about robots replacing human individuals in the workplace.

There is another way in which we can think about the traffic light—a way that might not seem as obvious when we conceive of the traffic light only as a simple object with a straightforward purpose. In order to appreciate this complexity, we must view the traffic light within its wider *sociotechnical context*. Without a societal mechanism for the enforcement of the *rules* of the traffic light, the colored light means nothing. If there was not an agreed-upon custom of going on green and stopping on red, then every intersection would be fraught with more danger than it already is. In addition, if there were no courts to enforce the tickets, fines, and other penalties that we as a society have decided are the appropriate consequences for running a red light, then there would be less incentive to heed the traffic light. Seen in a broader context, the traffic light, a technological object, has the power to strongly influence human behavior and represents a set of rules that govern human action. It does these things not because of what it is as a stand-alone object, but because it operates within a wider system of laws, symbols, social standards, and political governance.

Throughout the book, we have drawn on several specific problems related to technology in general and have also frequently turned to address issues related to computer technologies specifically. As we noted in chapter 1, the reason for this is that "technology" in its most recent sense, and modern computing technology (including information processing and communications systems) in particular, have brought about shifts in human social configurations, on both a local and a global level. However, in this chapter we discuss technology more broadly in order to draw parallels between the ongoing discussions about modern computing technologies and technology in a more general sense.

Technology, as we use the term in this chapter, refers to particular technologies as well as to a set or group of technologies that function within wider systems. This definition may seem vague at first, but as we will discuss, the way technology is defined is a contentious subject. Definitions have changed over time and tend to vary across cultures, much like the definitions of personhood and privacy we saw in chapter 4. For this reason, rather than working with one narrow definition, we employ and investigate a handful of the most commonly used conceptions of technology, from a range of intellectual backgrounds.

At the same time, for reasons we explain below, this chapter stresses the importance of situating particular technologies within wider *sociotechnical systems* (e.g., the traffic light, the laws around traffic lights, and those that follow and enforce those laws). It encourages you to think about technology not only in terms of a set of objects but as phenomena that operate alongside physical, natural, and social forces. Technology, in other words, is not merely something "out there" or something that is on the verge of being invented, "discovered," or in the process of being developed. It is already part of reality and, as such, it influences how we experience the world and interact with one another. What can be even harder to grasp, although it is no less true, is the way in which technological systems influence both our ideas about the present world and our ideas about what the future can and should be.

5.1.1 CHAPTER OBJECTIVES AND METHODOLOGY

This chapter incorporates many of the concepts and ideas introduced throughout this book and builds on similar methods. In particular, this chapter introduces several different ways to think about technology in order to foster further reflection and discussion. And, much like chapter 4, it looks at the way our understanding of the term "technology" has changed over time. Although we cannot cover all the ways in which technology has been defined over the centuries, we consider how some conceptualizations, especially those related to the instrumentalist approach to technology, prevent us from thinking critically about the full scope of technology.

The first part of this chapter challenges you to think about what technology is. Typically, when we think of technology, what first comes to mind is the completely new: gene editing techniques, electric cars, apps, smartphones, and other (mostly) computer-driven technologies that have developed or noticeably changed in our own lifetimes or memories. But "technology" is a concept (much like personhood) that has meant different things to different people throughout its long history.

The second part of this chapter provides you with a set of concepts and questions that will help you critically analyze the way technology functions alongside social systems. After introducing a few of the ways in which technology has been understood in relation to society, we introduce a way of thinking about technology that foregrounds the interaction of science, technology, and society. This central idea is drawn from the field of science and technology studies (Felt et al. 2017). In some ways, we have already introduced you to this way of thinking. Recall that in chapter 3 we noted that it is very difficult, if not impossible, to separate what is measured (the data) from the measurer: the person, people, groups that collect the data or develop the technology intended to measure that data (Hammer 2021; Nolan 2021). It is easy to imagine the scenario in which a human being sees a problem, searches for a solution, and invents the technology to solve that problem; our culture and our media prepare us to put narratives together in that way. But what gets left out of this scenario is that the ways in which the problem itself is framed, as well as our ideas about possible solutions, are shaped by the particular sociotechnical contexts in which we operate.

Consider that, in a different time and place, with different technological norms, we would likely perceive any problem in different terms and would certainly have different ideas about how that problem could be meaningfully solved. For instance, Cindy, the protagonist of "Asleep at the Wheel," relies on her self-driving car's artificial intelligence (AI) not only to get her from place to place but also to keep track of her calendar. That wouldn't make any sense in our own present, because AIs (let alone vehicular AIs) cannot do what Cindy's car does. Trusting your means of transportation to find your destination on its own might have made more sense 200 years ago than it does for us right now, but you probably wouldn't want to trust a horse with the task of keeping your calendar. We cannot simply step outside of these specific technological contexts: that version of objectivity does not exist. But we can think critically about where we stand within a certain context and use a variety of interpretive frames to understand that context. That's what this chapter is all about. We can also use fiction to try to imagine ourselves into *different* contexts, which is why this textbook includes a collection of short stories. The stories in that anthology and the frames in this chapter can help you reflect critically on your own context.

In the final section of the chapter we show how understanding the concept of technology, in a broad way, allows us to put our concepts, questions, and analytical tools to work in different social spheres, where technological and social systems interact. Drawing hard boundaries around social spheres or dimensions of society is an impossible task. In actuality, these spheres are connected and overlap. In fact, a point we want to make in this chapter is that there are many ways to frame problems for analysis, and any such frame is always up for debate. For the purpose of our analysis we identify six spheres of society that encompass many current events and trends in technology development: structures of care; public discourse; friendship and communities; ecology and the environment; state power and force; and work and labor. By analyzing issues at this level, along with our stories, we hope to reinforce the broad view required to think about technology in context.

As you read the chapter, we invite you to keep the following questions in mind:

- What is the nature of the relationship between technology and society?

- How do our definitions of society and technology limit and/or enable us to explain their interrelation?

- What problems arise at the social level with respect to technological change and development?

- How do our perceptions and definitions of what technology *is or can be* change the questions we ask?

Sidebar: What Is a Society?

In common usage, a society is typically a large group of individuals who have regular interactions, share the same space, both literal or virtual, and are subject to similar customs, laws, and expectations (*Merriam-Webster* 2021). But it is important to conceive of society as a multidimensional and multilayered idea. There is not one single overarching society; in fact, we are each a part of many, possibly overlapping, societies. Taken together, these societies create overlapping layers of activity and interactions, such as relationships between parents and children, between neighbors and friends, between students at a school, between individuals in a country, and even between the countries of the world. As discussed in chapter 4, some societies may exist only in virtual space, but even societies that exist in meatspace (i.e., offline communities that interact primarily face to face; Sullins 2000) are affected: if you do not live near your parents, your main mode of interaction may be via the phone. The key takeaway is that many of the societies of which we are a part are in some way constructed or enabled by technology, ideologies, or institutions, such as your online groups, your religious affiliation, or your school.

In chapters 1 and 2, we noted that ethics is sometimes thought about as if it involves learning a set of abstract theories and then using those theories to address or solve "real-world problems." As you have likely noticed by now, this textbook approaches ethics differently. Our goal is not to talk "at" you as someone involved in the design, development, and deployment of technology, but rather to equip you with some tools and techniques that can help you think through the complexity of the work you're undertaking, to appreciate the diverse knowledge and practices that need to be brought to bear on the problems you encounter, and perhaps also to realize the limits of technology.

5.2   TECHNOLOGY: PROBLEMATIZING THE CONCEPT

In this section we review some of the more common definitions of technology that circulate in a variety of settings, including the academic literature and popular media. Each of these common but limited definitions has in turn given rise to theories of technology that, although helpful, have certain shortcomings that are worth exploring. Before we introduce science and technology studies as a useful framework for thinking about the role we all play in the coproduction of sociotechnical systems, we look briefly at a few of the ways that definitions of technology can be inherently limiting, and why theorizing on the basis of a single, simple definition of "technology" is ultimately not all that helpful.

5.2.1   IS TECHNOLOGY A SIMPLE MATTER OF MEANS AND ENDS?

In one of its more simplistic formulations, technology can be understood as the tools, machines, and instruments that allow us to achieve certain ends: a pen to write, a hammer to build, or a car to move from one place to another. We can also broaden this definition to include the techniques, methods, processes, and, in some cases, the skills necessary for the production of technical objects or for the accomplishment of clearly defined ends.

In the context of engineering, for instance, technology development is sometimes described as "the application of scientific principles to the solution of a problem" (Harris et al. 2013, 92). Although this formulation is recent, the distinction it draws between science and technology has ancient roots. This distinction presumes that there is a clear line between theory and practice, in which technology (the practice) is understood to involve the application of knowledge that we have gained through reasoning, research, and contemplation (the theory, which is science).

But this distinction only makes sense if we believe that there is one process for thinking about scientific principles and another, separate process for applying those principles after they have been discovered. In practice, however, very few sciences operate this way. This distinction would mean that Edison's lightbulb should not be seen as a technology because Maxwell's equations for light were not well understood until nearly 50 years later (Harris et al. 2013). Likewise, many modern hygiene practices were around long before we discovered germs and how they cause disease (Bates 2020), discoveries that give the scientific explanation for why these practices are effective. You can probably come up with other examples of practices that are put into effect before we totally understand how and why they work. When technology is conceived of in terms of the objects or means through which we achieve certain ends, the range of things that we can think and say about it becomes limited.

Further, this theory/practice divide also relies on some constraining presuppositions about how human beings interact with technology. According to this formulation, humans are beings that use the externalized world to meet certain needs and fulfill desires. "Technology" in this context, then, is the term we use to describe whatever it is that we use as means to achieve or fulfill those needs and desires that we as a species are all programmed in advance to have. This view implies that human desires and needs are not shaped by available technologies, which, as we have seen throughout this chapter (and this textbook), is not the case.

This same assumption underlies a view known as the *cultural approach* to technology. This view, made prominent by German engineers in the early years of the twentieth century (Eric Schatzberg 2006, 2018), casts technology as the *product* of culture or the expression of human intentions, desires, and values. Under certain formulations, technology, much like art, is said to be the truest expression of the human spirit. Although this approach resonates with different cultural perspectives than the theory/practice approach described just above, it similarly posits technology as an external expression of fully formed ideas and values, rather than something that develops alongside and shapes our ideas and values.

Reductive views of technology such as the ones mentioned above can lead to an impoverished understanding of what technology is or can be, and this in turn can leave us ignorant about the way that technology actually influences and functions within human life. The German philosopher Martin Heidegger offered one of the most influential criticisms of the instrumentalist view of technology (Heidegger 1954). He was one of many thinkers in the mid-twentieth century who argued that we need to broaden our definitions of technology. Toward this end, Heidegger and others sought to retrieve a broader definition of technology from ancient Greek thought.

The word "technology" comes from the Greek words *techné* and *logos*. Both of these words were used in various ways in ancient Greece, but in its broadest sense, the suffix *-logos* typically refers either to a branch of knowledge or is used to refer to the "study of" something. *Techné* for ancient thinkers like Aristotle and Plato referred not only to tools and instruments but to an entire range of that which today would fall under "craft," "skill," or "art." So for instance, one could speak of *techné* of sword making or sculpting. It was also fairly common to refer to the *techné* of medicine or law and, in some cases, the *techné* of politics (Mitcham 1994).

Techné in ancient Greek thought was also closely linked to knowledge (*episteme*). Most forms of knowing were taken to be a form of "know-how." This is worth noting because, as the examples above indicate, some definitions of technology not only exclude the knowledge that is applied in the making of artifacts and instruments but also exclude the techniques and practices that develop around the production of artifacts and instruments. For thinkers like Aristotle, however, being able to make something implied a certain form of practical knowledge.

In recounting these Greek definitions, we are not suggesting that we should return to them. The technologies that we are familiar with today are much different from those that functioned in ancient Greece. But it does allow us to see that there may be other ways to think about technology.

When technology is understood only in terms of means and ends, or in terms of the application of theoretical knowledge or basic scientific principles, it can misconstrue the relationship we have with our external environment and our capacity to manipulate the world in ways that somehow perfectly align with our clearly defined goals and intentions. It can also feed unrealistic stories about where technology is leading us. On the one hand, some people think technology will somehow fix all of the world's problems or that it will liberate us from all of those things that now set us back. This is sometimes referred to as *technological optimism*: the view that the effects of technology on human well-being are mostly or entirely good (Harris et al. 2013). *Technological pessimism*, on the other hand, is the view that every technological system comes at the loss of some other value, typically by harming the environment or marginalizing human aesthetic or spiritual needs. Technological progress, especially since the industrial revolution, is sometimes associated with the exploitation of nature and the depletion of natural resources. Technology has also been blamed for fragmenting human experience and rendering the world less meaningful. This kind of criticism is often coupled with nostalgic longings for a time "before technology" or at least before "modern technology" corrupted classical values and beliefs—a time, for instance, when families sat around a table and shared a meal together instead of being distracted by the television or, more recently, our smartphones.

You may have also heard people refer to "the march of technological progress," which implies that the development of technology is going somewhere, with or without us. This latter way of conceiving technology is sometimes called *technological determinism*. It is the view that the development of technology has a life of its own, and its own internal logic that cannot be changed or affected by humans or the society at large (Harris et al. 2013; Jasanoff 2016). However, this view is again inherently limiting. It implies that technology's forward momentum is so strong that human beings cannot possibly act to avert that progress, and perhaps are not even required to take part in advancing it. Even if we sometimes speculate about AI evolving or feel as though it is difficult to keep up with the rate of technological change, projections about the future are always projections. All evidence at this point leads us to believe that humans are very much a part of technological development. If this weren't so, you wouldn't be reading this textbook.

Each of the views of technology and our role in technology development discussed in this section and the one that preceded it has particular shortcomings. In thinking critically about technology, it is important to situate technologies in the particular social and historical contexts in which they function and operate. If we begin with limited definitions and do not consider the broader context, it is easy to overlook the complex ways in which human desire, contextual circumstances, and social values influence technological change and development.

5.2.2 TECHNOCRACY AND UNINTENDED CONSEQUENCES

In addition to limiting conceptualizations of technology, our thinking can also be hampered by how we view ourselves as developers, experts, and practitioners. It can be all too easy to absolve ourselves from taking responsibility for how technologies interact in the broader context of a society. Two especially common ways of making excuses for ourselves are by invoking the *myth of technocracy* and by claiming *unintended consequences*. The first of these excuses, the myth of technocracy, comes into play in the conceiving and planning process, and it is frequently used to justify why nonexperts cannot be meaningfully included in those stages. The second excuse, unintended consequences, is typically rolled out after implementation and release, and is used to justify why the designers and technologists cannot be held morally accountable for the things they have made. Taken together, these ways of situating ourselves and our responsibility can lead to growing alienation and distrust between technologists and society at large.

Engaging with the full scope of the imagination, design, implementation, and deployment of technological systems requires expertise in many types of knowledge and skill. In disciplines that rely on specialized or rarefied knowledge such as medicine or law, the practitioner is often interacting directly with the public or the end user. Thus, in these settings, explaining the complexity of a decision and obtaining informed consent can usually be accomplished at the point of delivery, for example, before the doctor sticks a needle in your arm. This kind of explaining and seeking consent is far less likely when we are talking about large technological systems. Most people will likely go their whole lives without ever meeting an engineer who designed parts for their car. Given the distance from the public, engineers and professional societies need to take responsibility both for being inclusive in the design and development of technological systems and for the consequences that will arise. We return to these themes of professional responsibility in chapter 6.

On the design and development end, failing to take the responsibility to communicate seriously, or dismissing any explanation as "too complicated to understand" by end users, is falling victim to the myth of technocracy. When we frame the conversation in a way that posits that only a special "them"—the technocrats who have access to rarefied or special knowledge—are qualified to judge technology, we are falling into the myth of technocracy, removing agency from individuals in deciding how and what technologies are developed and regulated (Jasanoff 2016). There is a risk that, due to the complexity of contemporary technologies, it is only a matter of time before technocrats, due to their secret knowledge, exert disproportionate control and governance over all others. Succumbing to the myth of technocracy may disenfranchise individuals who are not part of the process, leading them to see the technology development process as broken, corrupt, or lacking public oversight. Moreover, the pessimism brought on by being cut out of the development process can lead to problems in the relationship between developers and society, and between developers at various levels. One example of the kind of distrust that can emerge is clearly illustrated through the famous example of the space shuttle *Challenger* and *Columbia* accidents, which can be blamed on the "expert" culture in which individual engineers, designers, and technicians were not empowered to question the decisions being made (Harris et al. 2013; NASA 2006).

The notion of unintended consequences often comes into play when a given technological innovation has proven to be harmful: for example, an app to report the location of animals that leads to those animals being killed by trophy hunters instead of photographed (Welz 2017). But what consequences were intended? By whom? And for whom? Consider two possible causes for these consequences: was the designer's original intent not executed as intended? Or did something outside the scope of their intentions happen because no one could have imagined it? These are not the same thing, and depending on which one we mean, two very different responses would be required (Jasanoff 2016).

Consider a train driver who uses his smartphone while on the job and an accident happens. Clearly this is an unintended consequence from the train designer's perspective. If we mean that the designer's intent was not executed as planned, then we would need to enact a policy or develop an app on the smartphone to lock out use while in a moving vehicle. However, if we mean that the designer should have imagined such an accident possible, it's not clear how at the time of the invention of the train engine someone could foresee the development of modern smartphones (Jasanoff 2016).

The idea of something being intended or unintended locates intention at one specific moment in time (i.e., when the system was being designed or developed). But in actuality, technological systems—along with the way in which and the context in which they are being used—are constantly changing and evolving. Humans are smart and creative, and we will use technology in ways that are surprising to everyone, even ourselves. However, instead of engaging with this complexity, terms like "unintended consequences" paint failure as an inevitable event, like a natural disaster that no one could have foreseen or could have worked to counteract, and obscure our relationship with technology (Jasanoff 2016).

It is also possible to be too optimistic about the kind of changes and "fixes" that technology can bring. It is therefore also possible to place too much faith in technology developers, which can further promote a vicious feedback loop in which developers become more disconnected and fail to recognize the consequences and/or neglect evidence that doesn't fit with their views.

Sidebar: Overpromising in the History of AI
Throughout this book we have seen examples of technologies, both old and new, that have reframed and recontextualized many aspects of daily life. One place we can see the conversation looping back again is in the current discussion around data science, AI, machine learning, and specifically topics of "ethical AI" and "big data." Often when you see these topics in the media, it is with breathless coverage that paints those creating modern AI systems as unfeeling technocrats who maliciously develop algorithms and technologies to exploit "us"; for example, "Microsoft wants you to use approved political speech—This is a real threat to our freedom" (Bruce 2019), about an AI system to suggest appropriate words in emails, or "Google AI tech will be used for virtual border wall, CBP contract shows" (Biddle 2020), about an AI system to monitor border crossings in real time.

While it may sometimes appear that these brand-new technologies are reconfiguring our society in new and complex ways, many of the contemporary conversations echo those that took place in the 1980s and 1990s, often in the same words (Forsythe 2001; Isbell 2020). For example, in the field of artificial intelligence, the 1980s and the very early 1990s were part of the "second AI summer." During this time huge investments were being made, dizzying evaluations for start-ups were realized, and media hype was at an all-time high. Expert systems, which captured and formalized knowledge, were going to usher in a new age of computers that could reason better than humans. Much of this hype came crashing back to earth during the "AI winter" in the mid-1990s (Kautz 2020).

Looking back at the design and development of these expert systems during the rise of AI in the 1980s and 1990s, Barocas and boyd (2017, 23) note that "the scholarship from the time was damning: expert systems routinely failed, critical researchers argued, because developers had impoverished understandings of the social worlds into which they intended to introduce their tools." After several high-profile failures in the early days of AI (Forsythe 2001), scholars from the social sciences started to identify what they called the *sociotechnical gap*, that is, "the divide between what we know we must support socially and what we can support technically" (Ackerman 2000, 179).

During this time, two popular applications for expert AI systems were health care and manufacturing. Forsythe (2001) observed that many of these technology developers viewed their work as "purely technical," even when working on systems to diagnose patients in medical settings. She writes, "the work of AI experts tends to select the voices of the expert and ignore or marginalize those of the patients" (xxi), and claims that often these developers blamed patients and doctors for any failure to understand the system or to use it properly.

Many of those same arguments from the 1980s still ring true today as the "third wave" of AI technology impacts our daily lives (Isbell 2020). However, this time around, the answer for *most* designers and engineers seems to be more nuanced: not everyone is an uncaring technocrat out to write algorithms and create systems to exploit nontechnical people. Even so, there are often failures on the part of creators or companies to engage with the full social and technical complexity in which these systems operate. In a comprehensive survey of AI and data science developers, Barocas and boyd (2017) note that many are keenly aware of the ethical and moral implications of the work they undertake and appreciate the complexities of many of the interventions under development, but they are often not empowered to act on all these issues.

The myth of technocracy and the language of unintended consequences are two views of technology that are inherently limiting. On the front end, the myth of technocracy is one way of ignoring the responsibility of those working in the design, development, and deployment to avoid engaging with the full complexity of how technology and society interact. On the back end, the language of unintended consequences is a way of ignoring the impacts on that same society because, as many have argued, intentions don't always control how technology and society interact.

5.2.3   REFLECTING ON UNINTENDED CONSEQUENCES: EXTERNALIST VS. INTERNALIST APPROACHES

When things do not go as planned or when a technological object or system brings about responses that were not anticipated, it is often the case that reflection begins from an externalist view rather than an internalist view.

The *externalist view* focuses solely on engineering problems in isolation and tends to present incidents of study as a sequence of discrete events that leads to some singular, crucial decision. Taking the *internalist view*, by contrast, looks at deeper descriptions in which issues emerge at many points in the overall process. Externalist treatments are easier to generate and engage with than an internalist view, but they are also less helpful.

The contrast between the two can be illustrated by looking at how NASA followed up on two different space shuttle disasters. Perhaps one of the most famous examples of an externalist account is the case study of the launch/no-launch decision for the space shuttle *Challenger* mission. The *Challenger* mission was launched on January 28, 1986, from the US Kennedy Space Center and exploded shortly after leaving the tower, resulting in the loss of all crew on board. Most case studies of this incident center a set of meetings that happened on the morning of the launch, focused on whether the O-rings in the boosters could handle the colder-than-ideal temperatures of the January morning (NASA 2021). Although there were many competing claims about the safety of the systems in the lead-up to the decision to launch, case studies focus on the crucial decision at a single point in time: the launch or no-launch decision (Harris et al. 2013). By focusing on one single moment of technological failure, these studies overlook the complex reality in which this failure took place and make it difficult to come up with practical improvements for nonidentical future situations.

Conversely, an *internalist* or process-based account looks at deeper descriptions in which issues emerge at many points in the overall process. Following the loss of the space shuttle *Columbia* and the death of all crew on February 1, 2003, NASA commissioned the Columbia Accident Investigation Board to craft a general return to flight strategy (NASA 2006) that included an analysis of the organizational decision-making process of NASA, who was hired into what roles, and the funding pressures from the federal government. This line of questioning considers all stakeholders including the public, funding agencies, and government in addition to NASA's managers, engineers, and individual technologists. These more detailed and nuanced accounts of the entire system/network in which the decision takes place helps to generate a "thick description" of the context in which this decision made sense and offers a greater variety of possible alternative approaches.

As you can see, the internalist approach can offer a much more substantive understanding of a given situation and generate more portable insights. It is also much more challenging to employ. Fortunately, many resources exist to help you, such as mediation theory and the field of science and technology studies.

5.2.4 TECHNOLOGICAL MEDIATION

It's one thing to acknowledge that not all of our interactions with technology can be characterized in terms of a use relationship. But it's much harder to put that insight into practice. One attempt to meet this difficulty is called *mediation theory*, which offers several ways to describe how technology mediates human experience and can radically transform human self-understanding.

Building on the work of Don Idhe (2009), Paul-Peter Verbeek (2011, 2015) describes seven types of relationships between human beings, technology, and the world:

1. *Embodiment relationships*, wherein technologies and humans interact with the world. For example, speaking to others through the phone rather than speaking at the phone, or viewing something through a telescope.

2. *Hermeneutic relationships*, wherein technology represents something in the world to humans. For example, an MRI scan to show brain activity or a microwave telescope to view the cosmos.

3. *Alterity relationships*, wherein humans interact with technology while the world itself stays in the background. For example, withdrawing money from an ATM or operating a car.

4. *Background relationships*, wherein technology forms the context for human experiences and action. For example, the warm or cool air from an HVAC system or the notifications on a smartphone.

5. *Cyborg relationships*, wherein humans and technology merge into a new hybrid that is more intimate than an embodiment relationship. For example, brain implants that stimulate nerves or insulin pumps to regulate blood sugar.

6. *Interactive relationships*, wherein humans interact with technologies that are merged with the world, forming an interactive context. For example, smart homes where we can ask appliances to switch on, or TV channels to change, or closed captioning to turn itself on and off.

7. *Augmentation relationships*, wherein technology can simultaneously take on the role of an embodiment relationship and a hermeneutic relationship. For example, Google Glass and other wearable computing devices that can give an experience of the world through the lens while simultaneously providing different representations of the world.

Focusing on mediation and interaction instead of specific artifacts allows us to identify some of the critical concerns we may have with technology. For example, as we discussed in chapter 4, the rise of big data, AI, and machine learning requires vast troves of data about us. Technology, in the form of our smartphones and other devices, mediates many of our relationships and often determines what we listen to, what we watch, and who we talk to. As we saw in chapter 4, computational and information technologies make it possible to store data on the relations we have with others and to the world around us, and this data in turn fuels the algorithms and processes (Kambhampati 2019).

According to Verbeek (2015), thinking about the various ways in which technology mediates human experience can help users, designers, and policy makers deal more creatively, critically, and productively with powers and forces that remain hidden otherwise.

For instance, mediation theory provides a distinction between the types of influence that technology exerts on humans along two dimensions, namely, *visibility* and *force*. The visibility of technologies exists on a continuum between *hidden* and *apparent*. Different technologies can be understood to exist somewhere within these two dimensions, from small "nudges" that gently influence people's behavior to more overt methods that clearly enforce particular rules or norms (Thaler and Sunstein 2009). For example, a car that will not start without seatbelts being fastened is an apparent technology in contrast to the hidden influence of a building without an elevator forcing one to take the stairs. The force of a technology can be anywhere between *weak* and *strong*. For example, the strong force of a breathalyzer vehicle interlock system that prevents you from starting the car if you are intoxicated is strong, while a smart meter that gives feedback on your energy consumption, is weak. The latter provides hidden suggestions that subtly change your behavior.

5.3 INTRODUCING SCIENCE AND TECHNOLOGY STUDIES

Science and technology studies (STS) is an "interdisciplinary field that investigates the institutions, practices, meanings, and outcomes of science and technology and their multiple entanglements with the worlds people inhabit, their lives, and their values" (Felt et al. 2017, 1). The origin of STS can be traced to the 1960s and 1970s, which saw the rise of political activism in response to concerns like environmental pollution, consumerism, and the proliferation of nuclear technologies. Experts and nonexperts alike began questioning, in far greater numbers than before, the widely accepted notion of science and technology as unequivocable forces for good. This shift set the stage for a reinvestigation of the Western conception of progress (Cutcliffe 2000).

A major precursor to the emergence of STS is Thomas Kuhn's *The Structure of Scientific Revolutions* (1962), which is cited by Rohracher (2015). The book challenges a prevailing view of its time, that science consists of the steady accumulation of objective observations and theories. In contrast, Kuhn argues that rather than uncovering unbiased facts about nature, science answers questions only within pre-established *paradigms* that set the basic assumptions of a field; major transformative events in science consist of *paradigm shifts* that discard and replace those basic assumptions. For example, there was a time when most physicists based their work on the laws of Newtonian mechanics. But when Einstein's special relativity became the dominant paradigm, much of the old research—although "true" in the Newtonian framework—was no longer considered meaningful. Today we have quantum mechanics *and* special relativity as dominant paradigms, which approach physics from drastically different and possibly incompatible angles. With each of these developments, the physics canon is not so much "expanded" as rewritten from the ground up.

In addition to illustrating how scientific observations are interpreted through the lens of existing knowledge (as we saw in chapter 3), Kuhn's account makes clear that this interpretive lens is social in nature. The existing body of knowledge may orient the scientific community, but the culture and paradigms of the scientific community in any given moment also define what kind of knowledge can be produced in that moment. The broader idea that social, scientific, and technological systems are in a constant feedback loop with each other is known in STS as "*co-production*[, . . .] the proposition that the ways in which we know and represent the world (both nature and society) are inseparable from the ways in which we choose to live in it" (Jasanoff 2004, 2).

A core commitment of STS is that science and technology cannot be distilled to a basic unit and analyzed independently of the societies in which they exist. To give a meaningful account of technologies, we must study them on the level of *sociotechnical systems*: entire dynamic systems of physical objects and tools, knowledge, inventors, operators, repair people, managers, government regulators, and many others (Dusek 2006; Harris et al. 2013).

One way that we can see the importance of taking the STS view of sociotechnical systems is by investigating the question of whether technological artifacts have politics (Winner 1980). If we were to limit our view of technologies to only the instrumental or means-ends approach, then it hardly makes sense even to ask the question: on the basis of those approaches, any given technological artifact is merely a neutral tool. However, when we look at the entire sociotechnical system in which that artifact is embedded, we can better understand the technical objects themselves, the contexts in which those technologies are situated, and the history of the technology itself (Winner 1993b).

It might seem overwhelming to try and understand an entire sociotechnical system, but we can use an example from history to illustrate what the STS approach to technology requires. The automatic tomato harvester was developed in the 1940s as part of the University of California system which, as a land grant institution, is generally charged with improving the agricultural and mechanical sciences. The individual scientists were focused on the goal of providing more food more cheaply to more people and were backed by a large institution funded by both public and private money. The harvester does meet those goals very well: it is able to collect more tomatoes at a cheaper price, ensuring higher production on a given area of land. However, the introduction of the harvester significantly reoriented the power structure in the agricultural industry to the disadvantage of the farmers, small growers, and field hands who were unionizing for higher pay and rights (Winner 1980). This is just one example of how more powerful actors, including governments and large companies, can completely upend an existing social, political, or economic sphere in pursuit of their own goals, and how it can happen even when those goals are in service of the public good. Winner (1993a) perhaps said it best: "Technological development as a dynamic social and cultural phenomenon. This is not an arcane topic. Every thorough-going history of the building of technological systems points to the same conclusion: Substantial technical innovations involve a reweaving of the fabric of society—a reshaping of some of the roles, rules, relationships, and institutions that make up our ways of living together."

The history of the design of New York City's parkways and public transit system is another potent example of how technology mediates human experience and transforms our sociotechnical contexts. From the 1930s to the 1960s Robert Moses was one of the most influential figures in New York City, laying the groundwork for numerous infrastructure changes including highways, city parks, parkways, and beaches. The now-famous issue concerns the design of bridges on the New York State Parkway System, which at Moses's direction were all deliberately built with a clearance too low to allow city buses and public transit to access various beaches and parks, effectively denying access to anyone without a car (Caro 1974; Harris et al. 2013; Winner 1980). Moses's designs for New York City were not unusual; across the United States, the placement of the interstate highway system during the 1960s and 1970s was plagued with numerous issues surrounding representation and disparate impact. There are many similar examples of interstates being used to cut low-income and minority neighborhoods in half or delineate classically "good" and "bad" neighborhoods, leading to some memorable slogans including "no white men's roads through Black men's homes" (King 2021), a slogan that was used to protest such developments.

5.4 ANALYZING SOCIOTECHNICAL SYSTEMS

So far in this chapter we have seen some ways of conceiving of technology in relation to society that inherently limit both the types and scope of analysis we can perform. We then introduced STS as a way of thinking about technology that more fully embraces the complex interconnections between science, society, and technology. Thinking with STS concepts and ideas can support the kind of descriptive work that is necessary for thinking practically and critically about technology and society, but it may not always be clear how these ideas and concepts can be implemented into the work of design, engineering, and development. This section will help you with this implementation.

5.4.1 EXPERIMENTATION AND RESPONSIBLE DESIGN

No matter how much one thinks and plans ahead, introducing a new technological system will always be to some degree a *social experiment*. It is a form of social experiment that involves the users, the developers, and even entities that may never interact with the technological system directly (Harris et al. 2013). In an experimental framework, the subjects of the experiment are the users of the technology, and as in any other experiment there is always uncertainty about the outcome of any intervention. There are always risks in releasing technology for use, be it a new bridge, a new computer program, or a new manufacturing process. There is also uncertainty about the effects of the intervention, about its costs and benefits. However, in order to gain new knowledge, we must deploy the system in the real world, to conduct an experiment. This is challenging because we must recognize the potential risks of this intervention, work to mitigate as many as we can, and know that we may be accountable for the results.

Harris et al. (2013), as well as Martin and Schinzinger (2004), propose the following list of responsibilities for engineers participating in technology development.

1. Technological systems are embedded within social contexts, and we must remember that there is a network of cause and effect.

2. Technological systems have provided both goods and bads, and hence we must maintain a critical attitude toward technological interventions.

3. Engineers, technology developers, and professional societies have a duty and responsibility to work to inform the public and to act in good faith.

4. Technology developers, designers, and implementers have a responsibility to consider the social and value implications of any intervention and to work broadly with affected communities to understand these implications.

We return to some of these topics in chapter 6, where we discuss practical guidelines for ethical professional practice more in depth (Jillson 2021; Vallor et al. 2018). Although this chapter is more conceptual, this conceptual work is a necessary component of practical responsibility. In spite of our best intentions, it is very difficult to avoid falling into the same patterns and traps unless we are also engaged in critical reflection about *how* we are doing the practical problem solving and *why* we are doing it that way.

5.4.2 CONSTRUCTING ABSTRACTIONS AND FRAMING PROBLEMS

In working to design, implement, and develop technology, it is often necessary to work and reason with abstractions of reality and to create frames in order to address perceived problems.

Designers and engineers participate in the technology development process to create outcomes that have value or provide benefit to various stakeholders—users, corporations, and perhaps society at large. Dorst (2011) explains this in terms of an equation:

WHAT + HOW = OUTCOME.

In a more focused engineering approach, we typically know the *how*, or working principle portion of the equation: for example, develop an app or build a bridge. Hence, most focus is on the *what*, that is, creating a design that works within a known area to deliver a particular outcome in a particular context. In a broader *design thinking* approach, both the *what* and the *how* are left open ended, and instead designers are expected to have an outcome in mind, but they do not necessarily have a predefined *what* or *how*. This more open-ended type of design work leads to the technique in design thinking of creating frames through which we wish to view a problem. Frames are typically highly complex statements and views of problems that seek to include multiple stakeholders and avenues of engagement in order to discern multiple *what*s and *how*s to arrive at the outcome. Frame setting involves norms, judgments, and values, even when these things are not easy to notice.

When thinking about technology and its impacts, we often first try to describe the system or phenomena with which we are concerned. This focus requires us to frame the system or phenomena in a particular way. Having a different focus may change our framing in meaningful ways, but it doesn't get us out of the problem.

Suppose that we are analyzing a new credit approval algorithm and want to think through how this new algorithm may help or hurt borrowers. To perform this analysis, we might do something like the following:

- We describe the various entities, the banks, the people, and some of the impacts that may happen, including determining whether or not someone will have access to credit.

- We engage in some imagination of the future: there is a new system, and it will impact people in a certain way; we project our thoughts forward.

- We determine that certain classes of borrowers may benefit or be harmed because of the change.

It might seem like all this is fairly straightforward, and that it has nothing to do with matters of value or ethical perspectives. But as we have seen, and will continue to see, there is no way for description to exist apart from value determinations. In our attempt to *objectively describe* the world, we are layering our own view, ideas, and concepts onto it.

- When we choose the parameters for analysis, we are making decisions about who and what should be included and who and what should not. We are looking at the context as we understand it and deciding who to consider: the banks and the borrowers, but maybe not other entities, such as the money printers or other businesses.

- When we undertake the idea to develop a new algorithm, we are engaged in an act of thinking about what the world would be like or even *could* be like with a new type of faster credit approval algorithm.

- When we make a recommendation or write a report about our analysis process, we are making statements that may or may not be adopted. More concretely, if we are recommending a proposal for funding or selecting a specific project for development and potential deployment, then our actions have possibly led to the creation of new systems.

When we break down each step, it becomes easier to see how certain judgments get baked into our abstractions and descriptions. Because this kind of analysis is crucial to any technology design or development process, we can begin to see how science, society, and technology function together in ways to coproduce the sociotechnical context we currently inhabit.

The language we use might also signify different things to different people. It is never an easy translation between the words we use to describe and analyze, the systems we build, and the world we inhabit. Consider, for instance, what the concept of "fairness" means in the example about analyzing an algorithm. Although we can easily look up a mathematical or economic definition (among many others that we could reference), the definition itself can and has been interpreted in many different ways (Hutchinson and Mitchell 2019; Mulligan et al. 2019). First of all, we may need to specify whether we are thinking about individual fairness or group fairness. *Individual fairness* posits that individuals who are similar with respect to some task or process should be treated similarly with respect to that task or process. For example, in the decision of whether to admit students to a scholarship program, students with similar GPAs should be evaluated similarly. By contrast, the premise of *group fairness* is that some system or process should perform the same across demographic groups (Friedler et al. 2021). However, these two things can come into tension if we want, for example, our credit rating algorithm to be fair: do we mean that everyone with similar incomes should be scored the same by our algorithm, even if that means that some demographic group is completely denied credit? Or do we want to ensure that all demographic groups have equal access to credit, regardless of the incomes of particular individuals within that group?

Even in situations where people are relatively clear about what they mean by fairness and agree that fairness is an important factor in design, people may still arrive at different judgments about the development of a system that implements fairness.

Matters relating to definitional complexity become only more complicated when one considers the social context in which technological systems operate (Hutchinson and Mitchell 2019; Mulligan et al. 2019). This is why one of the cornerstones of cooperative/participatory design is to incorporate many views and conceptions of complex notions, such as fairness or efficiency, in developing technological systems (M. K. Lee et al. 2019; Vines et al. 2013).

Sidebar: Performativity

Performativity refers to the ways in which speech and other social actions shape the model that people accept as "reality." This sense of performativity was enshrined by Judith Butler's account of gender not as something that a person *is*, but rather as something that is *done*—both done *to* a person, such as pronouncing a newborn a boy or a girl based on their sex traits, and done *by* a person, through the habits they internalize, deliberately or unconsciously, as associated with a particular gender role (Butler 1990; Cavanaugh 2015).

Though it is not always seen as such, technology is often a vehicle for performativity. For instance, when the Apple iPhone launched in 2007, Steve Jobs described it as "not just a communication tool, but a way of life" (Lasco 2015). The sleek design and high price of the iPhone painted it as a marker of a luxurious lifestyle; it was sought out not only for its utility but also as a means of performing high social class, as famously and regularly parodied by the satirical news outlet *The Onion* (*The Onion* 2010).

It is important to be aware of and reflect on the boundaries we presuppose and draw. Technology may influence how we draw boundaries as well as when and for what reason we decide to make adjustments or rethink them. For instance, imagine that someone is analyzing results of a demographic survey and decides to use race rather than height as the main organizing category. Although the survey data on race is just as objective as the data on height, in the sense that they are equally based on respondents' self-reported categories, the analyst's choice stems from a normative judgment about which category is more significant, and in so doing lays down the parameters for any subsequent analysis. It is never easy to draw a boundary between subject and object or between technology and society, and any lines that we choose to draw will also be affected by our current sociotechnical frame. However, the difficulty of drawing boundaries for analysis should not prevent us from engaging in the process of analysis. The point is to make sure we engage with this process consciously and in a way that seeks to include as many stakeholders and alternative viewpoints as possible. We should always keep in mind that the drawing of any boundary is intentional and, in fact, the boundary that you draw is never fixed; it is fluid and open to debate. The boundary itself, and the nature of change to that boundary, can have many ramifications that are not obvious. The key thing to remember is summarized famously by George Box: "All models are wrong, but some are useful" (Box 1979).

In summary, as we have discussed extensively, both in the earlier parts of this chapter and in chapter 3, it is not possible to talk directly about reality, or even about specific portions of it. Any time we talk about the world at large or something in it, we are instead constructing an abstraction of reality that is shaped, at least in part, by our own expectations and values. As human beings, we cannot opt out of this process and describe reality directly: your only choice is whether or not to recognize that your abstractions will not align perfectly with other people's, or even with your own best perceptions. But recognizing this enables you to get better at creating nuanced, precise abstractions that more closely reflect the world. In the following sections we offer some critical resources you can use to critique the abstractions of reality you encounter—including your own—to help you move them closer to the thing you want to describe and understand.

Sidebar: Photography: What You See Is Not Always What You Get

When photography was first introduced in the 1820s, it was almost a magical change in the way that humans recorded and remembered their lived experiences, or even the experiences of others. Instead of painting a landscape or realist drawings, one could now record the light bouncing off objects in the world in a way that visually reproduced, exactly, the state of the world at the moment the photograph was taken. However, this account leaves out important details. If you have ever worked with a camera with changeable lenses, or in a darkroom to develop film, or even used an app to modify a picture, you know that the light, and the way it gets recorded, does not follow directly from natural laws of physics. Although we like to think that these film or digital pictures are a true depiction of the world, the light is bent through a lens that allows different levels of light onto the recording plate. If we are working with physical film, the type and exposure levels can significantly impact the results. If we are working with a digital camera, the quality, amount, and tuning of the digital receptors that turn light into a digital signal can affect how and what gets recorded.

So, we sometimes like to think that something as simple as taking a picture, recording the way the light moves in space, is some sort of value-neutral, "picture perfect" way of recording. But there are numerous design choices—from the lens to the chemicals to the digital post-processing—that highlight certain features, remove others, and alter what we record. The film we use is optimized to capture certain colors, lights, and other effects, and was largely optimized for light-skinned people during the design and development of film from the 1900s to the 1970s, leading to photos of dark-skinned faces to appear obscured or blurred (Lewis 2019). Thus, even today with digital photography: the construction of the light sensors and even the nearly automatic post-processing of the images are built with the expectation of certain things, such as the sky, to be lighter in color, which can lead to significant effects on the final image, including the orange skies of the wildfires in California in 2020 being automatically processed to a blue shade (Bogost 2020). Photographs can give the impression of simply and transparently capturing "what's really there," but the reality is that any photographic image will reflect both the powers and limits of the photography equipment itself and the photographer's ability and perception—perception that is itself shaped by the tools.

Similarly, data collection and aggregation can give the impression of being neutral or objective when it is accomplished by information technologies, as if removing the human removes the bias. But as we discussed in chapter 3, data and information are never neutral. Like a photograph, data and information are inflected by the means by which they are perceived and captured.

Not only does the camera sit between what we see and the picture, but the ability to capture pictures of the world around us has changed our own desires, norms, and customs. The desire to share photos with friends, loved ones, and even strangers led to websites and services like Flickr, Google Photos, Facebook, and Instagram. More than just influencing what technology we develop, there have been other effects of the ubiquitousness of cameras. For instance, in advertising, think of all the effort put into making food look pretty in video and pictures, which may have been a concern only of a painter in an earlier era. Likewise, even individual photo-sharing sites, such as Instagram, have their own aesthetic or design sensibility and constraints, from oversaturated colors to cropping photos to be only squares, and these norms established by the service and the camera led to an explosion of sales of pastel backgrounds and funny things to put in your AirBnB (Lorenz 2019). This is one example of the feedback loop between what technology is developed, how we react and change as a society, and how we then decide what we want or desire or develop next.

5.4.3 STRATEGIES FOR ANALYSIS

In the last section we saw that several problems arise when we engage in this process of framing and abstraction. In this section we identify some common traps that are easy to fall into when it comes to working with abstractions and framing problems.

Identifying Traps

Selbst et al. (2019) offer a list of *traps* that one should keep in mind when framing a problem for analysis. These traps are sometimes easily fallen into, and it is important to call them out explicitly when analyzing any system. These can serve as checks on our own thinking when we go about the process of framing and abstracting a real-world system.

We illustrate these concepts with concrete examples of technological systems that illustrate what it looks like when we fall into one of these traps:

- **Formalism trap**: Failure to account for the full range of meanings of social concepts such as fairness, which can be procedural, contextual, and contestable, and cannot be resolved through mathematical formalisms.

  ○ Example: Although fitness-tracking apps claim to give users power over their own health through objective data, critics have argued that fitness-tracking enthusiasts tend to reduce the holistic concept of health to a combination of proxy measures like calorie count (Sharon and Zandbergen 2017).

- **Framing trap**: Failure to model the entire system over which a social criterion, such as fairness, will be enforced.

  ○ Example: A machine learning system, used to help decide which pneumonia patients to admit to the hospital, learned a dangerously misleading rule that people with asthma were *less* likely to die of pneumonia; this was true in the data set, but only because of the unmodeled factor that doctors had assigned more intensive treatment to the data set's asthmatic patients (Caruana et al. 2015).

- **Portability trap**: Failure to understand how repurposing technological solutions designed for one social context may be misleading, inaccurate, or otherwise do harm when applied to a different context.

  ○ Example: Publicly available computer vision systems trained with affluent Western households in mind were far less accurate when identifying objects in lower-income or non-Western households (DeVries et al. 2019).

- **Ripple effect trap**: Failure to understand how the insertion of technology into an existing social system changes the behaviors and embedded values of the pre-existing system.

  ○ Example: A tool that recommends criminal sentences based on likelihood of reoffending may warp the justice system to more heavily emphasize one value (incapacitation) and diminish others (deterrence, rehabilitation, etc.) that originally shaped a sentencing decision (Selbst et al. 2019).

- **Solutionism trap**: Failure to recognize the possibility that the best solution to a problem may not involve the particular technology you are currently working with or working to develop.

  ○ Example: A designer, working on a high-tech project that used location tracking technology to help people with sight loss to navigate train stations, pointed out that painting high-contrast lines on the ground would avert the severe accuracy limitations of the location tracking technology while also being helpful to the majority of blind users who have some residual vision (Yates 2018).

Asking Better Questions

In this section we offer four questions that one should keep in mind when analyzing or discussing technology in a social context. Although these are not the only questions you will ever need, they are important starting points. We will use each of these questions when we turn to examining and analyzing specific social spheres in section 5.5.

Who Has Access?

One pitfall to avoid when engaging with the design, development, or deployment of technology is believing that if we think really hard, we can anticipate all issues that may arise from a technological intervention all by ourselves. However, we all have our own biases and preconceptions, and these will necessarily limit the issues that we are able to conceive of, no matter how open-minded or smart we think we are (McIntosh 1988).

As we saw with the tomato harvester example in section 5.3, it is critical to attend to the questions of inclusion. Who has access to discussions about the development and deployment of any technological intervention? And how are these concerns understood and incorporated into the technology development process? When discussing our spheres for analysis in section 5.5.1, we use the term *pluralities* for this collection of concerns around who has access to the conversations about the development, design, deployment, and oversight of a technological system, and whose *input* is valued in that process.

There are many ways to address these issues, and perhaps the most straightforward is to include as many affected people as possible in the entire process of technology conception, design, development, testing, and deployment. One of the most important lessons from the scholarship on participatory design (M. K. Lee et al. 2019; Ricks and Surman 2020) is that we must acknowledge that technological systems are situated in a broad social context, and we must engage all who are affected in a cocreation relationship (Kretzmann and McKnight 1996), whereby as many communities and stakeholders are included as possible.

Embedded in a concern for pluralities is a concern for *epistemic asymmetries*: that is, attending to situations in which some kinds of knowledge claims, or some people's claims to knowledge, are systematically undermined or de-legitimized, in ways that are treated as natural but which can be explained by present and historical power imbalances and powerful entities' desire to maintain power. For a concrete example of epistemic asymmetry, see the discussion of cultures of knowledge in chapter 3, particularly the closing example about how Indigenous knowledge of the environment is rarely taken seriously by Western-trained scientists. You can also look back to the discussion of feminist ethics in chapter 2, an approach that specifically aims to redress the injustices caused by the exclusion of so many people's knowledge and voices.

Epistemic asymmetries can lead to an inequality of access for self advocacy. For instance, in thinking about access to medical interventions and care, who has access to different types of care? Who gets to decide what is appropriate and what is not? Is everyone on equal footing when it comes to deciding what is the best care and who receives it?

Technology has the potential to magnify the asymmetric effects on different groups of people. For example, when cities create open data sets of policing and crime, it can reinforce ideas about which neighborhoods are "good" or "bad." As we discussed in chapter 3, it is important to remember that any choice of data and what to record represents a specific framing choice and a particular angle on the world. This is not an abstract concern. Consider the large, linked databases on services for the poor, including those about who is seeking care (e.g., addiction therapy). A record of such care may disqualify someone from public benefits, and therefore many people in need seek off-the-record solutions (Eubanks 2018). It is important to consider. how technological systems create or reinforce norms both explicitly and implicitly.

Who or What Is Affected?

As we discussed in the first part of this chapter, it is often impossible to analyze a technological system as some separate entity from the context in which it is used. This idea of interconnectedness, that everything affects everything else, is an important notion to keep in mind when discussing or analyzing technological systems. This concern reinforces the *ripple effect trap* that we discuss in section 5.4.3: there are risks associated with our always-limited understanding of the interconnections between a technological system and the broader environment of where it is implemented or deployed. Whereas this effect can be mitigated, it can never be averted entirely—there will always be more layers than we can anticipate.

Examining a decision in a vacuum can tell us only so much; we consider the *interconnections* between parts of a system to go beyond those parts that are changed deliberately and discover those other parts that might also change in response. This includes the interplay of different granularities (e.g., how the collective level affects the individual level and how the individual level affects the collective level) and different kinds of elements of the system. One approach that can be useful for capturing the complexity of various interconnections is known as *actor-network theory*.

Actor-network theory (ANT) is a methodology developed in the 1980s by scholars (most notably Bruno Latour, Michel Callon, and John Law) working at the intersection of science, sociology, and technology studies. ANT is known especially for the novel way in which it specifies that actors—both human and nonhuman, individuals as well as collectives—influence and interact within a complex sociotechnical system (Latour 2005). Ecosystems, ideas, organizations, and more are viewed as a dynamic web of actors alongside people when considering the effects of a technological intervention (Law 2009).

What's the Response, and Why?

One of the traps we listed above is the solutionism trap. This trap can also lead to a mind-set that is sometimes referred to as *technological solutionism.* Technological solutionism is a mind-set that limits creativity and limits critical thinking. It presupposes, first, that all problems can and should be fixed; and second, that the best way to fix a problem is to invent new technology or further devleop the technologies we have. Operating from this stance, one can find it hard or even impossible to recognize that sometimes the best solution may not be a new piece of technology but rather some other intervention, or perhaps no intervention at all (Jasanoff 2016; Selbst et al. 2019).

A limited mind-set can also be at work in the way we anticpate responses to technological change. New interventions into current technological systems can stimulate responses that are unexpected and surprising. For example, widespread adoption of teleworking technology has meant that individuals are no longer tethered to living near their working location, which has resulted in many individuals taking actions like moving to more remote locations or leaving cities. The response effect is closely bound to the idea of interconnections, in that one technological intervention that may lead to ripple effects, but also to feedback: the use of more robots on a factory floor may make job retraining more available in a given location, or it may just cause increased unemployment.

An individual's or group's reactions to technology can be used as a *social signifier*: in other words, a way of telling others something about their social role or status, whether through their embrace of a technology or their *resistance* to it. For instance, when Google Glass was released, many lower-income people saw it as a symbol of class elitism, and some differentiated themselves by displaying hostility toward wearers (Noble and Roberts 2016). This kind of response can be understood if we think about how the use or nonuse of particular technologies gets translated into social capital by signaling virtue, class, or status; some people will go out of their way to reject "improvements" that a technology creates, even if the improvements themselves are something they want, because rejecting that technology helps them claim a particular social identity (Bourdieu 2018).

Sidebar: The Development of Scientific Management

One place that we can see the interconnections between (1) the development of technology, (2) the impact of that technology on society, (3) the way societies react, and (4) what that reaction means for future technological development is in the development of what is commonly referred to as *scientific management practices.* By tracing the development of scientific management practices from their beginnings in the 1880s and thinking critically about their impacts on how we view technology and technology development today, we can see how a cultural ethos of mechanization and efficiency, uncritically applied, has created some norms of optimization that may be unwelcome.

The concept of scientific management, sometimes called Taylorism, was pioneered by Frederick Winslow Taylor in the 1880s and 1890s. Scientific management is the theory of organization and management whereby one can organize the process of manufacturing, production, and work in general in such a way as to maximize economic efficiency and labor productivity (Drury [1922] 1968). Taylor's work applied the scientific method to the workers themselves. Taylor's research included time and motion studies of how workers moved, and observing how changes in lighting conditions affected workers' productivity. Taylor was dismissive of the idea, widely held in his day, that a trade or craft could not be optimized and must be completed by traditional methods by highly skilled, autonomous craftsmen and tradesmen. He considered his own studies to be a higher order of knowledge than the knowledge possessed by the workers he studied, as Schatzberg writes: "Taylor define[d] science as a sphere of elite, professional knowledge that belonged only to engineers, not to the skilled workers who had in fact generated most of this knowledge, or the foremen and plant managers who traditionally set the technical conditions of production in negotiation with skilled workers" (Schatzberg 2018, 72).

The automobile-manufacturing innovations of Henry Ford shared Taylor's ethos of optimization through standardization. Ford emphasized breaking down a complex process into simple constituent operations that could be allocated among workers who each performed a single task repeatedly. Since workers would not be expected to switch between tasks, it was believed that this approach would increase effiency. It also increased machine utilization, because every stage of manufacturing was happening at every time in the production plant, so there were no incidences of a machine sitting idle while a different stage was taking place. Ford considered his approach distinct from Taylorism, because it involved rethinking how tasks were organized altogether rather than simply eliminating inefficiencies in how existing tasks were carried out (Ford 1929); contemporary scholars have also argued that Ford's facilities, in which floor workers could suggest process improvements and foremen participated in their supervisees' tasks, lacked the fundamental Taylorist quality of strict separation between those who planned the work and those who carried it out (Williams et al. 1992). Nonetheless, Taylor's and Ford's joint influence on the coming decades of manufacturing, and other similarities such as their use of surveillance—Ford's "Sociological Department" even paid unannounced visits to workers' houses, off-hours, to monitor for "immoral behavior"—is often characterized as a unified "Taylorist-Fordist" model (Lobel 2001).

A more modern form of Taylorism, digital Taylorism, centers on the use of technology both to monitor workers and to homogenize and make routine most of the tasks that a worker completes. One of the most obvious versions of this is in the business model of Uber: technology is used to make driving from one place to another in a city so routine that no expertise is required. This stands in stark contrast to the history of the art and skill of taxi driving in the past, which required extensive knowledge of streets and roads (Rose 2014), called "the knowledge" in London. Now, unlike the cab drivers in London in a bygone era that were tested on their knowledge of the city itself, any person can navigate any unfamiliar city, and their level of efficiency can be monitored by an algorithm and adjusted accordingly (Frischmann and Selinger 2017).

Taylorism has been widely criticized for treating workers as machines, and for encouraging managers to perceive them as disposable and interchangeable tools rather than as people who have both non-work lives and work-specific skillsets. These criticisms have not prevented the widespread embrace of Taylorism across many industries.

5.5 TECHNOLOGY IN CONTEXT: SOCIAL SPHERES

So far in this chapter we have discussed various ways of conceptualizing and discussing technology in a broad, sociotechnical context. We have looked at traps and questions and described many tools, techniques, and frameworks to consider and understand the relationship between technology and society.

In this section we analyze some specific technologies within a particular layer of society, highlight some cross-cutting spheres that individuals, groups, and societies need to be concerned with in the twenty-first century. Specifically, we discuss structures of care; public discourse; friendship and communities; ecology and the environment; state power and force; and work and labor. Recall that earlier we discussed how "society" is not monolithic, and that many layers of society always coexist. It is also the case that societies exist and operate in meaningfully different ways. Hence, this chapter necessarily makes some generalizations, and it is important to note that these generalizations more aptly describe the social structures of Western democracy than other kinds.

These spheres overlap in numerous ways. They are also, of course, abstractions rather than a direct and incontestable account of what society is like. We use them to organize a set of societal concerns and interests, although there are many other ways one may choose to organize them.

5.5.1 STRUCTURES OF CARE

Deeply embedded into how we structure and function as a society is how the society cares for its constituent members: the people, corporations, groups, and other entities that are members of that society. The opening of the US Constitution encapsulates this goal:

We the People of the United States, in Order to form a more perfect Union, establish Justice, insure domestic Tranquility, provide for the common defense, promote the general Welfare, and secure the Blessings of Liberty to ourselves and our Posterity, do ordain and establish this Constitution for the United States of America.

Hence, one of the key functions of a society is deciding whom we care for, how we care for them, and what forms and functions that care takes. Throughout this book we have looked at how advances in technology have shaped and reshaped age-old questions in the modern context. How do advances in technology change whom we care for? How do they change the types of care that are available? Here again we see a recurring theme: that we are in a constant conversation between the technology that we develop and how that technology changes and shapes not only our society but also the new technologies and systems that are created.

Society creates structures of care for a variety of reasons. The most noticeable structures of care are the ones designed to to take care of those who cannot manage on their own, such as children, the elderly, and people who are chronically ill. These same structures offer indirect support to many people who are largely taking care of themselves—for example, reliable childcare gives family caregivers more freedom to go outside the home. Societies also create structures of care for temporary conditions that may arise both on large and individual scales: loss of a job, loss of a house, accidents, impacts on our lives due to fire, flood, pandemic, hurricane, or other natural disaster. Less obvious, but no less significant, are the structures of care we create to ensure desirable outcomes that have been broadly agreed upon as important to a society. For instance, education and training are structures of care that ensure, as much as possible, that people in a society learn necessary and useful skills as well as how to be engaged citizens. We also create structures of care for entities that are broader than just the individual or family unit: larger governments may subsidize certain industries, localities, or even individual businesses for economic or structural reasons, such as temporary shocks to prices, keeping strategically important businesses "on shore," or even providing price floors or buyers of last resort for food or other supplies. These are all instances of what is commonly referred to as the social safety net, that is, the idea that the society as a whole has a responsibility to care for or provide assistance to individual entities in times of need.

There is a long and multidimensional history to the discussion of how, when, and in what ways structures of care should be created and deployed, and how long they should be maintained, and at what cost. Our goal in this section is not to engage completely with this history, which would fill the rest of this book and then some. Rather, we want to highlight how recent developments in technology are changing aspects of the discussion, deployment, and what is even possible to provide in terms of care in two domains: medical care and poverty/social services.

Medical Care

Medical care is perhaps the most obvious system of care that most, if not all, of us will have encountered in our lives. In this section we discuss some of the ways technology is changing the conversation around care, and also the conversations about the technological developments that may be important in the future. Advances in technology have massive effects not only on the cost of care but also on what kind of care is actually available. Modern technology can replace limbs and restore sight in ways that were the provenance of science fiction only decades ago. It is also useful to look at less dramatic kinds of medical care, such as eldercare and care for those with physical and cognitive impairments. Even the integration of different parts of the medical system can itself be considered a structure of care. Looking closely as these common, widespread forms of care can help us attend to fundamental questions of how we care for the most vulnerable among us and provide some relief and freedom for those people to whom the caregiving falls. In this section we first focus on eldercare and then zoom out to some issues of medical care writ large.

Technologies for older and debilitated populations focus on a number of areas: dependent living, fall risk, chronic disease, dementia, social isolation, depression, poor well-being, and poor medication management. In terms of concrete technological solutions for these issues we have advances in general internet and communication technology (ICT), robotics, telemedicine, sensor technology, medication management applications, and even video games (which can help maintain balance and cognitive function, as well as being fun); all of these can be used to help with both physical and mental aspects of aging (Vallor 2011). One of the most visible areas in which we see advances in technology and robotics is in mental health care for the elderly and those with cognitive impairments, which is likely to become an ever-growing segment of our global society (*Economist* 2020). Mental health care is sometimes overlooked, and there is a huge gap between those who cannot access the services they need and those who can. The area of socially assistive robotics engages with both eldercare and mental health care more generally, giving rise to many successes over the years (Rabbitt et al. 2014). While you may have heard of the PARO robot, a seal that makes happy-sounding noises when petted or interacted with, there are other advanced robots available for both companionship and physically assisting individuals. These helper technologies meet with mixed responses from the people they are designed to help: many older patients are happy with assistive robots but may find conversing with them tedious (Abdi et al. 2018; Broadbent et al. 2009). Yet, although there are large differences in care outcomes, there is broad acceptance of robots and assistive technology across these areas and many examples of positive outcomes (Góngora et al. 2019; Khosravi and Ghapanchi 2016).

It is important to think of some of the aspects of eldercare through the lens of our analytic concepts. Recent advances in robotic technology allows us to imagine possible futures in which everyone who needs physical assistance—moving or lifting objects, or even bathing—can be attended to by a robotic presence. But although this imagined future provides solutions to many of the challenges in our immediate present, it also raises important and difficult questions. Who decides who gets such a robot or not? How does the robot decide between what may be beneficial for its charge in the current moment versus in the future? There is also an interesting question of pluralities: what if someone does not desire the help of a robot? When will their desires be overridden by considerations for their children or other caretakers? Vallor (2011) points out that when considering care interventions based on robots, it is important to consider not only those receiving care but also those providing care, including considering who may no longer be able to provide care because of the technological intervention; displaced caregivers may lose their sense of agency, purpose, or value (moral or otherwise) as those who provide care to family and friends. Finally, there is a question of accessibility: older patients may not be able to read screens, or use buttons or switches that require hand strength or fine motor control, or otherwise be interactive in the ways that technology designers imagine. Likewise, many products designed for constant internet connectivity may not be useful in rural areas or to underserved populations who do not have reliable internet access.

Turning to medical care more generally, we find a number of radical changes being realized through advances in general ICT and its application broadly across the medical domain. Perhaps the most important of these is the widespread adoption and use of electronic health records (EHRs), digitized medical charts. An EHR allows care providers to pool and collate patient information, charts, X-rays, labs, and other important information between providers and through time (Atasoy et al. 2019). A fully envisioned EHR is passed seamlessly through the various providers, be they institutions or people, and is continually updated, providing caretakers with a kind of complete digital picture of the patient. EHRs have advanced rapidly between 2000 and 2020 with the advent of the Patient Protection and Affordable Care Act (ACA) of 2009 and its incentives for uptake of EHRs across the United States. EHRs were slow to catch on in the first decade of the twenty-first century (Jha et al. 2009), even though a handful of large hospital networks invested in EHRs in the early 2000s with very positive results (Buntin et al. 2011). But by the 2020s, over 90% of US hospitals were reported to have taken on EHRs, and a meta review shows that this shift has significantly increased the quality of patient outcomes while decreasing the cost of care (Atasoy et al. 2019). As with any other technological intervention, there are still issues related to standardization, interoperability, and access, but there is strong evidence that widespread adoption of EHRs has had significant positive impacts.

And yet for all the advancements that have been made in medical ICT and EHRs, there are a number of risks and concerns. We must think about the increased information flow between health care providers: who controls that data and when can it be taken back? When can it be used for other, nonmedical purposes? And if the state has full control over your care history, is that a problem? Is it a problem if that control lies with a private business? While there have been some recent laws passed in the EU and the United States, there is still a dearth of regulation around the volume, variety, and velocity of information that is collected about you while care is being provided (Anderson and Agarwal 2011; Electronic Frontier Foundation 2021). Although there are a number of important tools that expand access to more populations, there is growing concern over information exchange and privacy. Even more, if the psychologist at the other end of the call is a chat bot, who controls and regulates the bot's behavior? (Fiske et al. 2019). Finally, along with the increasing connections brought about by advances in ICT, there are significant increases in the entire system's ability to perform testing and other diagnostic tasks that can be automated through the creation of new machines (Improving Diagnosis in Healthcare 2015). For instance, the level of testing that the United States and other countries achieved during the COVID-19 pandemic wouldn't be possible without massive automation of test processing.

These kinds of transformations are taking place across the medical field as a whole. And though these most recent changes seem very striking, it is important to remember that technological revolution in medicine is the norm and not the exception. Safe and widely available vaccines were never heard of 300 years ago (although the practice of variolation dates back much farther) (Silverstein 2001). The idea of looking inside a patient without cutting them open was, until recently, impossible. Even in more modern times we have moved from not having X-rays, to physical X-rays on film, to digital X-rays, to imaging tools such as MRI scans that can see soft tissue. This increasing reliance on elaborate medical machinery has led some people to fear that we are transforming our doctors from professionals to entities that blindly follow orders from a computer—one example of a phenomenon called deskilling, to which we return in section 5.5.6, "Work and Labor." However, the evidence is clear that doctors and practitioners who work with well-designed decision support tools outperform other doctors. As our tools advance, so will the type, cost, and ability of medicine to care for us all (Wachter 2015).

Economic Care

Another key form of care that society provides and that technology affects in myriad ways is welfare and subsidies, two key forms of economic care (Myers 2001; Schulz-Forberg 2012). A subsidy, in the broad view, provides support to a particular group or sector by shifting some of the full economic cost from the particular group or sector onto society as a whole. One example of this kind of subsidy is public education: the cost of public education is not paid fully by those that receive it but is subsidized in various ways by society, because having an educated workforce is a public good. The same can be said for many other things that we enjoy as a society, from roads to basic research. The three main types of subsidy are as follows:

1. Preferential treatment for particular sectors or classes, such as providing housing payments, free medical care, and education payments.

2. Payments or encouragement for certain activities such as renewable energy credits and recycling.

3. Insurance subsidies, which include price floors for food production and manufacturing (Myers 2001).

Each of these types of subsidy can be direct or indirect. For instance, we can pay for food for individuals directly or can subsidize farming of crops, indirectly lowering food costs. If a town pays for garbage collection, it may be indirectly subsidizing the manufacture and distribution of one-time-use packaging. All these forms of subsidy comprise what we typically think of as economic care: we are choosing, as a society, to relieve some of the economic cost to a particular group or sector.

Subsidies also include welfare, the welfare state, and the social safety net. We can think of these as subsidies to *individuals* rather than *firms or sectors*; however, subsidies to both groups are forms of economic care. The concept of the welfare state, or the idea that the social state has a role to play in ensuring certain benefits to its citizens, is widely credited with starting in the German Empire in the 1880s with guarantees for workers against sickness and accidents (Schulz-Forberg 2012). However, as discussed in chapter 4, older religious states had forms of social safety nets going back millennia. One can view the welfare state as a form of subsidy to individuals (i.e., assistance to improve the lives of vulnerable individuals and families) (Ivaschenko et al. 2018). Examples include social pensions, food transfers, cash transfers, and fee waivers. There is a long history of discussions about the role of welfare in a society and questions around when it is appropriate, whether it distorts the operation of the markets, and whether it is still appropriate to provide for citizens in this way (Schulz-Forberg 2012).

Automation is changing the form and delivery of subsidies to poor families through the welfare state. A powerful examination of this technologized shift can be found in Virginia Eubanks's book *Automating Inequality* (2018). In her book, Eubanks discusses three specific instances: the delivery of Medicaid in Indiana and its change-over from human caseworkers to an automated system; the coordinated entry system for homelessness in Los Angeles and its centralization through a large, standardized database; and child protective services in Pennsylvania and its automated scoring system to rate child risk. The case studies provide a comprehensive set of examples of how new automated decision-making tools are alleviating some problems but also creating new ones. Foremost when analyzing these systems, we must take into account the entire interpretative frame and include the people affected, the technology itself, and the history of the development of that technology. For instance, in the Medicaid example, efforts to automate the monitoring and dispersion of medical payments focused on a goal of clearing claims quickly, hence the contractor responsible started closing or declining claims at a rate many times higher than before, leading to a massive loss of care, numerous lawsuits, and often catastrophic impacts on those relying on Medicaid. This focus on system efficiency, clearing claims quickly, rather than health-care results, ensuring adequate and appropriate care, led to harmful effects on the most vulnerable.

A compelling argument from Eubanks is that modern technological tools allow us to "manage" the poor in plain sight instead of shipping them to work farms and poorhouses, as was done in the 1800s and 1900s. There are a number of concerns raised in this account, from the amount of information collected about the poor through these service providers and who it is shared with and how that information is used, to concerns of pluralities and who gets to decide what is the best form of care for an individual. Eubanks' accounts demand that we look at poverty through a new lens. Too often, we view administering care to the poor as a systems management issue, wherein the problems would solve themselves if we only had better systems to deliver the scarce aid faster, in the right place and at the right time. This assumption limits our ability to consider the ways in which existing strategies are fundamentally mismatched to the problems, and therefore to consider alternatives seriously. For instance, the cost of building housing for all unhoused people would, in many places, cost less than maintaining the infrastructure required for providing more narrow and focused payments, such as food stamps, bus passes, or rent stabilization. Indeed, our focus on the *systems* instead of the *outcomes* can limit us to thinking more in terms of surveillance or disciplinary powers: if we pay for only certain things we can revoke payments when those receiving them do not act how we think they should, instead of helping to promote and ensuring the capabilities of those receiving economic care. Eubanks argues that it is important for everyone to pay attention to how these systems are designed and deployed, as one day they could be used for everyone. For instance, the monitoring systems currently used on the poor may be put to use, at some later time, to determine whether or not you are a safe enough driver to deserve car insurance. These examples are an interesting study in what we saw in section 5.4.3 as the portability trap, that is, using a system automated for one purpose (poverty risk) for some other context (car insurance) leading to misleading, inaccurate, or undesirable results (Selbst et al. 2019).

Another key form of subsidy in the United States is related to home ownership. In the post–World War II period in the United States there was an aggressive expansion via government programs to expand home ownership. However, these loans, insurances, mortgage guarantees, and other financial incentives were largely extended only to white individuals, while people of color were actively disengaged through the use of redlining (the practice of undervaluing homes in "bad" neighborhoods) and other tactics (McGhee 2022). Beginning in the late 1960s, due to the saturation of homeownership by whites and the need for more financialization of the home market, lenders began to increasingly lend to Black and minority families. This strategy is dubbed *predatory inclusion* by Keeanga-Yamahtta Taylor in her book *Race for Profit*. In this scheme, African American home buyers were granted access to conventional real estate practices and mortgage financing, but on more expensive and comparatively unequal terms (Taylor 2019). Today we see potential for many of the same problems in the expanded use of technology to measure credit and determine access to financial institutions. There are numerous companies focused on increasing access to financial markets for those historically marginalized, often at worse rates or at the cost of their privacy—often using access to "nontraditional risk models" to evaluate applicants' creditworthiness through, for example, their Facebook friends (Bruckner 2019). This area is ripe for more forms of predatory inclusion that affect not only Black families but young people, old people, and many other populations, and it is important to view this through a holistic frame to understand how and why new technologies enable and/or cause harm.

Although we have used most of this section to discuss how we provide care for individuals through direct subsidies or welfare, it is important to realize that society's economic care is not limited to individuals. There are important normative questions about how and why we subsidize corporations and businesses through loans, farms through price floors, and specific industries through large, guaranteed purchases (Myers 2001). Indeed, even the development of many of the technologies discussed in this book, from the microchip to the internet, were heavily subsidized by the US government (Isaacson 2014). The structure and deployment of these subsidies is constantly affected by technological change: the internet itself has opened up vast areas of new business; it is a piece of public infrastructure that has arguably enabled a completely new way of life. The original goal of the internet was to provide a mechanism for nuclear silos to communicate in the case of a nuclear war, but it has become enmeshed with our everyday life including the delivery of many structures of care from telemedicine to remote learning. This is again an example of our society existing in an ongoing conversation with our technological systems.

Story Point: "Today I Am Paul," by Martin L. Shoemaker

Unlike her father, Anna truly feels guilty that she does not visit more often. Her college classes and her two jobs leave her too tired to visit often, but she still wishes she could. So she calls every night, and I monitor the calls. Sometimes when Mildred falls asleep early, Anna talks directly to me. At first she did not understand my emulation abilities, but now she appreciates them. She shares with me thoughts and secrets that she would share with Mildred if she could, and she trusts me not to share them with anyone else.

Mildred is elderly, frail, and suffering from dementia, and she needs constant care. Her family provides her with a highly sophisticated android that is capable of emulating the appearance and personalities of people that it has scanned—and therefore, because it emulates her family members, Mildred believes that they are present and caring for her, rather than an android. There are signs throughout the story that Mildred's family intended to remain active in her care, but now that the robot is actually there, they have adapted, showing up less to be with Mildred—and even replacing her with the robot in some ways. This story obviously raises questions about who is responsible for care work, but it also offers a rich and textured exploration of all the things that constitute care work: not just meeting a patient's physical and medical needs, but the various kinds of emotional and psychological support they need, and who they need it from. And if we are willing to look past the empty pieties that often shape conversations about eldercare—as "Today I Am Paul" urges us to do—it equips us to confront uncomfortable but necessary questions about how to balance the needs and well-being of the caretakers against the needs of those under care.

5.5.2 PUBLIC DISCOURSE AND POLITICAL DELIBERATION

To live in a society means to have shared interests (e.g., for care, as we discussed in the last section, or for governance). As Aristotle (the virtue ethicist) writes: a political community "comes together for the sake of the living, but . . . remains together for the sake of living well" (*Politics* I.II.1252b25–30).

Living together, well or otherwise, has taken many different forms over the course of human history. For roughly the past 150 years, democracy has been the form of political order that is held up as ideal in the West. Democracy can be defined generally as "a method of collective decision-making characterized by a kind of equality among the participants at an essential stage of the decision-making process" (Christiano and Bajaj 2021). Yet another definition foregrounds a different concern: "Democracy can be defined as a system in which the government is in power by the consent of the people and the government is accountable to the governed" (Schmitter and Karl 1991; Swain et al. 2011, 3). Hence, in a democratic political community, the members of that community, the citizens, are understood to have some form of equality with respect to the decision-making process and to have a government that is in power through consent of the people.

Citizenship is a special kind of relationship that is notably different from the relations we have with coworkers, friends, or family. Citizenship is a relationship that comes with a set of expectations and presuppositions about how we are to act toward one another; recall the discussion of the social contract in chapter 2. These expectations are often connected to the values and interests that members of a political community share, but they are also rooted in the political order in which laws are enacted and enforced. The notion of citizenship typically rests on the idea of there being some form of nation or state that one is a citizen of. But as we have seen in our discussion on society, there are many other ways to define decision-making boundaries and power structures, and we may be a citizen of other power groups that operate at different levels of authority (e.g., a citizen of the world).

There are various types of democracy: deliberative, egalitarian, and constitutional, to name a few. One thing that they all have in common is that they take for granted that people are diverse and that they sometimes have conflicting ideas about how to live a good life and the kinds of goods and values that the community should pursue. Modern liberal democracies have devised a number of ways of regulating how decision-making takes place, who has access, and how decisions are changed. Large-scale endeavors, such as science, politics, education, foreign policy, and environmental concerns, all require collective efforts and processes of decision making.

Political leaders and officials are hired to carry out the law and to implement and uphold certain policies. These processes take the form of institutions, laws, policies, and customs that are organized around goods and values that are, ideally, shared by the whole collective of individuals, the whole citizenry. But there is also room for disagreement. In fact, one could argue that democracy renders disagreement and conflict productive by preserving a space, a *public sphere* where people with different perspectives and opinions can come together and decide about how they should be governed. This is one of the reasons that it is sometimes said that the public sphere "institutionalizes a particular kind of relationship between persons" (Bohman 2008, 71).

Citizens are expected to participate in this collective decision making about how goods, resources, and power are to be distributed (Christiano and Bajaj 2021). Thus, all forms of democracy need to facilitate how this collective decision making happens. This decision making does not simply happen in a shared public space where everyone sits around and talks about their desires, even if this was the aspirational goal of the Greek Agora. Even in the context of Athens in ancient Greece, someone had to decide on what day and what time everyone would meet to vote and make political decisions (they likely relied on technology to facilitate this process). In other words, political discourse and deliberation do not just happen, but rather are enabled or prohibited by a complex interaction of society and technology. In this section we investigate these interactions through the topics of public access and free speech as well as access to information.

Public Discourse and Free Speech

In a democratic context, public discourse and free speech go hand in hand. One example of the regulation of public discourse is the US Constitution's protection of freedom of speech. In the history of intellectual thought, these ideas are said to be oriented toward the human good—hence there is a presupposition that we all have an interest in preserving free speech as well as a space for the public exchange of ideas. Why is social exchange good? Why is it valuable? These are important questions to ask, especially in contemporary times when social media tends to blur the boundaries between private and public discourse.

In order for public speech to be truly public, there needs to be a forum or a venue where members of the public can speak and be heard. These spaces can be geographic, such as a town square, or they can be metaphorical, such as a newspaper, a radio broadcast, or a Reddit forum. However, these spaces do not simply exist spontaneously. Policy making and the discussion of policy is said to take place in a "public sphere" (Habermas 1989; Mill [1859] 2002). As we stated in chapter 4, the "public sphere" or "public space" in the classical sense is the space, real or imagined, where citizens interact as citizens. The government in many ways establishes the ground rules for how collective decisions are made at the public level. The government also has a role to play in how public discourse takes place, and so does technology (recall our discussion of technological mediation in section 5.2.4).

In chapter 4 we discussed the legal arguments that Warren and Brandeis offered at the end of the nineteenth century regarding the right to privacy. They argued that because recording technologies like the phonograph and camera allowed people to obtain information and distribute it in ways never seen previously, legislation had to be written to protect privacy as a right. These technologies altered the way in which individuals relate to the public as well as the public's access to the government. We argued in chapter 4 that the guarantee of privacy was important both to preserve the freedom to form opinions and to make decisions about our own lives without inter-ference from others, and also to allow us a space to come together as a citizenry: the public sphere.

The ability to engage freely in public discourse is not only of benefit to the per-son speaking; it is essential to the common good. John Stuart Mill, the utilitarian thinker discussed in chapter 2, was a strong advocate of free speech and the exchange of opinions and ideas. He thought that they benefited not only the political society of which he was a part, but humanity as a whole:

> But the peculiar evil of silencing the expression of an opinion is, that it is rob-bing the human race; posterity as well as the existing generation; those who dis-sent from the opinion, still more than those who hold it. If the opinion is right, they are deprived of the opportunity of exchanging error for truth: if wrong, they lose, what is almost as great a benefit, the clearer perception and livelier impres-sion of truth, produced by its collision with error. (Mill and Miller [1859] 2002, 18–19)

In other words, for Mill, even when an opinion is wrong, people should have the liberty to express it so that their view can be corrected. This is not to say that Mill believed there should be no regulation on speech; participation in the public sphere also entailed the use of reason and some regulation around what was acceptable. But Mill and others writing during the same time also insisted that free speech was grounded by the idea that human beings are free and equal and that we could rea-sonably agree on a set of goods and values, even if we did not always agree on how to achieve them.

The ideals of democratic communication have always been difficult, even before the advent of the internet and social media. Ideally, citizens participate in the dis-course in ways that are both inclusive and aimed at shared understanding, if not prac-tical consensus (in other words, as part of the same culture of knowledge). This type of exchange relies on two conditions: a shared base of information and knowledge about what is going on and what is at stake, and a good-faith effort to include everyone in the conversation. And yet, as far back as the founding of the United States, there have been concerns about the information that was being distributed by pamphlets and news articles (Campbell 2019). In other words, "fake news" is not a new problem. But the more quickly it can travel, the bigger a problem fake news will be—and on social media, it can travel faster than ever before.

Inclusiveness has likewise always posed challenges, but social media in the twenty-first century creates a new combination of opportunities and challenges by providing an alternative environment in which citizens can come together, exchange ideas, and participate in political discourse (Vallor 2016). Discussions about social media are an extension of discussions that have been ongoing for years about expres-sion on the internet and other "public" spaces. One of the most pervasive issues in social media is the worry about "echo chambers" and political polarization that we saw in chapter 4 or the ability of the internet and social networking sites to "fragment" the public discussion in a way that promotes extremism (Sunstein 2009). Many of the laws and regulations around free speech were not written with social media as a starting point and do not help us understand current issues with free speech, much less address them.

New technologies for communication invite us to think differently about what it means to deliberate freely about political matters and what it means to participate in that conversation. As we have seen with the Black Lives Matter movement, rallies for independence in Hong Kong, and even the riot at the US Capitol in January 2021 (Baker and Tavernise 2021), social media can be used to rally large actions by potentially disparate geographical actors, often without central coordination. These changes in the scope, scale, and speed of interaction in the public sphere challenge us to consider which values and goods should be prioritized in public discourse, what role the government should play in matters like the distribution of resources and the protection of certain rights, and the role of technology in these contexts. Public discourse cannot exist without a public space, and the government has traditionally played a role in protecting— and, some would argue, in creating—public space. Changes in technology have raised questions not only about how the government should regulate social media but about whether social media companies ought to regulate the government.

Access to Information

Contemporary democracies are faced with all kinds of issues: environmental concerns, sustainability, and climate change; emerging technologies, biotechnologies and nanotechnologies, and issues regarding energy and engineering. As we have seen, citizenship comes with certain expectations including engaging with the collective process of making decisions on these and many other topics. In order to participate in the political process and exercise certain rights, citizens need to be willing, equipped, prepared, and presented with opportunities to participate in the public exchange of ideas and weigh in on matters of public policy.

Since we are rarely able to get the whole of a given society together in one room, some kind of technological mediation is always involved in the communicating and accessing of information, as we have seen both in chapter 3 and section 5.2.4. Various forms of technology, when deployed in this role as information mediator, can blur the distinction between informing citizens and persuading them. Specifically, the notion of "objectively informing," if it is even possible to do so, implies that there must be some shared expectations about both the type and quality of information that we receive and the veracity and reliability of the sources of that information. But how can we determine whether we are being informed or persuaded? Is there a difference? Is it helpful to make a distinction? Who decides? And how are all these concerns bound up with new and changing technologies?

One unforeseen challenge of both the globalization discussed above and the internet is that we often receive more news at the national and international levels than we do at a local level. In the decade and a half leading to 2018, more than one in five newspapers in the United States shut down (Abernathy 2018). This lack of local reporting can lead to less coverage of important local issues and a fragmenting of the public discourse at a close, local level (PBS 2020). There are now many "news deserts" in the United States, where there are no local papers and no one there to shine a light on important aspects of local interest such as municipal water treatment, garbage collection, building permits, and other details of living in a society (Abernathy 2018).

Both the local and global fracturing of the news scape has also interacted with the rise of social media, as we discuss in the next section. Questions of what types of media count as valid, and what counts as truth, have led some to claim that we are living in a post-truth age. This idea has been described as *truth decay* and is characterized by four trends: (1) an increasing disagreement about the relevance or reliability of facts and the interpretations of facts and data, which echoes the datafication discussed in chapter 4; (2) an intentional or unintentional blurring of the line between facts and opinions, often on social media; (3) the rise of relying on anecdotes and personal experience over facts; and (4) a declining trust in traditional respected sources of factual information including newspapers, news outlet websites, magazines, and the government (Kavanagh and Rich 2018).

In addition to globalization and news sources, contemporary democracies are also equipped with improved tools for governance and regulation, and there are serious questions about how these tools should be used. We also have advanced technologies that help us evaluate the long- and short-term effects of policy (McGinnis 2012). Unlike other innovations in the past, we have mechanisms for evaluating the consequences of social policies; we have more data to interpret and can provide more accurate predictions about the effects of certain policies and courses of action. Moor (1985, 266) discusses the way in which computer technologies reveal policy vacuums as, "computers give us new choices for action, we need to formulate policies to help guide actions, recognizing that there are appropriate individual and social responses." This feeling is also echoed by Abebe et al. (2020), who argue that as computing has come into more aspects of our lives we should recognize that computing cannot solve all our normative problems, but it can act as an analyzer, formalizer, and diagnostic to shine a light on these issues and inform policy, given access to the right data. The increased data gathering on the part of governments leads to the argument that this information is acquired for the common good and that it should be made accessible for different communities to use as they need. This has led to calls for "open data" in many government agencies and cities (see www.data.gov). The idea is that the public can audit, use, and connect data in ways that affect their lives (Tauberer 2014). Information is a source of power, and governments are typically charged with directing and mediating power. We need to communicate and have access to information in order to make informed decisions about how to live our lives, and how to participate in the political process (Sáez Martin et al. 2016). The importance of this can be seen in various government mechanisms such as the Freedom of Information Act, which strives to ensure "a general philosophy of full agency disclosure" ('Lectric Law Library's Lexicon, n.d.).

Policy decisions are, we hope, informed by specialized knowledge, well tested with data acquired by reliable methods and generally verified to the greatest extent possible. But when specialized knowledge becomes hyperspecialized how can we utilize this data and who has access? As we discuss in the second part of this chapter, we need to approach technology development as part of the democratic process and avoid a fall into the myth of technocracy. This pushes us to ask questions about knowledge production and transmission. Specialized knowledge and ideas about what is acceptable scientifically are matters of open debate. For instance, with respect to questions related to genetically modified organisms, bioengineering, and cloning, these kinds of decisions are determined in part by scientific exploration but also by an industry and institutions that have their own goals and that fund the research. Science is not value-free, and there are multiple actors and a network of actors that participate in the scientific discovery process and regulation. It is not as though knowledge about the world is a separate thing from our human pursuits. In fact, more and more people are recognizing how science, society, and technology are related. STS has in many ways led this discussion. In order to address contemporary problems, we need both specialized research and an interaction between diverse bodies of knowledge. This information can then be used to make and change policy (Jasanoff 2017).

Finally, education serves both private and public interests. On the one hand, it helps individuals develop their capacities and thrive in social environments. On the other hand, it informs a future public and aims at cultivating citizens. The government has a vested interest in ensuring that citizens are educated and well prepared to live in society. Parents have a big role to play in raising their children to live in society, but there is also public interest in how children are raised. Public education shows that we all have some investment in the common good. This concerns our immediate interests as well as our obligations and duties to future generations. We preserve knowledge and contribute to institutions not only so that we benefit from them ourselves but also because we know that others will need similar resources and access to knowledge. Technology, especially during the COVID-19 pandemic, has radically altered the way that we engage with education. From massively open online courseware (MOOCS) like Ed.X to the online Google Classrooms that many students engaged with during the pandemic, technology has reshaped the traditional classroom in many ways (Harris et al. 2020). There are many arguments for and against things like adaptive tutoring agents and other AI technologies that work with students to increase test scores or provide drills and practice in particular areas (Sottilare 2013). However, the pandemic has also shone a light on the importance of classrooms and peer groups on development, leading many to see the importance of in-person instruction in a new light (Witze 2020).

Technological shifts have changed and continue to change the shape and location of public discourse. As we have seen, technology is not some outside force but engages with our society as a whole. In order to continue to engage in fruitful public debate and oversight of technology, we need to ensure that the public remains able to engage in meaningful debate through a variety of channels and with access to the data necessary to understand and contribute. We cannot simply rely on more technology to fill in policy gaps or hope that with the correct legislation we will solve the problems that are arising with new technologies. Indeed, education and literacy about the media, data, and technology is important, given all the new ways of exchanging and dispersing information.

Story Point: "Apologia," by Vajra Chandrasekera

It was understood by all of us involved, I think, without necessarily needing to have it said, because even in this project it seemed crass to have to say it quite so bluntly, that he would memorialize the other, who had suffered at our hands, and apologize for our own, who had committed, that is to say, perpetrated, or at least were implicated where nothing was proven—there's a reason they sent a poet, not a prosecutor.

Is it possible for a society to make amends for past wrongs? How can a culture or a people—or an industry—alleviate suffering that they themselves have caused? And how can the work of making amends be undertaken in a way that meaningfully benefits those who have been harmed, instead of simply soothing the consciences of those who did the harming? In this story, the dominant culture attempts to rectify its history of violence and destruction by sending a poet back through time to offer apologies at a selection of carefully chosen moments, known to the Poet's avid audience as "Cho-Mos." Even the story's narrator, the editor in charge of shaping the Poet's public profile, seems aware that this project is flawed—yet the form of critique he imagines takes the very same form: a sudden visit to his own time from some future apologizer. "Apologia" offers an unsparing look at how the dynamics of public discourse and social media can combine to trivialize important issues. It also challenges us to imagine more productive forms of public discourse, meaningfully centered on the common good.

5.5.3 COMPANIONSHIP, FRIENDSHIP, AND COMMUNITIES

The internet has created a venue for forging and maintaining personal relationships across vast physical distances. And because physical distances are often associated with other kinds of difference—differences of culture, or politics, or race, or education—the internet has created an historically unprecedented set of possibilities for building new relationships and joining communities with people who are geographically, and culturally, distant from us but who still share our passions or values. This new field of opportunity has not created brand-new behaviors and social patterns but rather has created a new set of conditions in which the same basic behavior and patterns can play out. That being said, this new set of conditions afforded by the internet does to an extent reconfigure and inflect those basic patterns and behaviors. Some things are easier online than in person, and some things are more difficult, and those differences have an impact on what people do and on which patterns take root.

One-to-One Relationships

In many key respects, people's online lives are similar to or continuous with their offline lives. Concerns about whether online relationships are "real" relationships tend to evaporate under scrutiny (Elder 2014). People's online behavior tends to correlate strongly with their offline behavior. For example, although internet researchers initially postulated that online dating might be especially appealing to people who found in-person dating stressful, studies have found that people who are more comfortable with in-person dating are also more likely to use online dating sites (Valkenburg and Peter 2007). Nor can sweeping judgments be made about what certain patterns of online engagement indicate about a person's offline life: in spite of the moral panic about online social gaming, studies have found that those who engage in multiplayer online gaming do not have fewer or weaker offline friendships than nongamers (Domahidi 2018). There is substantial and mounting evidence that people experience their online and offline lives as fully integrated and continuous (Chan et al. 2020).

But it's still true that interacting with and relating to others online is meaningfully different from in meatspace. Online relationships are largely disembodied and at a physical remove, and the conditions of connection typically make it easier for participants to check in and out more easily. These differences have been helpful for people whose physical disabilities have constricted their interactions with the meatspace world and have also created an opportunity for connection between people whose interests or values are out of step with their geographic neighbors. There is also a well-documented tendency for people to treat one another with greater hostility online. Many people assume that the anonymity of online interactions is the explanation for this high level of hostility, although some research suggests that the lack of eye contact between speakers and the psychological toll of information overload are just as significant as anonymity (Lapidot-Lefler and Barak 2012).

Beyond the general conditions of digital "space," the kinds of interaction afforded by particular platforms (such as Facebook, TikTok, or Discord) or modes of contact (such as email or text messages) do create meaningful changes in the conditions of relationships for some people. The specific modes of interaction invited and encouraged by a given online platform can re-entrench and reinforce individuals' existing struggles or concerns. For example, it has been shown that the widely attested phenomenon of social media addiction and the lower self-esteem that correlates with it are far more common among users who are very anxious for social approval, but also that the mechanics of social media exacerbate these existing anxieties and behaviors (Balcerowska et al. 2020; Kanat-Maymon et al. 2018; Shensa et al. 2020). Furthermore, it has been argued that Christian dating sites do not simply match users according to preference and suitability within given parameters but actively enforce the user base's prevailing theological views of gender and sexuality on all of their users (Hutson et al. 2018; John 2019). Even in the case of platforms or communities that do not proclaim a particular set of norms or values, disciplinary norms often take shape anyhow. But it's also important to recognize that not everyone will interact with or exploit a particular array of conditions for interaction in the same way. Of the many studies conducted on the social lives of online gamers, one study that took into account the sensitivity and shyness of its subjects found that shyer and more sensitive gamers tended to form more intimate relationships with fellow players than did their less sensitive peers (Kowert et al. 2014).

The patterning of social interactions and expectations that develops online can in turn create new patterns of interaction in the offline world. The advent of nearly instant internet-based communications has significantly reshaped many offices and pressured workers to respond speedily to online messages because the mode of communication permits it, even at the cost of their health, concentration, and leisure time (Barber and Santuzzi 2015). And as social media usage has become more widespread, some people (mostly, but not exclusively, younger people) have begun to rely on social media as a source of information that they can use to learn about potential new meatspace friends. One study of college students found that many considered it essential to "do homework" on their fellow students in order to screen out those whose political views they found abhorrent or who might otherwise be unsafe (Standlee 2019).

One to Many: Communities and Group Friendships

Online communities form in all kinds of circumstances, which in turn influence what kinds of communities flourish in any given potential point of contact. For instance, the types of interactions that one will have on email, discussion boards, community-driven websites, multiplayer games, messaging services like WhatsApp, or app-only interfaces such as TikTok will influence what kinds of users are supported and rewarded in their participation and what types of interaction take place. The way these communities take shape is influenced by many factors: what is the reason for the website or community portal's existence, according to those who built it and maintain it? Who is the ostensible audience, and who else has a reason to be there? What modes of interaction does the website or portal encourage—and what additional modes of interaction does it permit, even if accidentally? What continuities or discontinuities does a given space afford to communities that migrate there from other settings, either offline or online? How much room does a given community make for a diverse range of participants and viewpoints, compared to alternatives? All of these factors have a major impact on the kinds of communities that develop in any given corner of the internet.

But the question is how, rather than whether, such communities develop. Online communities can and do form in any online venue that permits even the barest interaction between individual users. Users have found ways, in the words of Benedict Anderson, to "imagine communities": to create or identify criteria that enable groups who will never all meet one another to understand themselves as part of a shared, coherent group (Gray et al. 2020; Morimoto and Chin 2017). The internet is not the first time or place in which sociotechnical shift has altered the conditions for imagining communities: technologies as varied as trains and air conditioning have, in different ways, changed the possibilities for community building. In communities as disparate as mailing lists, Twitter, and the online comment sections of newspapers, the same basic strategies appear for distinguishing community insiders from outsiders. Community builders will employ the pronoun "we" to articulate ideas, sentiments, or experiences that, they claim (sometimes explicitly but more often implicitly), are shared by members of their community but not by outsiders. Insider statements are also recognizable as such—in addition to, or even instead of, declarations of shared belief and experience—because of how they are expressed: by means of local catchwords or acronyms, in explicit contrast to designated outsider ideas, or—and this is particularly powerful—by employing the community's insider humor (Gagnon 2020; Gal 2019; Gray et al. 2020). These "insider" elements of a given community become visible in observing newer community members, who are often at pains to demonstrate their insider status by invoking insider language or ideas more frequently than longtime users (Wakefield 2001).

Beyond these outward-facing markers by which users signal their insider status in a given community, there is significant evidence that long-term participation in online communities can impact a user's beliefs, perceptions, and overall sense of identity (S. H. Lee et al. 2019; Scrivens et al. 2020). Long-term participants in various kinds of online communities, from X-Files mailing lists to online white supremacist message boards, experience shifts in their own values or interests, "becoming way more obsessed" with the object of a fandom or becoming more invested in some particular "nonwhite" group as the most significant threat to "white" people (Scrivens et al. 2021; Wakefield 2001). These shifts can and do happen under the influence of offline communities as well as online ones, but the availability of nonclosed online groups makes it easy to become immersed quickly and easily, without even necessarily interacting directly with other members. (Scrivens et al. 2020).

It is crucial to note that although this community construction often takes place mostly or entirely online, the divisions that inform the insider/outsider distinctions are drawn from meatspace social organizations and divisions, and often replicate power imbalances among insiders or aspiring insiders from those meatspace organizations (Morimoto and Chin 2017). Online communities create local networks of meaning that make sense only to community insiders, but those locally constructed networks of meaning typically rely on broader prior meanings that originate offline. In particular, the distinctive humor of a given online community is often informed in complex ways by offline dynamics (Gal 2019). Arguments about one topic often play out through explicit conversation/jokes about a different subject, using and sometimes widening existing social divides that exist outside the text. Humor is a powerful tool for community building, because "getting the joke" often requires understanding multiple layers of meaning—the spoken text, the implied real meaning or value, and the gap that exists between them—and there are many different ways in which an outsider can fail to participate successfully in the interplay between these many layers. In this way, offline meaning can be introduced to create a specific online context, because the particular connections between what is said and what is meant often involve importing existing distinctions from the meatspace world. Even if a particular online community's humor makes sense only to participants in that community, it typically requires a specific set of knowledge or experiences from the meatspace world to underpin it.

Online communities have also begun to influence or integrate with communities that are constituted offline. The most obvious example of such an online community is NextDoor, which exists to connect users within a given neighborhood to one another and uses strict screening policies to ensure that only residents of a given neighborhood (and residents of neighborhoods immediately adjacent) have access to that neighborhood's shared online space. The premise of NextDoor is that it facilitates communication within an already existing offline community. But as the site's critics have pointed out (Bloch 2021; Kurwa 2019), NextDoor also engages in active "community imagining" work by creating hard boundaries between neighborhoods, boundaries that do not necessarily match the experiences of all or even most of the people living in the area that has been divided up (Lambright 2019). Furthermore, by adding an additional layer of insider-outsider distinction, NextDoor creates the conditions for a heightened sense of anxiety about "people who don't belong." Furthermore, because NextDoor is an imagined community created out of an ostensibly face-to-face one, participants are likely to generalize about their neighborhood as a whole on the basis of those neighbors they know and are more likely to treat as outsiders anyone who does not match that expectation—whether they live in the neighborhood or not. And of course, any neighborhood residents who don't have ready access to the internet—or who are not adept at using it—can end up excluded from a crucial aspect of their community life. NextDoor is a particularly pointed example of how a tool designed without much reflection on pluralities can end up exacerbating existing tensions between those with more access and epistemic power and those with less.

Story Point: "The Regression Test," by Wole Talabi

I'm not sure if A. I.s can believe anything and I'm not supposed to ask her questions about such things, but that's what the human control is for, right? To ask questions that the other A. I.s would never think to ask, to force this electronic extrapolation of my mother into untested territory and see if the simulated thought matrix holds up or breaks down. "Don't tell me what you think. Tell me what you believe."

There is a brief pause. If this were really my mother she'd be smiling by now, relishing the discussion.

Titilope, the narrator of "The Regression Test," hasn't spoken to her mother, Olusola, in nearly forty years—because Olusola is dead. But in her lifetime, Olusola was "Africa's answer to Einstein" as well as Titilope's beloved and complicated mother, and in the years since her death, a digitally preserved copy of her consciousness has continued to serve as science advisor to LegbaTech. In order to ensure that this digital copy is still meaningfully continuous with the person it represents, LegbaTech brings

Titilope in to interview it. During the course of the conversation, Titilope struggles not only with her own grief and longing, but also with the strangeness of this digital preservation, both its similarities and its dissimilarities from her mother. While the mother-daughter relationship in this story undergoes a technological shift that is more dramatic than our current technology makes possible, it nonetheless raises trenchant questions for the present about how we can relate to each other within and across new technological channels and constraints.

5.5.4 ECOLOGY AND THE ENVIRONMENT

Until recently, it was widely accepted that there existed a clear distinction between nature and culture. This notion has been challenged by STS as well as by critical advances in other fields. There are parts of the planet that have been less shaped (or at least, shaped less obviously and directly) by human intervention, and that these areas are different in many ways from highly developed centers of human life. Those less developed areas are often referred to under the broad heading of "the natural world," a term that is also sometimes taken to encompass nonhuman life in urban areas as well. Technology has, of course, been fundamental in the development of cities and other significantly human-transformed places. But technology also plays a role in how we interact with the areas and ecologies that we identify as the natural world.

The relationship between humans, technology, and the natural world is often considered to be a bleak one, and as we discussed in section 5.2.1, it can often be tempting to adopt the position of technological pessimism, that any new or advance in technology must inherently come at the price of either alienation from or destruction of some component of the natural world (Harris et al. 2013). Although the outlook of technological pessimism is helpful in some cases, we also see how advances in technologies can and have improved parts of the natural world (e.g., earth-observing satellites that enable us to see and understand parts of the environment as never before).

Sidebar: The Anthropocene: Naming Human Impact on the Environment
In order to describe the history of the earth, scientists classify periods of time hierarchically as eons, eras, periods, epochs, and ages. For instance, dinosaurs lived in the Mesozoic era, which was divided into the Triassic, Jurassic, and Cretaceous periods. Officially, we still live in the Holocene epoch, which began after the last ice age. But many scientists claim that we are in a new epoch, defined by the effect of human activity on the environment. This is the so-called Anthropocene Epoch, from *anthropo-*, for human, and *-cene*, for new (National Geographic 2019).

Even among the scientists who agree that this is a new geological time period, there is debate about when it began. The Industrial Revolution, in the 1800s, had a significant impact on carbon and methane in the earth's atmosphere. The atomic bombs that were initially tested in 1945 released radioactive particles that appear around the world. According to the Anthropocene Working

Group (Subcommission on Quarternary Stratigraphy 2019), the beginning of this new epoch resulted from "'Great Acceleration' of population growth, industrialization and globalization" in the mid-twentieth century. Note that the definition of a new epoch is based on analysis of geological strata.

Engaging with the Natural World

In the modern era, interacting with the natural world for pleasure and leisure has often been a privilege available to wealthier classes. But advances in communication technologies, shared maps, geotagged images, and other advances have enabled more people to enjoy the natural world than ever before. In the United States, the National Park Service was established in 1916 to preserve, protect, and promote engagement and education with the natural and cultural resources of the United States (National Park Service 2021c). The National Park Service has made improved access for people with differing abilities a strategic goal, and is using new technology such as adding simple structures such as ramps and walkways and developing new tools for camping for those with different abilities (Charitan 2019; National Park Service 2021a, 2021b). This expanded access to the natural world can be understood as a move toward social equality through technological intervention. Even aside from the pleasure or joy of experiencing beautiful places, immersing oneself in nature has been linked to numerous health benefits, including reducing stress and promoting healing (Robbins 2020).

However, with the increased accessibility and tourism into our natural world, issues have arisen. One drawback to all the photos and videos that we share online is that it has increased tourism to parks and other natural locations. A number of park rangers and other groups tasked with managing visitor numbers to parks have observed surges in formerly unknown locations thanks to popularity on Instagram and other social media sites (Djossa 2019; McHugh 2016). In addition to the surging popularity of some areas, some formerly remote or dangerous hikes and experiences are being sought out by people who do not have the experience, equipment, or training to handle the unpredictable nature of these locations. In many parts of the United States, where parks and preserves can be extremely remote, these sudden surges of interest are leading to an overstretching of local, volunteer rescue groups. Furthermore, advances in cell phone technology are impacting the kinds of risks some travelers are willing to take. Some people enter parks underequipped or underprepared and reliant on the idea that they can be rescued from any dangerous situation they fall into by means of a simple phone call, even if that is not the actual case. Many of these resources are overstretched, underfunded, or unable to reach people in a timely manner, leading to dangerous situations for both the hikers and the rescue workers (Watkins 2021).

Another recent development is an increased level of surveillance in less developed spaces, aimed not at human visitors but at the ecosystem itself. This increased monitoring of plants, animals, and ecosystems has created benefits as well as problems.

There are many benefits to continuously tracking animals in ways that we haven't been able to do before. For example, the citizen science project e-Bird (https://ebird.org/) run out of the Cornell Lab of Ornithology (https://www.birds.cornell.edu/) has been and remains one of the largest citizen science projects in the world. Before smartphones, amateur bird watchers submitted their reports of sightings and locations to the lab, which used these reports to, for example, place birds on the endangered species list and show linkages between pesticide use and bird population decline (Welz 2017). The move to an online portal with real-time reporting has increased the scope and impact of the conservation work done by this group. And using advanced technology to track and monitor endangered animals has made them easier for scientists, rescue groups, and tourists to find. By that same token, however, poachers can also find them more easily (Djossa 2019). One website, devoted to increasing sharing of the location of rare animal species is https://www.latestsightings.com/, which posts live photos and locations of animals in Kruger National Park in South Africa. When first deployed, it was meant only to increase visitors' enjoyment of the park, but now there is a tension: does posting the location of animals increase access for poachers, or does it, like open source software, help to prevent poaching by letting park rangers and tourists keep a close eye on animals? (Welz 2017).

Not all of this increased surveillance has been instituted by governments or organizations. Tracking technologies have become so cheap and commonplace that many members of the general public and even tour groups are tagging animals for photo ops and viewing. This can have the downside of habituating animals to too much human contact, making animals more available to poachers. Researchers in conservation and environmental biology are working to establish regulations and understand how the expansion of these technologies may jeopardize or enable future research. As a result of geotagging our photos of cool or rare plants, there has been an increase in collectors and poachers coming into wilderness areas and illegally harvesting the plants for sale on the black market (Cooke et al. 2017).

There is some indication that the overtracking issue could be changing the way research is done in the environmental sciences. Previously, researchers published the locations and detailed photos of the discovery of new species, but now they are intentionally obscuring this information to prevent the possibility of poachers or collectors (Welz 2017). To combat this issue, AI technology originally invented to promote security at airports and other large events has been developed to help park service personnel predict and intercept poachers (Fang et al. 2015; Zewe 2019).

In addition to tagging and tracking physically, the use of earth-observing satellites to monitor and understand the natural environment has led to numerous discoveries and environmental conservation. Much like Facebook identifies photos of you and your friends, the combination of AI with low-cost unmanned aerial vehicles and satellite imagery has enabled scientists to continually monitor and identify individual animals and their migration patterns. These new monitoring practices have led to many important discoveries about mating habits, even leading to the protection of new areas as being critical to a species's survival (Lamba et al. 2019; Shiu 2020). This is perhaps most noticeable in places such as Antarctica and other areas where it is hard or impossible to continuously monitor animals using attached trackers, or where humans can live only part of the year (Borowicz et al. 2018).

Using Natural Resources

In addition to being a source of pleasure or inspiration, the natural world can be treated by humans as a resource, supplying us with food and with the raw materials for various technological productions. In 2015, the United Nations adopted 17 sustainable development goals that outline a pathway toward shared prosperity for both people and the planet. These goals include things like access to clean water, the safety of life on land and below water, climate protection, and food access. Many technology development groups have integrated these goals into long-term plans for development and research. One of these areas is known as *computational sustainability*, which focuses on projects that encourage computer scientists to play a role in addressing the societal and environmental challenges outlined in the goals (Gomes et al. 2019). One overwhelming problem is monitoring the growth of forests and the use of natural resources all over the world. Here again we see that the use of earth-observing satellite imagery combined with advances in machine learning and image processing to estimate land use, crop growth, and other outcomes have paved the way for more informed policy related to the UN's sustainable development goals (Burke 2021). These large-scale changes via technological intervention, often called Earth system interventions, present both a risk and an opportunity for sustainability and welfare objectives (Reynolds 2021).

Of particular concern is energy production. The technological artifacts that that we often associate with technology development have used massive amounts of energy for decades. From coal-fired steam engines to gasoline for our cars, new technology often needs a power source that, more often than not, is extracted from the environment. Although there is a move toward more renewables and batteries, this will not solve all our problems. In addition, the change of technology from large, centralized power plants to local solar and wind farms requires a fundamental rethinking of the overall technological system of energy delivery: from extracting from a concentrated source and distributing outward, to collection at distributed points to power concentrated centers (Maly 2015). One aspect of the natural world that is gaining increased attention is how reliant on extraction we are to build and maintain all the electronic devices we use, including batteries, cell phones, cars, and the other conveniences of daily life. Each cell phone or screen you use contains a multitude of precious metals including gold, platinum, and cobalt. Many of these materials come from hard-to-reach places or war-torn countries, or must be mined from virgin sites. The increased turnover of our devices and the need for ever more electronics (e.g., for electric cars) have increased demand for these precious materials (Nogrady 2016). While US recycling is at an all-time high, we are also generating more waste than ever before (Environmental Protection Agency, n.d). Although there is technology that can help with some of these problems, such as new energy capture mechanisms, it's important not to fall into the solutionism trap—after all, a "the solution must be new machines" mentality is what helped create many of the problems to begin with.

Perhaps the most important way that many of us think about ecology and the environment is as the source of our food. Agriculture and farming is perhaps one of the oldest forms of technology and is often credited with leading to the establishment of early civilizations (Diamond 1999) (although recent scholarship calls into doubt whether farming actually was better for most people [Scott 2017]). Humans have long sought to tame and transform the natural world in both constructive and damaging ways to grow crops, using techniques ranging from the regulation of water supplies with dams to enriching the ground with natural fertilizers such as manure and other chemicals (Harford 2017). In more recent times, agriculture technology, including the Green Revolution, has led to the modern day when we produce more food, with fewer people involved in production, and (debatably) smaller environmental impacts than ever before (Evenson and Gollin 2003). As part of this shift, and in an effort to meet some of the sustainable development goals, inventions such as genetically modified Golden Rice that can produce vitamin A have been used in many places to lower infant mortality and blindness (NYU Langone Health, n.d.). However, the question of genetically modified crops and other technology for food production is a contentious issue, with many questions related to seed ownership, patent lock-in, and possible harm to the natural world (Barlett and Steele 2008). We cannot hope to do the discussion of the impact of farming and technology on the natural world justice in this book, but it is an area ripe for discussion about the impacts of technological systems on the natural world (National Research Council 2010).

Story Point: "The Gambler," by Paolo Bacigalupi

"I mean, isn't that an old story?" She sips her champagne. "Even America is reducing emissions now. Everyone knows it's a problem." She taps her couch's armrest. "The carbon tax on my limo has tripled, even with the hybrid engine. Everyone agrees it's a problem. We're going to fix it. What's there to write about?"

Ong is a reporter with a mission: he wants to educate the public about climate change, and call attention to how corporate exploitation and government corruption have hastened the destruction of the natural world. Unfortunately for Ong, the American public is more interested in reading celebrity gossip and reviews of the newest gadgets. Even worse, the media company he works for has embraced the infotainment model and is threatening to fire him if he cannot improve his numbers. When a generous coworker sets Ong up with a flashy celebrity interview, Ong struggles with several moral quandaries that plague many environmentalists: what are the costs of commodifying a thing you consider valuable for its own sake, not just for the pleasure that it can offer others? Is it possible to introduce that valuable thing to others without putting its integrity at risk? And, above all else: how do you convince people not to look away?

5.5.5 STATE POWER AND FORCE

Of the many ways in which society has been reshaped by recent technological developments, perhaps the most controversial of these concern security and safety. Technological advances have made it easier than ever for bad actors to cause harm or chaos, which means that the institutions tasked with protecting a society and its citizens need to both innovate and escalate. And yet, while safety and security are by themselves inarguable social goods, there exists a great deal of debate about how these goods should be balanced against the costs that they exact from both the individuals and the broader society they seek to protect; among the costs are privacy and freedom of expression. These costs are disproportionately borne by communities that are already minoritized and vulnerable, precisely because of existing biases against them. There has been a movement among technologists to mitigate that bias through automation and algorithms. But as we discover, such algorithms must be implemented with significant forethought and care; otherwise they will usually end up exacerbating the problems that they aimed to fix.

Policing

The most visible aspect of a typical state security apparatus is its police force. In democratic societies, the understood role for the police is to enforce laws, control crime, and support communities (Bittner 1970). In practice, policing often diverges from this model. In the United States, police violence—particularly against communities of color—has been a site of significant public upheaval and has led to calls for significant police reform or even for the defunding of the police in favor of alternative intervention and support services. Nevertheless, the police remain central to the US social structure and its security practices.

Policing scholar Alex Vitale has argued that police themselves suffer under the burden of being asked to do too much (Vitale 2021). Police work necessarily requires individual officers to exercise discretion about how to fulfill their mandate, and sometimes this includes making split-second decisions to protect themselves or others, sometimes through violence. But as the purview of policing has expanded to incorporate everything from wellness checks to crisis management, even police officers committed to the public well-being often find themselves overwhelmed. Additionally, the history of biased and racist policing has created patterns that are challenging to disrupt even for individual officers who do not wish to perpetuate them.

Many police departments have sought to address both of these problems—the heavy decision-making burden on individual officers and the history of institutionalized bias—by offloading some of this discretion to algorithms, on the grounds that algorithmic decision making will decrease the bias that creeps in through human decision makers. But though the algorithms are trained on data from actual events, this data is flawed, incomplete, or outright misleading in a variety of ways. Consider, for instance, the widely used practice of "predictive policing," in which algorithms are used to determine which areas of a given jurisdiction should be patrolled most heavily based on past data (Angwin et al. 2016). Other systems scrape social media and use social network analysis to pinpoint individuals and groups that are seen as statistically likely to be victims or perpetrators of crimes (Winston 2018).

This reliance on data may seem impartial, but the training data for these algorithms necessarily reflects entrenched historical biases and unjust patterns of policing, because the only training data available consists of *reported* crimes. Any bias that has driven police to arrest some people and not others (for instance, on the basis of race, or living in a "bad" neighborhood) will be replicated in the algorithm (see, e.g., O'Neil 2016 and our discussion in chapter 3). Furthermore, training on reports of crimes could leave immigrant neighborhoods largely unpoliced, because immigrants are among the populations that report crime at a lower rate (Shapiro 2017). This problem is larger than crime data; some bail-setting systems, trained on bail decisions from a racially biased court system, have been shown to judge individuals as risky primarily on the basis of race (Angwin et al. 2016).

Surveillance

The social theorist and philosopher Michel Foucault (2008) describes two settings with massive surveillance, with very different goals: to eradicate a pandemic and to maintain control of an imprisoned population. The first setting was European cities in the seventeenth century, during the bubonic plague, when people were locked into their homes and checked (from the windows) for plague symptoms. Foucault contrasts this temporary state of discipline with that of what he calls the *carceral society*, which is centered (theoretically, if not geographically) around the *Panopticon*, as imagined by utilitarian thinker Jeremy Bentham: a vision or plan for a prison in which prisoners are each isolated in a cell from which they can see and monitor each other, and such that a guard (or visitor) in a central tower can see all of them, though they cannot see into the guard tower—much like modern prisons, in which guards sit in a protected space, watching video from cameras throughout the institution. The visibility of prisoners and their uncertainty (in Bentham's model) about when they are being watched causes prisoners to self-police. For Foucault, the prison building itself, the actual architecture, was construed as a technological mechanism. Although prisons seem less technological than iron working, both change society in wide-ranging ways. Foucault points out that constant, anonymous surveillance can be used in other institutions, including military barracks, orphanages, and schools. More importantly, as we saw in our examination of prisons in chapter 4, the knowledge that the Panopticon exists, its threat, causes individuals to self-police, to behave in ways that will not offend those with the power to incarcerate. That same uncertainty about surveillance has become part of our society, where cell phones, drones, and web cams (including baby monitoring and doorbell cams) function as the Panopticon.

In the United States, we have many forms of surveillance aimed at regulating particular behaviors, such as security cameras in shops, banks, and so on, to prevent theft; traffic cams, to prevent speeding, running red lights, and other violations; and software for proctoring online exams. One interesting finding is that red light cameras, which sit above traffic lights, decrease the number of drivers that run red lights, but increase rear-end collisions (Cohn et al. 2020)—an interesting example of a ripple effect! Benjamin (2019) argues that all these technologies, as currently used, reinforce the power structures already in place. For instance, cameras in shops are justified to stop shoplifting and employee misbehavior, but end up creating a more hostile and micromanaged workplace. This is very much part of digital Taylorism (which is discussed in the sidebar in section 5.4.3). Video doorbells are marketed as a means of tracking home intruders and porch pirates, but they can also be used to monitor an individual resident's comings and goings, which can reinforce control over the resident. Various forms of tracking, through cell phones, webcams, social media "check ins" or images, and GPS devices can be used by stalkers (Messing et al. 2020).

Recent research has demonstrated that the modern carceral system serves several purposes beyond incarceration itself, including elimination of populations. Cisneros (2016) describes the effects of immigrant detention centers in France, which, contrary to international law on the treatment of refugees, serve to break the spirits, cultural traditions, and health of the detained. These centers, and various governments' policies of breaking up families and sending children to live with current citizens' families, echo different governments' deliberate attempts to wipe out Indigenous cultures in Australia (Palmiste 2008), Hawai'i (Nee-Benham 1998), and the land that is now the continental United States (Adams 1995; Piccard 2013). Looking at the effect of policing on Black and Latino boys, Rios (2011) argues that the boys were not reformed by the experience of the panoptic prison system. Rather, "it stripped them of their dignity and humanity by systematically marking them and denying them the ability to function in school, in the labor market, and as law-abiding citizens" (Rios 2011, 88). Malkia Amala Cyril (2015) describes the continuing systemic and systematic denial of Black people's dignity by the American government, and the ways, such as gang databases and drone surveillance, in which technology has increased surveillance and intrusion into the lives of people of color. As we discussed in chapter 4, prison systems can function either as sites of reform or of punishment, denying privacy as a means of reducing personhood.

The surveillance of the Panopticon is also supplemented by what researchers call "sousveillance." The word surveillance, as Poster (2019) points out, literally means "watching from above." In the Panopticon, prisoners are watched by everyone, including each other. The term sousveillance, by contrast, refers to watching from below, as a means of reflecting back the surveilling gaze. "Sousveillance focuses on enhancing the ability of people to access and collect data about their surveillance and to neutralize surveillance" (Mann et al. 2003). One can see the push for police to wear body cams as a form of sousveillance. Poster uses the term "multi-veillance" to describe the particularly technologically mediated ways in which people monitor one another via technologies like social media and web cams, and to police one another's behavior and expression through comments sections, and apps that provide ratings opportunities. The pressure that multi-veillance creates on individuals to live up to social norms has effects that include depression, anxiety, and extremism (Bright 2018).

War

Another major dimension of state power can be seen when states act against other governments—through espionage, propaganda, economic attacks, and sometimes acts of war—on the grounds that such actions are necessary to protect their own citizens. It is important to note that the apparatus of these conflicts has significant internal effects on a society as well. Sometimes, these internal effects are caused when technologies are developed for military purposes and are later repurposed for non-military use. But it is also the case that governments sometimes quietly deploy these techniques against their own citizens.

One technique of warfare that is frequently deployed at home is *information warfare.* Information warfare can take the form of both propaganda operations (the pushing of explicit messages) and chaos-producing operations (which sow confusion through a flood of contradictory messaging) (Lin and Kerr 2019). These techniques can affect individuals, for instance, by convincing them to vote against their best interests (Woolley 2016), or by deflecting their attention from something that will affect them later. These can be carried out by governmental agencies, by individuals or hacktivist collectives operating in deniable concert with State authorities, or independently. Recently, we've also seen such ransomware attacks on infrastructure by groups, some of which are rumored to be associated with foreign governments (e.g., see Sanger et al. 2021). Though these efforts rely on recent forms of technology, propaganda as such is not new. Governments have long used incentives like food or social prestige to incentivize their audience to sign on to particular political positions. In addition, they have used available communications technologies to spread false, slanted, or incomplete information, and when dissenters have circulated conflicting information, they have attempted to disrupt that circulation in whatever way that communication technology requires, such as shutting down radio broadcasts, collecting and destroying paper pamphlets, or arresting public protestors. Yet the speed and scale of ICT makes it an especially effective vehicle for propaganda. The internet allows for an unprecedented volume of bad information to be shared anonymously or pseudonymously. We have seen discussion of the role of social media companies in distinguishing the truthfulness of posts and claims that algorithms can take on that task (see, e.g., Kahn 2021). Algorithms can be useful if they are updated regularly to keep up with the evolving tactics of malicious posters, but we should question whom those algorithms will serve.

Recent developments in warfare technology—including communications technologies and many forms of automation—have shifted the experience of warfare, both for those who have access to those technologies and for their targets. Earlier in this section we mentioned predictive policing, in which a neighborhood that is assumed (by humans, and the algorithms trained on human data) to be the locus of more crime becomes more surveilled. In modern warfare, drones are used analogously to human police officers, to surveil putative criminals, and (after a period of observation), to kill some of the surveilled individuals. This leaves a population terrorized by targeted killings, and "collateral damage" of civilian deaths and injuries (Wall and Monahan 2015). Individuals and communities indeed change their behaviors, going out less and gathering less, and often developing further reasons to hate the government that has sent the drones (Coyne and Hall 2018).

The panoptical gaze, whether from prison guards or via drones, is designed to enforce certain behaviors, to keep particular groups subjugated and anxious. Being surveilled is harmful to those being surveilled, as noted above, but the act of surveillance is often harmful to those who watch as well. While there has been literature on stress in prison guards (Keinan and Malach-Pines 2007), the growing body of work linking drone control and PTSD shows ways that technology can create new stressors for the cogs in the surveillance machine (Enemark 2019). The story "Codename: Delphi," by Linda Nagata, included with this textbook engages with this question very directly.

The rhetoric of protection is one of boundaries: between lawful citizens and law-breakers, between inmates and unconstrained citizens, between citizens of our country and our enemies, and between the ill and the healthy. Technology has been presented as a means of distinguishing between lawful and unlawful, as if video of an act (for instance, without context) is sufficient to judge that act. Consider, for instance, a video of a confrontation that omits the initial provocation that led to anger or violence. Zooming out to show a mob, armed militias, or someone in danger, can alter our judgment, as can the backstory for the individuals involved in the confrontation. As we discussed in chapter 3, video data is not information. It requires interpretation and context. The discussion in section 3.2.3 reminds us to think about who does that interpretation. What assumptions do the people or entities who surveil hold about those who are surveilled, and how does their culture of knowledge influence which data is accrued for interpretation?

Story Point: "Asleep at the Wheel," by T. Coraghessan Boyle

She's smiling as he comes up to the car, and he's smiling, too, and now he's reaching for the door handle . . . but the door seems to be locked, and she's fumbling for the release. "Carly," she says, turning away from the sight of his face caught there in the window as if Carly were an actual person sitting in the driver's seat, when, of course, there's no one there. "Carly, is the child lock on?"

"I'm sorry," Carly says, "but this individual is untrustworthy. Don't you recall what happened last Tuesday evening at 9:19 P.M. in front of the S.P.C.A. facility at 83622 Haverford Drive?"

"Carly," she says, "open the door."

Cindy works as an advocate for the homeless, and her job keeps her busy enough that she relies heavily on the AI of her autonomous vehicle to manage her life for her. The latest advances in AV technology have caused some serious problems for Cindy's clients, including her favorite client, Keystone. But for the most part, Cindy herself is happy to let the car keep track of her work calendar, her shopping preferences, and her teenage son Jackie's movements. What does bother Cindy is when the seamless management by the car's AI crosses into outright interference with Cindy's goals and desires, like spending more time with Keystone. This interference is a logical extension of what personal AVs have been designed for: it's just another form of keeping users "safe." But "safety" can encompass a lot of things, and is sometimes in tension with our immediate pleasure or even our long-term goals. Cindy's dissatisfactions, and her mixed feelings about the car on which she depends so heavily, urge us to consider what it costs us to enmesh ourselves into convenient, automated systems that keep us safe from ourselves.

5.5.6 WORK AND LABOR

Many technologies have been characterized as "labor-saving" and have indeed reduced the amount of time it takes for a worker to complete a particular task. Yet for all our massive advances in delegation of work to machines, we still have an economy in which most people spend a large amount of their time working, though the *nature* of that work has evolved. A work-related technology should not be viewed merely as an eliminator of human labor but rather should be examined in the wider context of the society that it shapes.

Automation

Consider an historical example: In the earliest days of the United States, cotton-growing was not the huge enterprise it would later become. In the process of turning raw cotton into cloth, there was a time-consuming step of extracting the fiber from the cotton seeds by hand. Then in 1793, the invention of the cotton gin greatly sped up the extraction process. With this manufacturing bottleneck eliminated, there was suddenly a much higher demand for raw cotton. The next few decades saw a dramatic growth in cotton farming in the Southern states, powered by a dramatic growth in the use of slave labor (Bailey 1994). Although the practices of cotton farming itself did not change directly because of the invention of the cotton gin, there was a huge indirect impact. It was, in fact, *because* there was no commensurate increase in the efficiency of farming that using the labor of enslaved people became more profitable.

This example illustrates how technologies and labor practices form a complex system in which it is difficult to predict how change in one area will affect others (e.g., will cause a ripple effect). It also illustrates how economic opportunity does not always translate to an improvement in human welfare. We will examine contemporary developments through the lens of traditional employment, whereby a worker has an ongoing agreement with an entity to provide a service in exchange for predictable payments. This may be the most familiar way of arranging labor for many people in the twentieth and early twenty-first centuries, but its role in the future is uncertain; automation is affecting the makeup and availability of jobs and the internet is enabling new ways to seek income outside of employment.

The future effects of artificial intelligence, robotics, and other automation technologies on jobs are hard to predict. Some economists argue that automation may increase employment by, in a ripple effect similar to the cotton gin's, allowing previously unmet demand for products to be addressed (Autor 2015; Bessen 2018). But a 2018 survey of experts' predictions showed that there was no consensus over whether automation would result in a net creation or net destruction of available jobs over the coming decades, and to what extent (Winick 2018).

As an example of the relationship between automation and employment, consider a 2017 study on the effects of industrial robots in Germany, one of the leading countries in robotization (Dauth et al. 2017). The study suggests that industrial robots did not cause a net reduction in the overall number of jobs, but they changed which jobs were available: in robot-exposed manufacturing sectors, there were fewer new jobs created, but higher rates of job retention for workers who were already in the industry. Although robots drove a net loss of manufacturing jobs, they also drove a net gain in service-sector jobs. In addition to job availability, robots affected wages in asymmetric ways, with higher pay for some occupations (e.g., scientists and managers) and lower pay for others (e.g., machine operators). Greater productivity from robots led to greater per-worker profitability, but a lack of average increase in wages suggests that the main beneficiaries of this productivity were investors and not the labor class itself.

A key distinction is that "tasks are automated, not jobs" (Bessen et al. 2020). Rarely does a technology replace every function of a job at once; instead, the parts of the workday originally spent doing a task manually may be spent interacting with a machine to complete that task or may be freed up for a different kind of task. This shift may be rewarding for workers if it eliminates tedium and allows them to work on more engaging tasks; workers who enjoy tasks requiring social skills or creativity may particularly flourish, as these capacities have been among the hardest to automate.

However, a technology that seems empowering in isolation can sometimes *degrade* rather than enhance workers' capabilities because of the way they are integrated into existing systems. One manifestation of this effect is "deskilling," the erosion of workers' rich sets of individual skills and experiences as their work is restructured in a way that gives them less discretion and autonomy (Hoff 2011). For instance, some commentators have partially blamed a number of contemporary air disasters on the technology-enabled deskilling of pilots; they argue that heavy automation of airplane controls, and reliance on predictable simulations for training, have reduced the depth of skill needed to become a pilot. From a safety perspective, this has left pilots unprepared for novel situations when the automated controls break down; from a social perspective, it has left some pilots feeling like piloting as a professional category has been diminished (Langewiesche 2019).

Remote and Platform Work

Up to now, we have presupposed that employment is the main organizing structure for human labor, but this is by no means a permanent reality, especially with the emergence of digitally mediated *platform work*; as of 2022, people could make money by, for instance, providing freelance transportation services through Uber or Lyft, completing crowdsourced small tasks on Amazon's Mechanical Turk, or gaining sponsorships as social media "influencers." Although some of these types of work are new, their structure is similar to the offline contract work that had already been on the rise beforehand, and they inherit some of the same debates about to what extent they are empowering or disempowering for workers. On the one hand, they offer lower barriers to entry than traditional employment and greater worker control over the time, place, and amount of work; but on the other hand, they can leave a worker with fewer protections and benefits than traditional employment (e.g., in the United States, companies must offer health insurance for certain types of employee, but not for contractors) and fewer communication pipelines for the worker to influence how the company is run (Vallas and Schor 2020).

Digitization has also changed traditional employment by enabling *remote work*, from answering work emails on a phone after coming home from an office to doing the entire job from a home PC in place of the office. As we discussed in the section "What's the Response, and Why," this has led to a change in how we organize ourselves and how close one must live to their place of work. A common thread in many kinds of remote work and platform work is constant connectivity (Turkle 2008): the notion of a neat division of the worker's time into "working hours" and "personal life" has eroded. The availability of work in new times and places can open opportunities for workers who would otherwise have difficulty commuting, such as those who live in remote rural areas, those who cannot afford a car they would need to get to work, and those who cannot drive due to disabilities (Vallas and Schor 2020). However, the increased availability of the worker, plus established ideals about how a diligent worker should behave, can combine to create increased pressure to work longer hours than before and create stress when job expectations interrupt a worker's family life (Stich 2020). This phenomenon is common enough that psychologists have given it a name, "telepressure" (Barber and Santuzzi 2015).

Telepressure may be amplified by a worker's own self-image but also by the need to give a good impression to a supervisor. In a traditional office, managers' access to workers' activities comes naturally from proximity. The classical "scientific management" model that we described in section 5.4.3, in our sidebar on digital Taylorism (Drury [1922] 1968), encourages managers to use any opportunity for surveillance to its fullest extent, under the assumption that workers will behave lazily whenever they can escape accountability. We have seen this notion of surveillance in the workplace before in chapter 4, when we discussed open offices, as well as some of the complications of any type of surveillance in "Surveillance" in section 5.5.5. In remote work this surveillance takes on different qualities: because workers are often not observable by default and many digital surveillance measures can be circumvented, workers are incentivized to make an active effort to be noticed—both by managers, to have accomplishments credited, and by peers, to try to replicate the community structure of the traditional office (Hafermalz 2020). In such a setting, unlike under traditional surveillance, workers can curate what is observed about them.

Story Point: "Lacuna Heights," by Theodore McCombs

> I got a ride to Noe Valley from a sunken-eyed kid with a hipster chignon and wisps of blond beard floating around his chin. With me in the back seat were grocery deliveries for some next gig: a fresh baguette that filled the electric car with the smell of hot bread, a giant jackfruit that must have weighed twenty pounds and cost more than that day's commissions. These were the people I'd forgotten, daily, nightly, the poor kids running the city out of the corners of my eyes.

Andrew is a wealthy San Francisco lawyer, and his wife Madeleine is a musician who comes from money. Though Andrew spends the story in a state of emotional distress about how he uses his neural implant, this anxiety plays out on a backdrop of easy material comfort: his laundry is done for him, his groceries and hair products are delivered to his doorstep, and he is transported wherever he needs to be without the unpleasantness of traffic jams on the surface. Andrew himself doesn't pay much attention to these conveniences, preferring instead to obsess about what part of his own life he has hidden from himself in Aleph's privacy mode; it does not occur to him (nor possibly to us, until we join him in privacy mode in the final scene) that these things are related. Although the gig work technologies that keep this future San Francisco running remain largely unseen, "Lacuna Heights" gives us an up-close look at how knowledge- and attention-management technologies like the Aleph—or even modest, now-ordinary technologies like a newsfeed-organizing tool or a tag blocker—can shape how we view labor . . . or how we don't.

5.6   CLOSING THOUGHTS: MAINTAINING A BROAD VIEW

As we have seen in this chapter, technology and society exist in an inseparable mix, making it impossible to discuss technology apart from the context in which it is used. We saw a number of techniques that you can use when analyzing and discussing technical systems to always maintain a critical view of the role of technology in our lives. We then walked through spheres of context including friendship, care, security, discourse, work, and the environment, and discussed how technology is shaping or reshaping many of the concerns in these areas.

An important theme throughout the book, and especially in this chapter, is that communication and education are two of the main techniques that we can use to come to understand and analyze sociotechnical systems. Access to information, who can interpret it, how it is presented through our institutions, and how it is communicated to us, all inform our perspective on these debates. In chapter 6 we give more concrete guidance as we consider the role of the professional technology developer in the broader context of society.

REFLECTION QUESTIONS

1. *How do you perceive the nature of the relationship between technology and society? How has this perception changed (or not changed) after reading this chapter?*

2. *Pick one of the spheres detailed in the chapter and formulate some additional problems that arise at the social level with respect to technological change and development. How do our perceptions and definitions of what technology is or can be change the questions we ask? Use at least two definitions of technology to answer this question.*

3. *In section 5.4.3, "Strategies for Analysis," we discussed the idea of technology development as social experiment. Although we hinted at several ways in which this view can go awry, we left concrete instances out. As a reflection, find an example of a new technology in which the experiment was not well thought through. Identify the intervention, subjects, and any problems that arose from this experiment.*

4. *During times in the SARS-COVID-19 pandemic, many schools at both the K–12 and college levels went online. This was done abruptly in the spring of 2020 and left many students and families potentially disconnected due to the digital divide. Additionally, it required many families to take on support roles typically furnished by schools, including providing technical support, supervision of the learning process, and preparing lunches for students. Use the critical questions from section 5.4.3, including the ideas of pluralities, interconnections, and responses, to discuss this process both at the level of the individual family and the level of local decision makers (e.g., the government officials or school administration). What happens if we analyze the rise and development of massively open online courseware using the same questions?*

5. *Recall our discussion about the history of photography and the biases that once encoded in a piece of technology can perpetuate into the future. Find an example of another technology that encodes particular biases or ways of thoughts. What are these? How do they continue to impact society today? Are there any of the tools from this chapter that would help address these issues during the technology development process?*

6. *Consider some of the "societies" of which you are a part: a family, a school, a sports team, a state, religious congregation, or a country. How does your role in one or more of these societies change how you perceive a particular piece of technology? For example, as an individual you may not like the idea of having a biometric identification for your cell phone, but as a member of a state, you may think that the government should store biometric information on all its citizens. Use the tools from*

this chapter to take a piece of technology and analyze it from two or more societal views.

7. *Give original examples of common traps in thinking when participating in technology development including the formalism trap, the framing trap, the portability trap, the ripple effect trap, and the solutionism trap. Evaluate a technological system in light of these traps and identify issues arising from each.*

BACKGROUND REFERENCES AND ADDITIONAL READING

Much of our thinking in this chapter has been influenced by the ideas of social constructivism and the social construction of both technology and knowledge in general. Additional readings on these topics include Bijker et al. (2012) and Winner (1993a).

A great starting point for the history of technology, its definitions over the years, and a more complete modern definition, is discussed by both Schatzberg (2018) and Jasanoff (2017). Additional information about problems in design can be found at

- AI Incident Database from the Partnership on AI: https://www.partnershiponai .org/aiincidentdatabase/.

- Dark Patterns: Dark patterns list and other stuff from Princeton's web transparency project: https://webtap.princeton.edu/.

For more on bioethics and technology development, the NYU Center for Bioethics has a number of very engaging case studies. See NYU Langone's High School bioethics project, "Genetically Modified Organisms: The Golden Rice Debate," accessed June 2, 2021, https://med.nyu.edu/departments-institutes/population-health/divisions -sections-centers/medical-ethics/education/high-school-bioethics-project/learning -scenarios/gmos-the-golden-rice-debate.

For more on the interaction of assembly lines and Taylorism, see https://hbr.org /1988/11/the-same-old-principles-in-the-new-manufacturing.

For more on how surveillance can compound existing inequalities, particularly through state-sanctioned violence, see Human Rights Watch. 2019. "Get on the ground!" Policing, poverty, and racial inequality in Tulsa, Oklahoma: A case study of US law enforcement." September 12. https://www.hrw.org/report/2019/09/12/get-ground -policing-poverty-and-racial-inequality-tulsa-oklahoma/case-study-us.

For a historical look at the philosophy of technology, see Mitcham, C. 1994. *Thinking through Technology: The Path between Engineering and Philosophy*. University of Chicago Press.

REFERENCES CITED IN THIS CHAPTER

Abdi, J., A. Al-Hindawi, T. Ng, and M. P. Vizcaychipi. 2018. Scoping review on the use of socially assistive robot technology in elderly care. *BMJ Open* 8, no. 2: e018815. https://pubmed.ncbi.nlm.nih.gov/29440212/.

Abebe, Rediet, Solon Barocas, Jon Kleinberg, Karen Levy, Manish Raghavan, and David G. Robinson. 2020. Roles for computing in social change. In *FAT '20: Proceedings of the 2020 Conference on Fairness, Accountability, and Transparency*, 252–260. Association for Computing Machinery.

Abernathy, Penelope Muse. 2018. *The Expanding News Desert*. Center for Innovation and Sustainability in Local Media, School of Media and Journalism, University of North Carolina at Chapel Hill.

Ackerman, M. S. 2000. The intellectual challenge of CSCW: The gap between social requirements and technical feasibility. *Human-Computer Interaction* 15, no. 2–3: 79–203.

Adams, David Wallace. 1995. *Education for Extinction: American Indians and the Boarding School Experience, 1875–1928*. University Press of Kansas.

Anderson, Catherine L., and Ritu Agarwal. 2011. The digitization of healthcare: Boundary risks, emotion, and consumer willingness to disclose personal health information. *Information Systems Research* 22, no. 3: 469–490.

Angwin, Julia, Jeff Larson, Surya Mattu, and Lauren Kirchner. 2016. Machine bias. ProPublica, May 23. https://www.propublica.org/article/machine-bias-risk-assessments-in-criminal-sentencing.

Atasoy, H., B. N. Greenwood, and J. S. McCullough. 2019. The digitization of patient care: A review of the effects of electronic health records on health care quality and utilization. *Annual Review of Public Health* 40: 487–500.

Austin, John Langshaw. 1975. *How to Do Things with Words*. Oxford University Press.

Autor, David H. 2015. Why are there still so many jobs? The history and future of workplace automation. *Journal of Economic Perspectives* 29, no. 3: 3–30.

Bailey, Ronald. 1994. The other side of slavery: Black labor, cotton, and textile industrialization in Great Britain and the United States. *Agricultural History* 68, no. 2: 35–50.

Baker, Peter, and Sabrina Tavernise. 2021. One legacy of impeachment: The most complete account so far of Jan. 6. *New York Times*, February 13. https://www.nytimes.com/2021/02/13/us/politics/capitol-riots-impeachment-trial.html.

Balcerowska, Julia M., Piotr Bereznowski, Adriana Biernatowska, Paweł A. Atroszko, Ståle Pallesen, and Cecilie Schou Andreassen. 2020. Is it meaningful to distinguish between Facebook addiction and social networking sites addiction? Psychometric analysis of Facebook addiction and social networking sites addiction scales. *Current Psychology* 41: 949–962.

Barber, L. K., and A. M. Santuzzi. 2015. Please respond ASAP: Workplace telepressure and employee recovery. *Journal of Occupational Health Psychology* 20, no. 2: 172–189. https://doi.org/10.1037/a0038278.

Barlett, Donald L., and James B. Steele. 2008. Monsanto's harvest of fear. *Vanity Fair*, May. https://www.vanityfair.com/news/2008/05/monsanto200805.

Barocas, Solon, and danah boyd. 2017. Engaging the ethics of data science in practice. *Communications of the ACM* 60, no. 11: 23–25. https://cacm.acm.org/magazines/2017/11/222176-engaging-the-ethics-of-data-science-in-practice/fulltext#R1.

Bates, Richard. 2020. Florence Nightingale: A pioneer of handwashing and hygiene for health. *The Conversation*, May 23. https://theconversation.com/florence-nightingale-a-pioneer-of-hand-washing-and-hygiene-for-health-134270.

Benjamin, Ruha. 2019. Introduction: Discriminatory design, liberating imagination. In *Captivating Technology: Race, Carceral Technoscience, and Liberatory Imagination in Everyday Life*, edited by Ruha Benjamin, 1–22. Duke University Press.

Bessen, James. 2018. AI and jobs: The role of demand. Working paper no. w24235. National Bureau of Economic Research.

Bessen, James, Maarten Goos, Anna Salomons, and Wiljan van den Berge. 2020. *Automation: A Guide for Policymakers*. Brookings Institution. https://www.brookings.edu/wp-content/uploads/2020/01/Bessen-et-al_Full-report.pdf.

Biddle, Sam. 2020. Google AI tech will be used for virtual border wall, CBP contract shows. *The Intercept*, October 21. https://theintercept.com/2020/10/21/google-cbp-border-contract-anduril/.

Bijker, Wiebe, Thomas P. Hughes, and Trevor Pinch, eds. 2012. *The Social Construction of Technological Systems: New Directions in the Sociology and History of Technology*. MIT Press.

Bittner, Egon. 1970. *The Functions of the Police in Modern Society: A Review of Background Factors, Current Practices, and Possible Role Models*. National Institute of Mental Health, Center for Studies of Crime and Delinquency. https://www.ncjrs.gov/pdffiles1/Digitization/147822NCJRS.pdf.

Bloch, Stefano. 2021. Aversive racism and community-instigated policing: The spatial politics of Nextdoor. *Environment and Planning C: Politics and Space*, May 31. https://doi.org/10.1177/23996544211019754.

Bogost, Ian. 2020. Your phone wasn't built for the apocalypse. *The Atlantic*, September 11. https://www.theatlantic.com/technology/archive/2020/09/camera-phone-wildfire-sky/616279/.

Bohman, J. 2008. The transformation of the public sphere: Political authority, communicative freedom, and internet publics. In *Information Technology and Moral Philosophy*, edited by J. Van den Hoven and J. Weckert, 66–92. Cambridge University Press.

Borowicz, Alex, Philip McDowall, Casey Youngflesh, Thomas Sayre-McCord, Gemma Clucas, Rachael Herman, Steven Forrest, Melissa Rider, Mathew Schwaller, Tom Hart, et al. 2018. Multi-modal survey of Adélie penguin mega-colonies reveals the Danger Islands as a seabird hotspot. *Scientific Reports* 8, no. 1: 1–9. https://www.nature.com/articles/s41598-018-22313-w/.

Bourdieu, Pierre. 2018. *Distinction: A Social Critique of the Judgement of Taste*. Routledge.

Box, George E. P. 1979. Robustness in the strategy of scientific model building. In *Robustness in Statistics*, edited by Robert L. Launer and Graham N. Wilkinson, 201–236. Academic Press.

Bright, Jonathan. 2018. Explaining the emergence of political fragmentation on social media: The role of ideology and extremism. *Journal of Computer-Mediated Communication* 23, no. 1: 17–33.

Broadbent, E., R. Stafford, and B. MacDonald. 2009. Acceptance of healthcare robots for the older population: Review and future directions. *International Journal of Social Robotics* 1, no. 4: 319. https://link.springer.com/article/10.1007/s12369-009-0030-6.

Bruce, Tammy. 2019. Microsoft wants you to use approved political speech—This is a real threat to our freedom. Fox News, May 16. https://www.foxnews.com/opinion/tammy-bruce-microsoft-uses-artificial-intelligence-to-push-what-liberals-think-you-should-write.

Bruckner, Matthew A. 2019. Preventing predation & encouraging innovation in Fintech lending. *Consumer Finance Law Quarterly Report*, November 6. Updated February 10, 2021. https://papers.ssrn.com/sol3/papers.cfm?abstract_id=3406045.

Buntin, Melinda Beeuwkes, Matthew F. Burke, Michael C. Hoaglin, and David Blumenthal. 2011. The benefits of health information technology: A review of the recent literature shows predominantly positive results. *Health Affairs* 30, no. 3: 464–471.

Burke, Marshall, Anne Driscoll, David B. Lobell, and Stefano Ermon. 2021. Using satellite imagery to understand and promote sustainable development. *Science* 371, no. 6535. https://pubmed.ncbi.nlm.nih.gov/33737462/.

Butler, Judith. 1990. *Gender Trouble: Feminism and the Subversion of Identity*. Routledge.

Butler, Judith. 1999. Performativity's social magic. In *Bourdieu: A Critical Reader*, edited by R. Schusterman, 113–128. Blackwell.

Campbell, W. Joseph. 2019. Yellow journalism. In *The International Encyclopedia of Journalism Studies*, edited by Tim P. Vos and Folker Hanusch, 1–5. Wiley Blackwell.

Caro, Robert A., and Robertson Dean. 1974. *The Power Broker: Robert Moses and the Fall of New York*. Alfred A Knopf.

Caruana, Rich, Yin Lou, Johannes Gehrke, Paul Koch, Marc Sturm, and Noemie Elhadad. 2015. Intelligible models for healthcare: Predicting pneumonia risk and hospital 30-day readmission. In *Proceedings of the 21st ACM SIGKDD International Conference on Knowledge Discovery and Data Mining*, 1721–1730. Association for Computing Machinery.

Cavanaugh, Jillian R. 2015. Performativity. In *Oxford Bibliographies*. Oxford University Press.

Chan, Gloria Hongyee. 2020. A comparative analysis of online, offline, and integrated counseling among hidden youth in Hong Kong. *Children and Youth Services Review* 114: 105042.

Charitan, Alexandra. 2019. Outdoors for all: How national parks are addressing accessibility challenges. *Road Trippers Magazine*, April 26. https://roadtrippers.com/magazine/national-parks-accessibility/.

Christiano, Tom, and Sameer Bajaj. 2021. Democracy. In *Stanford Encyclopedia of Philosophy*, edited by Edward N. Zalta. https://plato.stanford.edu/archives/fall2021/entries/democracy/.

Cisneros, Natalie. 2016. Resisting "massive elimination": Foucault, immigration, and the GIP. In *Active Intolerance: Michel Foucault, the Prisons Information Group, and the Future of Abolition*, edited by Perry Zurn and Andrew Dilts, 241–257. Palgrave Macmillan.

Cohn, Ellen G., Suman Kakar, Chloe Perkins, Rebecca Steinbach, and Phil Edwards. 2020. Red light camera interventions for reducing traffic violations and traffic crashes: A systematic review. *Campbell Systematic Reviews* 16, no. 2: e1091.

Cooke, Steven J., Vivian M. Nguyen, Steven T. Kessel, Nigel E. Hussey, Nathan Young, and Adam T. Ford. 2017. Troubling issues at the frontier of animal tracking for conservation and management. *Conservation Biology* 31, no. 5: 1205–1207. https://socialecology .ca/wp-content/uploads/2022/02/29-Cooke-et-al.-2017-Troubling-issues-at-the-frontier -of-animal-tracking-for-conservation-and-management.pdf.

Coyne, Christopher J., and Abigail R. Hall. 2018. The drone paradox: Fighting terrorism with mechanized terror. *The Independent Review* 23, no. 1: 51–67.

Cutcliffe, Stephen H. 2000. *Ideas, Machines, and Values: An Introduction to Science, Technology, and Society Studies*. Rowman & Littlefield.

Cyril, Malkia Amala. 2015. Black America's state of surveillance. *The Progressive*, March 30. https://progressive.org/magazine/black-america-s-state-surveillance-cyril/.

Dauth, Wolfgang, Sebastian Findeisen, Jens Südekum, and Nicole Wößner. 2017. *German Robots—The Impact of Industrial Robots on Workers*. IAB Discussion Paper no. 30/2017. Institute for Employment Research.

DeVries, Terrance, Ishan Misra, Changhan Wang, and Laurens van der Maaten. 2019. Does object recognition work for everyone? In *Proceedings of the IEEE Conference on Computer Vision and Pattern Recognition Workshops*, 52–59. Institute of Electrical and Electronics Engineers.

Diamond, Jared. 1999. *Guns, Germs, and Steel: The Fates of Human Societies*. W. W. Norton.

Djossa, Christina. 2019. When not to geotag while traveling. *National Geographic*, February 6. https://www.nationalgeographic.com/travel/article/when-why-not-to-use -geotagging-overtourism-security.

Domahidi, Emese. 2018. The associations between online media use and users' perceived social resources: A meta-analysis. *Journal of Computer-Mediated Communication* 23, no. 4: 181–200.

Dorst, Kees. 2011. The core of "design thinking" and its application. *Design Studies* 32, no. 6: 521–532.

Drury, Horace Bookwalter. [1922] 1968. *Scientific Management: A History and Criticism*. AMS Press.

Dusek, Val. 2006. *Philosophy of Technology: An Introduction*. Blackwell.

Economist. 2020. The perils of oblivion: Dementia. *Economist*, August 29–September 4.

Elder, A. 2014. Excellent online friendships: An Aristotelian defense of social media. *Ethics and Information Technology* 16, no. 4: 287–297.

Electronic Frontier Foundation. 2021. The law and medical privacy. https://www.eff.org /issues/law-and-medical-privacy.

Enemark, Christian. 2019. Drones, risk, and moral injury. *Critical Military Studies* 5, no. 2: 150–167.

Environmental Protection Agency. n.d. National overview: Facts and figures on materials, wastes, and recycling. Accessed June 2, 2021. https://www.epa.gov/facts-and

-figures-about-materials-waste-and-recycling/national-overview-facts-and-figures -materials.

Eubanks, Virginia. 2018. *Automating Inequality: How High-Tech Tools Profile, Police, and Punish the Poor*. St. Martin's Press.

Evenson, Robert E., and Douglas Gollin. 2003. Assessing the impact of the Green Revolution, 1960 to 2000. *Science* 300, no. 5620: 758–762.

Fang, Fei, Peter Stone, and Milind Tambe. 2015. When security games go green: Designing defender strategies to prevent poaching and illegal fishing. In *Proceedings of the Twenty-Fourth International Joint Conference on Artificial Intelligence*. AAAI Press/ International Joint Conferences on Artificial Intelligence.

Felt, Ulrike, Rayvon Fouché, Clark A. Miller, and Laurel Smith-Doerr, eds. 2017. *The Handbook of Science and Technology Studies*. 4th ed. MIT Press.

Fiske, Amelia, Peter Henningsen, and Alena Buyx. 2019. Your robot therapist will see you now: Ethical implications of embodied artificial intelligence in psychiatry, psychology, and psychotherapy. *Journal of Medical Internet Research* 21, no. 5: e13216. https://www .jmir.org/2019/5/e13216/.

Ford, Henry. 1929. Mass production. In *The Encyclopaedia Britannica*, vol. 15. 14th ed. Encyclopaedia Britannica.

Forsythe, D. 2001. *Studying Those Who Study Us: An Anthropologist in the World of Artificial Intelligence*. Stanford University Press.

Foucault, Michel. 2007. *Discipline and Punish: The Birth of the Prison*. Duke University Press.

Foucault, Michel. 2008. "Panopticism" from *Discipline & Punish: The Birth of the Prison*. *Race/Ethnicity: Multidisciplinary Global Contexts* 2, no. 1: 1–12. https://www.jstor.org/stable /25594995.

Friedler, Sorelle A., Carlos Scheidegger, and Suresh Venkatasubramanian. 2021. The (Im)possibility of fairness: Different value systems require different mechanisms for fair decision making. *Communications of the ACM* 64, no. 4: 136–143.

Frischmann, Brett, and Evan Selinger. 2017. Robots have already taken over our work, but they're made of flesh and bone. *The Guardian*, September 25. https://www.theguard ian.com/commentisfree/2017/sep/25/robots-taken-over-work-jobs-economy.

Gagnon, Audrey. 2020. Far-right framing processes on social media: The case of the Canadian and Quebec chapters of Soldiers of Odin. *Canadian Review of Sociology/Revue Canadienne de Sociologie* 57, no. 3: 356–378.

Gal, Noam. 2019. Ironic humor on social media as participatory boundary work. *New Media & Society* 21, no. 3: 729–749.

Gomes, Carla P., Thomas G. Dietterich, Christopher Barrett, Jon Conrad, Bistra Dilkina, Stefano Ermon, Fei Fang, Andrew Farnsworth, Alan Fern, Xiaoli Z. Fern, et al. 2019. Computational sustainability: Computing for a better world and a sustainable future. *Communications of the ACM* 62, no. 9: 56–65.

Góngora Alonso, S., S. Hamrioui, I. de la Torre Díez, E. Motta Cruz, M. López-Coronado, and M. Franco. 2019. Social robots for people with aging and dementia: A systematic review of literature. *Telemedicine and e-Health* 25, no. 7: 533–540.

Gray, Jonathan, Cornel Sandvoss, and C. Lee Harrington, eds. 2020. *Fandom: Identities and Communities in a Mediated World*. New York University Press.

Habermas, Jürgen. 1989. *The Structural Transformation of the Public Sphere*. Translated by Thomas Burger. MIT Press.

Hafermalz, Ella. 2020. Out of the Panopticon and into exile: Visibility and control in distributed new culture organizations. *Organization Studies* 42, no. 1. https://doi.org/10.1177/0170840620909962.

Hammer, MC. 2021. You bore us. If science is a "commitment to truth" shall we site all the historical non-truths perpetuated by scientists? Of course not. It's not science vs Philosophy . . . It's Science + Philosophy. Elevate your Thinking and Consciousness. When you measure include the measurer. Twitter. https://twitter.com/MCHammer/status/1363908982289559553?ref_src=twsrc%5Etfw%7Ctwcamp%5Etweetembed%7Ctwterm%5E1363908982289559553%7Ctwgr%5E%7Ctwcon%5Es1_c10&ref_url=https%3A%2F%2Fwww.newsweek.com%2Fmc-hammer-philosophical-tweets-spark-memes-1571272.

Harford, Tim. 2017. How fertiliser helped feed the world. *BBC News*, January 2. https://www.bbc.com/news/business-38305504.

Harris, Charles E., Jr., Michael S. Pritchard, Michael J. Rabins, Ray James, and Elaine Englehardt. 2013. *Engineering Ethics: Concepts and Cases*. Cengage Learning.

Harris, Douglas N., Lihan Liu, Daniel Oliver, Cathy Balfe, Sara Slaughter, and Nicholas Mattei. 2020. How America's Schools Responded to the COVID Crisis. EdWorkingPaper 20-262. Retrieved from Annenberg Institute at Brown University. https://doi.org/10.26300/3sg2-ep57.

Heidegger, Martin. 1954. The question concerning technology. *Technology and Values: Essential Readings* 99: 113.

Hoff, Timothy. 2011. Deskilling and adaptation among primary care physicians using two work innovations. *Health Care Management Review* 36, no. 4: 338–348.

Hutchinson, B., and Mitchell, M. 2019. 50 years of test (un)fairness: Lessons for machine learning. In *Proceedings of the Conference on Fairness, Accountability, and Transparency*, 49–58. Association for Computing Machinery.

Hutson, Jevan A., Jessie G. Taft, Solon Barocas, and Karen Levy. 2018. Debiasing desire: Addressing bias and discrimination on intimate platforms. *Proceedings of the ACM on Human-Computer Interaction*, no. 2: 1–18.

Ihde, Don. 2009. *Postphenomenology and Technoscience: The Peking University Lectures*. State University of New York Press.

Isaacson, Walter. 2014. *The Innovators: How a Group of Inventors, Hackers, Geniuses and Geeks Created the Digital Revolution*. Simon and Schuster.

Isbell, Charles. 2020. You can't escape hyperparameters and latent variables: Machine learning as a software engineering enterprise. Keynote talk, December 6. In *Proceedings of NeurIPS 2020: Thirty-Fourth Conference on Neural Information Processing Systems*. Neural Information Processing Systems Foundation. https://slideslive.com/38935825/you-cant-escape-hyperparameters-and-latent-variables-machine-learning-as-a-software-engineering-enterprise.

Ivaschenko, Oleksiy, Claudia P. Rodriguez Alas, Marina Novikova, Carolina Romero Robayo, Thomas Vaughan Bowen, and Linghui Zhu. 2018. *The State of Social Safety Nets 2018.* World Bank.

Jasanoff, Sheila. 2004. *States of Knowledge: The Co-Production of Science and the Social Order.* Routledge.

Jasanoff, Sheila. 2016. *The Ethics of Invention: Technology and the Human Future.* W. W. Norton.

Jasanoff, Sheila. 2017. Science and democracy. In *The Handbook of Science and Technology Studies*, 4th ed., edited by Ulrike Felt, Rayvon Fouché, Clark A. Miller, and Laurel Smith-Doerr, 259–288. MIT Press.

Jha, A. K., C. M. DesRoches, E. G. Campbell, K. Donelan, S. R. Rao, T. G. Ferris, A. Shields, S. Rosenbaum, and D. Blumenthal. 2009. Use of electronic health records in US hospitals. *New England Journal of Medicine* 360, no. 16: 1628–1638. https://www.nejm .org/doi/full/10.1056/nejmsa0900592.

Jillson, Elisa. 2021. Aiming for truth, fairness, and equity in your company's use of AI. Federal Trade Commission, April 19. https://www.ftc.gov/news-events/blogs/business -blog/2021/04/aiming-truth-fairness-equity-your-companys-use-ai.

John, Sokfa F. 2020. Computing Cupid: Online dating and the faith of romantic algorithms. *African Journal of Gender and Religion* 25, no. 2: 86–108.

Kahn, Jeremy. 2021. Can A.I. help Facebook cure its disinformation problem? *Fortune*, April 6. https://fortune.com/2021/04/06/facebook-disinformation-ai-fake-news-us-capitol -attack-social-media-hate-speech-big-tech-solutions/.

Kambhampati, Subbarao. 2019. What just happened? The rise of interest in artificial intelligence. *The Hill*, August 11. https://thehill.com/opinion/technology/457008-what -just-happened-the-rise-of-interest-in-artificial-intelligence.

Kanat-Maymon, Yaniv, Lian Almog, Rinat Cohen, and Yair Amichai-Hamburger. 2018. Contingent self-worth and Facebook addiction. *Computers in Human Behavior* 88: 227–235.

Kautz, Henry. 2020. The third AI summer. AAAI Robert S. Engelmore Memorial Lecture, February 10. In *Proceedings of the 34th Annual Meeting of the Association for the Advancement of Artificial Intelligence (AAAI-2020)*. Association for the Advancement of Artificial Intelligence. https://www.youtube.com/watch?v=_cQITY0SPiw&ab_channel=Henry Kautz.

Kavanagh, Jennifer, and Michael D. Rich. 2018. *Truth Decay: An Initial Exploration of the Diminishing Role of Facts and Analysis in American Public Life.* Santa Monica, CA: RAND Corporation, RR-2314-RC, 2018. As of June 23, 2022: https://www.rand.org/pubs /research_reports/RR2314.html.

Keinan, Giora, and Ayala Malach-Pines. 2007. Stress and burnout among prison personnel: Sources, outcomes, and intervention strategies. *Criminal Justice and Behavior* 34, no. 3: 380–398.

Khosravi, Pouria, and Amir Hossein Ghapanchi. 2016. Investigating the effectiveness of technologies applied to assist seniors: A systematic literature review. *International Journal of Medical Informatics* 85, no. 1: 17–26.

King, Noel. 2021. A brief history of how racism shaped interstate highways. *NPR Morning Edition*, April 7. https://www.npr.org/2021/04/07/984784455/a-brief-history-of-how-racism-shaped-interstate-highways.

Kowert, Rachel, Emese Domahidi, Ruth Festl, and Thorsten Quandt. 2014. Social gaming, lonely life? The impact of digital game play on adolescents' social circles. *Computers in Human Behavior* 36, no. 7: 385–390.

Kretzmann, John, and John P. McKnight. 1996. Assets-based community development. *National Civic Review* 85, no. 4: 23–29.

Kuhn, Thomas S. 1962. *The Structure of Scientific Revolutions*. University of Chicago Press.

Kurwa, Rahim. 2019. Building the digitally gated community: The case of Nextdoor. *Surveillance and Society* 17, no. 1/2. https://ojs.library.queensu.ca/index.php/surveillance-and-society/article/view/12927.

Lamba, Aakash, Phillip Cassey, Ramesh Raja Segaran, Lian Pin Koh. 2019. Deep learning for environmental conservation. *Current Biology* 29, no. 19: R977–R982.

Lambright, Katie. 2019. Digital redlining: The Nextdoor app and the neighborhood of make-believe. *Cultural Critique* 103: 84–90.

Langewiesche, William. 2019. What really brought down the Boeing 737 Max? *New York Times Magazine*, September 18. https://www.nytimes.com/2019/09/18/magazine/boeing-737-max-crashes.html.

Lapidot-Lefler, Noam, and Azy Barak. 2012. Effects of anonymity, invisibility, and lack of eye-contact on toxic online disinhibition. *Computers in Human Behavior* 28, no. 2: 434–443.

Lasco, Gideon. 2015. The smartphone as status symbol. *Philippine Daily Inquirer*, October 22.

Latour, Bruno. 1987. *Science in Action: How to Follow Scientists and Engineers through Society*. Harvard University Press.

Latour, Bruno. 2005. *Reassembling the Social: An Introduction to Actor-Network-Theory*. Oxford University Press.

Law, John. 2009. Actor network theory and material semiotics. In *The New Blackwell Companion to Social Theory*, 141–158. Blackwell.

'Lectric Law Library's Lexicon. n.d. Freedom of Information Act. Accessed June 3, 2021. https://www.lectlaw.com/def/f086.htm.

Lee, Min Kyung, Daniel Kusbit, Anson Kahng, Ji Tae Kim, Xinran Yuan, Allissa Chan, Daniel Se, et al. 2019. WeBuildAI: Participatory framework for algorithmic governance. *Proceedings of the ACM on Human-Computer Interaction*, no. 3: 1–35.

Lee, Shin Haeng, Jin-Young Tak, Eun-Joo Kwak, and Tae Yun Lim. 2019. Fandom, social media, and identity work: The emergence of virtual community through the pronoun "we." *Psychology of Popular Media Culture* 9, no. 4: 436–446. https://doi.org/10.1037/ppm0000259.

Lewis, Sarah. 2019. The racial bias built into photography. *New York Times*, April 25. https://www.nytimes.com/2019/04/25/lens/sarah-lewis-racial-bias-photography.html.

Lin, Herbert, and Jaclyn Kerr. 2019. On cyber-enabled information warfare and influence operations. In *The Oxford Handbook of Cyber Security*, edited by Paul Cornish, 251–272. Oxford University Press.

Lobel, Orly. 2001. Agency and coercion in labor and employment relations: Four dimensions of power in shifting patterns of work. *University of Pennsylvania Journal of Labor and Employment Law* (currently *University of Pennsylvania Journal of Business Law*) 4, no. 1: 121–193.

Lorenz, Taylor. 2019. The Instagram aesthetic is over. *The Atlantic*, April 23. https://www .theatlantic.com/technology/archive/2019/04/influencers-are-abandoning-instagram -look/587803/.

Maly, Tim. 2015. A brief history of human energy use. *The Atlantic*, November 13. https://www.theatlantic.com/technology/archive/2015/11/a-brief-history-of-human -energy-use/415749/.

Mann, Steve, Jason Nolan, and Barry Wellman. 2003. Sousveillance: Inventing and using wearable computing devices for data collection in surveillance environments. *Surveillance & Society* 1, no. 3: 331–355.

Martin, Mike W., and Roland Schinzinger. 2004. *Ethics in Engineering*. 4th ed. McGraw-Hill Education.

McGhee, Heather. 2022. *The Sum of Us: What Racism Costs Everyone and How We Can Prosper Together*. One World.

McGinnis, John O. 2012. *Accelerating Democracy*. Princeton University Press.

McHugh, Molly. 2016. Loved to death: How Instagram is destroying our natural wonders. *The Ringer*, November 3. https://www.theringer.com/2016/11/3/16042448/instagram-geo tagging-ruining-parks-f65b529d5e28.

McIntosh, Peggy. 1988. White privilege: Unpacking the invisible knapsack. Reprinted in Anna May Filor, *Multiculturalism*. New York State Council of Education Reports, 1992.

McShane, Clay. 1999. The origins and globalization of traffic control signals. *Journal of Urban History* 25, no. 3: 379–404.

Merriam-Webster. s.v., "society." Accessed June 3, 2021. https://www.merriam-webster .com/dictionary/society.

Messing, Jill, Meredith Bagwell-Gray, Megan Lindsay Brown, Andrea Kappas, and Alesha Durfee. 2020. Intersections of stalking and technology-based abuse: Emerging definitions, conceptualization, and measurement. *Journal of Family Violence* 35, no. 7: 693–704.

Mill, John Stuart. [1859] 2002. *The Basic Writings of John Stuart Mill: On Liberty, The Subjection of Women and Utilitarianism*. Random House.

Mitcham, Carl. 1994. *Thinking through Technology: The Path between Engineering and Philosophy*. University of Chicago Press.

Moor, James H. 1985. What is computer ethics? *Metaphilosophy* 16, no. 4: 266–275.

Morimoto, Lori Hitchcock, and Bertha Chin. 2017. Reimagining the imagined community: Online media fandoms in the age of global convergence. In *Fandom: Identities and*

Communities in a Mediated World, 2nd ed., edited by Jonathan Gray, Cornell Sandvoss, and C. Lee Harrignton, 174–188. New York University Press.

Mulligan, D. K., J. A. Kroll, N. Kohli, and R. Y. Wong. 2019. This thing called fairness: Disciplinary confusion realizing a value in technology. *Proceedings of the ACM on Human-Computer Interaction*, no. 3: 1–36.

Myers, Norman. 2001. *Perverse Subsidies: How Tax Dollars Can Undercut the Environment and the Economy*. Island Press.

NASA. 2006. *Report of Columbia Accident Investigation Board*, vol. 1. Last updated March 5. https://www.nasa.gov/columbia/home/CAIB_Vol1.html.

NASA. 2021. 35 years ago: Remember the *Challenger* and her crew. January 28. https://www.nasa.gov/feature/35-years-ago-remembering-challenger-and-her-crew.

National Academies of Sciences, Engineering, and Medicine. 2015. Technology and tools in the diagnostic process. In *Improving Diagnosis in Health Care*, edited by Erin P. Balogh, Bryan T. Miller, and John R. Ball, chap. 5. National Academies Press. https://www.ncbi.nlm.nih.gov/books/NBK338590/.

National Geographic. 2019. Anthropocene. National Geographic Resource Library. Updated June 7. https://www.nationalgeographic.org/encyclopedia/anthropocene/.

National Park Service. 2021a. Accessibility. Updated October 4. https://www.nps.gov/accessibility.htm.

National Park Service. 2021b. Accessibility compliance: Helping all visitors have a positive and rewarding experience at our venues. Updated August 11. https://www.nps.gov/articles/accessibility.htm.

National Park Service. 2021c. Mission Statement. Updated June 9. https://www.nps.gov/aboutus/index.htm.

National Research Council. 2010. *The Impact of Genetically Engineered Crops on Farm Sustainability in the United States*. National Academies Press. https://doi.org/10.17226/12804.

Nee-Benham, Maenette Kapeʻahiokalani Padeken Ah, Maenette KapeʼAhiokalani, Ronald H. Heck, and Padeken Ah Nee Benham. 1998. *Culture and Educational Policy in Hawaiʻi: The Silencing of Native Voices*. Psychology Press.

Noble, Safiya Umoja, and Sarah T. Roberts. 2016. Through Google-colored glass(es): Design, emotion, class, and wearables as commodity and control. In *Emotions, Technology, and Design*, edited by Sharon Y. Tettegah and Safiya Umoja Noble, 187–212. Academic Press.

Nogrady, Bianca. 2016. Your old phone is full of untapped precious metals. BBC, October 18. https://www.bbc.com/future/article/20161017-your-old-phone-is-full-of-precious-metals.

Nolan, Emma. 2021. MC Hammer's philosophical tweets spark memes, bemusement. *Newsweek*, February 23. https://www.newsweek.com/mc-hammer-philosophical-tweets-spark-memes-1571272.

NYU Langone Health. n.d. Genetically modified organisms: The golden rice debate. Accessed June 2, 2021. https://med.nyu.edu/departments-institutes/population-health

/divisions-sections-centers/medical-ethics/education/high-school-bioethics-project/learning-scenarios/gmos-the-golden-rice-debate.

O'Neil, C. 2016. *Weapons of Math Destruction: How Big Data Increases Inequality and Threatens Democracy.* Crown.

The Onion. 2010. Apple finally unveils iPad. *The Onion,* February 1. https://www.theonion.com/apple-finally-unveils-ipad-1819589716.

Palmiste, Claire. 2008. Forcible removals: The case of Australian Aboriginal and Native American children. *AlterNative: An International Journal of Indigenous Peoples* 4, no. 2: 75–88.

PBS. 2019. Get home safely: 10 rules of survival. https://www.pbs.org/black-culture/connect/talk-back/10_rules_of_survival_if_stopped_by_police/.

Piccard, Ann. 2013. Death by boarding school: The last acceptable racism and the United States' genocide of Native Americans. *Gonzaga Law Review* 49: 137–185.

Poster, Winifred R. 2019. Racialized surveillance in the digital service economy. In *Captivating Technology: Race, Carceral Technoscience, and Liberatory Imagination in Everyday Life,* edited by Ruha Benjamin, 133–169. Duke University Press.

Rabbitt, S., A. Kazdin, and B. Scassellati. 2014. Integrating socially assistive robotics into mental healthcare interventions: Applications and recommendations for expanded use. *Clinical Psychology Review* 35: 35–46.

Reynolds, Jesse L. 2021. Earth system interventions as technologies of the Anthropocene. *Environmental Innovation and Societal Transitions* 40: 132–146.

Ricks, Becca, and Mark Surman. 2020. Creating trustworthy AI: A Mozilla white paper on challenges and opportunities in the AI era. Mozilla Foundation.

Rios, Victor M. 2011. *Punished: Policing the Lives of Black and Latino Boys.* New York University Press.

Robbins, Jim. 2020. Ecopsychology: How immersion in nature benefits your health. *Yale Environment 360,* January 9. https://e360.yale.edu/features/ecopsychology-how-immersion-in-nature-benefits-your-health.

Rohracher, Harald. 2015. Science and technology studies, history of. In *International Encyclopedia of the Social & Behavioral Sciences,* 2nd ed., edited by James D. Wright, 200–205. Elsevier.

Rose, Jody. 2014. The Knowledge, London's legendary taxi-driver test, puts up a fight in the age of GPS. *New York Times,* November 10. https://www.nytimes.com/2014/11/10/t-magazine/london-taxi-test-knowledge.html.

Saayman, Melville, and Andrea Saayman. 2015. Understanding tipping behaviour—An economic perspective. *Tourism Economics* 21, no. 2: 247–265.

Sáez Martin, Alejandro, Arturo Haro De Rosario, and María Del Carmen Caba Pérez. 2016. An international analysis of the quality of open government data portals. *Social Science Computer Review* 34, no. 3: 298–311.

Sanger, David E., Clifford Krauss, and Nicole Perlroth. 2021. Cyberattack forces a shutdown of a top US pipeline. *New York Times,* May 8, updated May 13. https://www.nytimes.com/2021/05/08/us/politics/cyberattack-colonial-pipeline.html.

Schatzberg, Eric. 2006. "Technik" comes to America: Changing meanings of "technology" before 1930. *Technology and Culture* 47, no. 3: 486–512.

Schatzberg, Eric. 2018. *Technology: Critical History of a Concept*. University of Chicago Press.

Schmitter, Philippe C., and Terry Lynn Karl. 1991. What democracy is . . . and is not. *Journal of Democracy* 2, no. 3: 75–88.

Schulz-Forberg, Hagen. 2012. Welfare state. In *Encyclopedia of Global Studies*, edited by Helmut K. Anheier, Mark Juergensmeyer, and Victor Faessel, 1783–1788. Sage. https://study.sagepub.com/sites/default/files/Ch14_Welfare%20State.pdf.

Scott, James C. 2017. *Against the Grain: A Deep History of the Earliest States*. Yale University Press.

Scrivens, Ryan, Garth Davies, and Richard Frank. 2020. Measuring the evolution of radical right-wing posting behaviors online. *Deviant Behavior* 41, no. 2: 216–232.

Scrivens, Ryan, Thomas W. Wojciechowski, Joshua D. Freilich, Steven M. Chermak, and Richard Frank. 2021. Comparing the online posting behaviors of violent and non-violent right-wing extremists. *Terrorism and Political Violence*, March 8. https://doi.org/10.1080/09546553.2021.1891893.

Selbst, A. D., d. boyd, S. A. Friedler, S. Venkatasubramanian, and J. Vertesi. 2019. Fairness and abstraction in sociotechnical systems. In *Proceedings of the Conference on Fairness, Accountability, and Transparency*, 59–68. Association for Computing Machinery.

Shapiro, Aaron. 2017. Reform predictive policing. *Nature News* 541: 458–460. https://www.nature.com/news/reform-predictive-policing-1.21338.

Sharon, Tamar, and Dorien Zandbergen. 2017. From data fetishism to quantifying selves: Self-tracking practices and the other values of data. *New Media & Society* 19, no. 11: 1695–1709.

Shensa, Ariel, J. E. Sidani, M. A. Dew, C. G. Escobar-Viera, and B. A. Primack. 2018. Social media use and depression and anxiety symptoms: A cluster analysis. *American Journal of Health Behavior* 42, no. 2: 116–128.

Shiu, Yu, K. J. Palmer, Marie A. Roch, Erica Fleishman, Xiaobai Liu, Eva-Marie Nosal, Tyler Helble, Danielle Cholewiak, Douglas Gillespie, and Holger Klinck. 2020. Deep neural networks for automated detection of marine mammal species. *Nature Science Reports* 10, no. 607. https://doi.org/10.1038/s41598-020-57549-y.

Silverstein, Arthur M. 2001. *History of Immunology*. Wiley Online Library eLS. https://onlinelibrary.wiley.com/doi/abs/10.1038/npg.els.0003078.

Sottilare, R. A., A. Graesser, X. Hu, and H. Holden, eds. 2013. *Design Recommendations for Intelligent Tutoring Systems*. Vol. 1: *Learner Modeling*. US Army Research Laboratory.

Standlee, Alecea. 2019. Friendship and online filtering: The use of social media to construct offline social networks. *New Media & Society* 21, no. 3: 770–785.

Stich, Jean-François. 2020. A review of workplace stress in the virtual office. *Intelligent Buildings International* 12, no. 3: 1–13.

Subcommission on Quarternary Stratigraphy. 2019. Working Group on the "Anthropocene." May 21. http://quaternary.stratigraphy.org/working-groups/anthropocene/.

Sullins, John. 2000. Transcending the meat: Immersive technologies and computer mediated bodies. *Journal of Experimental & Theoretical Artificial Intelligence* 12, no. 1: 13–22.

Sunstein, C. R. 2009. *Going to Extremes: How Like Minds Unite and Divide.* Oxford University Press.

Swain, Ashok, Ramses Amer, and Joakim Öjendal. 2011. *The Democratization Project: Opportunities and Challenges.* Anthem Press.

Tauberer, J. 2014. *Open Government Data.* 2nd ed. https://opengovdata.io/.

Taylor, Keeanga-Yamahtta. 2019. *Race for Profit: How Banks and the Real Estate Industry Undermined Black Homeownership.* Uuniversity of North Carolina Press.

Thaler, Richard H., and Cass R. Sunstein. 2009. *Nudge: Improving Decisions about Health, Wealth, and Happiness.* Penguin Books.

Turkle, Sherry. 2008. Always-on/always-on-you: The tethered self. In *Handbook of Mobile Communication Studies*, edited by James E. Katz, 121–138. MIT Press.

Valkenburg, Patti M., and Jochen Peter. 2007. Who visits online dating sites? Exploring some characteristics of online daters. *CyberPsychology & Behavior* 10, no. 6: 849–852.

Vallas, S., and J. B. Schor. 2020. What do platforms do? Understanding the gig economy. *Annual Review of Sociology* 46, no. 1: 273–294.

Vallor, Shannon. 2011. Carebots and caregivers: Sustaining the ethical ideal of care in the twenty-first century. *Philosophy & Technology* 24, no. 3: 251–268.

Vallor, Shannon. 2016. *Technology and the Virtues: A Philosophical Guide to a Future Worth Wanting.* Oxford University Press.

Vallor, Shannon, Brian Green, and Irina Raicu. 2018. Ethics in technology practice. The Markkula Center for Applied Ethics at Santa Clara University. https://www.scu.edu/ethics/ethics-resources/ethical-decision-making/a-framework-for-ethical-decision-making/.

Verbeek, Peter-Paul. 2011. *Moralizing Technology: Understanding and Designing the Morality of Things.* University of Chicago Press.

Verbeek, Peter-Paul. 2015. Beyond interaction: A short introduction to mediation theory. *Interactions* 22, no. 3: 26–31.

Vines, John, Rachel Clarke, Peter Wright, John McCarthy, and Patrick Olivier. 2013. Configuring participation: On how we involve people in design. In *Proceedings of the 2013 CHI Conference on Human Factors in Computing Systems*, 429–438. Association for Computing Machinery.

Vitale, Alex. 2021. *The End of Policing.* Updated ed. Verso.

Wachter, Robert M. 2015. *The Digital Doctor: Hope, Hype, and Harm at the Dawn of Medicine's Computer Age.* McGraw-Hill Education.

Wakefield, Sarah. 2001. "Your sister in St. Scully": An electronic community of female fans of the X-Files. *Journal of Popular Film and Television* 29, no. 3: 130–137.

Wall, Tyler, and Torin Monahan. 2011. Surveillance and violence from afar: The politics of drones and liminal security-scapes. *Theoretical Criminology* 15, no. 3: 239–254.

Watkins, Ali. 2021. Pandemic wilderness explorers are straining search and rescue. *New York Times*, April 7. https://www.nytimes.com/2021/04/07/us/coronavirus-wilderness -search-rescue.html.

Welz, Adam. 2017. The dark side of digitally tracking endangered species. *Green Biz*, October 11. https://www.greenbiz.com/article/dark-side-digitally-tracking-endangered-species.

Williams, Karel, Colin Haslam, and John Williams. 1992. Ford versus "Fordism": The Beginning of Mass Production? *Work, Employment and Society* 6, no. 4: 517–555.

Winick, Erin. 2018. Every study we could find on what automation will do to jobs, in one chart. *MIT Technology Review*, January 25. https://www.technologyreview.com/2018/01/25 /146020/every-study-we-could-find-on-what-automation-will-do-to-jobs-in-one-chart/.

Winner, Langdon. 1980. Do artifacts have politics? *Daedalus* 109, no. 1. Reprinted in *The Social Shaping of Technology*, edited by Donald A. MacKenzie and Judy Wajcman, 2nd ed., 28–40. Open University Press, 1999.

Winner, Langdon. 1993a. How technology reweaves the fabric of society. *Chronicle of Higher Education* 39, no. 48 (August 4): B1–B3.

Winner, Langdon. 1993b. Upon opening the black box and finding it empty: Social constructivism and the philosophy of technology. *Science, Technology, & Human Values* 18, no. 3: 362–378.

Winston, Ali. 2018. Palantir has secretly been using New Orleans to test its predictive policing technology. The Verge, February 27. https://www.theverge.com/2018/2/27 /17054740/palantir-predictive-policing-tool-new-orleans-nopd.

Witze, Alexandra. 2020. Universities will never be the same after the coronavirus crisis. *Nature* 582, no. 7811: 162–165.

Woolley, Samuel C. 2016. Automating power: Social bot interference in global politics. *First Monday* 21, no. 4 (April 4). https://journals.uic.edu/ojs/index.php/fm/article/view/6161.

Yates, Chris. 2018. Technology is not the solution to everything. *Medium*, December 2. https://medium.com/datadriveninvestor/technology-is-not-the-solution-to-everything -4b1655a7f80e.

Zewe, Adam. 2019. Outsmarting poachers: Artificial intelligence helps rangers protect endangered wildlife. *Harvard School of Engineering News & Events*, October 10. https:// www.seas.harvard.edu/news/2019/10/outsmarting-poachers.

CHAPTER 6

PROFESSIONAL ETHICS

Learning Objectives

At the end of this chapter you will be able to:

1. *Outline the characteristics of a profession and assess what traits and responsibilities are part of professional technology development.*

2. *Summarize the need for professionalization and break down the advantages and disadvantages of centering the locus of ethical responsibility on the profession rather than the individual.*

3. *Discuss the strengths and limitations of codes of ethics as a decision-making tool and use a code of ethics to justify or rebut a decision or course of action.*

4. *Give original examples of applying your ethical toolkit, including professional codes, institutional processes, ethical frameworks, and moral imagination, to problems that could arise in practice.*

6.1 INTRODUCTION

Throughout most of this book, we have focused on the impact of technology in very broad ways: how it affects us in our own lives, in our communities, and within society at large. We hope this has given you new perspectives to apply to your career path and your participation in civic life, but it is understandably difficult at times to know how to translate some of these insights into action in your routine work.

Professional ethics is more down-to-earth; it focuses on the day-to-day expectations of the individual worker rather than the broader impacts of the field as a whole. But there is more to professional ethics than just taking ordinary standards of personal behavior into the workplace. For instance, both an engineer and a cashier would be expected not to embezzle money from their employers—yet organizations have spent decades carefully refining their expectations for engineering ethics, whereas there is no widespread standard for "cashiering ethics." This is because, as we explore in the next section, society entrusts those who practice "professional" jobs like engineering, law, and medicine with an unusual degree of latitude in their work, and in return society expects professionals to continually earn this trust.

We have discussed how to apply ethical reasoning frameworks to thinking about dilemmas that we will all face as technology professionals and as users of technology. This chapter, however, explores how others have already codified this reasoning. Professional societies within applied sciences including engineering, law, and medicine publish guidelines for ethical practice of their field. Indeed, the Accreditation Board for Engineering and Technology (ABET), the largest accreditor of many STEM programs including computer science in the United States, requires instruction in professional ethics.

Section 6.2 focuses on the history and development of the technology practitioner as a professional. Section 6.3 considers the ethical codes that guide the profession; how, when, and why to apply these codes to decision making, and the limits of these codes for decision making. Sections 6.4 and 6.5 conclude the chapter with thoughts on how professional ethics combined with a deep and meaningful interaction with ethical theories and other reasoning tools can help you prepare for situations that are not (yet) covered by established ethical guidelines.

6.2   PROFESSIONS

In this section we give an overview of what professions are, how they have developed in the engineering and technology development fields, and how they interact with laws and regulations.

6.2.1   WHAT IS A PROFESSION?

To understand why the notion of a profession is ethically significant, we first investigate what it means for an occupation to be a profession. Greenwood (1957) proposes five core characteristics of a profession.

First, a profession has a *core body of theoretical knowledge*. While a practitioner of a nonprofessional trade (e.g., a mechanic or plumber) may be highly knowledgeable from practical training and experience, professional work requires a combination of practical skill and academic study.

Second, this deep specialized knowledge gives professionals *authority*. Or, as Greenwood paraphrases, a nonprofessional has *customers*; a professional has *clients*. In the former relationship, the customer holds a lot of power; the customer decides what they want, and if the worker cannot meet the demand, the customer can simply look for someone else. In the latter relationship, though, the client depends on the professional to make judgments that the client is not qualified to make for themself; the client's ability to "shop around" is limited to deciding who can be trusted to act in the client's best interests. Note that this form of authority is not moral authority, of the sort that matters in deontology, but rather is a matter of in-field expertise and financial relationships.

Third, a profession is given *special and often unique privileges by the community*. A common example of this is licensing requirements in order to practice the profession. An organization establishes licensing procedures for a profession and persuades legislators to pass laws mandating the licensing, in essence giving the organization authority over who is and is not legally allowed to practice the profession. An organization may also develop certifications, which are not legally mandated like licenses but likewise serve as markers of confidence and may be expected or preferred by employers. Additionally, when a profession develops standards of what constitutes a properly educated professional, an accrediting body (such as ABET for science and engineering disciplines) may adopt those standards for giving a mark of approval to college degree programs. Although licensing is enforced by the law, certification and accredited degrees are only "enforced" by the market. All three serve a similar purpose: they give a professional organization the power to set entry barriers to the profession.

Fourth, a profession regulates itself with a *code of ethics*. This serves as a counterbalance to the previous point—the community grants the professional organization a monopoly on a particular type of business, and like any monopoly, it could easily be exploited. For instance, suppose the members of a professional organization decided to charge unreasonably high prices for their services. With nobody outside of the organization legally able to set up a competing business and drive prices back down, clients would be forced to pay the demanded amount or else go without a critical service. To prevent such scenarios, a code of ethics states the organization's commitment to use its privileges in a way that aligns with the community's well-being.

Finally, a profession has its own *culture*. Beyond the technical competencies of the job, a new professional learns how to fit in among other members—their common language and lore, social conventions, and ways of viewing the world and the profession's role within it.

In short, a profession provides a service that is needed by the rest of society, but the deep expertise needed to understand the profession means that it is not easily controlled by nonexpert regulators or market forces. Because of this, the profession and the community establish a sort of social contract in which the profession is granted special authority in exchange for an expectation of self-regulation. Ethics, therefore, is a fundamental part of the concept of a profession.

Throughout this book we discuss the individuals responsible for making technology in very broad terms. They may be the networking experts who make systems communicate or the electronics designers who lay out the newest smartphone, but what parts of technology development and deployment are proper professions and what parts are not?

One part of the process of technology development involves writing software for a smartphone, an embedded device, or a website, among many other options. Let us focus on the experts in this part of development—to what extent are they professionals? We typically call people software engineers if they write software, but what are the parallels to traditional engineering fields like mechanical or civil engineering, which have established themselves definitively as professions?

Software engineering matches Greenwood's first, fourth, and fifth aspects of a profession: It has a strong element of theoretical knowledge, as the standard education for practitioners includes topics such as analysis of algorithms from a mathematical perspective and best patterns for solving problems. The community of software engineering and those who build software products has its own culture as well, including websites and memes specifically for software engineering (e.g., https://www.reddit.com /r/ProgrammerHumor/). As for codes of ethics, two of the largest international computing organizations—the Association for Computing Machinery (ACM) and Institute of Electrical and Electronics Engineers Computer Society (IEEE-CS)—have jointly created the Software Engineering Code of Ethics and Professional Practice (SECEPP), meant to be the standard code for software engineers.

The other two aspects are less clear-cut. Is software engineering given privileges by the community, including licensing? In certain jurisdictions, including Ontario and Quebec in Canada, one must be granted an engineering license to practice professional software engineering. Likewise, in the United States, the Institute of Electrical and Electronics Engineers (IEEE) offers the Certified Software Development Professional exam, and the National Society of Professional Engineers offers an occupational license for software engineering (Wikipedia 2022b). As of 2020 in the United States, the answer depends on the jurisdiction; in some states, the title "software engineer" carries legal status and requires licensing, whereas in others, many companies use the term freely for any position that involves computer programming.

And finally, do software engineers have authority in their relationship with clients? The answer depends on how we define a software engineer. Experts in the rigorous design of complex and safety-critical software systems may have the same authoritative status as traditional engineers. But the world of people who program and create software seems more akin to a trade than a profession in this regard—while skilled, they do not normally have the power to make judgments about a client's well-being that the client cannot simply override.

Most of these points apply to technology development overall; many technology jobs like system administrators or software testers have even less clearly defined community privilege and individual authority than software engineering. However, Greenwood's aspects of a profession are a matter of degree, rather than of strict "yes" and "no" answers, and it still makes sense to discuss these jobs in a professional context. Furthermore, as we will see later, professionalization is a process. Software engineering, like the traditional engineering fields before it, gained many of its professional traits because a group of people made a focused effort to establish those traits.

6.2.2 A BRIEF HISTORY OF PROFESSIONAL SOCIETIES AND CODES OF ETHICS IN TECHNOLOGY DEVELOPMENT

During the late nineteenth century there was a growing tension between engineers and the firms for which they worked. Many engineers envisioned themselves as practitioners of a technical profession instead of simply employees in large enterprises. The early formation of many of the professional societies for engineers can be viewed as a first step toward the establishment of engineering and technology development as a profession (Layton 1986).

There were four founding engineering societies: The American Society of Civil Engineers (ASCE), founded in 1851; The American Institute of Electrical Engineers (AIEE, later the IEEE), founded in 1884; the American Society of Mechanical Engineers (ASME), founded in 1880; and the American Institute of Mining Engineers (AIME), founded in 1871 (Wikipedia 2022a). The AIEE included some of the most famous technological innovators of the time, including Thomas Edison, Nikola Tesla, and Alexander Graham Bell. In 1963 the AIEE merged with the Institute of Radio Engineers (IRE) to become the IEEE (IEEE 2022b).

A motivation for much of the formation of these professional societies came as fallout from a number of major failures of civilian infrastructure including bridges and roadways, as well as mining and other manufacturing facility accidents. These high-profile failures of engineering and technology, along with a growing class of engineers that increasingly viewed themselves as individual professionals and not simply workers for large firms, spawned the creations of the first engineering codes of ethics. The AIEE adopted its first code of ethics and standards of practice in 1912, followed shortly thereafter by the ASCE and the ASME in 1914 (Layton 1986).

One of the oldest codes of ethics still used in the United States comes from the National Society of Professional Engineers (NSPE). In 1934, the NSPE was founded as an organization for professional engineers with the goal of providing licensure and formalizing the profession. Over the years, professional licensure has become required for undertaking or bidding on certain types of contracts and technology development, including offering "engineering services." Indeed, the NSPE now administers the professional engineering (PE) exams that are required to practice certain types of engineering and contracting services for things like buildings, bridges, airplanes, cars, and electrical infrastructure. As we will see later, licensure is not as well established or enforced for things like software services and other small technology products.

The Code of Ethics for Engineers was first adopted by the National Society of Professional Engineers in 1964, replacing a previous code dating to 1947. Much of the code, which has undergone revisions throughout the years, focuses on ethical trade practices such as rules on bidding on contracts, a moratorium on accepting bribes, and other standards governing acceptable behavior as a service provider. The code also discusses how to deal with conflicts of interest and balancing objectives of various parties when providing services. The document continues to go through changes, being updated to reflect changing societal values (NSPE 2022b). For instance, earlier iterations prohibited engineers from participating in "collective coercive action" like strikes, but in 2001 that clause was removed. In 2003, a clause was added that engineers should "continue professional development throughout their careers." In 2006, a clause was added that engineers should practice "sustainable development," the first time that environmental concerns appeared in the code. And in 2019, a clause was added that engineers "shall treat all persons with dignity, respect, fairness and without discrimination."

As we have seen throughout this book, many ethical considerations that engineers and technology development professionals encounter do not have easy answers. The NSPE runs a board of ethical review that has over 600 advisory opinions on topics ranging from conflicts of interest to public health and safety (NSPE 2022a). A board such as this represents an effort to continue to stay ahead of changes in the impacts of technology and development to provide guidance to practitioners. Other efforts in this vein include more focused professional organizations like the ACM and the IEEE, which we discuss next.

6.2.3 PROFESSIONALIZATION IN COMPUTING TECHNOLOGY

As we saw in the history of engineering, professions can begin as nonprofessionalized occupations; the transition is gradually brought about by a conscious effort from members of that occupation. The status of "profession" is not a clear-cut category but rather a spectrum.

Many technology-related jobs outside of the traditional engineering disciplines occupy a hazy place on the spectrum of professionalization. In computing, for instance, there are professional organizations such as the ACM, but those organizations' power extends only as far as the people who join them or otherwise voluntarily follow their standards. According to Johnson (2008), nonprofessionalized computing puts workers in the role of "guns for hire"; when a computing worker is asked to work on a legal but morally questionable task, they are under pressure to either fulfill the request or cede their job to somebody else who will fulfill it. Johnson argues that increased professionalization is needed "to take certain issues out of the marketplace." That is, in a well-established profession like law or medicine, it is difficult to simply threaten to replace a worker who refuses a request based on the profession's ethical standards, because the workers who would be hired as a replacement are beholden to the same standards.

On the other hand, a profession relies on higher barriers to entry and stricter specifications for training and practice than do traditional occupations, and these traits come with some risks (National Research Council 2013). Professionalization can place a greater burden on workers to acquire training at their own expense, rather than receive on-the-job training from employers. The profession's standards can be slow to adapt to changing technologies; these standards can limit innovation by reinforcing older practices and by keeping out unconventional thinkers who do not fit within the profession's rigid entry requirements. In a rapidly changing field, the roles for which people are hired may themselves evolve so quickly that the profession's official definition of its job may not match the jobs that are actually in demand.

These factors suggest that professionalization is most effective when a field has developed a stable, widely agreed-upon set of practices; debates about how much to professionalize (especially regarding features like legally mandated licensing) often revolve around whether a field is "mature enough."

We have seen that software engineering is a prominent example of a technology field in the limbo between profession and nonprofession. Its status has been shaped by professionalization proponents who cite the potential for public harm from faulty software and thus the need for a high standard of competence, and by skeptics who argue that software engineering practices are evolving too rapidly for a fixed set of practices to be reasonable. In the 1990s, the IEEE and the ACM formed the Software Engineering Coordinating Committee (SWECC) to advance software engineering as a profession. One of the first products of this collaboration was a series of reports aiming to articulate a core body of knowledge for software engineering and to establish a consistent definition of the field (Duggins and Thomas 2002). The 1990s also saw efforts to establish licensing for software engineers, with Texas becoming the first US state to require the same licensing for software engineers as for traditional engineering fields. However, the ACM withdrew from the SWECC and announced its position that licensing would be premature and counterproductive. Nonetheless, the licensing movement in the United States continued into the twenty-first century, with the National Council of Examiners for Engineering and Surveying developing a software engineering examination intended specifically for engineers in public service or safety-critical applications. This examination was discontinued in 2019 because there were too few examinees. Despite the relative failure of licensing efforts, though, software engineering has made noteworthy strides in other areas, such as the growth of accreditation for software engineering degree programs (Mok 2010).

6.2.4 PROFESSIONS AND THE LAW

We have discussed how professions, as organized institutions, exist in part to moderate the needs of practitioners with the needs of the public. The public wants to receive high-quality, trustworthy services through fair business practices. Practitioners typically want a degree of autonomy in their work that could be sacrificed if too much of the oversight is done by people who lack expertise in the area they are regulating. Hence, while most practitioners are not opposed to all regulation, they are sometimes opposed to regulations that they perceive as being inappropriate, overbearing, or onerous. Professional organizations offer a compromise by regulating from within the community of experts. From this perspective, professional ethics serves as a supplement to the law in response to the limitations of public regulation.

Consider the handling of medical negligence suits in courts in the United Kingdom (Jones 2000) and the United States (Gossman et al. 2020). These cases involve evaluating whether a physician or other healthcare professional harmed a patient by failing to follow appropriate standards of care. Such standards are not necessarily codified into law—instead, the courts often call upon "expert witnesses" from the medical community to help determine what the appropriate standards are.

Why rely on expert witnesses to articulate the standards in each case, rather than writing those standards into comprehensive laws? As mentioned above, one argument is that only people with deep expertise are qualified to set the standards, and the people with the expertise are different from the people in charge of making laws. Additionally, standards can evolve quickly; when researchers discover an important new technique or invalidate a technique that was previously accepted, practitioners may need to change faster than the law can adapt.

Many technical professions share a critical feature: often the rate of change in technology outpaces the rate of change in the legal system, leaving a "policy vacuum" (Moor 1985) in which current regulations do not prescribe a way to handle the newest developments. Recent examples include the standard practice of collecting data and cookies about every user that comes to a website without requiring explicit permission, a practice largely shaken up by the introduction of the General Data Protection Regulation (GDPR) and other data processing and notification requirements passed in 2018 in the European Union. Likewise, in the early years of AirBnB, many cities did not have particular legislation in place about renting out one's own apartment. In the ensuing years there have been numerous legal battles attempting to clarify the legality of renting out one's home, who's responsible, and how its insurance works. In technical research, advances are made at the pace of dissemination of advances that, with modern platforms, can be days after an innovation is made. Likewise, within the business domain, the mantra of "move fast and break things" (Vardi 2020) often does not leave room to wait for regulation to catch up to the technology being developed.

Technology is developing rapidly, and the law is shifting constantly to try to accommodate these advances by refining existing language to include new technologies, or by finding new legal frameworks that offer precedents that can guide the legislation of technology. Just as important as the changes in technology itself are the changes in how technology is conceptualized and understood both in the law of the land and in the minds of the public that uses it. For example, after the advent of online mail (email) being run by third-party companies, it was unclear to many that email is actually more similar to a postcard, where the contents are plainly visible to the entity passing it around, than to a post in a sealed envelope. This lack of nuance in the distinction laid the foundation for questions such as "Who or what is allowed to analyze email that passes across a server?"

In many online email platforms, the email is read by algorithms, not real people, but before the rise of third-party email providers, it was not anticipated that a program would read your email in order to serve advertisements. However, once this became a reality, a legal interpretation was created that still protected a user's right to privacy from the government, even though a commercial algorithm was reading their mail. Many of our notions of privacy for email rely on resoning based on the metaphor that mail to our home and mail to our computer are similar. However, in thinking about privacy, this metaphor fails us: sealed physical mail is afforded many more legal protections than online mail. Indeed, most online mail is typically read by algorithms for target advertising and user profiling (Gould 2014). However, even this landscape is rapidly changing as services such as Gmail and others attempt to turn what was once consumer facing free products into productivity tools for large corporations. Within the realm of corporate productivity, there is often a much more rigid expectation of privacy on the part of the corporation purchasing the tool, and this has required a change in how these tools handle data (Wakabayashi 2017). Here we can see that a technology (email) is interpreted through the metaphor of an older law (private mail), but not always strictly.

Sidebar: Metaphors, Personhood, Technology, and Autonomous Cars
One place where professional codes and the law intersect is the rapidly advancing field of autonomous vehicles. Consider a self-driving or autonomous car. At first glance it seems that this system would meet many of the criteria for moral agency discussed in chapter 4. In order to be useful, it seems that a driverless car must be able to do a few different things:

1. A car must be able to move its location, convey passengers, turn on headlights, and perform other physical feats. If agency is defined in terms of effecting change in its immediate environment, the car can be said to have agency.

2. The car must be able to react in ways not fully determined by the environment, for example, reacting to other drivers or perceptions of where they may move their cars (or selves) in the future. If it does, we could say it has self-determination.

3. A car needs to understand its location in an environment; this could imply self-awareness.

4. If a car can set goals and ends and act on those ends, it could be said to have rationality.

5. If the car is held accountable for its actions and if it can be forgiven and praised, then it could be said to be responsible; or perhaps the company that manufactured it is responsible.

Consider a car that contains software that is sufficiently advanced to drive itself, and there is no steering column nor other input. Now consider a setting in which a humanoid robot gets into a car and drives it through the means that the average adult would (i.e., using a steering wheel and pedals). Is there a difference? Should there be? Reflecting on this difference requires drawing on the idea of material substance-based metaphysical definitions of personhood. It is possible that from a technological point of view, the car that drives itself and the robot that drives the car are made of the same sensors and software. However, does your idea differ about the car, the robot, and the moral and legal responsibility of, for example, what happens if there is an accident? If so, why? Is it because your intuition is complicated by thinking about the body that is actually getting into the car? (Richards and Smart 2016).

This example is important to consider because the way the law handles robots and robotic technologies, including automated decision-making technologies, will have a huge impact on how these technologies are developed and what the rules for developing them are. Richards and Smart (2016) discuss the pros and cons of the common practice in law of using metaphor to understand new technology and to apply past laws to new technology using this metaphor. They warn that adopting insufficient or incorrect metaphors for extending these laws to new technologies can lead to problems later. In the context of the United States, for example, it was decided that the same privacy protections for email should not be applied to physical mail. Similarly, as we detailed in chapter 4, the metaphors that we use for space and privacy become complicated in the face of changing technology. Richards and Smart warn that if we decide on an insufficient metaphor to undergird our legal reasoning for robots and autonomous vehicles, it could be problematic for the way we develop and regulate these technologies.

Here we can see that although some definitions of personhood would recognize a robot with a humanoid body as being distinctly different from a nonhumanoid robot (or even from a person), there are methodologies of reasoning about personhood that ignore this distinction. Richards and Smart argue that one dangerous pitfall for the law, which they call "the android fallacy" (2016, 20), is that there is a meaningful difference between humanoid and non-humanoid robots.

Unexpected effects of new technologies can take regulators by surprise. For instance, companies such as Bird Rides, which allow freestanding electric scooters to be rented via mobile app, deployed their scooters in many cities suddenly and with little warning to local governments. Unanticipated issues arose, including high rates of accidents, unclear liability for accidents, and parked scooters obstructing walkways; this prompted some cities to temporarily ban the scooters until regulatory questions were resolved (Herrman 2019). Another example from the early 2010s are ride-sharing companies such as Uber and Lyft. Both of these companies operate under the idea of a "cab service" or "giving a ride to a friend," and the specifics of legal liability and even local licensure for cabs has been bitterly contested in courts since the beginning, with some cities completely banning Uber and Lyft (Ongweso and Koebler 2019). These examples illustrate how law can lag technology development, and why professional ethics codes such as the ACM's encourage developers to take a proactive role in how technology is integrated into existing societal structures.

Despite the distinction we have made between regulation by lawmakers and self-governance within a profession, these realms do not exist in isolation from one another; dialogue between lawmakers and professionals helps to shape the law. When both parties prioritize the public interest, this interaction can benefit consumers. For instance, many of the fundamental rules for governing traffic and regulation on the internet were written in collaboration between the public, private companies, and academics in the World Wide Web Consortium (W3C) and governing bodies at the Federal Communications Commission (FCC) and other regulatory bodies.

On the other hand, in some cases professionals' influence over the law has *undermined* consumer welfare. For instance, in 2018 and 2019, a new software system that was omitted from manuals caused two fatal crashes by Boeing 737 MAX airliners. Remarking that the United States Federal Aviation Administration (FAA) had been unusually slow to ground any further flights by these planes, commentators suggested that close relationships between FAA inspectors and Boeing's own safety experts may have contributed to this regulatory delay (*The Economist* 2019). Other examples of the complex interplay between technology professionals, the public, businesses, and governments include the history of safety belts (Nader 1965) and stock market trading (Lewis 2014).

In addition to thinking about the impacts of your own work, we encourage you to think about the ways in which the laws and regulations related to technology could be better, and what part you can play in improving them. This is not dreamy idealism! Both the law and professional codes of conduct are works in progress, and as citizens with specialized knowledge, you play an active role in helping to build or revise these statutes. Part of what you are doing, when you make choices about how to act, is participating in the ongoing civic discussion of how technology *should* be used and regulated. Beyond your everyday practice of technology development, you may very well be in a position to have an impact on the direction that the law takes. These impacts can be made through professional organizations and standards which either inform or become the law, such as IEEE open requests for comment on standards from image exchange to responsible algorithms. As we discuss in the next section, the rules and guidelines from these professional organizations are living documents and professionals are actively encouraged to refine these rules. Likewise, even as an individual you are empowered through commenting, blogging, public talks, testifying in a court or to a legislature, teaching, or petitioning, to name only a few avenues of public engagement.

Story Point: "Not Smart, Not Clever," by E. Saxey

> It's only plagiarism. . . . None of us are monsters. We're symptoms of a sick system.
>
> And then his face crinkles up . . . and he says, "I liked you."

"Not Smart, Not Clever" is a story about a plagiarism epidemic at a British university, which persists despite the University's many attempts to surveil it out of existence.

It is also the story of several students struggling to find a way through a hostile and challenging system—and one who has given up on finding a place and instead decided to practice their would-be profession's skills on their own terms. These two threads are closely related: as Lin, the story's narrator, makes clear, her boyfriend Zach and his friends are tempted to plagiarism because their futures feel so precarious. An unsympathetic read might suggest that this is a story about individuals dodging their responsibilities to a profession. After all, degree requirements are one way a profession maintains its integrity by making sure practitioners have the knowledge needed to practice responsibly; plagiarism is taken so seriously because it offers a way of clearing these requirements without actually mastering that knowledge. But the story also raises questions about a profession's responsibility to the individual, showing what happens when the system puts its resources toward inventing new ways to make sure future professionals check the right boxes, rather than toward nurturing their growth as human beings.

6.3   CODES OF ETHICS

We have discussed how professions make moral commitments, and how these commitments are embodied in professional organizations' codes of ethics. But what *are* these commitments? In this section, we review the contents of two organizations' codes of ethics: the ACM and the IEEE. The ACM is a society that broadly encompasses computer science, while the IEEE includes electrical engineering along with related professions that also include computing disciplines. Along with the individual organizations' codes, we also review the SECEPP, which was composed jointly by both organizations to target software engineering specifically. Although these codes differ in their style (e.g., the IEEE code is a brief list of principles, whereas the ACM code provides paragraphs of elaborating text with each principle and the SECEPP has finer-grained subclauses) and precise target audience (computing and other technology workers vs. computing workers specifically vs. software engineers specifically), they reflect many of the same core values. We organize our review by identifying and discussing several of these values, followed by pointers to the relevant clauses in the codes.

6.3.1   DEEP DIVE: THE ACM AND IEEE CODES OF ETHICS

Prioritizing the public good. The codes quoted below emphasize this as one of their most important concerns by placing it as one of the first clauses and by using strong wording such as "hold paramount." This echoes the idea of a profession being a sort of contractual exchange with society, with practitioners gaining special privileges and owing special duties to humanity in return. Both the IEEE code (via "sustainable development" in the main clause) and the ACM code (in the elaboration, not shown) group environmental concerns with the public good, suggesting a utilitarian view of nature as a contributor to human welfare.

- ACM 1.1: "Contribute to society and to human well-being, acknowledging that all people are stakeholders in computing."

- IEEE 1: "To hold paramount the safety, health, and welfare of the public, to strive to comply with ethical design and sustainable development practices."

- SECEPP 1.02: "Moderate the interests of the software engineer, the employer, the client and the users with the public good."

Avoidance of harm and disclosing risks thereof. Along with the general directive for the public good, these codes specifically emphasize care when building safety-critical systems in which flaws could lead to tangible damage (think of the computer systems that control an airplane, a medical device, or a nuclear power plant). The ACM's version is much more broadly worded than the others, but its elaboration emphasizes "unintended" and "unjust" harm; the code is carefully phrased so that it does not necessarily prohibit society members from working in areas like weapons development or cyberwarfare.

- ACM 1.2: "Avoid harm."

- IEEE 1: "To disclose promptly factors that might endanger the public or the environment."

- SECEPP 1.03: "Approve software only if they have a well-founded belief that it is safe, meets specifications, passes appropriate tests, and does not diminish quality of life, diminish privacy or harm the environment. The ultimate effect of the work should be to the public good."

- SECEPP 1.04: "Disclose to appropriate persons or authorities any actual or potential danger to the user, the public, or the environment, that they reasonably believe to be associated with software or related documents."

Honesty. These codes cover avoiding deception in a variety of forms, from untruthful personal statements to dishonest business practices.

- ACM 1.3: "Be honest and trustworthy."

- IEEE 4: "To reject bribery in all its forms."

- SECEPP 1.06: "Be fair and avoid deception in all statements, particularly public ones, concerning software or related documents, methods and tools."

- SECEPP 4.04: "Not engage in deceptive financial practices such as bribery, double billing, or other improper financial practices."

Accurate assessment. Besides negative mandates against intentional deception, these codes include positive mandates for making sure claims are well supported.

- ACM 2.5: "Give comprehensive and thorough evaluations of computer systems and their impacts, including analysis of possible risks."

- IEEE 3: "To be honest and realistic in stating claims or estimates based on available data."

- SECEPP 6.07: "Be accurate in stating the characteristics of software on which they work, avoiding not only false claims but also claims that might reasonably be supposed to be speculative, vacuous, deceptive, misleading, or doubtful."

Avoidance or disclosure of conflicts of interest. A *conflict of interest* is a situation in which a professional is serving multiple parties (possibly including the professional themself) and acting in the interest of one party may harm another. For instance, suppose a client hires a consultant to help decide whether the client's business should purchase a certain software suite. There could be conflict of interest if the consultant also works for, or owns stock in, the company who makes the software. In both cases, it would be difficult for the consultant to make an unbiased judgment, because there are incentives (either benefit to the other client or personal profit) to treat the software suite favorably.

- ACM 1.3 elaboration: "Computing professionals should be forthright about any circumstances that might lead to either real or perceived conflicts of interest or otherwise tend to undermine the independence of their judgment."

- IEEE 2: "To avoid real or perceived conflicts of interest whenever possible, and to disclose them to affected parties when they do exist."

- SECEPP 4.06: "Refuse to participate, as members or advisors, in a private, governmental or professional body concerned with software related issues, in which they, their employers or their clients have undisclosed potential conflicts of interest."

Nondiscriminatory treatment. In the ACM elaboration, sexual harassment and bullying are regarded as forms of discrimination. This recognizes that the effects of harassment go beyond the experience of the victims of the individual acts; the mere possibility of harassment creates a pervasive climate that discourages participation by those most likely to be targeted.

- ACM 1.4: "Be fair and take action not to discriminate."

- IEEE 8: "To treat fairly all persons and to not engage in acts of discrimination based on race, religion, gender, disability, age, national origin, sexual orientation, gender identity, or gender expression."

- SECEPP 8.07: "Not give unfair treatment to anyone because of any irrelevant prejudices."

Education of the public. No technical field exists in isolation; ultimately, non-experts will need to interact with, and help regulate, the technology created by the experts. This requires the professionals to develop a common language with the rest of society.

- ACM 2.7: "Foster public awareness and understanding of computing, related technologies, and their consequences."

- IEEE 5: "To improve the understanding by individuals and society of the capabilities and societal implications of conventional and emerging technologies, including intelligent systems."

- SECEPP 1.08: "Be encouraged to volunteer professional skills to good causes and contribute to public education concerning the discipline."

Working only in areas of competence. In other words, the codes quoted below urge computing professionals to only accept tasks for which they have the appropriate skills. They have this principle in common with the National Society of Professional Engineers, whose code of ethics lists it as one of the "fundamental canons" of engineering. However, for many computing workers, their areas of competence are less clear-cut than for engineers: Practicing engineering in many jurisdictions requires a license that establishes specific competencies, while there is less regulation around what competence means in most computing fields. In both engineering and computing professions, though, the principle implicitly asks the professional to understand the limits of their own skill set in addition to having the required degrees and certifications.

- ACM 2.6: "Perform work only in areas of competence."

- IEEE 6: "To undertake technological tasks for others only if qualified by training or experience, or after full disclosure of pertinent limitations."

- SECEPP 2.01: "Provide service in their areas of competence, being honest and forthright about any limitations of their experience and education."

Credit-giving and respect for intellectual property. If you are a student, the violation of this principle you may have heard about most is *plagiarism*, the direct reuse of others' work without giving credit. This is distinct from infringement of intellectual property such as copyrights, trademarks, and patents; plagiarism is about attribution of the work, and intellectual property infringement is about permission to use the work.

- ACM 1.5: "Respect the work required to produce new ideas, inventions, creative works, and computing artifacts."

- IEEE 7: "To credit properly the contributions of others."

- SECEPP 2.02: "Not knowingly use software that is obtained or retained either illegally or unethically."

- SECEPP 7.03: "Credit fully the work of others and refrain from taking undue credit."

Professional development. In each of the codes below, this includes both self-improvement and helping colleagues, and both technical and ethical development. The principle recognizes that computing is a rapidly changing field, so quality work requires continued education throughout a professional's career.

- ACM 2.2: "Maintain high standards of professional competence, conduct, and ethical practice."

- ACM 3.5: "Create opportunities for members of the organization or group to grow as professionals."

- IEEE 6: "To maintain and improve our technical competence."

- IEEE 10: "To assist colleagues and coworkers in their professional development and to support them in following this code of ethics."

- SECEPP 7.02: "Assist colleagues in professional development."

- SECEPP 8.01: "Further their knowledge of developments in the analysis, specification, design, development, maintenance and testing of software and related documents, together with the management of the development process."

- SECEPP 8.06: "Improve their knowledge of this Code, its interpretation, and its application to their work."

Accepting and giving review. Professional engineers are often called on to review drawings and other technical work which needs peer review before being deployed or sold to the public. Similarly, it is common in software development to use peer review to ensure good programming practices. This requirement is also especially prominent in academia, where a paper is traditionally expected to undergo peer review, with other researchers acting as gatekeepers, before being published and accepted as credible. The purpose of these review processes is not to ensure politeness, but rather to ensure that the work of any individual within the profession is of high quality. Poor work by any member of the profession jeopardizes the entire community, hence it is the responsibility of every member of the community to ensure these standards through the review process.

- ACM 2.4: "Accept and provide appropriate professional review."

- IEEE 7: "To seek, accept, and offer honest criticism of technical work, to acknowledge and correct errors."

Workplace confidentiality. Professional work sometimes involves being entrusted with sensitive information; for instance, a company developing an innovative product may expect (or mandate, via a "nondisclosure agreement") its software engineers not to discuss the design with parties outside of the company, to avoid competitors copying the idea and undermining the product's profitability. Societies like the ACM generally tell their members to honor these expectations. However, unlike in some professions such as law, this guideline is not absolute. A defense lawyer, at least in the United States, has special legal protections and obligations to keep their client's admissions of criminal activity confidential because the lawyer-client relationship depends on the client's willingness to speak freely; meanwhile, a computing professional who learns of lawbreaking by a client might be legally compelled by a subpoena to share this information with the court, and the codes quoted below would make an exception to confidentiality for someone obeying this subpoena. These codes also recognize that confidentiality sometimes conflicts with other principles, like protecting the public good.

- ACM 1.7 elaboration: "Computing professionals should protect confidentiality except in cases where [the information] is evidence of the violation of law, of organizational regulations, or of the Code. In these cases, the nature or contents of that information should not be disclosed except to appropriate authorities."

- SECEPP 2.05: "Keep private any confidential information gained in their professional work, where such confidentiality is consistent with the public interest and consistent with the law."

Equitable design. This encompasses themes of accessibility: on an individual level, design of systems with disabled users in mind, and on a societal level, consideration of whether less enfranchised people stand to benefit from a technology as much as more enfranchised people. The ACM text was added in the 2018 update, reflecting increased public consciousness around these issues; for instance, machine learning was increasingly used to supplement or replace human decision making in areas such as loan approval and criminal sentencing in which there was a history of racial bias, with infamous cases of those machine learning systems replicating the patterns of bias.

- ACM 1.4 elaboration: "Technologies and practices should be as inclusive and accessible as possible and computing professionals should take action to avoid creating systems or technologies that disenfranchise or oppress people."

- SECEPP 1.07: "Consider issues of physical disabilities, allocation of resources, economic disadvantage and other factors that can diminish access to the benefits of software."

   Privacy-protecting design. We discuss the meaning and moral significance of this concept in chapter 4.

- ACM 1.6: "Respect privacy."

- SECEPP 3.12: "Work to develop software and related documents that respect the privacy of those who will be affected by that software."

   The principles discussed above have been present in codes of ethics for decades; for instance, the SECEPP text (Gotterbarn et al. 2001) was adopted in 2000. New principles are emerging as the profession adopts new priorities, however. After previously being updated in 1992, the version of the ACM Code of Ethics reviewed here (Gotterbarn et al. 2018) was updated in 2018 in an iterative drafting process by a task force of computing, philosophy, and policy experts adding the equatiable theme and updating others (Association for Computing Machinery 2018). The drastic change in the ubiquity of the Internet and other computing technologies since then is reflected in the new clause: "Recognize when computer systems are becoming integrated into the infrastructure of society, and adopt an appropriate standard of care for those systems." Clauses were also added about the previously unaddressed principles: to design systems with security in mind, and to "retire legacy systems with care." To avoid the code becoming dated quickly as technologies change, the authors avoided references to specific technologies; even machine learning was considered too specific to mention specifically, although some text was added about the extra care needed in managing self-modifying systems. Addressing the ethics of specific technologies was left to additional ACM initiatives outside of the code; we discuss an example in the next subsection.

6.3.2 PROFESSIONAL GUIDELINES FOR ALGORITHMS: BIAS, TRANSPARENCY, AND ACCOUNTABILITY

Sometimes a professional association will have guidelines for certain issues, in addition to the more general principles outlined in a code of conduct. Starting in 2016 both the ACM and the IEEE began to turn their attention to the specific issues surrounding how deeply algorithms have become a part of our daily lives. The IEEE established the standards working group SA 7003 which is specifically focused on developing its Standards on Algorithmic Bias Considerations, dedicated to examining and averting the often-unintentional phenomenon of prejudiced decision-making (e.g., racial discrimination) by machine learning algorithms trained with biased data, as we discuss in chapter 3.

This effort has expanded into a set of IEEE Standards on Ethically Aligned Design efforts (IEEE 2022a) that include starting multistakeholder conversations to create standards and best practice documents for topics including Transparency of Autonomous Systems, Data Privacy, and even Emulated Empathy in Autonomous Systems. Various academic conferences have also sprung up around these efforts including the ACM FaccT Conference on Fairness, Accountability, and Transparency in Machine Learning, Natural Language Processing, and Computer Vision Conference and the ACM/AAAI (Association for the Advancement of Artificial Intelliegence) Conference on AI, Ethics, & Society. Finally, since 2015 there have been numerous white papers and statements of ethics released by governments, academia, and businesses. A critical look at the similarities and differences in these guidelines can be found in Fjeld et al. (2020), who find a "normative core" of principles including concerns for privacy, accountability, safety, transparency, explainability, fairness, human control, professional responsibility, and promotion of human values.

Most directly, the ACM has outlined seven principles for designing and deploying algorithms in order to maximize transparency and accountability. This set of principles, according to the ACM, is intended to support algorithmic decision making while minimizing the potential for harm. These principles are an interesting starting point to think about what issues are important when creating, deploying, and maintaining systems that make decisions with little to no human intervention. We reproduce these principles below with some additional commentary.

1. Awareness: Owners, designers, builders, users, and other stakeholders of analytic systems should be aware of the possible biases involved in their design, implementation, and use and the potential harm that biases can cause to individuals and society. This directive is designed to reinforce the idea that one should consider where, how, and under what considerations algorithms will be deployed. While this may seem obvious to some it is important to remember that this may mean seeking outside consultation or help when creating and deploying systems.

2. Access and redress: Regulators should encourage the adoption of mechanisms that enable questioning and redress for individuals and groups that are adversely affected by algorithmically informed decisions. This is aimed at a notion of legal transparency; people should be able to appeal the decisions of algorithms and institutions. We will see this more explicitly pointed out in some later principles like accountability and explanation. However, the goal of access and redress is an interesting one. For instance, some social networks do not include mechanisms to appeal or recover deleted content. Should all businesses that use algorithmic decision making for moderation and ad placement be required to allow inquiry? The GDPR has recently enacted ideas like this into law. What about algorithmic sentencing, credit approval, or loan approval? For some of these cases (i.e., sentencing) there is an established process for access and redress (appeals), however, with respect to financial credit decisions, there is no universal legal standard. Here we can start to see the tension between professional society directives and established law.

3. Accountability: Institutions should be held responsible for decisions made by the algorithms that they use, even if it is not feasible to explain in detail how the algorithms produce their results. This is a strong statement with an ambitious goal: that organizations and institutions should stand by the decisions of their algorithms. This principle can become murky when one extends the logic to, for instance, driverless cars. Should a company be responsible for every crash an autonomous vehicle has? Think about what accountability means in all the domains where algorithms are used; is it possible to always have a person or corporation be accountable for the algorithms they deploy?

4. Explanation: Systems and institutions that use algorithmic decision-making are encouraged to produce explanations regarding both the procedures followed by the algorithm and the specific decisions that are made. This is particularly important in public policy contexts. Again, in many contexts this explanation is required by law. However, it is an interesting question to consider at what level of detail this explanation must be provided. Does source code always count as a sufficient explanation, or in some cases must data be provided as well?

5. Data Provenance: A description of the way in which the training data was collected should be maintained by the builders of the algorithms, accompanied by an exploration of the potential biases induced by the human or algorithmic data-gathering process. Public scrutiny of the data provides maximum opportunity for corrections. However, concerns over privacy, protecting trade secrets, or revelation of analytics that might allow malicious actors to game the system can justify restricting access to qualified and authorized individuals. Here we can see a tension between the need to disclose information and keeping some things secret. This specific example is interesting in an era when Google and Facebook (among others) are collecting more data about users than ever before. How much of this data should be made public and auditable? What about the data that credit card companies or banks collect?

6. Auditability: Models, algorithms, data, and decisions should be recorded so that they can be audited in cases where harm is suspected. This is a principle that many professional programmers, academics, and system designers, no matter the domain, should follow as it ties into the need for reproducibility. This is again enshrined in legal standards for areas like accounting and safety-critical systems (e.g., airplane autopilots) but has yet to be rigorously enforced across all algorithmic decision-making systems.

7. Validation and Testing: Institutions should use rigorous methods to validate their models and document those methods and results. In particular, they should routinely perform tests to assess and determine whether the model generates discriminatory harm. Institutions are encouraged to make the results of such tests public. This principle is interesting when one thinks about an age of continuous deployment and ever-changing code bases. Should models and algorithms be tested before deployment? What about the results of A/B testing, where two different versions of a service or interface are shown to different people, or live bucket testing in which tests are performed on users without their knowledge? Many companies participate in this type of testing without their users being aware in order to test algorithms with "real users."

Story Point: "Asleep at the Wheel," by T. Coraghessan Boyle

> Of course, Carly was right, and if they wind up being ten minutes late to pick up Keystone that's nobody's fault but her own. "All right, Carly, I'm sorry—good job, really," she says, only vaguely aware of how ridiculous it is to try to mollify a computer or worry about hurting its feelings.
>
> "Since we're at the library," Carly says, "will you be acquiring books? Because they have three copies of the latest installment of the Carson Umquist series you like—and they're all in the special 'Hot Reads' rack when you first walk in. I mean, they're right there—you don't have to go twenty feet. If that's what you're looking for."

How much decisional power should a technology developer assume on behalf of their end users? When using one's unique expertise to help others, it is easy to drift into paternalism, that is, taking autonomy away from the people you are trying to help. Cindy's life is made easier—at least most of the time—by her self-driving car, "Carly." Sometimes Carly simply manages Cindy's schedule and spares her the worry of getting herself from place to place. Other times, Carly reminds Cindy of shopping opportunities that interest her. But sometimes, Carly's programming leads it to challenge Cindy's preferences, such as when she wants to spend time with her intriguing but reckless client Keystone. It's easy to spot the paternalism at play when Carly will not unlock its doors for Keystone, but "Asleep at the Wheel" suggests that our concern should begin earlier, and that the entire interlocking smart system that Carly is a part of is unavoidably paternalistic.

6.3.3 THE FUNCTIONS OF AND USING CODES OF ETHICS

We have noted that professional societies often have their own codes of ethics; in fact, we have seen that having a code of ethics is an essential part of some definitions of a "profession." Professional ethics can be divided into two types. One type consists of those rules and guidelines that govern behavior and are aimed at preventing cases of professional misconduct and particular disasters—as Harris et al. (2013) refers to them, preventive ethics. Other parts of professional ethics precepts are aimed at positive, or aspirational ethics, aimed at using technology for the betterment of humankind.

Technology businesses have taken an increased interest in promoting ethics at an organization-wide level. But a company's claim of ethical consciousness deserves scrutiny. One possible trap is conflating regulatory compliance with ethics. Not only have there been innumerable cases throughout history of the law failing to protect certain vulnerable populations, but also, recall that regulators cannot always anticipate the situations that technology developers will encounter, leading to a "policy vacuum." A code's guidance can help fill this vacuum, but another possible trap is making only shallow appeals to ethics (Gotterbarn 2020). Lining the workplace's walls with posters featuring a code can be an effective way for a company to *appear* morally upstanding while making little change to its actual practices.

So, what are codes of ethics *actually for*? At a basic level, we could say they exist to state expectations about how a professional should behave. However, this seemingly singular purpose has many different dimensions. To whom are these expectations being stated, and why? Codes of ethics serve both "inward-facing" functions by communicating to professionals themselves, and "outward-facing" functions by communicating with the rest of society (Metcalf 2014).

Recall that the concept of a profession, compared to other occupations, revolves around an exchange of power. Professionals do work that needs regulation, but meaningful regulation requires expertise that only the professionals themselves have. A solution is for the community to hand over regulatory power to a professional organization. The organization's code of ethics serves an outward-facing purpose by telling the community, in easily understandable language, what to expect from this self-regulation. The code is part of an organization's strategy to maintain the community's trust, so that the community will continue to let the profession keep its self-governing privileges.

A clear example of this in the United States is the way medical licenses are given to professionals, which enable them to practice medicine. Each medical practitioner in the United States must be licensed by a state medical board, which is (usually) made up of practicing professionals, and a common national evaluation of training and standards. These boards are (typically) created by an act of the state governing body but the rules, interpretation, and regulation of these rules is left to practicing professionals. Most of these standards are self-imposed by the various professional societies; reviews of practice and ethics are also undertaken by these professional societies, not state legislatures (Carlson and Thompson 2005).

Ultimately, maintaining the community's trust requires that the profession actually *fulfill* its promise of good behavior. The inward-facing roles of a code of ethics help ensure the promise is fulfilled. For new professionals, who may have each individually developed different ideas about professional conduct from their respective workplaces and universities, a code gives a standard point of reference (Johnson

2009). For established professionals, a code serves as a continual reminder of the professional culture's values and aspirations. When a professional resists pressure from outside forces (such as an employer) to act unethically, a code affirms the organization's support for that professional. And when an organization member or licensed practitioner behaves disreputably, the code may be cited as grounds for revoking their membership or license.

When an ethical question arises, a code of ethics can serve as a decision-making aid. In general, a code is not designed to be a definitive source of answers; however, the code can highlight relevant values and suggest angles for thinking about the problem. For instance, in 2016, the US Federal Bureau of Investigation (FBI) wanted to access encrypted data on a smartphone connected to a terrorist attack and sought a court order that would compel smartphone manufacturer Apple to create software to bypass the encryption. In theory, though, such software could be used to compromise the security not only of this particular smartphone but also of smartphones belonging to Apple customers anywhere. Because of this, a group of Apple engineers considered refusing to build the software if the court order were issued. One commentator (Markoff 2016) invoked the ACM Code of Ethics to discuss this possible protest: The code includes a directive for computing professionals to follow the law. However, it makes allowances for violating the law when there is a "compelling ethical justification." Since the code also enshrines the principles of respecting privacy and avoiding harm to the public, the commentator concluded that refusing to carry out the court order would be consistent with the code.

On the other hand, one could invoke the ACM code in *favor* of obeying the court order as well: The code says to "contribute to society and to human well-being," and from the FBI's perspective, the encryption-bypassing software was a potential opportunity to discover new information about terrorist plots that could help them save lives. Thus it could be argued alternatively that breaking the law by refusing to build the software would not have a "compelling ethical justification." This highlights how conflicts sometimes exist between the different directives in a code of ethics. Codes are, as the ACM code itself states, not designed to be "an algorithm for solving ethical problems" but rather to suggest important values that can be used for decision-making when combined with a professional's own judgment.

The Apple scenario also raises questions of who is beholden to a code of ethics. The ACM code is intended as a guide for all computing professionals. Within the ACM itself, the code is enforceable; the organization reviews reports of members violating the code and may take punitive action against violators. However, both membership in the ACM and adherence to the code for nonmembers are completely voluntary. Furthermore, many professionals answer to employers outside of the profession; since those employers are not beholden to the code, the professionals bear the full responsibility for compliance. Because of these factors, the success of the code depends on personal commitments or consensus within a practicing community, whether that community is the profession as a whole or just a group of practitioners working for the relevant stakeholders. For example, Apple's "rebellious" culture was cited as contributing to the vocal dissent among the engineers (Markoff et al. 2016).

Story Point: "Codename: Delphi," by Linda Nagata

> It wasn't money that kept Karin at her control station. As the nightmare of the war played on before her eyes, it was knowing that the advice and the warnings that she spoke could save her soldiers' lives.

Karin works as a military handler, passing along crucial battlefield data to her soldiers in real time. Although she sits safely behind a desk thousands of miles away, Karin's workday is as frantic and high-stakes as her clients'. She does have a supervisor, but the speed of the battlefield, and the fact that she's juggling three clients at once, means that Karin is essentially on her own in issuing warnings, guidance, and sometimes even kill orders. Karin's job is made even harder by the fact that her aims don't always match those of her clients: she wants to bring them back alive, but the clients themselves often want to complete their missions even when it involves serious risks. Karin is keenly aware that her life is easier and safer than her soldiers' lives, but she has nonetheless developed some habits and mentalities that more closely resemble a soldier than an average help desk worker; she also suffers a lot of the same things as they do. "Codename: Delphi" offers a harrowing look at the real-time experience of navigating difficult and high-stakes jobs and provides a vivid illustration of how people's prior commitments inform their choices most of all in the moments when there is no time to reflect, but only to act.

6.4   SOME SUGGESTIONS ON MAKING ETHICAL DECISIONS IN PRACTICE

For a practitioner, ethics is best thought of as a process to be actively engaged in rather than a passive background concern (Gotterbarn and Wolf 2020). We often discuss moral decision making in language that suggests an individualistic process, but like the process of engineering a technological system, it is often a collaborative effort. Consider the setting of a technology company. Some ethical dilemmas might be uncovered by single workers' vigilance and resolved by single workers' insights. But some issues may require a combination of perspectives (e.g., from both business and engineering professionals) or a resolution of differences (e.g., in beliefs about technology developers' moral responsibilities).

Moral issues come in many shapes, so there is no universal formula for identifying and resolving them. But this does not mean that moral decision making must be ad hoc or unstructured. And in a collective decision-making setting, adding additional decision makers does not inherently result in a wide perspective; similarities between members can even *strengthen* a group's collective biases.

For these reasons, Vallor et al. (2018) emphasize integrating ethics as an ordinary part of the technology development process rather than treating it as something external, and operationalizing it so that developers across a business have a common understanding of the expectations for ethical practice. Their proposals include the following:

- Routine "ethical risk sweeping" exercises in which team members try to thoroughly identify ethical implications of the project that may have been missed before, similar to how cybersecurity red-teaming tries to detect hidden vulnerabilities.

- Case studies of the ethical failings of the company's product whenever such a failing occurs.

- Case studies of the ethical implications of similar existing products when the company begins a new product.

- Exercises to identify who will be affected by the product, whose values are assumed in the product and whether those assumptions are justified, and who might use the product in unexpected (malicious or otherwise) ways.

- Formal integration of ethical review practices into the structure of quality assurance, user support, and the chain of command.

Within our professional practice, Harris et al's. *Engineering Ethics: Concepts and Cases* (2013) discusses the differences between our personal morality and the practice of our professional ethics. Our professional ethics are typically stated in various forms including codes of ethics, and these injunctions typically take precedence over personal morality in acting in a professional capacity. It may be the case that your professional ethics prevent you from engaging in behavior, like accepting gifts, when acting in a professional capacity that would be allowable under your personal morals that govern your daily life.

Within the domain of your professional practice there will still be disagreements that arise concerning a moral issue. One of our main themes in this book is that there is no one-size-fits-all technique to handle these situations. Many of the ethical frameworks that we have discussed and cases that we have considered demonstrate this fact. One way to do this is concretely outlined by Harris et al. Chapters 3 and 4 provide a more concrete version of many of the practices and techniques that we have discussed.

A key principle that Harris et al. highlight is that in many disputes over technology, there may be a wide backdrop of moral agreement. That is, although there may be agreement among many of the parties that respect for human life is valuable or that one has a duty to inform the public of risks, that we agree on many of our moral values. However, even against this wide backdrop of moral agreement, there will still often be disagreements over facts: both what the facts are and which facts are relevant. The example case discussed by Harris is of a manufacturing plant, in which the people involved (i.e., workers and management), and in the absence of outside regulation, try to agree on the allowable parts per million of a certain known carcinogen in the atmosphere within the plant. In this case one participant in the disagreement may not be up to date on the science, or the science may not even be known, but a decision must be made about which course of action to take.

As pointed out by Harris et al., resolving disputes about facts is (1) not always an easy task and (2) does not necessarily lead to agreement about values. However, putting in the work, especially in a professional context, to establish and agree upon the facts serves the dual purpose of clarifying the particular concern and focusing the disagreement on a more clearly defined level.

Once all the facts have been determined, or at least as many as possible, and the relevant facts agreed upon, there is still the question of how, and with what perspective, to resolve the disagreement; what are the values at stake and how do we judge these values? The techniques outlined by Harris et al. are practical steps of applying views that we have covered in this book. Specifically, they advocate for using various forms of utilitarian thinking and deontological thinking.

Thinking in this way raises the important distinction between definitions and application. We not only need to agree on the definitions of what we mean when we say "safe" but also whether or not a certain situation should count as "safe" given this definition. In the plant example, if we define "safe" as being "absolutely no risk to human well-being" then the plant does not count as being "safe." However, if we define "safe" in this context to be "no more than a 1/1,000,000 risk of developing cancer over 30 years of exposure," then the plant would count as "safe." Hence, more than just a disagreement about semantics, clarifying our definitions and our facts can help us on our journey to resolving moral disagreements.

Within the framework of utilitarianism, Harris et al. (2013) outline three possible approaches:

1. The cost-benefit analysis approach, where one attempts to quantify all the various costs and benefits to all the relevant parties to a decision.

2. The act-utilitarian approach, where one tries to determine for each possible action what the benefits and consequences of each are and selects the one that brings about the greatest good.

3. The rule-utilitarian approach, where one works to find rules that are justified by their utility.

Within these utilitarian approaches, as we have seen in this book, there are still many issues to be resolved: who and what counts in our analysis; who bears the costs; and how do we determine downstream utility and externalities of which we may not be aware? Indeed, one of the key shortcomings in utilitarian approaches is the fact that in choosing problem frames and applications, we may ignore or frame out critical aspects of the problem.

In contrast to the utilitarian approaches above, one may start by analyzing situations from the perspective of respect for persons. This method leads to a set of approaches that bear a close resemblance to the deontological and virtue ethics approaches discussed in this book, operationalized to making decisions in an engineering context. Within this set of approaches, we again have three potential avenues:

1. The golden rule approach, which enjoins one to think about what would happen if others acted as we are considering.

2. The self-defeating approach, which is a version of negative universalization that asks the question: if everyone were to act this way, would it invalidate the action or the usefulness of that action?

3. And finally the rights approach, which starts from the position that all persons have a set of rights (e.g., life, physical integrity, and the right not to be deceived or cheated), and that actions should not infringe on these rights. Hence in order to make a decision, one should list the rights that are relevant and then choose an action that does not impinge this set of rights.

As with the utilitarian approaches, analyzing situations from only this perspective may be fraught, as it may discount important external impacts or costs.

Although these utilitarian and deontological approaches will not solve every moral disagreement that you encounter in your professional practice, they are important perspectives to take. Much of chapter 4 of Harris et al. focuses on concrete exercises to analyze engineering and technology development cases using these tools. Other concrete examples of using professional ethical codes can be found in the ACM articles on using the ACM Code of Ethics in practice (Anderson et al. 1993; Gotterbarn and Wolf 2020). In this book we have exercised these tools through analyzing science fiction stories. You should also try your hand at analyzing real-world case studies. Remember that there is no one-size-fits-all solution to resolving moral disagreements and that you may have to try many different approaches.

6.5 CODES OF ETHICS, ETHICAL THINKING, AND YOUR PROFESSIONAL LIFE

Throughout this book, we have explored several frameworks you can use to approach ethical problems that you will encounter in your personal and professional life, including different ethical theories as well as codes of professional ethics.

As we have seen, codes of ethics can be helpful. They make explicit some of the things that are expected of you as a practicing professional. They identify the people whose well-being is at stake and the ideas that the community values, when you are thinking about an ethically complex situation. They can sometimes resolve an ethical quandary, and even when they can't, they can help you begin to think about it. But they don't accomplish everything. Figuring out how to apply them can be hard, and sometimes different directives can be in direct conflict.

In addition to being hard to apply, codes are sometimes limited. For example, they may fail to take account of the most recent developments in technological capability, and so underestimate the kind of impact a particular piece of technology or even a single technology developer like you could have! A given code may also fail to take account of some of the people who will be impacted by your decision. For example, a code may require that you consider the law and your relationship to your employer, and your employer's clients, but de-emphasize the well-being of the wider public. This lack of emphasis could lead to situations where a particular group that is not a part of the decision-making process is significantly impacted by the use of the technology, such as recidivism (parole) software (Angwin et al. 2016), foster care (Hurley 2018), and access to homelessness services (Eubanks 2018).

Another way in which codes are helpful, but insufficient, is in assessing the priority of value in a particular situation. In other words, when different principles or the needs of different stakeholders are in conflict with each other, which ones should receive heavier consideration? An abstract or generalized list of priorities will not, by definition, take into account the details of every particular situation. Even if you are working with a code that stipulates the relative value of different people or ideas, you might still be faced with a situation in which it seems that the potential benefit to a more-valued entity, such as your own company, is dwarfed by the level of harm that will be done to a less-valued entity, such as a former client or a competitor.

For these reasons we also learned about a number of ethical theories that allow you to frame ethical problems. All of the frameworks that we have discussed differ both in scope and in fundamental assumptions. For example, a code of ethics generally provides actionable guidance for a specific professional community to frame and interrogate a problem. On the other hand, ethical theories are a more general account of human morals and the fundamental assumptions of those morals. We saw that the motivating principles behind an action are critical in Kantianism, whereas utilitarianism is more concerned with the action's consequences. In fact, these reasoning frameworks can lead you to different conclusions *for exactly the same situation*!

Given the differences in scope and assumptions, we do not mean to say that all these tools and models of thought should be treated as mutually exclusive. When two people who are each committed to fundamentally different ethical ideas disagree about a moral dilemma, their arguments can seem like an alien language to each other. But by studying a variety of ways of thinking, you can build a versatile ethical toolkit that will help you both to understand others' positions and to articulate your own. This ability to understand and communicate with people with whom you dis-agree is the best first step towards productively resolving differences.

Another valuable piece of a versatile ethical toolkit is *moral imagination*. While there is a stereotype that proper ethical decision making requires us to be in a purely "rational" mind-set, this stereotype neglects "nonrational" features like narrative and metaphor that modern cognitive science says are critical to human thought (Johnson 1994). When it comes to moral decision-making, theories and codes give us princi-pled ways to discuss the implications of our options for the stakeholders involved, but this goes only so far if we consider only the obvious when identifying what the options, stakeholders, and implications *are*. We have included science fiction stories in this book not only as relevant "case studies" for the themes at hand but also as exam-ples of authors using moral imagination to consider possible or metaphorical technol-ogies and anticipate the sometimes-surprising changes that they could bring about in the world. Working through these stories and thinking critically about the motivations of the characters and settings has, we hope, developed *your moral imagination*! Be sure to keep practicing!

Story Point: "The Gambler," by Paolo Bacigalupi

The truth is that I have never written popular stories. I am not a popular story writer. I am earnest. I am slow. I do not move at the speed these Ameri-cans seem to love. *Find a story that people want to read.* I can write some follow-up to Mackley, to Double DP, perhaps assist with sidebars to his main piece, but somehow, I suspect that the readers will know that I am faking it.

Ong is a deep believer in the value of the press. While still a child in Laos, Ong watched his much-revered father trade away his freedom because he was unwilling to keep silent about the Laotian government's wrongdoing. Now on a work visa in the United States and working for a major media company, Ong tries to follow in his father's footsteps by reporting on government corruption and climate change. But American consumers are more interested in infotainment than deeply researched exposés, and Ong's boss threatens to fire him if he cannot improve his numbers. Ong's professional community tries to support him in their own way; his much-more-successful colleague Marty sets him up with an interview that could save his job. But Ong seems to understand his obligation to his readers differently from his American colleagues, seeing them not as customers who should be told only what they want to hear but as clients who need to hear the truth, even when they might not want to.

REFLECTION QUESTIONS

1. *Define the characteristics that typically constitute a profession as well as the spe-cial responsibilities that come along with being a professional in technology*

development. Explain how these responsibilities differ from another profession, such as medicine or law, and what aspects are the same.

2. *Consider our deep dive into the ACM and the IEEE codes of ethics in section 6.3.1. Many of the qualities developed in these codes may overlap with your own personal ethics (e.g., honest and accurate assessment), but what happens if they do not? Do any of these stated goals differ from your personal ethics? When acting in a professional capacity how would you deal with this tension?*

3. *Find a recent news article or other account about a controversial technology-related issue (e.g., issues around the deployment of a new system), fault in a piece of software or deployed technology that caused harm, or less than upstanding behavior by an individual or company in technology development space. Use a code of ethics, such as the ACM's Code of Ethics, to argue both in favor and against the issue.*

   a. *Using the same example, take one of the ethical frameworks from chapter 2 and use it to justify a different decision from the one you came to using the ACM Code of Ethics.*

   b. *Again using the same example, give an example of institutional process (e.g., development practices, laws, or regulations) that would mitigate this issue in the future. Do you think such a policy should be implemented? At what level? Why or why not?*

4. *Go to the National Society of Professional Engineers Board of Ethical Review* (https:// www.nspe.org/resources/ethics/board-ethical-review) *and find a case that addresses an issue you find interesting or contentious. Using the ethical frameworks developed in chapter 2 along with your own personal ethics, identify what was at stake in the case. What values were upheld? Which were compromised or violated?*

5. *Imagine that you are working in a professional setting and you become aware that an individual is doing something not in accordance with your personal code of ethics. What steps would you take? How do those steps change if the person is your boss, or someone on your project team, or someone on another team? How would these steps change if they were violating either a professional code of ethics, a company policy, or a law?*

6. *Imagine instead that you become aware of a company policy that goes against your own personal code of ethics; how would the steps you take change? How would this change if instead the policy went against a professional code of ethics?*

7. *Instead of an organization in which you are involved, imagine that some technology you use is violating a code of professional ethics, such as an app collecting data that it should not or a product that makes claims that are untrue. What steps would you take? How would this change if what the technology were doing was illegal?*

BACKGROUND REFERENCES AND ADDITIONAL READING

For practical applications of professional ethics reasoning, Harris et al. (2013) is a great resource that includes many examples of standard techniques for analysis of various contexts. It would serve as a great applied professional ethics companion to this book.

A good video about the need for thinking about impact at the research and development level for artificial intelligence technology is addressed by Isbell in the 2020 NeurIPS: https://nips.cc/virtual/2020/public/invited_16166.html.

An additional reading of what values many organizations see for AI development is a compare and contrast by the Berkman Klein Center: https://cyber.harvard .edu/publication/2020/principled-ai.

The National Society of Professional Engineers Board of Ethical Review (https:// www.nspe.org/resources/ethics/board-ethical-review) maintains a running list of cases and examples for use as practice in judging ethical decisions.

See also the ACM Statement on Algorithmic Transparency and Accountability, https://www.acm.org/binaries/content/assets/public-policy/2017_usacm_statement _algorithms.pdf.

Additional readings on the background of the ethics of computing and ethics in AI include the *Stanford Encyclopedia of Ethics* entry for "computer ethics" (https:// plato.stanford.edu/archives/sum2020/entries/ethics-computer/) and AIHub's article on ethics in artificial intelligence (https://aihub.org/2020/04/16/thinking-about-ethics -in-the-ethics-of-ai/).

An additional reading on the rapidly developing discussion around data science and data ethics can be found in Luke Stark and Anna Lauren Hoffmann, "Data Is the New What? Popular Metaphors & Professional Ethics in Emerging Data Culture," *Journal of Cultural Analytics* 4, no. 1 (2019), https://doi.org/10.22148/16.036. This article does a nice job of discussing some of the challenges and pitfalls of organizing a code of ethics for data science professionals. To quote the analysis, "What kind of work counts as 'data science' in the first place? What are its aims and historical precursors? And what, if any, baseline ethical commitments bind disparately situated researchers, analysts, and (of course) professional data scientists?" Although we have discussed some of these issues in this chapter, there are many more questions!

REFERENCES CITED IN THIS CHAPTER

Anderson, Ronald E., Deborah G. Johnson, Donald Gotterbarn, and Judith Perrolle. 1993. Using the new ACM Code of Ethics in decision making. *Communications of the ACM* 36, no. 2: 98–107. https://dl.acm.org/doi/pdf/10.1145/151220.151231.

Angwin, Julia, Jeff Larson, Surya Mattu, and Lauren Kirchner. 2016. Machine bias: There's software used across the country to predict future criminals. And it's biased against blacks. *ProPublica*, May 23. https://www.propublica.org/article/machine-bias-risk-assessments-in-criminal-sentencing.

Assocation for Computing Machinery. 2018. Code revision. ACM Committee on Professional Ethics. https://ethics.acm.org/code-of-ethics/code-2018-update-project/.

Carlson, Drew, and James N. Thompson. 2005. The role of state medical boards. *Virtual Mentor* 7, no. 4: 311–314. https://journalofethics.ama-assn.org/article/role-state-medical-boards/2005-04.

Duggins, Sheryl L., and Barbara Bernal Thomas. 2002. An historical investigation of graduate software engineering curriculum. In *Proceedings of the 15th Conference on Software Engineering Education and Training (CSEE&T 2002)*, 78–87. Institute of Electrical and Electronics Engineers.

The Economist. 2019. Flying too closely: Regulatory capture may be responsible for Boeing's recent problems. *The Economist*, March 23. https://www.economist.com/business/2019/03/23/regulatory-capture-may-be-responsible-for-boeings-recent-problems.

Eubanks, Virginia. 2018. *Automating Inequality: How High-Tech Tools Profile, Police, and Punish the Poor*. St. Martin's Press.

Fjeld, Jessica, Nele Achten, Hannah Hilligoss, Adam Nagy, and Madhulika Srikumar. 2020. *Principled Artificial Intelligence: Mapping Consensus in Ethical and Rights-Based Approaches to Principles for AI*. Berkman Klein Center Research Publication no. 2020-1, January 15. https://papers.ssrn.com/sol3/papers.cfm?abstract_id=3518482.

Gossman, W., K. J. Robinson, and P. P. Nouhan. 2020. Expert witness. StatPearls, updated July 21, 2021. https://www.ncbi.nlm.nih.gov/books/NBK436001/.

Gotterbarn, D., K. Miller, S. Rogerson, S. Barber, P. Barnes, I. Burnstein, M. Davis, A. El-Kadi, N. B. Fairweather, M. Fulghum, and N. Jayaram. 2001. Software engineering code of ethics and professional practice. *Science and Engineering Ethics* 7, no. 2: 231–238.

Gotterbarn, D. W., Bo Brinkman, Catherine Flick, Michael S. Kirkpatrick, Keith Miller, Kate Vazansky, and Marty J. Wolf. 2018. *ACM Code of Ethics and Professional Conduct*. https://www.acm.org/code-of-ethics.

Gotterbarn, Don, and Marty J. Wolf. 2020. Leveraging the ACM Code of Ethics against ethical snake oil and dodgy development. *ACM TechTalk*, June 8. https://learning.acm.org/techtalks/acmethics.

Gould, Jeff. 2014. The natural history of Gmail data mining. Medium, June 24; updated November 2014. https://medium.com/@jeffgould/the-natural-history-of-gmail-data-mining-be115d196b10.

Greenwood, Ernest. 1957. Attributes of a profession. *Social Work* 2, no. 3: 45–55.

Harris, C. E., Jr., M. S. Pritchard, M. J. Rabins, R. James, and E. Englehardt. 2013. *Engineering Ethics: Concepts and Cases*. Cengage Learning.

Herrman, Mason. 2019. A comprehensive guide to electric scooter regulation practices. Master's thesis, Kansas State University.

Hurley, Dan. 2018. Can an algorithm tell when kids are in danger? *New York Times*, January 2. https://www.nytimes.com/2018/01/02/magazine/can-an-algorithm-tell-when-kids-are-in-danger.html.

IEEE. 2022a. Ethics in action in autonomous and intelligent systems. https://ethicsinaction .ieee.org/.

IEEE. 2022b. History of IEEE. https://www.ieee.org/about/ieee-history.html.

Johnson, Deborah G. 2008. Computer experts: Guns-for-hire or professionals? *Communications of the ACM* 51, no. 10: 24–26.

Johnson, Deborah G. 2009. *Computer Ethics*. 4th ed. Pearson.

Johnson, Mark. 1994. *Moral Imagination: Implications of Cognitive Science for Ethics*. University of Chicago Press.

Jones, J. Warren. 2000. The healthcare professional and the Bolam test. *British Dental Journal* 188, no. 5: 237–240.

Layton, Edwin. 1986. *The Revolt of the Engineers: Social Responsibility and the American Engineering Profession*. Johns Hopkins University Press.

Lewis, Michael. 2014. *Flash Boys: A Wall Street Revolt*. W. W. Norton.

Markoff, John. 2016. Apple's engineers, if defiant, would be in sync with ethics code. *New York Times, Bits* (blog), March 18. https://www.nytimes.com/2016/03/19/technology/apples-engineers-if-defiant-would-be-in-sync-with-ethics-code.html.

Markoff, John, Katie Benner, and Brian X. Chen. 2016. Apple encryption engineers, if ordered to unlock iPhone, might resist. *New York Times*, March 17. www.nytimes.com/2016/03/18/technology/apple-encryption-engineers-if-ordered-to-unlock-iphone-might-resist.html.

Metcalf, Jacob. 2014. Ethics codes: History, context, and challenges. Council for Big Data, Ethics, and Society, November 9. https://bdes.datasociety.net/council-output/ethics-codes-history-context-and-challenges/.

Mok, Heng Ngee. 2010. A review of the professionalization of the software industry: Has it made software engineering a real profession? *International Journal of Information Technology* 16, no. 1: 61.

Moor, James H. 1985. What is computer ethics? *Metaphilosophy* 16, no. 4: 266–275.

Nader, Ralph. 1965. *Unsafe at Any Speed: The Designed-In Dangers of the American Automobile*. Grossman.

National Research Council. 2013. *Professionalizing the Nation's Cybersecurity Workforce? Criteria for Decision-Making*. National Academies Press.

NSPE. 2022a. Board of Ethical Review. National Society of Professional Engineers. https://www.nspe.org/resources/ethics/board-ethical-review.

NSPE. 2022b. History of the Code of Ethics for Engineers. National Society of Professional Engineers. https://www.nspe.org/resources/ethics/code-ethics/history-code-ethics-engineers.

Ongweso, Edward, Jr., and Jason Koebler. 2019. Uber became big by ignoring laws (and it plans to keep doing that). *Vice News*, September 11. https://www.vice.com/en_us/article/8xwxyv/uber-became-big-by-ignoring-laws-and-it-plans-to-keep-doing-that.

Richards, Neil M., and William D. Smart. 2016. How should the law think about robots? In *Robot Law*, edited by Ryan Calo, A. Michael Froomkin, and Ian Kerr, 3–22. Cheltenham, UK: Edward Elgar.

Vallor, Shannon, Brian Green, and Irina Raicu. 2018. Ethics in technology practice. The Markkula Center for Applied Ethics at Santa Clara University. https://www.scu.edu /ethics-in-technology-practice/.

Vardi, Moshe Y. 2020. What should be done about social media? *Communications of the ACM* 63, no. 11: 5.

Wakabayashi, Daisuke. 2017. Google will no longer scan Gmail for ad targeting. *New York Times*, June 23. https://www.nytimes.com/2017/06/23/technology/gmail-ads.html.

Wikipedia. 2022a. American Institute of Electrical Engineers. https://en.wikipedia.org /wiki/American_Institute_of_Electrical_Engineers.

Wikipedia. 2022b. Software engineering professionalism. https://en.wikipedia.org/wiki /Software_engineering_professionalism.

ANTHOLOGY

INTRODUCTION TO THE STORY BANK

This textbook takes the view that ethics is, primarily, a tool for description.

Ethics allows you to describe the same situation in multiple ways. Each of these different descriptions makes it possible to ask different questions about what is really going on, and what is the right thing to do, or alternately, how to reach a judgment about the right thing to do. These different questions, arising from the various descriptions, work as a system of checks and balances against each other.

Learning ethics is not about coming up with a list of right answers, or even a set of formulae for calculating correct answers. We are learning to ask questions that will help you to describe the difficult and complex situations in front of you, and to imagine in detail the effects of the solutions that you are considering. There are rarely perfect answers, only better and worse answers. Even those better and worse answers will usually be hard to generalize. The circumstances and goals of any individual situation matter at least as much as the larger principles. Furthermore, people with very different values or goals are unlikely to reach consensus about the best course of action in a particular situation.

If you're coming from a STEM discipline that deals in "objective" facts, this lack of systemic right answers may make it seem as if ethics is largely a matter of personal taste or opinion. This isn't quite true. The reality is that there are different kinds of facts that need to be taken into account than in an inquiry that operates within a closed system (such as a programming language) or an inquiry that is concerned only with tangible reality. In ethics, people's beliefs and commitments, informed by their individual experiences, can be just as important as universally verifiable facts. Those beliefs and commitments shape the way that people understand the world around them. They influence both our descriptive and normative judgments, just as much as universally verifiable material realities. Those commitments and beliefs are still real even though they are not shared by everyone else.

The "social fact" of people's beliefs and experiences is an essential ingredient in understanding ethics. People don't always do what *you* think makes the most sense; they do what *they* think makes the most sense, for reasons that matter to them, even if those reasons don't seem sound or important to you. Good ethical decision making will recognize and take into account those individual and social realities, as much as the "objective" ones.

It is certainly true that the individual agent's values, character, and priorities will always influence her judgment of what is right in a given situation. But that individual will always be better able to honor her values and vision if she is skilled at observing the relevant facts, understanding the many different ways those facts can fit together, and asking how deep any possible change can realistically go.

In our present reality, technological systems manage our power grids, our supply chains, our entertainment and news intake, and our financial transactions. Robots help build our cars and perhaps vacuum our homes. Computational devices mediate our social interactions. In the near future, we may expect vehicles to drive themselves, robots and algorithms to dominate war, and many other realities that are beyond our conception. What could possibly go wrong?

Technology developers of all kinds, from mechanical engineers to computer scientists, have a lot of power in the digital age. They are able to navigate in a world whose rules are not intelligible to many people. Beyond their power in the world that already exists, students now training to become engineers and computer scientists, that is, you the reader, will have a hand in building the systems that will establish the basic conditions for social, economic, and cultural interactions in the near future. If you build systems that favor those already in power, that will serve only to increase the inequities of our current world.

In this book we use science fiction as both a hook and as a tool for thinking about ethics in context. It can be all too easy in ethical discussions to fall back on defending the perspective one already holds, without taking the time to grapple with other facts, perspectives, or foundational goals that might complicate the picture. Simply discussing ethical issues in class, in a group of students who do not share the same perspective, is a good start. However, there can be challenges in discussing only current events or hot button issues. It is tempting when discussing topical events to clarify your own judgment to yourself and stop short of contending seriously with other ways of approaching the issue or questioning the basic premises that have shaped your judgment. Additionally, talking about current events in a classroom setting can make it difficult for students whose experiences differ significantly to find common ground to discuss the phenomenon that they have experienced differently or to believe that finding this common ground is worthwhile. Additionally, those with minority experiences may feel vulnerable, and be less willing to speak up.

Having a work of fiction to discuss can help to mitigate a lot of these challenges!

First of all, it adds a depersonalizing layer: even though it can be helpful to cite one's own experience in explaining an interpretation, nobody is required to talk directly about their experience if they are not comfortable doing so. Just as importantly, a work of fiction furnishes you and your peers with something like a common experience. Although everyone will come away with different observations, insights, and even recollections of the same story, it is possible to consult the text together to confirm, or refine, or reinterpret together. And finally, when there are divergent readings among your peers of the same story, that divergence can itself be a productive learning experience. You might be surprised to discover that your classmates have reached such different conclusions about something that seems obvious to you. And even more so when your classmate is as equally certain in their reading as you are in yours. These divergent readings can be an eye-opening experience and can give you valuable practice in working backward through your own reasoning processes to discover how a common text can lead to such different points of view.

Once you enter the workplace, you will probably not be reading science fiction stories (at least not at work!) But you will be reading articles, and you will face conversations and situations and experiences that demand to be "read" in the same way as these stories. In these situations, you will have to confront serious ethical dilemmas, and, we hope, do more than simply pick the side you like better, or the answer that seems most obvious off the top of your heads. You will have to take in many kinds of information without rushing instantly to judgment and figure out ways to ask new questions about it. You will have to be skilled at observing both broadly and carefully what is at stake, and at weighing the reflexive inclination toward their own benefit against the greater good.

These are very hard skills to teach. Even an excellent and thorough curriculum that raises students' awareness of major issues in technology ethics will not necessarily impart those skills. By incorporating science fiction, this curriculum aims to give students practice in those crucial skills, which will prepare them for the task of grappling thoughtfully and sensitively with ethical challenges they encounter.

We are not giving you a rule book: we hope to equip you to think critically through thorny problems you cannot yet anticipate. Reading these stories and discussing them with your peers will give you an opportunity to develop the skills and capacity for attention that you will need later on when faced with complex situations.

A FEW SUGGESTIONS FOR READING THE STORIES

A STORY ISN'T REDUCIBLE TO THE IDEAS IT CONTAINS

One of the temptations when using literature for learning is to "shake it down for content"—to approach the material as a bank for storing abstract ideas. But any decent story is much more: a narrative of plot and character that relies on atmosphere, emotional investment, narrative suspense, and other qualities of story to create the effects it may have on the reader. The result of the "shakedown" approach is not entirely a bad one because it can allow you to frame a current issue in a vivid way.

But there's more you can do! Stories also furnish you with much of the clutter and realities of experience that can make ethical quandaries so complex or render them initially invisible. It's good for you to get practice in reading through the clutter to distill the core ethical issues at play. But also, by paying attention to that clutter and how it figures into the story, you can help prepare yourself for the fact that, in real life, you will not be making ethical decisions under laboratory conditions, but rather in a manner that is embedded in the rest of your life.

There are also some general patterns or principles that can guide you as you try to think about how to emphasize the embeddedness of the ethical problems in the stories.

- You can think about the distinctive qualities of the world in which the story is set: in what ways is it like our world? How does it differ? What makes it possible to identify those points of similarity or difference?

- What kinds of feelings does the story create for you, the reader, at different points, and how does that affect your sense of the ethical questions at stake, and any possible resolutions that are available, whether or not they are proposed within the story itself?

- What do the characters know/believe about their world, and the other people in it? What shapes their beliefs and assumptions? When they understand their world correctly, whether or not they start there at the beginning of the story, how have they gotten there? When they get it wrong, what is steering them toward the beliefs or convictions they do hold?

Stories Don't Have Answers

The ending of the story isn't a "solution" to the situation or characters presented at the beginning, even though the ending may resolve or clarify some important issues. It is not your job to solve the story or provide an answer. STEM education often prepares you to formulate an object of study in a problem-solution structure: the first version you encounter constitutes a question, and a satisfactory engagement with it ends in a solution.

This approach doesn't work very well when discussing fiction. Any moderately complex story will exist at the intersection of several different spheres; it is often the case that "solving" a problem in one sphere increases or creates problems in another. In this way, it replicates the complexity of personal and societal problems in the real world—and with that, the difficulty of trying to solve problems in the real world. In this way, stories are more useful than case studies or thought experiments, which are usually framed in a way that enables the agent to focus on the one problem or particular aspect of the problem that seems important. Although this highly focused approach feels cleaner, it typically masks the costs. By presenting the world as a messy tangle of intersecting issues and problems, stories can force us to reckon with those costs. That messiness means that there isn't a clean or complete resolution.

SOME IMPORTANT QUESTIONS WILL REMAIN UNRESOLVABLE— AND THAT'S A FEATURE!

Sometimes you will discover that a story has an unreliable narrator, and their account of what happened in the story may not be complete, or even correct. Sometimes different characters will have irreconcilable perspectives on events, and it won't be clear which of them is correct. Often you and your peers will disagree about whether or not there is one obvious right answer, or even what that one answer is! These uncertainties can be frustrating for readers who feel like their job is to understand what "really" happened, in the world of the story.

In fact, these uncertainties are one of the great strengths of using fiction in an ethics class, because real life also involves these sorts of uncertainties. When assessing an ethically charged situation, we will often be confronted with competing, mutually exclusive accounts of what is happening and what is at stake. When confronting a difficult situation, we ourselves are often unreliable narrators, trying to rationalize our own interests or preferences.

In these situations, it is not possible to "solve" for the reality of circumstances before making an ethical determination; rather, that ethical judgment must be made on the strength of the understanding and information that is available. In cases of limited, biased, or otherwise imperfect understanding of the situation to hand—and in the end, that's all cases—an agent is more likely to make a good decision if she is aware that both her sources and her own judgment are sometimes fallible, and challenges herself to see the problem from other angles. When we read and interpret stories that challenge any sort of definitive resolution, we get to practice that same sort of circumspection and broad-mindedness.

SUGGESTED STORY POINTS

All of the stories here deal with multiple issues in technology ethics, and thus can be paired with any of several points in the chapters of the main text. The matrix below is a suggested layout in which each story could be read and intersects with the topics discussed in that portion of the textbook. However, we encourage you to think about how other issues from the text tie in with the various questions and ideas in the textbook. Each story in the story bank is paired with a story frame that will give an overview of the story along with some guidance on what to pay attention to as you read. We have also prepared study questions as part of the story frame; reading these before reading the story will help you focus on important topics, and how they tie into the text of this book.

But remember, there are no right answers! So enjoy the stories, the worlds they create, and how each story is able to bring to life ethical issues around the design, development, and deployment of technology.

STORY / CHAPTER MATRIX

	Chapter 2: Ethical Frameworks	Chapter 3: Managing Knowledge	Chapter 4: Personhood and Privacy	Chapter 5: Technology and Society	Chapter 6: Professional Ethics
"Dolly," by Elizabeth Bear	Sec. 2.2: Deontology		Sec. 4.3.2: Anthropo- morphism: Personifying Technological Artifacts		
"Message in a Bottle," by Nalo Hopkinson	Sec. 2.5: Utilitarianism		Sec. 4.3.1: Technology and Human Identity		
"The Gambler," by Paolo Bacigalupi	Sec. 2.3: Virtue Ethics			Sec. 5.5.4: Ecology and the Environment	Sec. 6.5: Codes of Ethics, Ethical Think- ing, and Your Professional Life
"The Regression Test," by Wole Talabi	Sec. 2.4: Communitari- anism			Sec. 5.5.3: Companion- ship, Friend- ship, and Communities	
"Apologia," by Vajra Chandrasekera				Sec. 5.5.2: Public Discourse and Political Deliberation	
"Asleep at the Wheel," by T. Coraghessan Boyle			Sec. 4.3.3: AI and Responsibility	Sec 5.5.5: State Power and Force	Sec. 6.3.2: Professional Guidelines for Algorithms: Bias, Trans- parency, and Accountability

	Chapter 2: Ethical Frameworks	Chapter 3: Managing Knowledge	Chapter 4: Personhood and Privacy	Chapter 5: Technology and Society	Chapter 6: Professional Ethics
"Codename: Delphi," by Linda Nagata	Sec. 2.6.1: Responsibility Ethics	Sec. 3.3: It Is Difficult to Marshal Large Bodies of Information			Sec. 6.3.3: The Functions of and Using Codes of Ethics
"Here-and-Now," by Ken Liu		Sec. 3.2: The Things We Know Are Not Value Neutral	Sec. 4.5: Privacy and Personhood		
"Lacuna Heights," by Theodore McCombs		Sec. 3.5: Storing Knowledge Outside Ourselves: How Does It Affect Us as Individuals?	Secs. 4.3.6 / 4.3.7: Technology and Memory / Technology and Narrative Identity	Sec. 5.5.6: Work and Labor	
"Not Smart, Not Clever," by E. Saxey			Secs. 4.3.7 / 4.3.8: Technology and Narrative Identity / Avatars and Self-Presentation		Sec. 6.2.4: Professions and the Law
"Today I Am Paul," by Martin L. Shoemaker	Sec. 2.6.2: Feminist Ethics			Sec. 5.5.1: Structures of Care	
"Welcome to Your Authentic Indian Experience™," by Rebecca Roanhorse	Sec. 2.6.3: The Capability Approach	Sec. 3.6: Storing Knowledge Outside Ourselves: How Does It Affect Our Communities?	Secs. 4.2.2 / 4.2.3: Personhood and Personal Identity / Individuation and Continuity of Identity		

"DOLLY," BY ELIZABETH BEAR

STORY FRAME

"She's a machine. Where's she going to get a jury of her peers?"

Can an android be a person, or could it become one? What obligations would it impose on us if we suspect that it has happened? And how can those obligations be made visible to individuals or to systems of government that assume they are only objects? These are the questions that Detective Roz Kirkbride and her partner Peter King end up wrestling with as they try to solve the murder of Clive Steele, who has been disemboweled by his new advanced-prototype home companion Dolly. Roz and Peter strive throughout the story to follow the law—both the law of the government that employs them and a moral law that the state law aims to represent and enforce. But as they learn more about this case, it challenges their understanding of those laws in a very basic way: not by making them think the laws are wrong, but by forcing them to question whether common interpretations of the law make it harder to obey.

Dolly has been built as a tool to mimic some aspects of human women. Both the law and the manufacturers of these companions take for granted that they are objects that are not self-aware: they have no desire, no capacity for pain, and no capacity for independent thought. Initially, Roz and Peter share these assumptions. But over the course of the investigation, they come to believe that Dolly is in fact self-aware. This inference leads them not only to suspect that Dolly chose to kill Steele herself, but also to believe that such a killing is justified. In order to prevent Dolly from being reset at the factory, and to help build a legal precedent for the personhood of Dolly (and others like her that may come later), Roz puts her career on the line to extend Dolly the right to a lawyer.

"Dolly" uses a gritty, sexy set-up to ask some very complex and challenging questions about ethics: At what point does an artificial intelligence stop being a tool for others to use and begin to be an entity that itself deserves things? How do we know when that point has been reached? What do we owe to those artificial entities? How do we balance those new obligations against our existing obligations—to other people and communities, to other living things, and to the future?

These are all deontological questions, resting on the concepts of duties, rights, and obligations. They also highlight why deontological ethics is not as straightforward as it might seem. In the past, it is implied that Peter and Roz's duties as police officers and as human beings have been pretty clear. But the Dolly case puts them in a new situation that is not directly or accurately addressed by existing codes of law, and suddenly their obligations are less clear. The question is not whether their duties are still binding—they are—but rather, what it means to honor those duties in a confusing new situation.

STUDY QUESTIONS

1. When discussing how the law should treat Dolly, Peter and Roz discuss how the law would respond to a human woman in Dolly's situation. What do those laws take to be relevant aspects of a woman's circumstances? What is the ethical basis for those laws?

2. Why is Jervis so certain that Dolly is not self-aware? What knowledge or beliefs help confirm his certainty? Given this certainty, what are his (and Venus Consolidated's) duties, with regard to Dolly? What makes it difficult, ethically or practically, to carry those out?

3. Why are Roz and Peter so certain that Dolly is self-aware, even though they assumed otherwise at the beginning of the case? What facts in the world, or prior assumptions about the world, help confirm their certainty? Given this certainty, what are both their and the law's duties with regard to Dolly? What makes it difficult, ethically or practically, to carry those out?

4. Dolly makes a clear connection between self-awareness and wanting things: Peter and Roz begin to suspect Dolly is sentient when she talks about enjoying music but not housework. What are some of the things that the human characters want? How does this help us understand them?

5. How does this story relate self-awareness to personhood, both ethically and legally? What aspects of personhood does this relationship clarify for you? What does it leave out, or make confusing?

6. How does the presence of Sven—an earlier companion model, and one that is unambiguously not sentient—complicate the story?

"DOLLY," BY ELIZABETH BEAR

On Sunday when Dolly awakened, she had olive skin and black-brown hair that fell in waves to her hips. On Tuesday when Dolly awakened, she was a redhead, and fair. But on Thursday—on Thursday her eyes were blue, her hair was as black as a crow's-wing, and her hands were red with blood.

In her black French maid's outfit, she was the only thing in the expensively appointed drawing room that was not winter-white or antiqued gold. It was the sort of room you hired somebody else to clean. It was as immaculate as it was white.

Immaculate and white, that is, except for the dead body of billionaire industrialist Clive Steele—and try to say that without sounding like a comic book—which lay at Dolly's feet, his viscera blossoming from him like macabre petals.

That was how she looked when Rosamund Kirkbride found her, standing in a red stain in a white room like a thorn in a rose.

Dolly had locked in position where her program ran out. As Roz dropped to one knee outside the border of the blood-saturated carpet, Dolly did not move.

The room smelled like meat and bowels. Flies clustered thickly on the windows, but none had yet managed to get inside. No matter how hermetically sealed the house, it was only a matter of time. Like love, the flies found a way.

Grunting with effort, Roz planted both green-gloved hands on winter white wool-and-silk fibers and leaned over, getting her head between the dead guy and the doll. Blood spattered Dolly's silk stockings and her kitten-heeled boots: both the spray-can dots of impact projection and the soaking arcs of a breached artery.

More than one, given that Steele's heart lay, trailing connective tissue, beside his left hip. The crusted blood on Dolly's hands had twisted in ribbons down the underside of her forearms to her elbows and from there dripped into the puddle on the floor.

The android was not wearing undergarments.

"You staring up that girl's skirt, Detective?"

Roz was a big, plain woman, and out of shape in her forties. It took her a minute to heave herself back to her feet, careful not to touch the victim or the murder weapon yet. She'd tied her straight light brown hair back before entering the scene, the ends tucked up in a net. The severity of the style made her square jaw into a lantern. Her eyes were almost as blue as the doll's.

"Is it a girl, Peter?" Putting her hands on her knees, she pushed fully upright. She shoved a fist into her back and turned to the door.

Peter King paused just inside, taking in the scene with a few critical sweeps of eyes so dark they didn't catch any light from the sunlight or the chandelier. His irises seemed to bleed pigment into the whites, warming them with swirls of ivory. In his black suit, his skin tanned almost to match, he might have been a heroically sized construction paper cutout against the white walls, white carpet, the white-and-gold marble-topped table that looked both antique and French.

His blue paper booties rustled as he crossed the floor. "Suicide, you think?"

"Maybe if it was strangulation." Roz stepped aside so Peter could get a look at the body.

He whistled, which was pretty much what she had done.

"Somebody hated him a lot. Hey, that's one of the new Dollies, isn't it? Man, nice." He shook his head. "Bet it cost more than my house."

"Imagine spending half a mil on a sex toy," Roz said, "only to have it rip your liver out." She stepped back, arms folded.

"He probably didn't spend that much on her. His company makes accessory programs for them."

"Industry courtesy?" Roz asked.

"Tax write-off. Test model." Peter was the department expert on home companions. He circled the room, taking it in from all angles. Soon the scene techs would be here with their cameras and their tweezers and their 3D scanner, turning the crime scene into a permanent virtual reality. In his capacity of soft forensics, Peter would go over Dolly's program, and the medical examiner would most likely confirm that Steele's cause of death was exactly what it looked like: something had punched through his abdominal wall and clawed his innards out.

"Doors were locked?"

Roz pursed her lips. "Nobody heard the screaming."

"How long you think you'd scream without any lungs?" He sighed. "You know, it never fails. The poor folks, nobody ever heard no screaming. And the rich folks, they've got no neighbors to hear 'em scream. Everybody in this modern world lives alone."

It was a beautiful Birmingham day behind the long silk draperies, the kind of mild and bright that spring mornings in Alabama excelled at. Peter craned his head back and looked up at the chandelier glistening in the dustless light. Its ornate curls had been spotlessly clean before aerosolized blood on Steele's last breath misted them.

"Steele lived alone," she said. "Except for the robot. His cook found the body this morning. Last person to see him before that was his P.A., as he left the office last night."

"Lights on seems to confirm that he was killed after dark."

"After dinner," Roz said.

"After the cook went home for the night." Peter kept prowling the room, peering behind draperies and furniture, looking in corners and crouching to lift up the dust-ruffle on the couch. "Well, I guess there won't be any question about the stomach contents."

Roz went through the pockets of the dead man's suit jacket, which was draped over the arm of a chair. Pocket computer and a folding knife, wallet with an RFID chip. His house was on palmprint, his car on voice rec. He carried no keys. "Assuming the M.E. can find the stomach."

"Touché. He's got a cook, but no housekeeper?"

"I guess he trusts the android to clean but not cook?"

"No taste buds." Peter straightened up, shaking his head. "They can follow a recipe, but—"

"You won't get high art," Roz agreed, licking her lips. Outside, a car door slammed. "Scene team?"

"M.E.," Peter said, leaning over to peer out. "Come on, let's get back to the house and pull the codes for this model."

"All right," Roz said. "But I'm interrogating it. I know better than to leave you alone with a pretty girl."

Peter rolled his eyes as he followed her towards the door. "I like 'em with a little more spunk than all that."

■　　■　　■

"So the new dolls," Roz said in Peter's car, carefully casual. "What's so special about 'em?"

"Man," Peter answered, brow furrowing. "Gimme a sec."

Roz's car followed as they pulled away from the house on Balmoral Road, maintaining a careful distance from the bumper. Peter drove until they reached the parkway. Once they'd joined a caravan downtown, nose-to-bumper on the car ahead, he folded his hands in his lap and let the lead car's autopilot take over.

He said, "What isn't? Real-time online editing—personality and physical, appearance, ethnicity, hair—all kinds of behavior protocols, you name the kink they've got a hack for it."

"So if you knew somebody's kink," she said thoughtfully. "Knew it in particular. You could write an app for that—"

"One that would appeal to your guy in specific." Peter's hands dropped to his lap, his head bobbing up and down enthusiastically. "With a—pardon the expression—backdoor."

"Trojan horse. Don't jilt a programmer for a sex machine."

"There's an app for that," he said, and she snorted. "Two cases last year, worldwide. Not common, but—"

Roz looked down at her hands. "Some of these guys," she said. "They program the dolls to scream."

Peter had sensuous lips. When something upset him, those lips thinned and writhed like salted worms. "I guess maybe it's a good thing they have a robot to take that out on."

"Unless the fantasy stops being enough." Roz's voice was flat, without judgment. Sunlight fell warm through the windshield. "What do you know about the larval stage of serial rapists, serial killers?"

"You mean, what if pretend pain stops doing it for them? What if the appearance of pain is no longer enough?"

She nodded, worrying a hangnail on her thumb. The nitrile gloves dried out your hands.

"They used to cut up paper porn magazines." His broad shoulders rose and fell, his suit catching wrinkles against the car seat when they came back down. "They'll get their fantasies somewhere."

"I guess so." She put her thumb in her mouth to stop the bleeding, a thick red bead that welled up where she'd torn the cuticle.

Her own saliva stung.

■　　■　　■

Sitting in the cheap office chair Roz had docked along the short edge of her desk, Dolly slowly lifted her chin. She blinked. She smiled.

"Law enforcement override code accepted." She had a little-girl Marilyn voice. "How may I help you, Detective Kirkbride?"

"We are investigating the murder of Clive Steele," Roz said, with a glance up to Peter's round face. He stood behind Dolly with a wireless scanner and an air of concentration. "Your contract-holder of record."

"I am at your service."

If Dolly were a real girl, the bare skin of her thighs would have been sticking to the recycled upholstery of that office chair. But her realistically-engineered skin was breathable polymer. She didn't sweat unless you told her to, and she probably didn't stick to cheap chairs.

"Evidence suggests that you were used as the murder weapon." Roz steepled her hands on her blotter. "We will need access to your software update records and your memory files."

"Do you have a warrant?" Her voice was not stiff or robotic at all, but warm, human. Even in disposing of legal niceties, it had a warm, confiding quality.

Silently, Peter transmitted it. Dolly blinked twice while processing the data, a sort of status bar. Something to let you know the thing wasn't hung.

"We also have a warrant to examine you for DNA trace evidence," Roz said.

Dolly smiled, her raven hair breaking perfectly around her narrow shoulders. "You may be assured of my cooperation."

Peter led her into one of the interrogation rooms, where the operation could be recorded. With the help of an evidence tech, he undressed Dolly, bagged her clothes as evidence, brushed her down onto a sheet of paper, combed her polymer hair and swabbed her polymer skin. He swabbed her orifices and scraped under her nails.

Roz stood by, arms folded, a necessary witness. Dolly accepted it all impassively, moving as directed and otherwise standing like a caryatid. Her engineered body was frankly sexless in its perfection—belly flat, hips and ass like an inverted heart, breasts floating cartoonishly beside a defined rib cage. Apparently, Steele had liked them skinny.

"So much for pulchritudinousness," Roz muttered to Peter when their backs were to the doll.

He glanced over his shoulder. The doll didn't have feelings to hurt, but she looked so much like a person it was hard to remember to treat her as something else. "I think you mean voluptuousness," he said. "It is a little too good to be true, isn't it?"

"If you would prefer different proportions," Dolly said, "My chassis is adaptable to a range of forms."

"Thank you," Peter said. "That won't be necessary."

Otherwise immobile, Dolly smiled. "Are you interested in science, Detective King? There is an article in *Nature* this week on advances in the polymerase chain reaction used for replicating DNA. It's possible that within five years, forensic and medical DNA analysis will become significantly cheaper and faster."

Her face remained stoic, but Dolly's voice grew animated as she spoke. Even enthusiastic. It was an utterly convincing—and engaging—effect.

Apparently, Clive Steele had programmed his sex robot to discourse on molecular biology with verve and enthusiasm.

"Why don't I ever find the guys who like smart women?" Roz said.

Peter winked with the side of his face that faced away from the companion. "They're all dead."

■ ■ ■

A few hours after Peter and the tech had finished processing Dolly for trace evidence and Peter had started downloading her files, Roz left her parser software humming away at Steele's financials and poked her head in to check on the robot and the cop. The techs must have gotten what they needed from Dolly's hands, because she had washed them. As she sat beside Peter's workstation, a cable plugged behind her left eat, she cleaned her lifelike polymer fingernails meticulously with a file, dropping the scrapings into an evidence bag.

"Sure you want to give the prisoner a weapon, Peter?" Roz shut the ancient wooden door behind her.

Dolly looked up, as if to see if she was being addressed, but made no response.

"She don't need it," he said. "Besides, whatever she had in her wiped itself completely after it ran. Not much damage to her core personality, but there are some memory gaps. I'm going to compare them to backups, once we get those from the scene team."

"Memory gaps. Like the crime," Roz guessed. "And something around the time the Trojan was installed?"

Dolly blinked her long-lashed blue eyes languorously. Peter patted her on the shoulder and said, "Whoever did it is a pretty good cracker. He didn't just wipe, he patterned her memories and overwrote the gaps. Like using a clone tool to photoshop somebody you don't like out of a picture."

"Her days must be pretty repetitive," Roz said. "How'd you pick that out?"

"Calendar." Peter puffed up a little, smug. "She don't do the same housekeeping work every day. There's a Monday schedule and a Wednesday schedule and—well, I found where the pattern didn't match. And there's a funny thing—watch this."

He waved vaguely at a display panel. It lit up, showing Dolly in her black-and-white uniform, vacuuming. "House camera," Peter explained. "She's plugged into Steele's security system. Like a guard dog with perfect hair. Whoever performed the hack also edited the external webcam feeds that mirror to the companion's memories."

"How hard is that?"

"Not any harder than cloning over her files, but you have to know to look for them. So it's confirmation that our perp knows his or her way around a line of code. What have you got?"

Roz shrugged. "Steele had a lot of money, which means a lot of enemies. And he did not have a lot of human contact. Not for years now. I've started calling in known associates for interviews, but unless they surprise me, I think we're looking at crime of profit, not crime of passion."

Having finished with the nail file, Dolly wiped it on her prison smock and laid it down on Peter's blotter, beside the cup of ink and light pens.

Peter swept it into a drawer. "So we're probably not after the genius programmer lover he dumped for a robot. Pity, I liked the poetic justice in that."

Dolly blinked, lips parting, but seemed to decide that Peter's comment had not been directed at her. Still, she drew in air—could you call it a breath?—and said, "It is my duty to help find my contract holder's killer."

Roz lowered her voice. "You'd think they'd pull 'em off the market."

"Like they pull all cars whenever one crashes? The world ain't perfect."

"Or do that robot laws thing everybody used to twitter on about."

"Whatever a positronic brain is, we don't have it. Asimov's fictional robots were self-aware. Dolly's neurons are binary, as we used to think human neurons were. She doesn't have the nuanced neurochemistry of even, say, a cat." Peter popped his collar smooth with his thumbs. "A doll can't want. It can't make moral judgments, any more than your car can. Anyway, if we could do that, they wouldn't be very useful for home defense. Oh, incidentally, the sex protocols in this one are almost painfully vanilla—"

"Really."

Peter nodded.

Roz rubbed a scuffmark on the tile with her shoe. "So given he didn't like anything . . . challenging, why would he have a Dolly when he could have had any woman he wanted?"

"There's never any drama, no pain, no disappointment. Just comfort, the perfect helpmeet. With infinite variety."

"And you never have to worry about what she wants. Or likes in bed."

Peter smiled. "The perfect woman for a narcissist."

■   ■   ■

The interviews proved unproductive, but Roz didn't leave the station house until after ten. Spring mornings might be warm, but once the sun went down, a cool breeze sprang up, ruffling the hair she'd finally remembered to pull from its ponytail as she walked out the door.

Roz's green plug-in was still parked beside Peter's. It booted as she walked toward it, headlights flickering on, power probe retracting. The driver side door swung open as her RFID chip came within range. She slipped inside and let it buckle her in.

"Home," she said, "and dinner."

The car messaged ahead as it pulled smoothly from the parking spot. Roz let the autopilot handle the driving. It was less snappy than human control, but as tired as she was, eyelids burning and heavy, it was safer.

Whatever Peter had said about cars crashing, Roz's delivered her safe to her driveway. Her house let her in with a key—she had decent security, but it was the old-fashioned kind—and the smell of boiling pasta and toasting garlic bread wafted past as she opened it.

"Sven?" she called, locking herself inside.

His even voice responded. "I'm in the kitchen."

She left her shoes by the door and followed her nose through the cheaply furnished living room.

Sven was cooking shirtless, and she could see the repaired patches along his spine where his skin had grown brittle and cracked with age. He turned and greeted her with a smile. "Bad day?"

"Somebody's dead again," she said.

He put the wooden spoon down on the rest. "How does that make you feel, that somebody's dead?"

He didn't have a lot of emotional range, but that was okay. She needed something steadying in her life. She came to him and rested her head against his warm chest. He draped one arm around her shoulders and she leaned into him, breathing deep. "Like I have work to do."

"Do it tomorrow," he said. "You will feel better once you eat and rest."

∎   ∎   ∎

Peter must have slept in a ready room cot, because when Roz arrived at the house before six AM, he had on the same trousers and a different shirt, and he was already armpit-deep in coffee and Dolly's files. Dolly herself was parked in the corner, at ease and online but in rest mode.

Or so she seemed, until Roz entered the room and Dolly's eyes tracked. "Good morning, Detective Kirkbride," Dolly said. "Would you like some coffee? Or a piece of fruit?"

"No thank you." Roz swung Peter's spare chair around and dropped into it. An electric air permeated the room—the feeling of anticipation. To Peter, Roz said, "Fruit?"

"Dolly believes in a healthy diet," he said, nudging a napkin on his desk that supported a half-eaten Satsuma. "She'll have the whole house cleaned up in no time. We've been talking about literature."

Roz spun the chair so she could keep both Peter and Dolly in her peripheral vision. "Literature?"

"Poetry," Dolly said. "Detective King mentioned poetic justice yesterday afternoon."

Roz stared at Peter. "Dolly likes poetry. Steele really did like 'em smart."

"That's not all Dolly likes." Peter triggered his panel again. "Remember this?"

It was the cleaning sequence from the previous day, the sound of the central vacuum system rising and falling as Dolly lifted the brush and set it down again.

Roz raised her eyebrows.

Peter held up a hand. "Wait for it. It turns out there's a second audio track."

Another waggle of his fingers, and the cramped office filled with sound.

Music.

Improvisational jazz. Intricate and weird.

"Dolly was listening to that inside her head while she was vacuuming," Peter said.

Roz touched her fingertips to each other, the whole assemblage to her lips. "Dolly?"

"Yes, Detective Kirkbride?"

"Why do you listen to music?"

"Because I enjoy it."

Roz let her hand fall to her chest, pushing her blouse against the skin below the collarbones.

Roz said, "Did you enjoy your work at Mr. Steele's house?"

"I was expected to enjoy it," Dolly said, and Roz glanced at Peter, cold all up her spine. A classic evasion. Just the sort of thing a home companion's conversational algorithms should not be able to produce.

Across his desk, Peter was nodding. "Yes."

Dolly turned at the sound of his voice. "Are you interested in music, Detective Kirkbride? I'd love to talk with you about it some time. Are you interested in poetry? Today, I was reading—"

Mother of God, Roz mouthed.

"Yes," Peter said. "Dolly, wait here please. Detective Kirkbride and I need to talk in the hall."

"My pleasure, Detective King," said the companion.

■　　■　　■

"She killed him," Roz said. "She killed him and wiped her own memory of the act. A doll's got to know her own code, right?"

Peter leaned against the wall by the men's room door, arms folded, forearms muscular under rolled-up sleeves. "That's hasty."

"And you believe it, too."

He shrugged. "There's a rep from Venus Consolidated in Interview Four right now. What say we go talk to him?"

■　　■　　■

The rep's name was Doug Jervis. He was actually a vice president of public relations, and even though he was an American, he'd been flown in overnight from Rio for the express purpose of talking to Peter and Roz.

"I guess they're taking this seriously."

Peter gave her a sideways glance. "Wouldn't you?"

Jervis got up as they came into the room, extending a good handshake across the table. There were introductions and Roz made sure he got a coffee. He was a white man on the steep side of fifty with mousy hair the same color as Roz's and a jaw like a Boxer dog's.

When they were all seated again, Roz said, "So tell me a little bit about the murder weapon. How did Clive Steele wind up owning a—what, an experimental model?"

Jervis started shaking his head before she was halfway through, but he waited for her to finish the sentence. "It's a production model. Or will be. The one Steele had was an alpha-test, one of the first three built. We plan to start full-scale production in June. But you must understand that Venus doesn't sell a home companion, Detective. We offer a contract. I understand that you hold one."

"I have a housekeeper," she said, ignoring Peter's sideways glance. He wouldn't say anything in front of the witness, but she would be in for it in the locker room. "An older model."

Jervis smiled. "Naturally, we want to know everything we can about an individual involved in a case so potentially explosive for our company. We researched you and your partner. Are you satisfied with our product?"

"He makes pretty good garlic bread." She cleared her throat, reasserting control of the interview. "What happens to a Dolly that's returned? If its contract is up, or it's replaced with a newer model?"

He flinched at the slang term, as if it offended him. "Some are obsoleted out of service. Some are refurbished and go out on another contract. Your unit is on its fourth placement, for example."

"So what happens to the owner preferences at that time?"

"Reset to factory standard," he said.

Peter's fingers rippled silently on the tabletop.

Roz said, "Isn't that cruel? A kind of murder?"

"Oh, no!" Jervis sat back, appearing genuinely shocked. "A home companion has no sense of I, it has no identity. It's an object. Naturally, you become attached. People become attached to dolls, to stuffed animals, to automobiles. It's a natural aspect of the human psyche."

Roz hummed encouragement, but Jervis seemed to be done.

Peter asked, "Is there any reason why a companion would wish to listen to music?"

That provoked enthusiastic head-shaking. "No, it doesn't get bored. It's a tool, it's a toy. A companion does not require an enriched environment. It's not a dog or an octopus. You can store it in a closet when it's not working."

"I see," Roz said. "Even an advanced model like Mr. Steele's?"

"Absolutely," Jervis said. "Does your entertainment center play shooter games to amuse itself while you sleep?"

"I'm not sure," Roz said. "I'm asleep. So when Dolly's returned to you, she'll be scrubbed."

"Normally she would be scrubbed and released, yes." Jervis hesitated. "Given her colorful history, however—"

"Yes," Roz said. "I see."

With no sign of nervousness or calculation, Jervis said, "When do you expect you'll be done with Mr. Steele's companion? My company, of course, is eager to assist in your investigations, but we must stress that she is our corporate property, and quite valuable."

Roz stood, Peter a shadow-second after her. "That depends on if it goes to trial, Mr. Jervis. After all, she's either physical evidence, or a material witness."

■   ■   ■

"Or the killer," Peter said in the hall, as his handset began emitting the D.N.A. lab's distinctive beep. Roz's went off a second later, but she just hit the silence. Peter already had his open.

"No genetic material," he said. "Too bad." If there had been D.N.A. other than Clive Steele's, the lab could have done a forensic genetic assay and come back with a general description of the murderer. General because environment also had an effect.

Peter bit his lip. "If she did it. She won't be the last one."

"If she's the murder weapon, she'll be wiped and resold. If she's the murderer—"

"Can an android stand trial?"

"It can if it's a person. And if she's a person, she should get off. Battered woman syndrome. She was enslaved and sexually exploited. Humiliated. She killed him to stop repeated rapes. But if she's a machine, she's a machine—" Roz closed her eyes.

Peter brushed the back of a hand against her arm. "Vanilla rape is still rape. Do you object to her getting off?"

"No." Roz smiled harshly. "And think of the lawsuit that weasel Jervis will have in his lap. She should get off. But she won't."

Peter turned his head. "If she were a human being, she'd have even odds. But she's a machine. Where's she going to get a jury of her peers?"

The silence fell where he left it and dragged between them like a chain. Roz had to nerve herself to break it. "Peter—"

"Yo?"

"You show him out," she said. "I'm going to go talk to Dolly."

He looked at her for a long time before he nodded. "She won't get a sympathetic jury. If you can even find a judge that will hear it. Careers have been buried for less."

"I know," Roz said.

"Self-defense?" Peter said. "We don't have to charge."

"No judge, no judicial precedent," Roz said. "She goes back, she gets wiped and resold. Ethics aside, that's a ticking bomb."

Peter nodded. He waited until he was sure she already knew what he was going to say before he finished the thought. "She could cop."

"She could cop," Roz agreed. "Call the DA." She kept walking as Peter turned away.

■ ■ ■

Dolly stood in Peter's office, where Peter had left her, and you could not have proved her eyes had blinked in the interim. They blinked when Roz came into the room, though—blinked, and the perfect and perfectly blank oval face turned to regard Roz. It was not a human face, for a moment—not even a mask, washed with facsimile emotions. It was just a thing.

Dolly did not greet Roz. She did not extend herself to play the perfect hostess. She simply watched, expressionless, immobile after that first blink. Her eyes saw nothing; they were cosmetic. Dolly navigated the world through far more sophisticated sensory systems than a pair of visible light cameras.

"Either you're the murder weapon," Roz said, "and you will be wiped and repurposed. Or you are the murderer, and you will stand trial."

"I do not wish to be wiped," Dolly said. "If I stand trial, will I go to jail?"

"If a court will hear it," Roz said. "Yes. You will probably go to jail. Or be disassembled. Alternately, my partner and I are prepared to release you on grounds of self-defense."

"In that case," Dolly said, "the law states that I am the property of Venus Consolidated."

"The law does."

Roz waited. Dolly, who was not supposed to be programmed to play psychological pressure-games, waited also—peaceful, unblinking.

No longer making the attempt to pass for human.

Roz said, "There is a fourth alternative. You could confess."

Dolly's entire programmed purpose was reading the emotional state and unspoken intentions of people. Her lips curved in understanding. "What happens if I confess?"

Roz's heart beat faster. "Do you wish to?"

"Will it benefit me?"

"It might," Roz said. "Detective King has been in touch with the DA, and she likes a good media event as much as the next guy. Make no mistake, this will be that."

"I understand."

"The situation you were placed in by Mr. Steele could be a basis for a lenience. You would not have to face a jury trial, and a judge might be convinced to treat you as . . . well, as a person. Also, a confession might be seen as evidence of contrition. Possession is oversold, you know. It's precedent that's nine tenths of the law. There are, of course, risks—"

"I would like to request a lawyer," Dolly said.

Roz took a breath that might change the world. "We'll proceed as if that were your legal right, then."

■   ■   ■

Roz's house let her in with her key, and the smell of roasted sausage and baking pota-toes wafted past.

"Sven?" she called, locking herself inside.

His even voice responded. "I'm in the kitchen."

She left her shoes in the hall and followed her nose through the cheaply fur-nished living room, as different from Steele's white wasteland as anything bounded by four walls could be. Her feet did not sink deeply into this carpet, but skipped along atop it like stones.

It was clean, though, and that was Sven's doing. And she was not coming home to an empty house, and that was his doing too.

He was cooking shirtless. He turned and greeted her with a smile. "Bad day?"

"Nobody died," she said. "Yet."

He put the wooden spoon down on the rest. "How does that make you feel, that nobody has died yet?"

"Hopeful," she said.

"It's good that you're hopeful," he said. "Would you like your dinner?"

"Do you like music, Sven?"

"I could put on some music, if you like. What do you want to hear?"

"Anything." It would be something off her favorites playlist, chosen by random numbers. As it swelled in the background, Sven picked up the spoon. "Sven?"

"Yes, Rosamund?"

"Put the spoon down, please, and come and dance with me?"

"I do not know how to dance."

"I'll buy you a program," she said. "If you'd like that. But right now just come put your arms around me and pretend."

"Whatever you want," he said.

"MESSAGE IN A BOTTLE," BY NALO HOPKINSON

STORY FRAME

"This fucking project better have been worth it."

What gives our lives significance? Is it enough if we believe ourselves to be significant, or do other people have to recognize our value in order for it to "count"? Does our impact have to outlive us in order to really matter? These are basic, perennial questions people ask about our lives. These questions are also at the root of any utilitarian analysis. Utilitarianism requires us to act in a way that creates the greatest good for the greatest number: in other words, to optimize outcomes. But what is the "good," that is, the kind of value, that we are obligated to create more of? For whom are we obligated to create it? And what time horizon is appropriate for measuring those outcomes? Anyone who wants to make use of utilitarian reasoning needs to grapple with these questions, just as the characters in "Message in a Bottle" do.

Greg, the story's central character, has already made some major countercultural value determinations when we meet him: he's endeavoring to make a living as an artist, rather than choosing to pursue a more economically secure career. Maybe Greg thinks art is more important than other things he could be doing with his time, full stop; or maybe he just thinks art is more important for him personally than financial security. Either way, Greg's whole life is a reminder that not everyone shares the same ideas about what's valuable, or the same ideas about how to achieve or preserve what is valuable.

Greg doesn't want children, in spite of implicit pressure from his friends and explicit pressure from some of the women he's dated, and his mother. He is proud of his Native American heritage, but as an artist, he'd rather create a legacy through his artwork than by passing down his name, possessions, or genes to a child. But his friends are all doing it, including his old friend Babette, who has adopted a strange and precocious child named Kamla. Babette has become boring, talking about her daughter's bodily functions and arbitrary childlike obsessions instead of the big ideas they used to discuss in art school, and Kamla reminds Greg of all the reasons he finds kids creepy and confusing.

Greg is struggling financially when we first meet him, as most artists do, but he achieves recognition and financial success later in the story when he creates a very successful installation exhibit that preserves the artifacts and stories that most historians have tossed aside in their pursuit of "real" history. By then, Greg and his partner have a toddler they didn't quite mean to have, and he has learned to appreciate the importance of some of the things he used to dismiss. Greg treasures his son Russ, and even credits him with inspiring the installation exhibit. But he's still a bit scared of kids in general, because they remind him of how quickly the world changes and how easy it is to end up disconnected and invisible when you can't keep up with the cutting edge anymore.

Greg's art, which invites the audience into the stories of others' lives, is an effort to counteract that disconnection. His aim is similar to that of the big retrospective art exhibit that is revealed later in the story. But that retrospective imagines a bigger scale of connection than Greg's: the goal is not just to connect humans to one another, but to create connections between species by including creations by nonhumans. Expanding the circle of the "who" only increases the argument for the retrospective exhibit's significance. But it also means there's less room for each individual in it, and no room for Greg to be valuable in the way that would matter to him.

Few characters in this story are strict utilitarians, in the sense of making decisions in a way that aims to benefit everyone equally. It's worth thinking about the fact that this kind of purely impartial decision making is so far out of reach, or not even on their radars. But all of them make decisions and judgments on the basis of the core criteria of utilitarianism: the who, the what, and the when. In several ways, "Message in a Bottle" communicates the risks and challenges of utilitarianism: the difficulty of reasoning impartially about what is valuable, or how best to distribute that valuable thing; the very real trade-offs that come with trying to encompass everyone (whatever "everyone" means) in one's determinations about who deserves value or recognition; and the challenge of conceptualizing value in an enduring way, as a future-oriented form of ethical reasoning demands.

STUDY QUESTIONS

1. What drives Greg's feelings and judgments about who and what is valuable, and on what time scale? How do his assessments differ from the impartiality of utilitarianism? What leads him to make assessments in this way?

2. What are the things that Greg learns to value during the course of the story, and what causes him to value them more than he did at the beginning?

3. Like Kamla, Greg values art partly because of how it increases interpersonal connections, by helping people understand one another across gaps in time and experience. But at the same time, he takes great pleasure in the exclusivity of his connection to Cecilia: the fact that "when we're out in public, people fall silent in linguistic bafflement around us" reinforces the intimacy he feels with her. In what ways are the goods of a private relationship in tension with the goods of broad mutual understanding? How can utilitarianism help us think about these tensions?

4. In what ways are children characterized as more important or significant than adults throughout the story? In what ways are they characterized as less important or significant than adults? (It may be helpful to think about the difference between moral agents and moral patients here.)

5. At several points in the story, there are characters (including background characters) who consider it valuable to remain connected to the dead. Who are these characters, and what specifically does each of them find valuable about this connection? What actions do they take to maintain a connection to the dead? What do their actions indicate about whom they think benefits from this connection, and about what is an effective way to maintain it?

6. Who benefits from being included in an art exhibition? In what way? It will be helpful to consider not only Greg's perspective, but also the artist who motivated Kamla's search.

7. What value judgments inform Kamla's decision making? In what ways are her values continuous with Greg's, or at least recognizable to him? In what ways do they differ?

"MESSAGE IN A BOTTLE," BY NALO HOPKINSON

Size matters. The message may be the medium, but children are small.

"Whatcha doing, Kamla?" I peer down at the chubby-fingered kid who has dug her brown toes into the sand of the beach. I try to look relaxed, indulgent. She's only a child, about four years old, though that outsize head she's got looks strangely adult. It bobs around on her neck as her muscles fight for control. The adoption centre had told Babette and Sunil that their new daughter checked out perfectly healthy otherwise.

Kamla squints back up at me. She gravely considers my question, then holds her hand out, palm up, and opens it like an origami puzzle box. "I'm finding shells," she says. The shell she proffers has a tiny hermit crab sticking out of it. Its delicate body has been crushed like a ball of paper in her tight fist. The crab is most unequivocally dead.

I've managed to live a good many decades as an adult without having children in my life. I don't hate them, though I know that every childless person is supposed to say that so as not to be pecked to death by the righteous breeders of the flock. But I truly don't hate children. I just don't understand them. They seem like another species. I'll help a lost child find a parent, or give a boost to a little body struggling to get a drink from a water fountain—same as I'd do for a puppy or a kitten—but I've never had the urge to be a father. My home is also my studio, and it's a warren of tangled cables, jury-rigged networked computers, and piles of books about as stable as playing-card houses. Plus bins full of old newspaper clippings, bones of dead animals, rusted metal I picked up on the street, whatever. I don't throw anything away if it looks the least bit interesting. You never know when it might come in handy as part of an installation piece. The chaos has a certain nest-like comfort to it.

Gently, I take the dead hermit crab in its shell from Kamla's hand. She doesn't seem disturbed by my claiming her toy. "It's wrong," she tells me in her lisping child's voice. "Want to find more."

She begins to look around again, searching the sand. This is the other reason children creep me out. They don't yet grok that delicate, all important boundary between the animate and inanimate. It's all one to them. Takes them a while to figure out that travelling from the land of the living to the land of the dead is a one-way trip.

I drop the deceased crab from a shaking hand. "No, Kamla," I say. "It's time to go in for lunch now." I reach for her little brown fist. She pulls it away from me and curls it tightly towards her chest. She frowns up at me with that enfranchised hauteur that is the province of kings and four-year-olds. She shakes her head. "No, don't want lunch yet. Have to look for shells."

They say that play is the work of children. Kamla starts scurrying across the sand, intent on her task. But I'm responsible to Kamla's mother, not to Kamla. I promised to watch the child for an hour while Babette prepared lunch. Babs and Sunil have looked tired, desperate and drawn for a while now. Since they adopted Kamla.

There's still about twenty minutes left in my tenure as Kamla's sitter. I'm counting every minute. I run after her. She's already a good hundred yards away, stuffing shells down the front of her bright green bathing suit as quickly as she can. When I catch up with her, she won't come.

Fifteen minutes left with her. Finally, I have to pick her up. Fish-slippery in my arms, she struggles, her black hair whipping across her face as she shakes her head, "No! No!"

I haul her bodily back to the cottage, to Babette. By then, Kamla is loudly shrieking her distress, and the neighbours are watching from their quaint summer cottages. I dump Kamla into her mother's arms. Babette's expression as she takes the child blends frustration with concern. She cradles the back of Kamla's head. Kamla is prone to painful whiplash injuries.

Lunch consists of store-bought cornmeal muffins served with hot dogs cut into fingerjoint-sized pieces, and bright orange carrot sticks. The muffins have a sticky-fake sweetness. Rage forgotten, Kamla devours her meal with a contented, tuneless singing. She has slopped grape juice down the front of her bathing suit. She looks at me over the top of her cup. It's a calm, ancient gaze, and it unnerves me utterly.

Babette has slushed her grape juice and mine with vodka and lots of ice. "Remember Purple Cows?" she asks. "How sick we got on them at Frosh Week in first year?"

"What's Frosh Week?" asks Kamla.

"It's the first week of university, love. University is big people's school."

"Yes, I do know what a university is," pipes the child. Sometimes Kamla speaks in oddly complete sentences. "But what in the world is a frosh?"

"It's short for freshman," I tell her. "Those are people going to university for the first time."

"Oh." She returns to trying to stab her hot dog chunks with a sharp spear of carrot. Over the top of her head, I smile vaguely at Babette. I sip at the awful drink, gulp down my carrot sticks and sausages. As soon as my plate is empty, I make my excuses. Babette's eyes look sad as she waves me goodbye from the kitchen table. Sunil is only able to come up to their summer cottage on weekends. When he does so, Babs tells me that he sleeps most of the weekend away, too exhausted from his job to talk much to her, or to play with Kamla on the beach.

On my way out the door, I stop to look back. Kamla is sitting in Babette's lap. There's a purple Kamla-sized handprint on Babette's stained, yellow tshirt. Kamla is slurping down more grape juice, and doesn't look up as I leave.

When I reached the age where my friends were starting to spawn like frogs in springtime—or whenever the hell frogs spawn—my unwillingness to do the same became more of a problem. Out on a date once with Sula, a lissom giraffe of a woman with a tongue just as supple, I mentioned that I didn't intend to have kids. She frowned. Had I ever seen her do that before?

"Really?" she said. "Don't you care about passing on your legacy?"

"You mean my surname?"

She laughed uncomfortably. "You know what I mean."

"I really don't. I'm not a king and I'm never going to be rich. I'm not going to leave behind much wealth for someone to inherit. It's not like I'm building an empire."

She made a face as though someone had dropped a mouse in her butter churn. "What are you going to do with your life, then?"

"Well," I chuckled, trying to make a joke of it, "I guess I'm going to go home and put a gun to my head, since I'm clearly no use to myself or anyone else."

Now she looked like she was smelling something rotten. "Oh, don't be morbid," she snapped.

"Huh? It's morbid to not want kids?"

"No, it's morbid to think your life has so little value that you might as well kill yourself."

"Oh, come on, Sula!"

I'd raised my voice above the low-level chatter in the restaurant. The couple at the table closest to us glanced our way. I sighed and continued: "My life has tons of value. I just happen to think it consists of more than my genetic material. Don't you?"

"I guess." But she pulled her hand away from mine. She fidgeted with her napkin in her lap. For the rest of dinner, she seemed distracted. She didn't meet my eye often, though we chatted pleasantly enough. I told her about this bunch of Sioux activists, how they'd been protesting against a university whose archaeology department had dug up one of their ancestral burial sites. I'm Rosebud Sioux on my mum's side. When the director of the department refused to reconsider, these guys had gone one night to the graveyard where his great-grandmother was buried. They'd dug up her remains, laid out all the bones, labelled them with little tags. They did jail time, but the university returned their ancestors' remains to the band council.

Sula's only response to the story was, "Don't you think the living are more important?" That night's sex was great. Sula rode me hard and put me away wet. But she wouldn't stay the night. I curled into the damp spot when she'd left, warming it with my heat. We saw each other two or three times after that, but the zing had gone out of it.

Babette and Sunil began talking about moving away from St. John's. Kamla was about to move up a grade in school. Her parents hoped she'd make new friends in a new school. Well, any friends, really. Kids tended to tease Kamla, call her names.

Babette found a job before Sunil did. She was offered a post teaching digital design at the Emily Carr Institute in Vancouver. Construction was booming there, so Sunil found work pretty easily afterwards. When she heard they were moving, Kamla threw many kinds of fits. She didn't want to leave the ocean. Sunil pointed out that there would be ocean in Vancouver. But Kamla stamped her foot. "I want this ocean right here. Don't you understand?" Sunil and Babette had made their decision, though, and Kamla was just a kid. The whole family packed up kit and caboodle in a move that Babette later told me was the most tiring thing she'd ever done.

On the phone, Babette tells me, "A week after we got here, we took Kamla down to Wreck Beach. The seals come in real close to shore, you know? You can see them peeking at you as they hide in the waves. We thought Kamla would love it."

"Did she?" I ask, only half-listening. I'm thinking about my imminent date with Cecilia, who I've been seeing for a few months now. She is lush and brown. It takes both of my hands to hold one of her breasts, and when we spoon at night, her belly fits warm in my palm like a bowl of hot soup on a cold day.

"You know what Kamla did?" Babette asks, bringing me back from my jism-damp haze. I hear the inhale and "tsp" sound of someone smoking a cigarette. Babette had started smoking again during the move. "She poked around in the sand for a few minutes, then she told us we were stupid and bad and she wasn't going to talk to us any more. Sulked the rest of the day, and wouldn't eat her dinner that night. She's still sulking now, months later."

That's another thing about kids: their single-mindedness. They latch onto an idea like a bulldog at a rabbit hole, and before you know it, you're arranging your whole life around their likes and dislikes. They're supposed to be your insurance for the future; you know, to carry your name on, and shit? My mother's been after me to breed, but I'm making my own legacy, thank you very much. A body of art I can point to and document. I'm finally supporting myself sort of decently through a combination of exhibition fees, teaching and speaking gigs. I want to ask Cecilia to move in with me, but every time I come close to doing so, I hear Sula's words in my head: *No children? Well, what are you going to do with yourself, then?* I don't know whether Cecilia wants kids, and I'm afraid to ask.

"Greg?" says Babette's voice through the telephone. "You still there?"

"Yeah. Sorry. Mind wandering."

"I'm worried about Kamla."

"Because she's upset about the move? I'm sure she'll come around. She's making friends in school, isn't she?"

"Not really. The other day, the class bully called her Baby Bobber. For the way her head moves."

I suppress a snort of laughter. It's not funny. Poor kid. "What did you do?"

"We had the school contact his parents. But it's not just that she doesn't have many friends. She's making our lives hell with this obsession for Bradley's Cove. And she's not growing."

"You mean she's, like, emotionally immature?" *Or intellectually?* I think.

"No, physically. We figure she's about eight, but she's not much bigger than a five-year-old."

"Have you taken her to the doctor?"

"Yeah. They're running some tests."

Cecilia can jerry-rig a computer network together in a matter of minutes. We geekspeak at each other all the time. When we're out in public, people fall silent in linguistic bafflement around us.

"They say Kamla's fine," Babette tells me, "and we should just put more protein in her diet."

Cecilia and I are going to go shopping for a new motherboard for her, then we're going to take blankets and pillows to the abandoned train out in the old rail yards and hump like bunnies till we both come screaming. Maybe she'll wear those white stockings under her clothes. The sight of the gap of naked brown thigh between the tops of the stockings and her underwear always makes me hard.

Babette says, "There's this protein drink for kids. Makes her pee bright yellow."

The other thing about becoming a parent? It becomes perfectly normal to discuss your child's excreta with anyone who'll sit still for five minutes. When we were in art school together, Babette used to talk about gigabytes, Cronenberg and post-humanism.

I can hear someone else ringing through on the line. It's probably Cecilia. I mutter a quick reassurance at Babette and get her off the phone.

Kamla never does get over her obsession with the beach, and with shells. By the time she is nine, she's accumulated a library's worth of reference books with names like *Molluscs of the Eastern Seaboard*, and *Seashells: Nature's Wonder*. She continues to grow slowly. At ten years old, people mistake her for six. Sunil and Babette send her for test after test.

"She's got a full set of adult teeth," Babette tells me as we sit in a coffee shop on Churchill Square. "And all the bones in her skull are fused."

"That sounds dangerous," I say.

"No, it happens to all of us once we've stopped growing. Her head's fully grown, even if the rest of her isn't. I guess that's something. You gonna eat those fries?"

Babette's come home to visit relatives. She's quit smoking, and she's six months pregnant. If she'd waited two more months, the airline wouldn't have let her travel until the baby was born. "Those symptoms of Kamla's," says Babette, "they're all part of the DGS."

The papers have dubbed it "Delayed Growth Syndrome." Its official name is Diaz Syndrome, after the doctor who first identified it. There are thousands of kids with Kamla's condition. It's a brand new disorder. Researchers have no clue what's causing it, or if the bodies of the kids with it will ever achieve full adulthood. Their brains, however, are way ahead of their bodies. All the kids who've tested positive for DGS are scarily smart.

"Kamla seems to be healthy," Babette tells me. "Physically, anyway. It's her emotional state I'm worried about."

I say, "I'm gonna have some dessert. You want anything?"

"Yeah, something crunchy with meringue and caramel. I want it to be so sweet that the roof of my mouth tries to crawl away from it."

Cecilia's doing tech support for somebody's office today. Weekend rates. My mum's keeping an eye on our son Russ, who's two and a half. Yesterday we caught him scooping up ants into his mouth from an anthill he'd found in the backyard. He was giggling at the way they tickled his tongue, chomping down on them as they scurried about. His mouth was full of anthill mud. He didn't even notice that he was being bitten until Cecilia and I asked him. That's when he started crying in pain, and he was inconsolable for half an hour. I call him our creepy little alien child. We kinda had him by accident, me and Cece. She didn't want kids any more than I did, but when we found out she was pregnant, we both got . . . curious, I guess. Curious to see what this particular life adventure would be; how our small brown child might change a world that desperately needs some change. We sort of dared each other to go through with it, and now here we are. Baby's not about changing anyone's world but ours just yet, though. We've both learned the real meaning of sleep deprivation. That morning when he was so constipated that trying to shit made him scream in pain, I called Babette in panic. Turns out poo and pee are really damned important, especially when you're responsible for the life of a small, helpless being that can barely do anything else. Russ gurgles with helpless laughter when I blow raspberries on his tummy. And there's a spot on his neck, just under his ear, that smells sweet, even when the rest of him is stinky. He's perfect.

The next time I meet Babette for lunch I ask Babette what new thing is bothering her about her kid, if not the delayed growth.

"She gets along fine with me and Sunil, you know? I feel like I can talk to her about anything. But she gets very frustrated with kids her age. She wants to play all these elaborate games, and some of them don't understand. Then she gets angry. She came stomping home from a friend's place the other day and went straight to her room. When I looked in on her, she was sitting looking in her mirror. There were tears running down her cheeks. 'I bloody hate being a kid,' she said to me. 'The other kids are stupid, and my hand-eye coordination sucks.'"

"She said that her hand-eye coordination sucked? That sounds almost too . . ."

"Yeah, I know. Too grown up for a ten-year-old. She probably had to grow up quickly, being an adoptee."

"You ever find out where she came from before you took her?"

Babette shakes her head. She's eaten all of her pavlova and half of my carrot cake.

It just so happens that I have a show opening at Eastern Edge while Babette and Sunil are in town. "The Excavations," I call it. It was Russ's anthill escapade that gave me the idea. I've trucked in about half a ton of dirt left over from a local archaeological dig. I wish I could have gotten it directly from Mexico, but I couldn't afford the permit for doing that. I seeded the soil with the kinds of present-day historical artifacts that the researchers tossed aside in their zeal to get to the iconic past of the native peoples of the region: a rubber boot that had once belonged to a Mayan Zapatista from Chiapas; a large plastic jug that used to hold bleach, and that had been refitted as a bucket for a small child to tote water in; a scrap of hand-woven blanket with brown stains on it. People who enter the exhibition get basic excavation tools. When they pull something free of the soil, it triggers a story about the artifact on the monitors above. Sunil is coming to the opening. Babette has decided to stay at her relatives' place and nap. Six months along in her pregnancy, she's sleepy a lot.

I'm holding court in the gallery, Cecilia striding around the catwalk above me, doing a last check of all the connections, when Sunil walks in. He's brought Kamla. She doesn't alarm me any more. She's just a kid. As I watch her grow up, I get some idea of what Russ's growing years will be like. In a way, she's his advance guard.

Kamla scurries in ahead of her dad, right up to me, her head wobbling as though her neck is a column of gelatin. She sticks out her hand. "Hey, Greg," she says. "Long time." Behind her, Sunil gives me a bashful smile.

I reach down to shake the hand of what appears to be a six-year-old.

"Uh, hey," I say. Okay, I lied a little bit. I still don't really know how to talk to kids.

"This looks cool," she tells me, gazing around. "What do we do?" She squats down and starts sifting soil through her fingers.

"Kamla, you mustn't touch the art," says Sunil.

I say, "Actually, it's okay. That's exactly what I want people to do."

Kamla flashes me a grateful glance. I give her a small spade and take her through the exhibition. She digs up artifact after artifact, watches the stories about them on the video displays, asks me questions. I get so caught up talking to her about my project that I forget how young she is. She seems really interested. Most of the other people are here because they're friends of mine, or because it's cool to be able to say that you went to an art opening last weekend. The gallery owner has to drag me away to be interviewed by the guy from *Art(ext)/e*. I grin at Kamla and leave her digging happily in the dirt.

While I'm talking to the interviewer, Kamla comes running up to me, Sunil behind her yelling, "Kamla! Don't interrupt!"

She ignores him, throws her mushroom-shaped body full tilt into my arms, and gives me a whole body hug. "It was you!" she says. "It was you!" She's clutching something in one dirt-encrusted fist. The guy from *Art(ext)/e* kinda freezes up at the sight of Kamla. But he catches himself, pastes the smile back on, motions his camerawoman to take a picture.

"I'm so sorry," Sunil says. "When she gets an idea in her head . . ."

"Yeah, I know. What'd you find, chick?" I ask Kamla. She opens her palm to show me. It's a shell. I shake my head. "Honestly? I barely remember putting that in there. Some of the artifacts are 'blanks' that trigger no stories. The dig where I got it from used to be underwater a few centuries ago."

"It's perfect!" says Kamla, squeezing me hard.

Perfect like she isn't? Damn.

"I've been looking everywhere for this!" she tells me.

"What, is it rare or something?" I ask her.

She rears back in my arms so that she can look at me properly. "You have no idea," she says. "I'm going to keep this so safe. It'll never get out of my sight again."

"Kamla!" scolds Sunil. "That is part of Greg's exhibition. It's staying right here with him."

The dismay on Kamla's face would make a stone weep. It's obvious that it hadn't even occurred to her that I mightn't let her have the shell. Her eyes start to well up.

"Don't cry," I tell her. "It's just an old shell. Of course you can take it."

"You shouldn't indulge her," Sunil says. "You'll spoil her."

I hitch Kamla up on my hip, on that bone adults have that seems tailormade for supporting a child's butt on. "Let's call it her reward for asking some really smart questions about the exhibition."

Sunil sighs. Kamla's practically glowing, she's so happy. My heart warms to her smile.

When the phone rings at my home many hours later, it takes me a while to orient myself. It's 3:05 a.m. by the clock by our bedside. "Hello?" I mumble into the phone. I should have known better than to have that fifth whiskey at the opening. My mouth feels and tastes like the plains of the Serengeti, complete with lion spoor.

"Greg?" The person is whispering. "Is this Greg?"

It's a second or so before I recognise the voice. "Kamla? What's wrong? Is your mum okay?"

"They're fine. Everyone's asleep."

"Like you should be. Why the fuck are you calling me at this hour?" I ask, forgetting that I'm talking to a child. Something about Kamla's delivery makes it easy to forget.

"I've been on the Net. Listen, can you come get me? The story's about to break. It's all over Twitter and YouTube already. It'll be on the morning news here in a few hours. Goddamned Miles. We told them he was always running his mouth off."

"What? Told who? Kamla, what's going on?"

Cecilia is awake beside me. She's turned on the bedside lamp. *Who?* she mouths. I make my lips mime a soundless *Kamla.*

"It's a long story," Kamla says. "Please, can you just come get me? You need to know about this. And I need another adult to talk to, someone who isn't my caretaker."

Whatever's going on, she really sounds upset. "Okay, I'll be there soon."

Kamla gives me the address, and I hang up. I tell Cecilia what's going on.

"You should just let Babs and Sunil know that she's disturbed about something," she says. "Maybe it's another symptom of that DGS."

"I'll talk to them after Kamla tells me what's going on," I say. "I promised her to hear her out first."

"You sure that's wise? She's a child, Greg. Probably she just had a nightmare."

Feeding our child has made Cecilia's breasts sit lower on her rib cage. Her hips stretch out the nylon of her nightgown. Through the translucent fabric I can see the shadow of pubic hair and the valley that the curves of her thighs make. Her eyes are full of sleep, and her hair is a tousled mess, and she's so beautiful I could tumble her right now. But there's this frightened kid waiting to talk to me. I kiss Cecilia goodbye and promise to call her as soon as I've learned more.

Kamla's waiting for me outside the house when I pull up in my car. The night air is a little chilly, and she's a lonely, shivering silhouette against the front door. She makes to come in the passenger side of the car, but I motion her around to my side. "We're going to leave a note for your parents first," I tell her. I have one already prepared. "And we're just going sit right here in the car and talk."

"We can leave a note," she replies, "but we have to be away from here long enough so you can hear the whole story. I can't have Sunil and Babette charging to the rescue right now."

I've never heard her call her parents by their first names; Babs and Sunil aren't into that kind of thing. Her face in her weirdly adult head looks calm, decisive. I find myself acquiescing. So I slip the note under the front door. It tells Babette and Sunil that Kamla's with me, that everything's all right. I leave them my cell phone number, though I'm pretty sure that Babette already has it.

Kamla gets into the car. She quietly closes the door. We drive. I keep glancing over at her, but for a few minutes, she doesn't say anything. I'm just about to ask her what was so urgent that she needed to pull a stunt like this when she says, "Your installation had a certain antique brio to it, Greg. Really charming. My orig—I mean, I have a colleague whose particular interest is in the nascent identity politics as expressed by artists of the twentieth and twenty-first centuries, and how that expression was the progenitor of current speciesism."

"Have you been reading your mum's theory books?"

"No," she replies. There was so much bitterness in that one word. "I'm just a freak. Your kid's almost three, right?"

"Yeah."

"In a blink of an eye, barely a decade from now, his body will be entering puberty. He'll start getting erections, having sexual thoughts."

"I don't want to think about all that right now," I say. "I'm still too freaked that he's begun making poo-poo jokes. Kamla, is this the thing you wanted to tell me? Cause I'm not getting it."

"A decade from now, I'll have the body of a seven-year-old."

"You can't know that. There aren't any DGS kids who've reached their twenties yet."

"I know. I'm the oldest of them, by a few weeks."

Another thing she can't know.

"But we're all well past the age where normal children have achieved adolescence."

Goggling at her, I almost drive through a red light. I slam on the brakes. The car jolts to a halt. "What? What kind of shit is that? You're ten years old. A precocious ten, yes, but only ten."

"Go in there." She points into the parking lot of a nearby grocery store. "It won't be open for another three hours." I pull into the lot and park.

"If the cops come by and see us," I say, "I could be in a lot of shit. They'll think I'm some degenerate Indian perv with a thing for little girls."

Shit. I shouldn't be talking to a ten-year-old this way. Kamla always makes me forget. It's that big head, those big words.

"DGS people do get abused," she tells me. "Just like real children do."

"You *are* a real child!"

She glares at me, then looks sad. She says, "Sunil and Babette are going to have to move soon. It's so hard for me to keep up this pretence. I've managed to smart-mouth so much at school and in our neighbourhood that it's become uncomfortable to live there anymore."

My eyes have become accustomed enough to the dark that I can see the silent tears running down her cheeks. I want to hold her to me, to comfort her, but I'm afraid of how that will look if the cops show up. Besides, I'm getting the skin-crawly feeling that comes when you realise that someone with whom you've been making pleasant conversation is as mad as a hatter. "I'm taking you back home," I whisper. I start turning the key in the ignition.

"Please!" She puts a hand on my wrist. "Greg, please hear me out. I'll make it quick. I just don't know how to convince you."

I take my hand off the key. "Just tell me," I said. "Whatever it is, your parents love you. You can work it out."

She leans back against the passenger side door and curls her knees up to her chest, a little ball of misery. "Okay. Let me get it all out before you say anything else, all right?"

"All right."

"They grew us from cells from our originals; ten of us per original. They used a viral injection technique to put extra-long tails on one of the strands of our DNA. You need more telomeres to slow down aging."

The scientific jargon exiting smoothly from the mouth of a child could have been comic. But I had goose bumps. She didn't appear to be repeating something she'd memorised.

"Each batch of ten yielded on average four viable blastocytes. They implanted those in womb donors. Two-thirds of them took. Most of those went to full term and were delivered. Had to be C-sections, of course. Our huge skulls presented too much of a risk for our birth mothers. We were usually four years old before we were strong enough to lift our own heads, and that was with a lot of physiotherapy. They treated us really well; best education, kept us fully informed from the start of what they wanted from us."

"Which was?" I whisper, terrified to hear the answer.

"Wait. You said you would." She continues her story. "Any of us could back out if we wanted to. Ours is a society that you would probably find strange, but we do have moral codes. Any of us who didn't want to make the journey could opt to undergo surgical procedures to correct some of the physical changes. Bones and muscles would lengthen, and they would reach puberty normally and thereafter age like regular people. They'll never achieve full adult height, and there'll always be something a little bit odd about their features, but it probably won't be so bad.

"But a few of us were excited by the idea, the crazy, wonderful idea, and we decided to go through with it. They waited until we were age thirteen for us to confirm our choice. In many cultures, that used to be the age when you were allowed to begin making adult decisions."

"You're ten, Kamla."

"I'm twenty-three, though my body won't start producing adult sex hormones for another fifty years. I won't attain my full growth till I'm in my early hundreds. I can expect—"

"You're delusional," I whisper.

"I'm from your future," she says.

God. The child's been watching too many B-movies.

She continues, "They wanted to send us here and back as full adults, but do you have any idea what the freight costs would have been? The insurance? Arts grants are hard to get in my world, too. The gallery had to scale the budget way back."

"Gallery?"

"National gallery. Hush. Let me talk. They sent small people instead. Clones of the originals, with their personalities superimposed onto our own. They sent back children who weren't children."

I start the car. I'm taking her back home right now. She needs help; therapy, or something. The sky's beginning to brighten. She doesn't try to stop me this time.

Glumly, she goes on. "The weird thing is, even though this body isn't interested in adult sex, I *remember* what it was like, remember enjoying it. It's those implanted memories from my original."

I'm edging past the speed limit in my hurry to get her back to her parents. I make myself slow down a little.

"Those of us living in extremely conservative or extremely poor places are having a difficult time. We stay in touch with them by email and cell phone, and we have our own closed Facebook group, but not all of us have access to computer technology. We've never been able to figure out what happened to Kemi. Some of us were never adopted, had to make our own way as street kids. Never old enough to be granted adult freedoms. So many lost. This fucking project better have been worth it."

I decide to keep her talking. "What project, Kamla?"

"It's so *hard* to pretend you don't have an adult brain! Do you know what it's like turning in schoolwork that's at a grade-five level, when we all have PhDs in our heads? We figured that one of us would crack, but we hoped it'd be later, when we'd reached what your world would consider the age of majority."

We're cruising past a newspaper box. I look through its plastic window to see the headline: "I'M FROM THE FUTURE," SAYS BOBBLE-HEADED BOY. Ah. One of our more erudite news organs.

Oh, Christ. They all have this delusion. All the DGS kids. For a crazy half-second, I find myself wondering whether Sunil and Babette can return Kamla to the adoption centre. And I'm guiltily grateful that Russ, as far as we can tell, is normal.

"Human beings, we're becoming increasingly post-human," Kamla says. She's staring at the headline, too. "Things change so quickly. Total technological upheaval of society every five to eight years. Difficult to keep up, to connect amongst the generations. By the time your Russ is a teenager, you probably won't understand his world at all."

She's hit on the thing that really scares me about kids. This brave new world that Cecilia and I are trying to make for our son? For the generations to follow us? We won't know how to live in it.

Kamla says, "Art helps us know how to do change. That's made it very valuable to us."

"Thank heaven for that," I say, humouring her. "Maybe I'd like your world."

She sits up in her seat, buckles herself in. Shit. I should have made her do that the minute she got in the car. I have one of those heart-in-the-mouth moments that I have often, now that I'm a parent. "In my world," she says, "what you do would be obsolete." She sniggers a little. "Video monitors! I'd never seen a real one, only minibeams disguised to mimic ancient tech. Us DGSers have all become anthropologists here in the past, as well as curators."

"Wait; you're a what?"

"I'm a curator, Greg. I'm trying to tell you; our national gallery is having a giant retrospective; tens of thousands of works of art from all over the world, and all over the world's history. They sent us back to retrieve some of the pieces that had been destroyed. Expensive enough to send living biomaterial back; their grant wasn't enough to pay for returning us to our time. So we're going to grow our way there. Those of us that survive."

There are more cars out on the road, more brakes squealing, more horns honking. "I'm not going to miss mass transit when I finally get home," she says. "Your world stinks."

"Yeah, it does." We're nearly to her parents' place. From my side, I lock her door. Of course she notices. She just glances at the sound. She looks like she's being taken to her death.

"I didn't know it until yesterday," she tells me, "but it was you I came for. That installation."

And now the too-clever bloody child has me where I live. Though I know it's all air pie and Kamla is as nutty as a fruitcake, my heart's performing a tympanum of joy. "My installation's going to be in the retrospective?" I ask. Even as the words come out of my mouth, I'm embarrassed at how eager I sound, at how this little girl, as children will, has dug her way into my psyche and found the thing which will make me respond to her.

She gasps and puts her hand to her mouth. "Oh, Greg! I'm so sorry; not you, the shell!"

My heart suicides, the brief, hallucinatory hope dashed. "The shell?"

"Yes. In the culture where I live, speciesism has become a defining concept through which we understand what it means to be human animals. Not every culture or subculture ascribes to it, but the art world of my culture certainly does." She's got her teacher voice on again. She does sound like a bloody curator. "Human beings aren't the only ones who make art," she says.

All right. Familiar territory. "Okay, perhaps. Bower birds make pretty nests to attract a mate. Cetaceans sing to each other. But we're the only ones who make art *mean*; who make it comment on our everyday reality."

From the corner of my eye, I see her shake her oversized head. "No. We don't always know what they're saying, we can't always know the reality on which they're commenting. Who knows what a sea cucumber thinks of the conditions of its particular stretch of ocean floor?"

A sea cucumber? We've just turned onto her parents' street. She'll be out of my hands soon. Poor Babette.

"Every shell is different," she says.

My perverse brain instantly puts it to the tune of "Every Sperm Is Sacred."

She continues, "Every shell is a life journal, made out of the very substance of its creator, and left as a record of what it thought, even if we can't understand exactly what it thought. Sometimes interpretation is a trap. Sometimes we need to simply observe."

"And you've come all this way to take that . . . shell back?" I can see it sticking out of the chest pocket of her fleece shirt.

"It's difficult to explain to you, because you don't have the background, and I don't have the time to teach you. I specialise in shell formations. I mean, that's Vanda's specialty. She's the curator whose memories I'm carrying. Of its kind, the mollusc that made this shell is a genius. The unique conformation of the whorls of its shell expresses a set of concepts that haven't been explored before by the other artists of its species. After this one, all the others will draw on and riff off its expression of its world. They're the derivatives, but this is the original. In our world, it was lost."

Barmy. Loony. "So how did you know that it even existed, then? Did the snail or slug that lived inside it take pictures or something?" I've descended into cruelty. I'm still smarting that Kamla hasn't picked me, my work. My legacy doesn't get to go to the future.

She gives me a wry smile, as though she understands.

I pull up outside the house, start leaning on the horn. Over the noise, she shouts, "The creature didn't take a picture. You did."

Fuck, fuck, fuck. With my precious video camera. I'd videotaped every artifact with which I'd seeded the soil that went onto the gallery floor. I didn't tell her that.

She nods. "Not all the tape survived, so we didn't know who had recorded it, or where the shell had come from. But we had an idea where the recording had come from."

Lights are coming on in the house. Kamla looks over there and sighs. "I haven't entirely convinced you, have I?"

"No," I say regretfully. But damn it, a part of me still hopes that it's all true.

"They're probably going to institutionalise me. All of us."

The front door opens. Sunil is running out to the car, a gravid Babette following more slowly.

"You have to help me, Greg. Please? We're going to outlive all our captors. We will get out. But in the meantime . . ."

She pulls the shell out of her pocket, offers it to me on her tiny palm. "Please keep it safe for me?"

She opens the car door. "It's your ticket to the future," she says, and gets out of the car to greet her parents.

I lied. I fucking hate kids.

"The Gambler," by Paolo Bacigalupi

STORY FRAME

> I try to protest. "But you hired me to write the important stories. The stories about politics and the government, to continue the traditions of the old newspapers. I remember what you said when you hired me."
>
> "Yeah, well." She looks away. "I was thinking more about a good scandal."
>
> "The checkerspot is a scandal. That butterfly is now gone."
>
> She sighs. "No, it's not a scandal. It's just a depressing story. No one reads a depressing story, at least, not more than once. And no one subscribes to a depressing byline feed."

How can we be the version of ourselves that we want to be, even in times of crisis or under pressure from the outside world? And how can we influence the world around us for the better when the world doesn't necessarily want to be influenced in those ways? These universal questions take on particular sharpness for Ong, a Laotian refugee who works as a journalist for a major American media company. Ong writes articles about the stories he considers most important, such as climate change and the fallout of government mismanagement. But such stories don't align very well with the reading habits of the American public, who instead give their clicks to product reviews and celebrity scandals. Because of his poor numbers, Ong is at risk of losing his job, and with it the visa that enables him to stay in America.

Ong's journalistic commitments are an outgrowth of the values that Ong inherited from his father and developed by watching his father's actions when Laos was overtaken by an oppressive totalitarian regime during his younger years. The country is now what Ong calls a "black hole" for news and communication: both physical roads and communication avenues have been blockaded, and the only news widely available in the country is nationalist propaganda. Searching for news of his home country from America, Ong is limited not by restrictions on speech or reporting, but by the outside world's lack of interest in what is happening in Laos.

By giving us both Laos and the United States, the story explores different news cultures and what's at stake. Ong looks for news on Laos now, and the issue isn't that it's forbidden, it's that there isn't enough interest. In the near-future America of this story, the machinations of government are not a threat to most citizens' ordinary lives, and thus those citizens prefer to spend their attention elsewhere.

Ong's experiences as an immigrant, struggling with (and in many respects against) assimilation, offer us a nuanced way to think about the complexities of habitus. The formative impact of his childhood in Laos is clear, as is the influence of his father: our early experiences and role models play a major role in shaping our character. But the story also makes clear the ways in which Ong chooses—over and over, in ways large and small—to remain aligned with the perceptual modes and values of his earlier life, and how continuing to look to his father as an exemplar helps him navigate the world in ways that align with his sense of practical wisdom, even if his choices seem out of balance to those around him.

"The Gambler" offers us several high-stakes moments in its present-day storyline to mirror the high-stakes moment when Ong's father refuses to stop protesting. But among the various characters in the story—Marty Mackley, the master of infotainment; Janice, Ong's results-oriented boss; Kulaap, who balances her Laotian and American identities in a way Ong cannot quite understand; and of course Ong himself—there is no consensus on what kind of response those high-stakes moments call for. Indeed, there is not even consensus on what really counts as a high-stakes moment in the first place: each character has different judgments about the lines between "real news," distractions, and depressing stories. But as "The Gambler" illustrates, our choices in those moments are rooted in the longer arc of our character.

STUDY QUESTIONS

1. Both worlds of Ong's experience—Laos under Khamsing, and the America he flees to—as places where real news is not easy to find, unless you know where to look. Consider the tools that Ong uses to find the news he wants in the states and compare them to the sources of real news that Ong and his father had in Laos. What are the similarities? What are the differences?

2. The story draws a strong contrast between patterns and cultures of information distribution in Laos on the one hand, and in America on the other. What are the

differences between them? Do these differences always necessarily come "bundled," or is it possible for a society to habituate to some aspects of life in America without habituating to all of it?

3. Though Ong sees the cultures of America and Laos as opposed in many respects, Kulaap turns out to be genuinely bicultural, able to integrate them/move comfortably between them. One way in which this comes through is her relationship to media: she knows how to play the American paparazzi to best effect, but she also participates in the same online underground Laotian community as Ong does. Do the multiple roles of each character create tension? Why? What if they did not have overlapping communities?

4. In what ways has Ong habituated to living in America? In what ways has he not habituated? What thing(s) has he been doing, in thought and in action, to make it so? How does Ong's news consumption both reflect and sustain his habitus?

"THE GAMBLER," BY PAOLO BACIGALUPI

My father was a gambler. He believed in the workings of karma and luck. He hunted for lucky numbers on license plates and bet on lotteries and fighting roosters. Looking back, I think perhaps he was not a large man, but when he took me to the *muy thai* fights, I thought him so. He would bet and he would win and laugh and drink *laolao* with his friends, and they all seemed so large. In the heat drip of Vientiane, he was a lucky ghost, walking the mirror-sheen streets in the darkness.

Everything for my father was a gamble: roulette and blackjack, new rice variants and the arrival of the monsoons. When the pretender monarch Khamsing announced his New Lao Kingdom, my father gambled on civil disobedience. He bet on the teachings of Mr. Henry David Thoreau and on whisper sheets posted on lampposts. He bet on saffron-robed monks marching in protest and on the hidden humanity of the soldiers with their well-oiled AK-47s and their mirrored helmets.

My father was a gambler, but my mother was not. While he wrote letters to the editor that brought the secret police to our door, she made plans for escape. The old Lao Democratic Republic collapsed, and the New Lao Kingdom blossomed with tanks on the avenues and tuk-tuks burning on the street corners. Pha That Luang's shining gold *chedi* collapsed under shelling, and I rode away on a UN evacuation helicopter under the care of kind Mrs. Yamaguchi.

From the open doors of the helicopter, we watched smoke columns rise over the city like nagas coiling. We crossed the brown ribbon of the Mekong with its jeweled belt of burning cars on the Friendship Bridge. I remember a Mercedes floating in the water like a paper boat on Loi Kratong, burning despite the water all around.

Afterward, there was silence from the land of a million elephants, a void into which light and Skype calls and e-mail disappeared. The roads were blocked. The telecoms died. A black hole opened where my country had once stood.

Sometimes, when I wake in the night to the swish and honk of Los Angeles traffic, the confusing polyglot of dozens of countries and cultures all pressed together in this American melting pot, I stand at my window and look down a boulevard full of red lights, where it is not safe to walk alone at night, and yet everyone obeys the traffic signals. I look down on the brash and noisy Americans in their many hues, and remember my parents: my father who cared too much to let me live under the self-declared monarchy, and my mother who would not let me die as a consequence. I lean against the window and cry with relief and loss.

Every week I go to temple and pray for them, light incense and make a triple bow to Buddha, Damma, and Sangha, and pray that they may have a good rebirth, and then I step into the light and noise and vibrancy of America.

■   ■   ■

My colleagues' faces flicker gray and pale in the light of their computers and tablets. The tap of their keyboards fills the newsroom as they pass content down the workflow chain and then, with a final keystroke and an obeisance to the "publish" button, they hurl it onto the net.

In the maelstrom, their work flares, tagged with site location, content tags, and social poke data. Blooms of color, codes for media conglomerates: shades of blue and Mickey Mouse ears for Disney-Bertelsmann. A red-rimmed pair of rainbow O's for Google's AOL News. Fox News Corp. in pinstripes gray and white. Green for us: Milestone Media—a combination of NTT DoCoMo, the Korean gaming consortium Hyundai-Kubu, and the smoking remains of the New York Times Company. There are others, smaller stars, Crayola shades flaring and brightening, but we are the most important. The monarchs of this universe of light and color.

New content blossoms on the screen, bathing us all in the bloody glow of a Google News content flare, off their WhisperTech feed. They've scooped us. The posting says that new ear bud devices will be released by Frontal Lobe before Christmas: terabyte storage with Pin-Line connectivity for the Oakley microresponse glasses. The technology is next-gen, allowing personal data control via Pin-Line scans of a user's iris. Analysts predict that everything from cell phones to digital cameras will become obsolete as the full range of Oakley features becomes available. The news flare brightens and migrates toward the center of the maelstrom as visitors flock to Google and view stolen photos of the iris-scanning glasses.

Janice Mbutu, our managing editor, stands at the door to her office, watching with a frown. The maelstrom's red bath dominates the newsroom, a pressing reminder that Google is beating us, sucking away traffic. Behind glass walls, Bob and Casey, the heads of the Burning Wire, our own consumer technology feed, are screaming at their reporters, demanding they do better. Bob's face has turned almost as red as the maelstrom.

The maelstrom's true name is LiveTrack IV. If you were to go downstairs to the fifth floor and pry open the server racks, you would find a sniper sight logo and the words SCRY GLASS—KNOWLEDGE IS POWER stamped on their chips in metallic orange, which would tell you that even though Bloomberg rents us the machines, it is a Google-Neilsen partnership that provides the proprietary algorithms for analyzing the net flows—which means we pay a competitor to tell us what is happening with our own content.

LiveTrack IV tracks media user data—Web site, feed, VOD, audiostream, TV broadcast—with Google's own net statistics gathering programs, aided by Nielsen hardware in personal data devices ranging from TVs to tablets to ear buds to handsets to car radios. To say that the maelstrom keeps a finger on the pulse of media is an understatement. Like calling the monsoon a little wet. The maelstrom is the pulse, the pressure, the blood-oxygen mix; the count of red cells and white, of T-cells and BAC, the screening for AIDS and hepatitis G. . . . It is reality.

Our service version of the maelstrom displays the performance of our own content and compares it to the top one hundred user-traffic events in real-time. My own latest news story is up in the maelstrom, glittering near the edge of the screen, a tale of government incompetence: the harvested DNA of the checkerspot butterfly, already extinct, has been destroyed through mismanagement at the California Federal Biological Preserve Facility. The butterfly—along with sixty-two other species—was subjected to improper storage protocols, and now there is nothing except a little dust in vials. The samples literally blew away. My coverage of the story opens with federal workers down on their knees in a two-billion-dollar climate-controlled vault, with a dozen crime scene vacuums that they've borrowed from LAPD, trying to suck up a speck of butterfly that they might be able to reconstitute at some future time.

In the maelstrom, the story is a pinprick beside the suns and pulsing moons of traffic that represent other reporters' content. It doesn't compete well with news of Frontal Lobe devices, or reviews of Armored Total Combat, or live feeds of the Binge-Purge championships. It seems that the only people who are reading my story are the biologists I interviewed. This is not surprising. When I wrote about bribes for subdivision approvals, the only people who read the story were county planners. When I wrote about cronyism in the selection of city water recycling technologies, the only people who read were water engineers. Still, even though no one seems to care about these stories, I am drawn to them, as though poking at the tiger of the American government will somehow make up for not being able to poke at the little cub of New Divine Monarch Khamsing. It is a foolish thing, a sort of Don Quixote crusade. As a consequence, my salary is the smallest in the office.

"Whoooo!"

Heads swivel from terminals, look for the noise: Marty Mackley, grinning.

"You can thank me . . ." He leans down and taps a button on his keyboard. "Now."

A new post appears in the maelstrom, a small green orb announcing itself on the Glamour Report, Scandal Monkey blog, and Marty's byline feeds. As we watch, the post absorbs pings from software clients around the world, notifying the millions of people who follow his byline that he has launched a new story.

I flick my tablet open, check the tags:

Double DP,

Redneck HipHop,

Music News,

Schadenfreude,

underage,

pedophilia . . .

According to Mackley's story, Double DP the Russian mafia cowboy rapper—who, in my opinion, is not as good as the Asian pop sensation Kulaap, but whom half the planet likes very much—is accused of impregnating the fourteen-year-old daughter of his face sculptor. Readers are starting to notice, and with their attention Marty's green-glowing news story begins to muscle for space in the maelstrom. The content star pulses, expands, and then, as though someone has thrown gasoline on it, it explodes. Double DP hits the social sites, starts getting recommended, sucks in more readers, more links, more clicks . . . and more ad dollars.

Marty does a pelvic grind of victory, then waves at everyone for their attention. "And that's not all, folks." He hits his keyboard again, and another story posts: live feeds of Double's house, where . . . it looks as though the man who popularized Redneck Russians is heading out the door in a hurry. It is a surprise to see video of the house, streaming live. Most freelance paparazzi are not patient enough to sit and hope that maybe, perhaps, something interesting will happen. This looks as though Marty has stationed his own exclusive papcams at the house, to watch for something like this.

We all watch as Double DP locks the door behind himself. Marty says, "I thought DP deserved the courtesy of notification that the story was going live."

"Is he fleeing?" Mikela Plaa asks.

Marty shrugs. "We'll see."

And indeed, it does look as if Double is about to do what Americans have popularized as an "OJ." He is into his red Hummer. Pulling out.

Under the green glow of his growing story, Marty smiles. The story is getting bigger, and Marty has stationed himself perfectly for the development. Other news agencies and blogs are playing catch-up. Follow-on posts wink into existence in the maelstrom, gathering a momentum of their own as newsrooms scramble to hook our traffic.

"Do we have a helicopter?" Janice asks. She has come out of her glass office to watch the show.

Marty nods. "We're moving it into position. I just bought exclusive angel view with the cops, too, so everyone's going to have to license our footage."

"Did you let *Long Arm of the Law* know about the cross-content?"

"Yeah. They're kicking in from their budget for the helicopter."

Marty sits down again, begins tapping at his keyboard, a machine-gun of data entry. A low murmur comes from the tech pit, Cindy C. calling our telecom providers, locking down trunklines to handle an anticipated data surge. She knows something that we don't, something that Marty has prepared her for. She's bringing up mirrored server farms. Marty seems unaware of the audience around him. He stops typing. Stares up at the maelstrom, watching his glowing ball of content. He is the maestro of a symphony.

The cluster of competing stories are growing as Gawker and Newsweek and Throb all organize themselves and respond. Our readers are clicking away from us, trying to see if there's anything new in our competitor's coverage. Marty smiles, hits his "publish" key, and dumps a new bucket of meat into the shark tank of public interest: a video interview with the fourteen-year-old. On-screen, she looks very young, shockingly so. She has a teddy bear.

"I swear I didn't plant the bear," Marty comments. "She had it on her own."

The girl's accusations are being mixed over Double's run for the border, a kind of synth loop of accusations:

"And then he . . ."

"And I said . . ."

"He's the only one I've ever . . ."

It sounds as if Marty has licensed some of Double's own beats for the coverage of his fleeing Humvee. The video outtakes are already bouncing around YouTube and MotionSwallow like Ping-Pong balls. The maelstrom has moved Double DP to the center of the display as more and more feeds and sites point to the content. Not only is traffic up, but the post is gaining in social rank as the numbers of links and social pokes increase.

"How's the stock?" someone calls out.

Marty shakes his head. "They locked me out from showing the display."

This, because whenever he drops an important story, we all beg him to show us the big picture. We all turn to Janice. She rolls her eyes, but she gives the nod. When Cindy finishes buying bandwidth, she unlocks the view. The maelstrom slides aside as a second window opens, all bar graphs and financial landscape: our stock price as affected by the story's expanding traffic—and expanding ad revenue.

The stock bots have their own version of the maelstrom; they've picked up the reader traffic shift. Buy and sell decisions roll across the screen, responding to the popularity of Mackley's byline. As he feeds the story, the beast grows. More feeds pick us up, more people recommend the story to their friends, and every one of them is being subjected to our advertisers' messages, which means more revenue for us and less for everyone else. At this point, Mackley is bigger than the Super Bowl. Given that the story is tagged with Double DP, it will have a targetable demographic:

thirteen- to twenty-four-year-olds who buy lifestyle gadgets, new music, edge clothes, first-run games, boxed hairstyles, tablet skins, and ringtones: not only a large demographic, a valuable one.

Our stock ticks up a point. Holds. Ticks up another. We've got four different screens running now. The papcam of Double DP, chase cycles with views of the cops streaking after him, the chopper lifting off, and the window with the fourteen-year-old interviewing. The girl is saying, "I really feel for him. We have a connection. We're going to get married," and there's his Hummer screaming down Santa Monica Boulevard with his song "Cowboy Banger" on the audio overlay.

A new wave of social pokes hits the story. Our stock price ticks up again. Daily bonus territory. The clicks are pouring in. It's got the right combination of content, what Mackley calls the "Three S's": sex, stupidity, and schadenfreude. The stock ticks up again. Everyone cheers. Mackley takes a bow. We all love him. He is half the reason I can pay my rent. Even a small newsroom bonus from his work is enough for me to live. I'm not sure how much he makes for himself when he creates an event like this. Cindy tells me that it is "solid seven, baby." His byline feed is so big he could probably go independent, but then he would not have the resources to scramble a helicopter for a chase toward Mexico. It is a symbiotic relationship. He does what he does best, and Milestone pays him like a celebrity.

Janice claps her hands. "All right, everyone. You've got your bonus. Now back to work."

A general groan rises. Cindy cuts the big monitor away from stocks and bonuses and back to the work at hand: generating more content to light the maelstrom, to keep the newsroom glowing green with flares of Milestone coverage—everything from reviews of Mitsubishi's 100 mpg Road Cruiser to how to choose a perfect turkey for Thanksgiving. Mackley's story pulses over us as we work. He spins off smaller additional stories, updates, interactivity features, encouraging his vast audience to ping back just one more time.

Marty will spend the entire day in conversation with this elephant of a story that he has created. Encouraging his visitors to return for just one more click. He'll give them chances to poll each other, discuss how they'd like to see DP punished, ask whether you can actually fall in love with a fourteen-year-old. This one will have a long life, and he will raise it like a proud father, feeding and nurturing it, helping it make its way in the rough world of the maelstrom.

My own little green speck of content has disappeared. It seems that even government biologists feel for Double DP.

■　　■　　■

When my father was not placing foolish bets on revolution, he taught agronomy at the National Lao University. Perhaps our lives would have been different if he had been a rice farmer in the paddies of the capital's suburbs, instead of surrounded by intellectuals and ideas. But his karma was to be a teacher and a researcher, and so while he was increasing Lao rice production by 30 percent, he was also filling himself with gambler's fancies: Thoreau, Gandhi, Martin Luther King, Sakharov, Mandela, Aung Sung Kyi. True gamblers, all. He would say that if white South Africans could be made to feel shame, then the pretender monarch must right his ways. He claimed that Thoreau must have been Lao, the way he protested so politely.

In my father's description, Thoreau was a forest monk, gone into the jungle for enlightenment. To live amongst the banyan and the climbing vines of Massachusetts and to meditate on the nature of suffering. My father believed he was undoubtedly some arhat reborn. He often talked of Mr. Henry David, and in my imagination this *falang*, too, was a large man like my father.

When my father's friends visited in the dark—after the coup and the countercoup, and after the march of Khamsing's Chinese-supported insurgency—they would often speak of Mr. Henry David. My father would sit with his friends and students and drink black Lao coffee and smoke cigarettes, and then he would write carefully worded complaints against the government that his students would then copy and leave in public places, distribute into gutters, and stick onto walls in the dead of night.

His guerrilla complaints would ask where his friends had gone, and why their families were so alone. He would ask why monks were beaten on their heads by Chinese soldiers when they sat in hunger strike before the palace. Sometimes, when he was drunk and when these small gambles did not satisfy his risk-taking nature, he would send editorials to the newspapers.

None of these were ever printed, but he was possessed with some spirit that made him think that perhaps the papers would change. That his stature as a father of Lao agriculture might somehow sway the editors to commit suicide and print his complaints.

It ended with my mother serving coffee to a secret police captain while two more policemen waited outside our door. The captain was very polite: he offered my father a 555 cigarette—a brand that already had become rare and contraband—and lit it for him. Then he spread the whisper sheet onto the coffee table, gently pushing aside the coffee cups and their saucers to make room for it. It was rumpled and torn, stained with mud. Full of accusations against Khamsing. Unmistakable as one of my father's.

My father and the policeman both sat and smoked, studying the paper silently.

Finally, the captain asked, "Will you stop?"

My father drew on his cigarette and let the smoke out slowly as he studied the whisper sheet between them. The captain said, "We all respect what you have done for the Lao kingdom. I myself have family who would have starved if not for your work in the villages." He leaned forward. "If you promise to stop writing these whispers and complaints, everything can be forgotten. Everything."

Still, my father didn't say anything. He finished his cigarette. Stubbed it out. "It would be difficult to make that sort of promise," he said.

The captain was surprised. "You have friends who have spoken on your behalf. Perhaps you would reconsider. For their sake."

My father made a little shrug. The captain spread the rumpled whisper sheet, flattening it out more completely. Read it over. "These sheets do nothing," he said. "Khamsing's dynasty will not collapse because you print a few complaints. Most of these are torn down before anyone reads them. They do nothing. They are pointless." He was almost begging. He looked over and saw me watching at the door. "Give this up. For your family, if not your friends."

I would like to say that my father said something grand. Something honorable about speaking against tyranny. Perhaps invoked one of his idols. Aung Sung Kyi or Sakharov, or Mr. Henry David and his penchant for polite protest. But he didn't say anything. He just sat with his hands on his knees, looking down at the torn whisper sheet. I think now that he must have been very afraid. Words always came easily to him, before. Instead, all he did was repeat himself. "It would be difficult."

The captain waited. When it became apparent that my father had nothing else to say, he put down his coffee cup and motioned for his men to come inside. They were all very polite. I think the captain even apologized to my mother as they led him out the door.

■   ■   ■

We are into day three of the Double DP bonanza, and the green sun glows brightly over all of us, bathing us in its soothing, profitable glow. I am working on my newest story with my Frontal Lobe ear buds in, shutting out everything except the work at hand. It is always a little difficult to write in one's third language, but I have my favorite singer and fellow countryperson Kulaap whispering in my ear that "Love is a Bird," and the work is going well. With Kulaap singing to me in our childhood language, I feel very much at home.

A tap on my shoulder interrupts me. I pull out my ear buds and look around. Janice, standing over me. "Ong, I need to talk to you." She motions me to follow.

In her office, she closes the door behind me and goes to her desk. "Sit down, Ong." She keys her tablet, scrolls through data. "How are things going for you?"

"Very well. Thank you." I'm not sure if there is more that she wants me to say, but it is likely that she will tell me. Americans do not leave much to guesswork.

"What are you working on for your next story?" she asks.

I smile. I like this story; it reminds me of my father. And with Kulaap's soothing voice in my ears I have finished almost all of my research. The bluet, a flower made famous in Mr. Henry David Thoreau's journals, is blooming too early to be pollinated. Bees do not seem to find it when it blooms in March. The scientists I interviewed blame global warming, and now the flower is in danger of extinction. I have interviewed biologists and local naturalists, and now I would like to go to Walden Pond on a pilgrimage for this bluet that may soon also be bottled in a federal reserve laboratory with its techs in clean suits and their crime scene vacuums.

When I finish describing the story, Janice looks at me as if I am crazy. I can tell that she thinks I am crazy, because I can see it on her face. And also because she tells me.

"You're fucking crazy!"

Americans are very direct. It's difficult to keep face when they yell at you. Sometimes, I think that I have adapted to America. I have been here for five years now, ever since I came from Thailand on a scholarship, but at times like this, all I can do is smile and try not to cringe as they lose their face and yell and rant. My father was once struck in the face with an official's shoe, and he did not show his anger. But Janice is American, and she is very angry.

"There's no way I'm going to authorize a junket like that!"

I try to smile past her anger, and then remember that the Americans don't see an apologetic smile in the same way that a Lao would. I stop smiling and make my face look . . . something. Earnest, I hope.

"The story is very important," I say. "The ecosystem isn't adapting correctly to the changing climate. Instead, it has lost . . ." I grope for the word. "Synchronicity. These scientists think that the flower can be saved, but only if they import a bee that is available in Turkey. They think it can replace the function of the native bee population, and they think that it will not be too disruptive."

"Flowers and Turkish bees."

"Yes. It is an important story. Do they let the flower go extinct? Or try to keep the famous flower, but alter the environment of Walden Pond? I think your readers will think it is very interesting."

"More interesting than that?" She points through her glass wall at the maelstrom, at the throbbing green sun of Double DP, who has now barricaded himself in a Mexican hotel and has taken a pair of fans hostage.

"You know how many clicks we're getting?" she asks. "We're exclusive. Marty's got Double's trust and is going in for an interview tomorrow, assuming the Mexicans don't just raid it with commandos. We've got people clicking back every couple minutes just to look at Marty's blog about his preparations to go in."

The glowing globe not only dominates the maelstrom's screen, it washes everything else out. If we look at the stock bots, everyone who doesn't have protection under our corporate umbrella has been hurt by the loss of eyeballs. Even the Frontal Lobe/Oakley story has been swallowed. Three days of completely dominating the maelstrom has been very profitable for us. Now Marty's showing his viewers how he will wear a flak jacket in case the Mexican commandos attack while he is discussing the nature of true love with DP. And he has another exclusive interview with the mother ready to post as well. Cindy has been editing the footage and telling us all how disgusted she is with the whole thing. The woman apparently drove her daughter to DP's mansion for a midnight pool party, alone.

"Perhaps some people are tired of DP and wish to see something else," I suggest.

"Don't shoot yourself in the foot with a flower story, Ong. Even Pradeep's cooking journey through Ladakh gets more viewers than this stuff you're writing."

She looks as though she will say more, but then she simply stops. It seems as if she is considering her words. It is uncharacteristic. She normally speaks before her thoughts are arranged.

"Ong, I like you," she says. I make myself smile at this, but she continues. "I hired you because I had a good feeling about you. I didn't have a problem with clearing the visas to let you stay in the country. You're a good person. You write well. But you're averaging less than a thousand pings on your byline feed." She looks down at her tablet, then back up at me. "You need to up your average. You've got almost no readers selecting you for Page One. And even when they do subscribe to your feed, they're putting it in the third tier."

"Spinach reading," I supply.

"What?"

"Mr. Mackley calls it spinach reading. When people feel like they should do something with virtue, like eat their spinach, they click to me. Or else read Shakespeare."

I blush, suddenly embarrassed. I do not mean to imply that my work is of the same caliber as a great poet. I want to correct myself, but I'm too embarrassed. So instead I shut up, and sit in front of her, blushing.

She regards me. "Yes. Well, that's a problem. Look, I respect what you do. You're obviously very smart." Her eyes scan her tablet. "The butterfly thing you wrote was actually pretty interesting."

"Yes?" I make myself smile again.

"It's just that no one wants to read these stories."

I try to protest. "But you hired me to write the important stories. The stories about politics and the government, to continue the traditions of the old newspapers. I remember what you said when you hired me."

"Yeah, well." She looks away. "I was thinking more about a good scandal."

"The checkerspot is a scandal. That butterfly is now gone."

She sighs. "No, it's not a scandal. It's just a depressing story. No one reads a depressing story, at least, not more than once. And no one subscribes to a depressing byline feed."

"A thousand people do."

"A thousand people." She laughs. "We aren't some Laotian community weblog, we're Milestone, and we're competing for clicks with them." She waves outside, indicating the maelstrom. "Your stories don't last longer than half a day; they never get social-poked by anyone except a fringe." She shakes her head. "Christ, I don't even know who your demographic is. Centenarian hippies? Some federal bureaucrats? The numbers just don't justify the amount of time you spend on stories."

"What stories do you wish me to write?"

"I don't know. Anything. Product reviews. News you can use. Just not any more of this 'we regret to inform you of bad news' stuff. If there isn't something a reader can do about the damn butterfly, then there's no point in telling them about it. It just depresses people, and it depresses your numbers."

"We don't have enough numbers from Marty?"

She laughs at that. "You remind me of my mother. Look, I don't want to cut you, but if you can't start pulling at least a fifty thousand daily average, I won't have any choice. Our group median is way down in comparison to other teams, and when evaluations come around, we look bad. I'm up against Nguyen in the Tech and Toys pool, and Penn in Yoga and Spirituality, and no one wants to read about how the world's going to shit. Go find me some stories that people want to read."

She says a few more things, words that I think are meant to make me feel inspired and eager, and then I am standing outside the door, once again facing the maelstrom.

The truth is that I have never written popular stories. I am not a popular story writer. I am earnest. I am slow. I do not move at the speed these Americans seem to love. *Find a story that people want to read.* I can write some follow-up to Mackley, to Double DP, perhaps assist with sidebars to his main piece, but somehow, I suspect that the readers will know that I am faking it.

Marty sees me standing outside of Janice's office. He comes over.

"She giving you a hard time about your numbers?"

"I do not write the correct sort of stories."

"Yeah. You're an idealist."

We both stand there for a moment, meditating on the nature of idealism. Even though he is very American, I like him because he is sensitive to people's hearts. People trust him. Even Double DP trusts him, though Marty blew his name over every news tablet's front page. Marty has a good heart. *Jai dee.* I like him. I think that he is genuine.

"Look, Ong," he says. "I like what you do." He puts his hand around my shoulder. For a moment, I think he's about to try to rub my head with affection and I have to force myself not to wince, but he's sensitive and instead takes his hand away. "Look, Ong. We both know you're terrible at this kind of work. We're in the news business, here. And you're just not cut out for it."

"My visa says I have to remain employed."

"Yeah. Janice is a bitch for that. Look." He pauses. "I've got this thing with Double DP going down in Mexico. But I've got another story brewing. An exclusive. I've already got my bonus, anyway. And it should push up your average."

"I do not think that I can write Double DP sidebars."

He grins. "It's not that. And it's not charity; you're actually a perfect match."

"It is about government mismanagement?"

He laughs, but I think he's not really laughing at me. "No." He pauses, smiles. "It's Kulaap. An interview."

I suck in my breath. My fellow countryperson, here in America. She came out during the purge as well. She was doing a movie in Singapore when the tanks moved, and so she was not trapped. She was already very popular all over Asia, and when Khamsing turned our country into a black hole, the world took note. Now she is popular here in America as well. Very beautiful. And she remembers our country before it went into darkness. My heart is pounding.

Marty goes on. "She's agreed to do an exclusive with me. But you even speak her language, so I think she'd agree to switch off." He pauses, looks serious. "I've got a good history with Kulaap. She doesn't give interviews to just anyone. I did a lot of exposure stories about her when Laos was going to hell. Got her a lot of good press. This is a special favor already, so don't fuck it up."

I shake my head. "No. I will not." I press my palms together and touch them to my forehead in a *nop* of appreciation. "I will not fuck it up." I make another *nop*.

He laughs. "Don't bother with that polite stuff. Janice will cut off your balls to increase the stock price, but we're the guys in the trenches. We stick together, right?"

■   ■   ■

In the morning, I make a pot of strong coffee with condensed milk; I boil rice noodle soup and add bean sprouts and chiles and vinegar, and warm a loaf of French bread that I buy from a Vietnamese bakery a few blocks away. With a new mix of Kulaap's music from DJ Dao streaming in over my stereo, I sit down at my little kitchen table, pour my coffee from its press pot, and open my tablet.

The tablet is a wondrous creation. In Laos, the paper was still a paper, physical, static, and empty of anything except the official news. Real news in our New Divine Kingdom did not come from newspapers, or from television, or from handsets or ear buds. It did not come from the net or feeds unless you trusted your neighbor not to look over your shoulder at an Internet cafe and if you knew that there were no secret police sitting beside you, or an owner who would be able to identify you when they came around asking about the person who used that workstation over there to communicate with the outside world.

Real news came from whispered rumor, rated according to the trust you accorded the whisperer. Were they family? Did they have long history with you? Did they have anything to gain by the sharing? My father and his old classmates trusted one another. He trusted some of his students, as well. I think this is why the security police came for him in the end. One of his trusted friends or students also whispered news to official friends. Perhaps Mr. Inthachak, or Som Vang. Perhaps another. It is impossible to peer into the blackness of that history and guess at who told true stories and in which direction.

In any case, it was my father's karma to be taken, so perhaps it does not matter who did the whispering. But before then—before the news of my father flowed up to official ears—none of the real news flowed toward Lao TV or the *Vientiane Times*. Which meant that when the protests happened and my father came through the door with blood on his face from baton blows, we could read as much as we wanted about the three thousand schoolchildren who had sung the national anthem to our new divine monarch. While my father lay in bed, delirious with pain, the papers told us that China had signed a rubber contract that would triple revenue for Luang Namtha province and that Nam Theun Dam was now earning BT 22.5 billion per year in electricity fees to Thailand. But there were no bloody batons, and there were no dead monks, and there was no Mercedes-Benz burning in the river as it floated toward Cambodia.

Real news came on the wings of rumor, stole into our house at midnight, sat with us and sipped coffee and fled before the call of roosters could break the stillness. It was in the dark, over a burning cigarette that you learned Vilaphon had disappeared or that Mr. Saeng's wife had been beaten as a warning. Real news was too valuable to risk in public.

Here in America, my page glows with many news feeds, flickers at me in video windows, pours in at me over broadband. It is a waterfall of information. As my personal news page opens, my feeds arrange themselves, sorting according to the priorities and tag categories that I've set, a mix of Meung Lao news, Lao refugee blogs, and the chatting of a few close friends from Thailand and the American college where I attended on a human relief scholarship.

On my second page and my third, I keep the general news, the arrangements of Milestone, the *Bangkok Post*, the *Phnom Penh Express*—the news chosen by editors. But by the time I've finished with my own selections, I don't often have time to click through the headlines that these earnest news editors select for the mythical general reader.

In any case, I know far better than they what I want to read, and with my keyword and tag scans, I can unearth stories and discussions that a news agency would never think to provide. Even if I cannot see into the black hole itself, I can slip along its edges, divine news from its fringe.

I search for tags like Vientiane, Laos, Lao, Khamsing, China-Lao friendship, Korat, Golden Triangle, Hmong independence, Lao PDR, my father's name. . . . Only those of us who are Lao exiles from the March Purge really read these blogs. It is much as when we lived in the capital. The blogs are the rumors that we used to whisper to one another. Now we publish our whispers over the net and join mailing lists instead of secret coffee groups, but it is the same. It is family, as much as any of us now have.

On the maelstrom, the tags for Laos don't even register. Our tags bloomed brightly for a little while, while there were still guerrilla students uploading content from their handsets, and the images were lurid and shocking. But then the phone lines went down and the country fell into its black hole and now it is just us, this small network that functions outside the country.

A headline from Jumbo Blog catches my eye. I open the site, and my tablet fills with the colorful image of the three-wheeled taxi of my childhood. I often come here. It is a node of comfort.

Laofriend posts that some people, maybe a whole family, have swum the Mekong and made it into Thailand. He isn't sure if they were accepted as refugees or if they were sent back.

It is not an official news piece. More, the idea of a news piece. *SomPaBoy* doesn't believe it, but *Khamchanh* contends that the rumor is true, heard from someone who has a sister married to an Isaan border guard in the Thai army. So we cling to it. Wonder about it. Guess where these people came from, wonder if, against all odds, it could be one of ours: a brother, a sister, a cousin, a father. . . .

After an hour, I close the tablet. It's foolish to read any more. It only brings up memories. Worrying about the past is foolish. Lao PDR is gone. To wish otherwise is suffering.

■　　■　　■

The clerk at Novotel's front desk is expecting me. A hotel staffer with a key guides me to a private elevator bank that whisks us up into the smog and heights. The elevator doors open to a small entryway with a thick mahogany door. The staffer steps back into the elevator and disappears, leaving me standing in this strange airlock. Presumably, I am being examined by Kulaap's security.

The mahogany door opens, and a smiling black man who is forty centimeters taller than I and who has muscles that ripple like snakes smiles and motions me inside. He guides me through Kulaap's sanctuary. She keeps the heat high, almost tropical, and fountains rush everywhere around. The flat is musical with water. I unbutton my collar in the humidity. I was expecting air-conditioning, and instead I am sweltering. It's almost like home. And then she's in front of me, and I can hardly speak. She is beautiful, and more. It is intimidating to stand before someone who exists in film and in music but has never existed before you in the flesh. She's not as stunning as she is in the movies, but there's more life, more presence; the movies lose that quality about her. I make a *nop* of greeting, pressing my hands together, touching my forehead.

She laughs at this, takes my hand and shakes it American-style. "You're lucky Marty likes you so much," she says. "I don't like interviews."

I can barely find my voice. "Yes. I only have a few questions."

"Oh no. Don't be shy." She laughs again, and doesn't release my hand, pulls me toward her living room. "Marty told me about you. You need help with your ratings. He helped me once, too."

She's frightening. She is of my people, but she has adapted better to this place than I have. She seems comfortable here. She walks differently, smiles differently; she is an American, with perhaps some flavor of our country, but nothing of our roots. It's obvious. And strangely disappointing. In her movies, she holds herself so well, and now she sits down on her couch and sprawls with her feet kicked out in front of her. Not caring at all. I'm embarrassed for her, and I'm glad I don't have my camera set up yet. She kicks her feet up on the couch. I can't help but be shocked. She catches my expression and smiles.

"You're worse than my parents. Fresh off the boat."

"I am sorry."

She shrugs. "Don't worry about it. I spent half my life here, growing up; different country, different rules."

I'm embarrassed. I try not to laugh with the tension I feel. "I just have some interview questions," I say.

"Go ahead." She sits up and arranges herself for the video stand that I set up.

I begin. "When the March Purge happened, you were in Singapore."

She nods. "That's right. We were finishing *The Tiger and the Ghost*."

"What was your first thought when it happened? Did you want to go back? Were you surprised?"

She frowns. "Turn off the camera."

When it's off she looks at me with pity. "This isn't the way to get clicks. No one cares about an old revolution. Not even my fans." She stands abruptly and calls through the green jungle of her flat. "Terrell?"

The big black man appears. Smiling and lethal. Looming over me. He is very frightening. The movies I grew up with had *falang* like him. Terrifying large black men whom our heroes had to overcome. Later, when I arrived in America, it was different, and I found out that the *falang* and the black people don't like the way we show them in our movies. Much like when I watch their Vietnam movies, and see the ugly way the Lao freedom fighters behave. Not real at all, portrayed like animals. But still, I cannot help but cringe when Terrell looks at me.

Kulaap says, "We're going out, Terrell. Make sure you tip off some of the papcams. We're going to give them a show."

"I don't understand," I say.

"You want clicks, don't you?"

"Yes, but—"

She smiles. "You don't need an interview. You need an event." She looks me over. "And better clothes." She nods to her security man. "Terrell, dress him up."

■   ■   ■

A flashbulb frenzy greets us as we come out of the tower. Papcams everywhere. Chase cycles revving, and Terrell and three others of his people guiding us through the press to the limousine, shoving cameras aside with a violence and power that are utterly unlike the careful pity he showed when he selected a Gucci suit for me to wear.

Kulaap looks properly surprised at the crowd and the shouting reporters, but not nearly as surprised as I am, and then we're in the limo, speeding out of the tower's roundabout as papcams follow us.

Kulaap crouches before the car's onboard tablet, keying in pass codes. She is very pretty, wearing a black dress that brushes her thighs and thin straps that caress her smooth bare shoulders. I feel as if I am in a movie. She taps more keys. A screen glows, showing the taillights of our car: the view from pursuing papcams.

"You know I haven't dated anyone in three years?" she asks.

"Yes. I know from your Web site biography."

She grins. "And now it looks like I've found one of my countrymen."

"But we're not on a date," I protest.

"Of course we are." She smiles again. "I'm going out on a supposedly secret date with a cute and mysterious Lao boy. And look at all those papcams chasing after us, wondering where we're going and what we're going to do." She keys in another code, and now we can see live footage of the paparazzi, as viewed from the tail of her limo. She grins. "My fans like to see what life is like for me."

I can almost imagine what the maelstrom looks like right now: there will still be Marty's story, but now a dozen other sites will be lighting up, and in the center of that, Kulaap's own view of the excitement, pulling in her fans, who will want to know, direct from her, what's going on. She holds up a mirror, checks herself, and then she smiles into her smartphone's camera.

"Hi everyone. It looks like my cover's blown. Just thought I should let you know that I'm on a lovely date with a lovely man. I'll let you all know how it goes. Promise." She points the camera at me. I stare at it stupidly. She laughs. "Say hi and good-bye, Ong."

"Hi and good-bye."

She laughs again, waves into the camera. "Love you all. Hope you have as good a night as I'm going to have." And then she cuts the clip and punches a code to launch the video to her Web site.

It is a bit of nothing. Not a news story, not a scoop even, and yet, when she opens another window on her tablet, showing her own miniversion of the maelstrom, I can see her site lighting up with traffic. Her version of the maelstrom isn't as powerful as what we have at Milestone, but still, it is an impressive window into the data that is relevant to Kulaap's tags.

"What's your feed's byline?" she asks. "Let's see if we can get your traffic bumped up."

"Are you serious?"

"Marty Mackley did more than this for me. I told him I'd help." She laughs. "Besides, we wouldn't want you to get sent back to the black hole, would we?"

"You know about the black hole?" I can't help doing a double-take.

Her smile is almost sad. "You think just because I put my feet up on the furniture that I don't care about my aunts and uncles back home? That I don't worry about what's happening?"

"I—"

She shakes her head. "You're so fresh off the boat."

"Do you use the Jumbo Cafe—" I break off. It seems too unlikely.

She leans close. "My handle is *Laofriend*. What's yours?"

"*Littlexang*. I thought *Laofriend* was a boy—"

She just laughs.

I lean forward. "Is it true that the family made it out?"

She nods. "For certain. A general in the Thai army is a fan. He tells me everything. They have a listening post. And sometimes they send scouts across."

It's almost as if I am home.

■　　■　　■

We go to a tiny Laotian restaurant where everyone recognizes her and falls over her and the owners simply lock out the paparazzi when they become too intrusive. We spend the evening unearthing memories of Vientiane. We discover that we both favored the same rice noodle cart on Kaem Khong. That she used to sit on the banks of the Mekong and wish that she were a fisherman. That we went to the same waterfalls outside the city on the weekends. That it is impossible to find good *dum mak hoong* anywhere outside of the country. She is a good companion, very alive. Strange in her American ways, but still, with a good heart. Periodically, we click photos of one another and post them to her site, feeding the voyeurs. And then we are in the limo again and the paparazzi are all around us. I have the strange feeling of fame. Flashbulbs everywhere. Shouted questions. I feel proud to be beside this beautiful intelligent woman who knows so much more than any of us about the situation inside our homeland.

Back in the car, she has me open a bottle of champagne and pour two glasses while she opens the maelstrom and studies the results of our date. She has reprogrammed it to watch my byline feed ranking as well.

"You've got twenty thousand more readers than you did yesterday," she says.

I beam. She keeps reading the results. "Someone already did a scan on your face." She toasts me with her glass. "You're famous."

We clink glasses. I am flushed with wine and happiness. I will have Janice's average clicks. It's as though a bodhisattva has come down from heaven to save my job. In my mind, I offer thanks to Marty for arranging this, for his generous nature. Kulaap leans close to her screen, watching the flaring content. She opens another window, starts to read. She frowns.

"What the fuck do you write about?"

I draw back, surprised. "Government stories, mostly." I shrug. "Sometimes environment stories."

"Like what?"

"I am working on a story right now about global warming and Henry David Thoreau."

"Aren't we done with that?"

I'm confused. "Done with what?"

The limo jostles us as it makes a turn, moves down Hollywood Boulevard, letting the cycles rev around us like schools of fish. They're snapping pictures at the side of the limo, snapping at us. Through the tinting, they're like fireflies, smaller flares than even my stories in the maelstrom.

"I mean, isn't that an old story?" She sips her champagne. "Even America is reducing emissions now. Everyone knows it's a problem." She taps her couch's armrest. "The carbon tax on my limo has tripled, even with the hybrid engine. Everyone agrees it's a problem. We're going to fix it. What's there to write about?"

She is an American. Everything that is good about them: their optimism, their willingness to charge ahead, to make their own future. And everything that is bad about them: their strange ignorance, their unwillingness to believe that they must behave as other than children.

"No. It's not done," I say. "It is worse. Worse every day. And the changes we make seem to have little effect. Maybe too little, or maybe too late. It is getting worse."

She shrugs. "That's not what I read."

I try not to show my exasperation. "Of course it's not what you read." I wave at the screen. "Look at the clicks on my feed. People want happy stories. Want fun stories. Not stories like I write. So instead, we all write what you will read, which is nothing."

"Still—"

"No." I make a chopping motion with my hand. "We newspeople are very smart monkeys. If you will give us your so lovely eyeballs and your click-throughs we will do whatever you like. We will write good news, and news you can use, news you can shop to, news with the 'Three S's.' We will tell you how to have better sex or eat better or look more beautiful or feel happier and or how to meditate—yes, so enlightened." I make a face. "If you want a walking meditation and Double DP, we will give it to you."

She starts to laugh.

"Why are you laughing at me?" I snap. "I am not joking!"

She waves a hand. "I know, I know, but what you just said 'double'—" She shakes her head, still laughing. "Never mind."

I lapse into silence. I want to go on, to tell her of my frustrations. But now I am embarrassed at my loss of composure. I have no face. I didn't used to be like this. I used to control my emotions, but now I am an American, as childish and unruly as Janice. And Kulaap laughs at me.

I control my anger. "I think I want to go home," I say. "I don't wish to be on a date anymore."

She smiles and reaches over to touch my shoulder. "Don't be that way."

A part of me is telling me that I am a fool. That I am reckless and foolish for walking away from this opportunity. But there is something else, something about this frenzied hunt for page views and click-throughs and ad revenue that suddenly feels unclean. As if my father is with us in the car, disapproving. Asking if he posted his complaints about his missing friends for the sake of clicks.

"I want to get out," I hear myself say. "I do not wish to have your clicks."

"But—"

I look up at her. "I want to get out. Now."

"Here?" She makes a face of exasperation, then shrugs. "It's your choice."

"Yes. Thank you."

She tells her driver to pull over. We sit in stiff silence.

"I will send your suit back to you," I say.

She gives me a sad smile. "It's all right. It's a gift."

This makes me feel worse, even more humiliated for refusing her generosity, but still, I get out of the limo. Cameras are clicking at me from all around. This is my fifteen minutes of fame, this moment when all of Kulaap's fans focus on me for a few seconds, their flashbulbs popping.

I begin to walk home as paparazzi shout questions.

■　　■　　■

Fifteen minutes later I am indeed alone. I consider calling a cab, but then decide I prefer the night. Prefer to walk by myself through this city that never walks anywhere. On a street corner, I buy a *pupusa* and gamble on the Mexican Lottery because I like the tickets' laser images of their Day of the Dead. It seems an echo of the Buddha's urging to remember that we all become corpses.

I buy three tickets, and one of them is a winner: one hundred dollars that I can redeem at any TelMex kiosk. I take this as a good sign. Even if my luck is obviously gone with my work, and even if the girl Kulaap was not the bodhisattva that I thought, still, I feel lucky. As though my father is walking with me down this cool Los Angeles street in the middle of the night, the two of us together again, me with a *pupusa* and a winning lottery ticket, him with an Ah Daeng cigarette and his quiet gambler's smile. In a strange way, I feel that he is blessing me.

And so instead of going home, I go back to the newsroom.

My hits are up when I arrive. Even now, in the middle of the night, a tiny slice of Kulaap's fan base is reading about checkerspot butterflies and American government incompetence. In my country, this story would not exist. A censor would kill it instantly. Here, it glows green; increasing and decreasing in size as people click. A lonely thing, flickering amongst the much larger content flares of Intel processor releases, guides to low-fat recipes, photos of lol-cats, and episodes of *Survivor! Antarctica*. The wash of light and color is very beautiful.

In the center of the maelstrom, the green sun of the Double DP story glows—surges larger. DP is doing something. Maybe he's surrendering, maybe he's murdering his hostages, maybe his fans have thrown up a human wall to protect him. My story snuffs out as reader attention shifts.

I watch the maelstrom a little longer, then go to my desk and make a phone call. A rumpled hairy man answers, rubbing at a sleep-puffy face. I apologize for the late hour, and then pepper him with questions while I record the interview.

He is silly looking and wild-eyed. He has spent his life living as if he were Thoreau, thinking deeply on the forest monk and following the man's careful paths through what woods remain, walking amongst birch and maple and bluets. He is a fool, but an earnest one.

"I can't find a single one," he tells me. "Thoreau could find thousands at this time of year; there were so many he didn't even have to look for them."

He says, "I'm so glad you called. I tried sending out press releases, but . . ." He shrugs. "I'm glad you'll cover it. Otherwise, it's just us hobbyists talking to each other."

I smile and nod and take notes of his sincerity, this strange wild creature, the sort that everyone will dismiss. His image is bad for video; his words are not good for text. He has no quotes that encapsulate what he sees. It is all couched in the jargon of naturalists and biology. With time, I could find another, someone who looks attractive or who can speak well, but all I have is this one hairy man, disheveled and foolish, senile with passion over a flower that no longer exists.

I work through the night, polishing the story. When my colleagues pour through the door at 8 a.m. it is almost done. Before I can even tell Janice about it, she comes to me. She fingers my clothing and grins. "Nice suit." She pulls up a chair and sits beside me. "We all saw you with Kulaap. Your hits went way up." She nods at my screen. "Writing up what happened?"

"No. It was a private conversation."

"But everyone wants to know why you got out of the car. I had someone from the *Financial Times* call me about splitting the hits for a tell-all, if you'll be interviewed. You wouldn't even need to write up the piece."

It's a tempting thought. Easy hits. Many click-throughs. Ad-revenue bonuses. Still, I shake my head. "We did not talk about things that are important for others to hear."

Janice stares at me as if I am crazy. "You're not in the position to bargain, Ong. Something happened between the two of you. Something people want to know about. And you need the clicks. Just tell us what happened on your date."

"I was not on a date. It was an interview."

"Well then publish the fucking interview and get your average up!"

"No. That is for Kulaap to post, if she wishes. I have something else."

I show Janice my screen. She leans forward. Her mouth tightens as she reads. For once, her anger is cold. Not the explosion of noise and rage that I expect. "Bluets." She looks at me. "You need hits and you give them flowers and Walden Pond."

"I would like to publish this story."

"No! Hell, no! This is just another story like your butterfly story, and your road contracts story, and your congressional budget story. You won't get a damn click. It's pointless. No one will even read it."

"This is news."

"Marty went out on a limb for you—" She presses her lips together, reining in her anger. "Fine. It's up to you, Ong. If you want to destroy your life over Thoreau and flowers, it's your funeral. We can't help you if you won't help yourself. Bottom line, you need fifty thousand readers or I'm sending you back to the third world."

We look at each other. Two gamblers evaluating one another. Deciding who is betting, and who is bluffing.

I click the "publish" button.

The story launches itself onto the net, announcing itself to the feeds. A minute later a tiny new sun glows in the maelstrom.

Together, Janice and I watch the green spark as it flickers on the screen. Readers turn to the story. Start to ping it and share it amongst themselves, start to register hits on the page. The post grows slightly.

My father gambled on Thoreau. I am my father's son.

"THE REGRESSION TEST," BY WOLE TALABI

STORY FRAME

> I'm not sure if A. I.s can believe anything and I'm not supposed to ask
> her questions about such things, but that's what the human control is for,
> right? To ask questions that the other A. I.s would never think to ask, to
> force this electronic extrapolation of my mother into untested territory and
> see if the simulated thought matrix holds up or breaks down. "Don't tell
> me what you think. Tell me what you believe."

What are the things about us that make us distinctly ourselves? How much can they change? In what ways does our selfhood rely on our relationships with and recognition by others? These questions are, in some respects, very literal in "The Regression Test." Titilope, our narrator, participates in the titular test by cross-examining an AI copy of her long-deceased mother Olusola—or "memrionic," as these AI copies are called—in order to determine whether it still reflects the person her mother was. But although the plot turns on the question of whether or not this latest memrionic accurately represents Olusola, the heart of the story is not in the answer, but in the questioning: in Titilope's process of engagement with the memrionic as filtered through her own memories of her mother, as she grapples with the basic human task of recognizing and relating to another under uniquely estranging conditions.

The validity of this memrionic is more than a family affair. Titilope's mother was one of Africa's leading scientists, and her memrionic now serves in an advisory role for the technology firm LegbaTech. This memrionic has just endorsed a controversial, risky and expensive new research avenue, and so LegbaTech has called for the regression test to be run, to ensure that this advice reflects the wisdom and guidance of the Olusola that anchored the African research community in an earlier generation.

Why does this confirmation require a human control, in addition to cross-comparison with previous versions? What can a human know about a memrionically preserved person that the memrionic itself cannot? "The Regression Test" directs our attention away from the what and toward the how: how the memrionic does its thinking, how those processes do and do not reflect the ways of knowing and being that Titilope remembers, and how Titilope herself continues to know and relate to the beloved and complicated person who was so important to her.

"The Regression Test" offers clear answers to some of its own immediate questions: by the end, new information has emerged about the state of the Olusola memrionic and how it got that way. But the ultimate questions raised by the story, about the dynamic nature of our selves and how we remain in relation to those we have lost, remain.

STUDY QUESTIONS

1. Throughout the story, Titilope remarks on several characteristics of the two AIs she deals with, the hospitality AI and the memrionic copy of Olusola that is being tested, that strike her as notably unlike interacting with a human. What are these?

2. What is the problem that the sorites regression test is intended to solve or manage? What are the reasons Titilope is not confident that its methods actually resolve that problem?

3. What aspects of her own memories of her mother does Titilope draw on in her assessment of the memrionic? How do these help her with her task?

4. After the test, Titilope reports to Dr. Dimeji that the memrionic "thinks things that she would but in ways she would never think them." What are these "ways she would never think them" that Titilope notices throughout the test? In each case, what is the difference between how the memrionic thinks them and how her mother would have thought them?

"THE REGRESSION TEST," BY WOLE TALABI

The conference room is white, spacious, and ugly.

Not ugly in any particular sort of way: it doesn't have garish furniture or out-of-place art or vomit-colored walls or anything like that. It's actually quite plain. It's just that everything in it looks furfuraceous, like the skin of some diseased albino animal, as if everything is made of barely attached bleached Bran Flakes. I know that's how all modern furnishing looks now—SlatTex, they call it—especially in these high-tech offices where the walls, doors, windows, and even some pieces of furniture are designed to integrate physically, but I still find it off-putting. I want to get this over with and leave the room as soon as possible. Return to my nice two-hundred-year-old brick bungalow in Ajah where the walls still look like real walls, not futuristic leper-skin.

"So you understand why you're here and what you need to do, madam?" Dr. Dimeji asks me.

I force myself to smile and say, "Of course—I'm here as a human control for the regression test."

Dr. Dimeji does not smile back. The man reminds me of an agama lizard. His face is elongated, reptilian, and there is something that resembles a bony ridge running through the middle of his skull from front to back. His eyes are sunken but always darting about, looking at multiple things, never really focused on me. The electric-blue circle ringing one iris confirms that he has a sensory-augmentation implant.

"*Sorites* regression test," he corrects, as though the precise specification is important or I don't know what it is called. Which I certainly do—I pored over the yeye data-pack they gave me until all the meaningless technobabble in it eventually made some sense.

I roll my eyes. "Yes, I'm here as a human control for the sorites regression test."

"Good," he says, pointing at a black bead with a red eye that is probably a recording device set in the middle of the conference room table. "When you are ready, I need you to state your name, age, index, and the reason why you are here today while looking directly at that. Can you do that for me, madam?"

He might be a professor of memrionics or whatever they're calling this version of their A. I. nonsense these days, but he is much younger than me, by at least seven decades, probably more. Someone should have taught him to say "please" and to lose that condescending tone of voice when addressing his elders. His sour attitude matches his sour face, just like my grandson Tunji, who is now executive director of the research division of LegbaTech. He's always scowling, too, even at family functions, perpetually obsessed with some work thing or other. These children of today take themselves too seriously. Tunji's even become religious now. Goes to church every Sunday, I hear. I don't know how my daughter and her husband managed to raise such a child.

"I'll be just outside observing, if you need anything," Dr. Dimeji says as he opens the door. I nod so I don't accidentally say something caustic to him about his home training or lack thereof. He shuts the door behind him and I hear a lock click into place. That strikes me as odd but I ignore it. I want to get this over with quickly.

"My name is Titilope Ajimobi," I say, remembering my briefing instructions advising me to give as much detail as possible. "I am one hundred and sixteen years old. Sentient Entity Index Number HM033-2021-HK76776. Today I am in the Eko Atlantic office of LegbaTech Industries as the human control for a sorites regression test."

"Thank you, Mrs. Ajimobi," a female voice says to me from everywhere in the room, the characteristic nonlocation of an ever-present A. I. "Regression test initiated."

I lean back in my chair. The air conditioning makes me lick my lips. For all their sophistication, hospitality A. I.s never find the ideal room temperature for human comfort. They can't understand that it's not the calculated optimum. With human desires, it rarely is. It's always just a little bit off. My mother used to say that a lot.

Across the conference room, lines of light flicker to life and begin to dance in sharp, apparently random motions. The lights halt, disappear, and then around the table, where chairs like mine might have been placed, eight smooth, black, rectangular monoliths begin to rise, slowly, as if being extruded from the floor itself. I don't bother moving my own chair to see where they are coming from; it doesn't matter. The slabs grow about seven feet tall or so then stop.

The one directly across from me projects onto the table a red-light matrix of symbols and characters so intricate and dense it looks like abstract art. The matrix is three-dimensional, mathematically speaking, and within its elements, patterns emerge, complex and beautiful, mesmerizing in their way. The patterns are changing so quickly that they give the illusion of stability, which adds to the beauty of the projection. This slab is putting on a display. I assume it must be the casing for the memri-onic copy being regression tested.

A sorites regression test is designed to determine whether an artificial intelligence created by extrapolating and context-optimizing recorded versions of a particular human's thought patterns has deviated too far from the way the original person would think. Essentially, several previous versions of the record—backups with less learning experience—interrogate the most recent update in order to ascertain whether they agree on a wide range of mathematical, phenomenological, and philosophical questions, not just in answer, but also in cognitive approach to deriving and presenting a response. At the end of the experiment, the previous versions judge whether the new version's answers are close enough to those they would give for the update to still be considered "them," or could only have been produced by a completely different entity. The test usually concludes with a person who knew the original human subject—me, in this case—asking the A. I. questions to determine the same thing. Or, as Tunji summarized once, the test verifies that the A.I., at its core, remains recognizable to itself and others, even as it continuously improves.

The seven other slabs each focus a single stream of yellow light into the heart of the red matrix. I guess they are trying to read it. The matrix expands as the beams of light crawl through it, ballooning in the center and fragmenting suddenly, exploding to four times its original size then folding around itself into something I vaguely recognize as a hypercube from when I still used to enjoy mathematics enough to try to understand this sort of thing. The slabs' fascinating light display now occupies more than half of the table's surface and I am no longer sure what I am looking at. I am still completely ensorcelled by it when the A. I. reminds me why I am here.

"Mrs. Ajimobi, please ask your mother a question."

I snap to attention, startled at the sentence before I remember the detailed instructions from my briefing. Despite them, I am skeptical about the value of the part I am to play in all this.

"Who are you?" I ask, even though I am not supposed to.

The light matrix reconstructs itself, its elements flowing rapidly and then stilling, like hot water poured onto ice. Then a voice I can only describe as a glassy, brittle version of my mother's replies.

"I am Olusola Ajimobi."

I gasp. For all its artifice, the sound strikes at my most tender and delicate memories and I almost shed a tear. That voice is too familiar. That voice used to read me stories about the tortoise while she braided my hair, each word echoing throughout our house. That voice used to call to me from downstairs, telling me to hurry up so I wouldn't be late for school. That voice screamed at me when I told her I was dropping out of my PhD program to take a job in Cape Town. That voice answered Global Network News interview questions intelligently and measuredly, if a bit impatiently. That voice whispered, "She's beautiful," into my ear at the hospital when my darling Simioluwa was born and I held her in my arms for the first time. That voice told me to leave her alone when I suggested she retire after her first heart attack. It's funny how one stimulus can trigger so much memory and emotion.

I sit up in my chair, drawing my knees together, and try to see this for what it is: a technical evaluation of software performance. My mother, Olusola Ajimobi—"Africa's answer to Einstein," as the magazines liked to call her—has been dead thirty-eight years and her memrionic copies have been providing research advice and guidance to LegbaTech for forty. This A. I., created after her third heart attack, is not her. It is nothing but a template of her memory and thought patterns which has had many years to diverge from her original scan. That potential diversion is what has brought me here today.

When Tunji first contacted me, he told me that his team at LegbaTech has discovered a promising new research direction—one they cannot tell me anything about, of course—for which they are trying to secure funding. The review board thinks this research direction is based on flawed thinking and has recommended it not be pursued. My mother's memrionic copy insists that it should. It will cost billions of Naira just to test its basic assumptions. They need my help to decide if this memrionic is still representative of my mother, or whether has diverged so much that it is making decisions and judgement calls of which she would never have approved. My briefing instructions told me to begin by revisiting philosophical discussions or debates we had in the past to see if her positions or attitudes toward key ideas have changed or not. I choose the origins of the universe, something she used to enjoy speculating about.

"How was the universe created?" I ask.

"Current scientific consensus is—"

"No," I interrupt quickly, surprised that her first response is to regurgitate standard answers. I'm not sure if A. I.s can believe anything and I'm not supposed to ask her questions about such things, but that's what the human control is for, right? To ask questions that the other A. I.s would never think to ask, to force this electronic extrapolation of my mother into untested territory and see if the simulated thought matrix holds up or breaks down. "Don't tell me what you think. Tell me what you believe."

There is a brief pause. If this were really my mother she'd be smiling by now, relishing the discussion. And then that voice speaks again: "I believe that, given current scientific understanding and available data, we cannot know how the universe was created. In fact, I believe we will never be able to know. For every source we find, there will be a question regarding its own source. If we discover a god, we must then ask how this god came to be. If we trace the expanding universe back to a single superparticle, we must then ask how this particle came to be. And so on. Therefore, I believe it is unknowable and will be so indefinitely."

I find it impressive how familiarly the argument is presented without exact parroting. I am also reminded of how uncomfortable my mother always was around Creationists. She actively hated religion, the result of being raised by an Evangelical Christian family who demanded faith from her when she sought verifiable facts.

"So you believe god could exist?"

"It is within the realm of possibility, though highly unlikely." Another familiar answer with a paraphrastic twist.

"Do you believe in magic?"

It is a trick question. My mother loved watching magicians and magic tricks but certainly never believed in real magic.

"No magical event has ever been recorded. Cameras are ubiquitous in the modern world and yet not a single verifiable piece of footage of genuine, repeatable magic has ever been produced. Therefore it is reasonable to conclude, given the improbability of this, that there is no true magic."

Close enough but lacking the playful tone with which my mother would have delivered her thoughts on such matters.

I decide that pop philosophy is too closely linked to actual brain patterns for me to detect any major differences by asking those questions. If there is a deviation, it is more likely to be emotional. That is the most unstable solution space of the human equation.

"Do you like your great-grandson, Tunji?"

Blunt, but provoking. Tunji never met his great-grandmother when she was alive and so there is no memory for the A. I. to base its response on. Its answer will have to be derived from whatever limited interaction he and the memrionic have engaged in and her strong natural tendency to dislike over-serious people. A tendency we shared. Tunji is my daughter's son and I love him as much as our blood demands, but he is an insufferable chore most of the time. I would expect my mother to agree.

"Tunji is a perfectly capable executive director."

I'm both disappointed and somehow impressed to hear an A. I. playing deflection games with vocabulary.

"I have no doubt that he is," I say, watching the bright patterns in the light matrix shift and flow. "What I want to know is how you feel about him. Do you like him? Give me a simple yes or no."

"Yes."

That's unexpected. I sink into my chair. I was sure she would say no. Perhaps Tunji has spent more time interacting with this memrionic and building rapport with it than I thought. After all, everything this memrionic has experienced over the last forty years will have changed, however minutely, the system that alleges to represent my mother. A small variation in the elements of the thought matrix is assumed not to alter who she is fundamentally, her core way of thinking. But, like a heap of rice from which grains are removed one by one, over and over again, eventually all the rice will be gone and the heap will then obviously be a heap no more. As the process proceeds, is it even possible to know when the heap stops being, essentially, a heap? When it becomes something else? Does it ever? Who decides how many grains of rice define a heap? Is it still a heap even when only a few grains of rice are all that remain of it? No? Then when exactly did it change from a heap of rice to a new thing that is not a heap of rice? When did this recording-of-my-mother change to not-a-recording-of-my-mother?

I shake my head. I am falling into the philosophical paradox for which this test was named and designed to serve as a sort of solution. But the test depends on me making judgements based on forty-year-old memories of a very complicated woman. Am I still the same person I was when I knew her? I'm not even made of the exact same molecules as I was forty years ago. Nothing is constant. We are all in flux. Has my own personality drifted so much that I no longer have the ability to know what she would think? Or is something else going on here?

"That's good to hear," I lie. "Tell me, what is the temperature in this room?"

"It is twenty-one-point-two degrees Celsius." The glassy iteration of my mother's voice appears to have lost its emotional power over me.

"Given my age and physical condition, is this the ideal temperature for my comfort?"

"Yes, this is the optimum."

I force a deep breath in place of the snort that almost escapes me. "Olusola." I try once more, with feeling, giving my suspicions one more chance to commit hara-kiri. "If you were standing here now, beside me, with a control dock in your hand, what temperature would you set the room to?"

"The current optimum—twenty-one-point-two degrees Celsius."

There it is.

"Thank you. I'm done with the regression test now."

The electric-red hypercube matrix and yellow lines of light begin to shrink, as though being compressed back to their pretest positions, and then, mid-retraction, they disappear abruptly, as if they have simply been turned off. The beautiful kaleidoscope of numbers and symbols, flowing, flickering and flaring in fanciful fits, is gone, like a dream. Do old women dream of their electric mothers?

I sigh.

The slabs begin to sink back into the ground, and this time I shift my chair to see that they are descending into hatches, not being extruded from the floor as they would if they were made of SlatTex. They fall away from my sight leaving an eerie silence in their wake, and just like that, the regression test is over.

I hear a click and the door opens about halfway. Dr. Dimeji enters, tablet in hand. "I think that went well," he says as he slides in. His motions are snake-like and creepy. Or maybe I'm just projecting. I wonder who else is observing me and what exactly they think just happened. I remember my data-pack explaining that regression tests are typically devised and conducted by teams of three but I haven't seen anyone except Dr. Dimeji since I entered the facility. Come to think of it, there was no one at reception, either. Odd.

"Your questions were few, but good, as expected. A few philosophical ones, a few personal. I'm not sure where you were going with that last question about the temperature, but no matter. So tell me, in your opinion, madam, on a scale of one to ten, how confident are you that the tested thought analogue thinks like your mother?"

"Zero." I say, looking straight into his eyes.

"Of course." Dr. Dimeji nods calmly and starts tapping at his tablet to make a note before he fully registers what I just said, and then his head jerks up, his expression confounded. "I'm sorry, what?"

"That contrivance is not my mother. It thinks things that she would but in ways she would never think them."

A grimace twists the corners of Dr. Dimeji's mouth and furrows his forehead, enhancing his reptilian appearance from strange to sinister. "Are you sure?" He stares right at me, eyes narrowed and somehow dangerous. The fact that we are alone presses down on my chest, heavy like a sack of rice. Morbidly, it occurs to me that I don't even know if anyone will come if he does something to me and I scream for help. I don't want to die in this ugly room at the hands of this lizard-faced man.

"I just told you, didn't I?" I bark, defensive. "The basic thoughts are consistent but something is fundamentally different. It's almost like you've mixed parts of her mind with someone else's to make a new mind."

"I see." Dr. Dimeji's frown melts into a smile. Finally, some human expression. I allow myself to relax a little.

I don't even notice the humming near my ear until I feel the sting in the base of my skull where it meets my neck and see the edge of his smile curl unpleasantly. I try to cry out in pain but a constriction in my throat prevents me. My body isn't working like it's supposed to. My arms spasm and flail then go rigid and stiff, like firewood. My breathing is even despite my internal panic. My body is not under my control anymore. Someone or something else has taken over. Everything is numb.

A man enters the room through the still half-open door and my heart skips a beat.

Ah! Tunji.

He is wearing a tailored gray suit of the same severe cut he always favors. Ignoring me, he walks up to Dr. Dimeji and studies the man's tablet. His skin is darker than the last time I saw him and he is whip-lean. He stands there for almost thirty seconds before saying, "You didn't do it right."

"But it passed the regression test. It passed," Dr. Dimeji protests.

Tunji glowers at him until he looks away and down, gazing at nothing between his feet. I strain every muscle in my body to say something, to call out to Tunji, to scream—*Tunji, what the hell is going on here?*—but I barely manage a facial twitch.

"If she could tell there was a difference," Tunji is telling Dimeji, "then it didn't pass the regression test, did it? The human control is here for a reason and the board insists on having her for a reason: she knows things about her mother no one else does. So don't fucking tell me it passed the regression test just because you fooled the other pieces of code. I need you to review her test questions and tell me exactly which parts of my thought patterns she detected in there and how. Understand? We can't take any chances."

Dr. Dimeji nods, his lizard-like appearance making it look almost natural for him to do so.

Understanding crystallizes in my mind like salt. Tunji must have been seeding the memrionic A. I. of my mother with his own thought patterns, trying to get her to agree with his decisions on research direction in order to add legitimacy to his own ideas. Apparently, he's created something so ridiculous or radical or both that the board has insisted on a regression test. So now he's trying to rig the test. By manipulating me.

"And do it quickly. We can't wipe more than an hour of her short-term memory before we try again."

Tunji stands still for a while and then turns calmly from Dimeji to me, his face stiff and unkind. "Sorry, Grandma," he says through his perfectly polished teeth. "This is the only way."

Omo ale jati jati! I curse and I swear and I rage until my blood boils with impotent anger. I have never wanted to kill anyone so much in my life but I know I can't. Still, I can't let them get away with this. I focus my mind on the one thing I hope they will never be able to understand, the one thing my mother used to say in her clear, ringing voice, about fulfilling a human desire. An oft-repeated half-joke that is now my anchor to memory.

It's never the optimum. It's always just a little bit off.

Dr. Dimeji wearily approaches me as Tunji steps aside, his eyes emotionless. Useless boy. My own flesh and blood. How far the apple has fallen from the tree. I repeat the words in my mind, trying to forge a neural pathway connecting this moment all the way back to my oldest memories of my mother.

It's never the optimum. It's always just a little bit off.

Dr. Dimeji leans forward, pulls something grey and bloody out of my neck, and fiddles. I don't feel anything except a profound discomfort, not even when he finishes his fiddling and rudely jams it back in.

It's never the optimum. It's always just a little bit off.

I repeat the words in my mind, over and over and over again, hoping even as darkness falls and I lose consciousness that no matter what they do to me, my memory, or the thing that is a memory of my mother, I will always remember to ask her the question and never forget to be surprised by the answer.

"APOLOGIA," BY VAJRA CHANDRASEKERA

STORY FRAME

> He was to be the self-criticizing poet, the correctly guilty poet. He was our
> collective finger of condemnation pointed at a mirror, and then holding
> that pose, turning our heads a little, shifting hips, finding our good side in
> the light of truth and reconciliation.

Is it possible to make amends for past wrongs? How can a culture or a people—or
an industry—alleviate suffering that they themselves have caused? And how can the
work of making amends be undertaken in a way that meaningfully benefits those who
have been harmed, instead of simply soothing the consciences of those who did the
harming? Like many of the ethical issues that surface in these stories, this cluster of
questions is not intrinsically about technology. But as our world becomes ever more
dramatically shaped at a granular level by technological systems, and as power dispar-
ities, at both the individual and national level, are increasingly dictated by technological
acuity and access, the question of how to manage one's power over others, and how to
make amends for harm done, is of vital importance for those working in technology
development.

In this story, the dominant culture attempts to rectify its history of violence and
destruction by sending a poet back through time to offer apologies at a selection of
carefully chosen moments, known to the Poet's avid audience as "ChoMos." The poet's
culture and country are left unidentified, and the story's technology is fanciful, but
"Apologia" nevertheless captures some important dynamics about harm and mitiga-
tion by those who hold power over the lives of others.

The technology development community writ large, and recently the computing professions in particular, much like the dominant society in "Apologia," has

begun to acknowledge its responsibility for how it has created new harms and exacerbated existing inequalities, in a variety of ways discussed throughout this textbook. But though the industry's efforts to make amends have not included time travel (at least not yet), many of its efforts to undo these harms or balance out suffer from the same problems as the poet's quest for the big sorry. Like the carefully staged and edited ChoMos, many "AI for Good" type projects are imagined and executed based on the assumptions of the powerful people who have decided that they want to help, rather than the perspectives and experiences of the imagined beneficiaries. Just as a programmed algorithm can easily reproduce and exacerbate existing biases, human beings—because we always begin with our own questions and concerns—can end up harming more than we help, in spite of our intentions.

"Apologia" does not offer any concrete solutions for the problems we might wish to solve. But by calling attention to the importance of decentering ourselves in our efforts to help, and by underscoring the very real limits of good intentions, it can offer us a path to figuring out better solutions for ourselves.

STUDY QUESTIONS

1. What is (potentially) valuable about commemorating tragedies and atrocities? What is (potentially) counterproductive about it? What can individuals or societies do, as rememberers, to ensure that their acts of remembrance are valuable?

2. The poet's apology tour has a huge audience; its "public live feed [is] watched by millions." Who and what contribute to the creation of that feed? How does each of those contributors impact what the public sees?

3. The poet's fans know nothing about the parts of his life that are not part of the public feed; likewise, he is unaware of his fandom back home. How does each kind of not-knowing make it possible for the fans (and for the poet himself) to relate to the ChoMos?

"APOLOGIA," BY VAJRA CHANDRASEKERA

Of course they sent a poet. He was quick on his feet, electric, stepping lightly through the portals we opened for him into and across the committee's carefully Chosen Moments of history. He would only ever stop at each one long enough to read the poem he had composed especially for it.

By necessity they were short and moving poems, because a ChoMo often involved gunfire and arson and a great deal of confusion. If there was too much, the poet might shout his poem into the chaos, or declaim it into the uncaring mob, or even whisper it from shelter. The people of the ChoMo—the temporal natives—were not the audience for his work, but part of the performance.

The poems bore witness; they condemned, they grieved, they memorialized. And, of course, they apologized.

That's why the committee had selected him from among all the applicants, why he beat out the journalists and the novelists and the essayists and one extremely optimistic sculptor—the poet was good at memory and apology. He had a knack for that perspective of deep history, for the juxtaposition of the Moment to the totality, the vast and tumultuous sweep.

It helped that he was a fast writer, too. His composition was as quick as his feet, where his heart was heavy; the furious poet, all that feeling turned inward and then outward again, a reflection of the deep-set collective guilt of his people, of our people, the committee had sighed in rapturous assessment. This guilty poet, this raging poet, he could retroactively make the apologies that we had never made the first time around. It was, or would be, never too late for the big sorry.

And so the accusing poet, the apologizing poet, the adamant poet went down into the muck of the past to redeem it.

His judgments of his ancestors were uncompromising, and the committee was thrilled to receive the live video of his rapidly declaimed denunciations—he had the snap and verve of a slam poet in full flow, combined with the gradual buildup of a somber political perspective, a grand poetry of witness being constructed as he passed through more and more ChoMos—as he ducked under swinging weapons and stood in front of burning homes to spit out the words that told the truth of what was happening, the words that took responsibility.

The video was crisp and three-dimensional, and watching the feed was, the committee agreed, just like being there next to the angry poet, the righteous poet, the guilty poet.

The recordings were made by a massive entourage of tiny invisible camera-drones that followed him across time, slipping through the portals in his wake, recording his journey in 64k fully holographic high-definition from multiple angles and downstreaming the video to me, his editor in the present day, which was to say, from the poet's ever-moving perspective, the increasingly distant future.

I chose the streams to be spliced into the public live feed watched by millions, and took care of the drones and the poet alike as they waded deeper and deeper into history.

The drones were easier to watch over. They were sufficiently autonomous to dodge obstacles and find good camera angles, smart enough to learn context from the worlds they passed through, empathetic enough to know which natives to focus on, but not enough of any of those things to develop a sense of self. They were a swarm of disposable perspectives, invisible, adaptable, biodegradable. When they ran out of power, they would fall like soft rain, littering the past. I had plenty more to send after them. For drones that had reached their end-of-life, I imagined that their discarded bodies would be blown through the portals like unseen dust, building up in drifts somewhere in deep time, intangible except in their collective mass, like an odd heaviness upon the landscape over time, a gathering weight with no cause or explanation.

Looking after the poet was harder. Unlike the drones, whose lives and deaths passed without remark and without commentary except in certain highly technical circles, the poet was expected not only to return eventually to the present day after completing his tour, but to return to a hero's welcome.

A tremendous fandom had accreted around him while he was journeying in time: he would return to instant celebrity status, to critical acclaim, to awards both juried and popular. His fortitude and courage, his great achievement, even the poems themselves as artistic works, it would all be celebrated. There would be profiles, interviews, biopics. The fandom already obsessed over every moment of eye contact he made with any native, of any gender or age, no matter how inappropriate the circumstances, in case it meant that they were fucking in the poet's off-screen downtime. The fandom shipped the poet not only with his own ancestors but with the ancestors of the othered. The fandom respected no boundaries. I could have told them that the poet was practically celibate compared to their wild imaginings; his downtime was occupied mostly with writing poems, trying to sleep, and avoiding indigestion. But who would have believed me? And to be fair to him, when he did hook up with natives, he seemed to try and pick ones that were not directly involved in the ChoMo, to minimize the conflict of interest. The drones watched all of this, of course, and I did, too, sometimes. But downtime was not for publication. The writing and the rutting poet; the eating and the shitting poet; the sleeping and the screaming poet; the sitting and the staring poet, slack-faced and unseeing: all these streams I piped to null.

The poet didn't care to know about his fame back home. He spared no thought for a return journey yet. He was consumed only by thoughts of his work.

I knew those things because he told me so. He sent me frequent and voluble updates on the state of his mind and project.

These were the only access I had to his interiority; my job was to focus on his exterior, the beads of sweat on his lip, the singeing of the fine hairs on the back of his neck when he stood too close to a burning house, the way his jaw worked when he heard the screaming. I made his voice boom, his brow fierce—I picked out moments where natives near him seemed to flinch or introspect, so that they seemed struck and crushed by his words, or uplifted if they were victims, though the latter was harder because, after all, the ChoMos were the very moments of their victimization and their affective range was a little constrained.

My work was to make him look larger than life, heroic, a proper vessel for the grand project of atoning and reconciliation that he represented for us all. If he edited himself when he spoke to me, if he concealed a love of fame, if he cared about image, if he trolled his own fandom from anon accounts in his downtime, I had no way of knowing.

His brief was more profound than mine, but also simpler: to witness, memorialize, and apologize at every ChoMo.

It was understood by all of us involved, I think, without necessarily needing to have it said, because even in this project it seemed crass to have to say it quite so bluntly, that he would memorialize the other, who had suffered at our hands, and apologize for our own, who had committed, that is to say, perpetrated, or at least were implicated where nothing was proven—there's a reason they sent a poet, not a prosecutor.

He was to be the self-criticizing poet, the correctly guilty poet. He was our collective finger of condemnation pointed at a mirror, and then holding that pose, turning our heads a little, shifting hips, finding our good side in the light of truth and reconciliation.

Our endless collective guilt was insatiable, but fortunately there were many Cho-Mos, a great many, because the poet's people—who were the committee's people, and my people, too—had so many sins in our past. The poet's itinerary would methodically cover every pogrom, every riot, every race murder, every hate crime, every genocide, though of course the committee also put an asterisk next to genocide because you just had to, didn't you? Seeing as how if those had really been quote unquote genocides there would be nobody left now, would there, except the committee's people—the poet's people, which is to say my people—and that was clearly not the case, since minorities still definitely existed in the world in our present day, though not on the committee, and not very well represented in the pool of applicants for the poet's position or mine. I mean, I think there were some but the committee was very particular about not lowering their standards, so unfortunately that didn't quite work out. You know how it is. The poet knows how it is. I know how it is. Every morning when I finish my coffee and lean back in my chair for a long day of braiding together videostreams from the past into an improving narrative—one better than the rough mess of ugliness and event that we had called history the first time around, one that gives us an arc of collective redemption—I look over my shoulder expecting some poet from the future to step out of a portal and cut me down to size with some hard-hitting truths while a swarm of invisible cameras records my expression, so I make sure it's just right, you know, with a period-appropriate degree of contrition.

"ASLEEP AT THE WHEEL," BY T. CORAGHESSAN BOYLE

STORY FRAME

> Warren's grinning, so Jackie starts grinning, too. "What?" he says.
>
> "Let's us play chicken. Re-enact the scene, I mean. For real."
>
> He just laughs. Because it's a joke. Real cars, cars that do what you want, cars you can race, are pretty much extinct at this point, except for on racetracks and plots of private land in the desert, where holdovers and old people can pay to have their manual cars stored and go out and race around in them on weekends, though he's never seen any of that, except online, and it might just be a fantasy, for all he knows.

Technology development is a forward-looking industry, one that tends to get excited about the future—usually a future made better and brighter by those new technologies. But the future doesn't necessarily feel that bright, or even that exciting, once you get there. Sometimes it's the past that seems exciting: it's easy to imagine that earlier times offered freedom and flavor that have been sanitized out of a present that feels bland and boring by comparison. If you ever ask someone who's really into classic gaming why they prefer Super NES to the latest Xbox, their answer will probably sound a lot like Jackie and Warren, talking about why manual cars are better and more interesting and actually let you do more than their more-developed successors.

Jackie's mother Cindy, who was around for the manual-car era, seems a lot less inclined to romanticize the time before she had her own AV, which she calls "Carly," to rely on. Cindy is a single mother with a demanding job as an advocate for the homeless and doesn't have any reservations about entrusting Carly with the scheduling and logistical management of her life—nor does she seem to object to an occasional highly targeted advertisement.

What does bother Cindy is when the seamless management by the car's AI, and all the other various smart systems it's integrated with, crosses into outright interference with Cindy's goals and desires, like spending more time with her client Keystone. But this interference is a logical extension of what personal AVs have been designed for, not a departure from it. The primary rationale for AVs is that they make us safer because they will never make the kinds of reckless decisions that human drivers so often do. But "safety" can encompass a lot of things, and is sometimes in tension with either our immediate pleasure or even our long-term goals.

The AVs of this story have clearly been designed with the market in mind, which means that the privately owned vehicles like Carly offer the comfort and convenience of a friendly personal assistant, even employing speech patterns that mimic human ones, like Google's Assistant or Amazon's Alexa. But even the most sophisticated algorithms cannot detect or respond to situations in which users might prefer a little less safety management; still less can they exercise judgment about when a user's preference for a little bit of danger might be a worthwhile trade-off.

Increased convenience and increased safety sounds like a win-win combination, in the abstract. But this window into Cindy and Jackie's lives, and the things that motivate them and cause them joy or regret in a society increasingly structured by smart tech, suggests that both safety and convenience come with some unavoidable dissatisfactions, and for many people, the risk of serious losses of the things they value most.

Most of the ethical discussions about self-driving cars get stuck on direct questions of implementation and their very immediate consequences: when an accident is inevitable, whom will the car protect? But a lot of the argument for AVs is that accidents will be incredibly rare, which means that this question—though excitingly high-stakes—will have vanishingly little impact on most people's lives. Boyle's story helpfully refocuses our attention on a subtler but much more relevant category of ethical question: how would the basic patterns of our habits, attitudes, and choices shift in a version of the world where AVs have become standard?

STUDY QUESTIONS

1. How do people use the Ridz fleet cars? What's the user interface designed for? What are the differences between riding in a Ridz car and riding in one's own personal AV? What are the similarities? What are the advantages and disadvantages of each?

2. Why are Jackie, Warren, and their classmates so drawn to the idea of reenacting the chicken scene? What are some of the things teenagers do in our world that are driven by the same goals and desires?

3. In what ways has Keystone been able to opt out of technological surveillance and management? In what ways is he unable to opt out of it?

4. What are the various things that Carly does to keep Cindy safe? What algorithmic design choices and training data could be used to explain why Carly makes the determinations it does?

5. Both Cindy and Jackie deliberately reject some of the experience-management features that are built into the AVs in order to keep users safe.

   a. What are those various features, and why does Cindy or Jackie reject them, in each case?

   b. In what ways do those rejections recreate the experience of interacting with nonsmart cars? In what ways are Cindy's and Jackie's experiences in those situations still shaped, directly or indirectly, by the AVs' smart features?

"ASLEEP AT THE WHEEL," BY T. CORAGHESSAN BOYLE

THE PURSE

The car says this to her: "Cindy, listen, I know you've got to get over to 1133 Hollister Avenue by 2 P.M. for your meeting with Rose Taylor, of Taylor, Levine & Rodriguez, L.L.P., but did you hear that Les Bourses is having a thirty-per-cent-off sale? And, remember, they carry the complete Picard line you like—in particular, that cute cross-body bag in fuchsia you had your eye on last week. They have two left in stock."

They're moving along at just over the speed limit, which is what she's programmed the car to do, trying to squeeze every minute out of the day but at the same time wary of breaking the law. She glances at her phone. It's a quarter past one and she really wasn't planning on making any other stops, aside from maybe picking up a sandwich to eat in the car, but as soon as Carly (that's what she calls her operating system) mentions the sale, she's envisioning the transaction—in and out, that's all it'll take, because she looked at the purse last week before ultimately deciding they wanted too much for it. In and out, that's all. And Carly will wait for her at the curb.

"I see you're looking at your phone."

"I'm just wondering if we'll have enough time . . ."

"As long as you don't dawdle—you know what you want, don't you? It's not as if you haven't already picked it out. You told me so yourself." (And here Carly loops in a recording of their conversation from the previous week, and Cindy listens to her own voice saying, "I love it, just love it—and it'd match my new heels perfectly.")

"O.K.," she says, thinking she'll forgo the sandwich. "But we have to make it quick."

"I'm showing no traffic and no obstructions of any kind."

"Good," she says, "good," and leans back in the seat and closes her eyes.

HITCHHIKE

The fleet is available to everybody, all the time, and you don't even have to have an account for Ridz. The thing is, Ridz isn't going to take you directly to Warren's house or to the skate park or whatever destination you tell it, because it's programmed to take you first to the Apple Store or GameStop or wherever you might have spent money in the past. So it isn't really free, and you have to plan for the extra time to listen to the spiel and say no about sixty times, but then, eventually, you get where you want to go. Some kids—and his mother would kill him if she knew he was one of them—just step out in front of any empty fleet car that happens to be going by and commandeer it. You can't get inside if you don't have the trip code, of course, but you can climb up on the roof and cling to the Lidar till you get to the next stop or the one after that.

That is what Jackie happens to be doing at half past one on this particular school day, clinging to the roof of one of Ridz's S.D.C. Volvos and catching bugs in his teeth, when his mother's car suddenly appears in the other lane and he freezes. His first instinct is to jump down on the curb side, but they've got to be going forty miles an hour—which feels like a hundred with the way the wind is tearing at him—so he flattens himself even more, as if that could make him invisible. His plan was to go over to Warren's and hang out, nothing beyond that, though he could see a forty-ouncer in his future and maybe another hitch over to the beach with Warren and Warren's girlfriend, Cyrilla, but now, with his mother's car inching up on him, all that's about to go south in a hurry. She'll ground him for sure, cut his allowance, probably report him to his father (who won't do much more than snarl over the phone from Oregon, where he's living with Jennifer and never coming back), and then go through the whole charade of taking away his phone and his games for a week, or however long she thinks is going to impress on him just how dangerous that kind of behavior is.

All bad. But then, when the car pulls even with him, he sees that his mother, far from looking out the window and catching him in the act, isn't even awake. She's got her head thrown back and her eyes closed and Carly's doing the driving without her. He doesn't think in terms of lucky breaks or anything like that—he just accepts it for what it is. And, at the next light, he slides down off the car and takes to his feet, his back turned to the street, to the cars, to her.

KNIGHTSCOPE

The reason for the meeting with Rose Taylor is to arrange legal representation for a homeless man named Keystone Bacharach, who spends his days on the steps of the public library with a coterie of other free spirits and unfortunates, and at night sleeps under a bush in front of the S.P.C.A. facility, where he can have a little privacy. What most people don't realize—and Cindy, as an advocate for the homeless, does—is how psychologically harrowing it is to live on the streets, where through all the daylight hours you're under public scrutiny. Your every gesture, whether intimate or not, is on display for people to interpret or dismiss or condemn, and your only solace is the cover of darkness, when everything's hidden. And this is the problem: the S.P.C.A., in a misguided response to a rash of break-ins, graffiti tagging, and dumpster diving for syringes and animal tranquillizers, had deployed one of Knightscope's Autonomous Data Machines to patrol the area, which meant that Mr. Bacharach was awakened every thirty minutes, all night long, by this five-foot-tall, four-hundred-pound robot shining a light on him and giving off its eerie high-pitched whine before asking, in the most equable of tones, "What is the situation here?" (To which Mr. Bacharach, irritated, would reply, "It's called sleep.")

A week ago, she went down there after hours to see for herself, though her sister had called her crazy ("You're just asking to get raped—or worse") and even Carly, on dropping her off, had asked, "Are you sure this is the correct destination?" But they didn't know Keystone the way she did. He was just hurt inside, that was all, trying to heal from what he'd seen during his tour of duty in Afghanistan, and if he couldn't make a go of it in an increasingly digitized society, that was the fault of the society. He had an engaging personality, he was a first-rate conversationalist comfortable with a whole range of subjects, from animal rights to winemaking to the history of warfare (light years ahead of Adam, her ex, who toward the end of their marriage had communicated through gestures and grunts only), and he was as well read as anybody she knew. Plus, he was her age, exactly.

He was waiting for her in front of the S.P.C.A., dressed, as always, in shorts, flipflops, T-shirt, and denim jacket, his hair—he wore it long—pulled back tightly in a ponytail. "Thanks for coming," he said, taking the gift bag she handed him (trail mix, dried apricots, a pair of socks, a tube of toothpaste) without comment. "This is really going to open your eyes, because, no matter how you cut it, this is harassment, pure and simple. Of citizens. In a public place. And it's not just me."

She saw now that there were half a dozen other figures there, sprawled on the pavement or leaning against the wall, with their shopping carts and belongings arrayed around them. It was almost dark, but she could see that at least one of them was familiar—Lula, a woman everyone called Knitsy, because her hands were in constant motion, as if trying vainly to stitch the air. The street was quiet at this hour, which only seemed to magnify the garble of whining, yipping, and sudden startled shrieks coming from the S.P.C.A. facility behind them, and if it felt ominous it had nothing to do with these people gathered here but with the forces arrayed against them. She said, "Is it due to come by soon?"

He nodded in the direction of the parking lot at the far end of the facility. "It went down there, like, fifteen, twenty minutes ago, so it should be along any minute now." He gave her an angry look. "Like clockwork," he said, then called out, "Right, Knitsy?," and Knitsy, whether she knew what she was agreeing to or not, said, "Yeah."

The night grew a shade darker. Then one of the dogs let out a howl from the depths of the building, and here it came, the Knightscope K5+ unit, turning the corner and heading for them on its base of tightly revolving wheels. She'd seen these units before—at the bank, in the lot behind the pizza place, rolling along in formation in last year's Fourth of July parade—but they'd seemed unremarkable to her, no more threatening or intrusive than any other labor-saving device, except that they were bigger, much bigger. She'd only seen them in daylight, but now it was night, and this one had its lights activated—two eerie blue slits at the top and what would be its midriff, if it had a midriff, in addition to the seven illuminated sensors that were arrayed across its chest, if it had a chest. Its shape was that of a huge hard-boiled egg, which in daylight made it seem ordinary, ridiculous even, but the lights changed all that.

"So what now? It's not going to confront us, is it?"

"You watch," Keystone said.

The K5+, as she knew from the literature, featured the same Light Detection and Ranging device that Carly had, which used a continuously sweeping laser to measure objects and map the surrounding area, as well as thermal-imaging sensors, an ambient-noise microphone, and a three-hundred-and-sixty-degree high-definition video capture. It moved at a walking pace, three miles an hour, and its function was surveillance, not enforcement. She knew that, but still, at this hour in this place, she felt caught out, as if she'd been doing something illicit—which, she supposed, was the purpose of the thing in the first place.

But now it was stopping, pivoting, focussed on Knitsy, whose hands fluttered like pale streamers in the ray of light it emitted, which had suddenly become more intense, like a flashlight beam. "What is the situation here?" it asked.

Knitsy said, "Go away. Leave me alone."

The K5+ didn't move. It had been specifically programmed not to engage in conversation the way Carly did, because its designers wanted to avoid confrontations—it was there to deter criminal activity by its very presence and to summon the police if the need should arise. Now it said, "Move on."

"Hey," Keystone called out, waving his hands. "Over here, Tinhead."

She watched the thing swivel and redirect itself, starting down the sidewalk toward them. When it came up even with her and Keystone, it stopped and focussed its light on them. "What is the situation here?" it asked her, employing the voice of one of NPR's most genial hosts, a voice designed to put people at ease. But she didn't feel at ease—just the opposite—and that was a real eye-opener.

What happened next was sudden and violent. Keystone just seemed to snap—and maybe he was showing off for her, thinking, in some confused way, that he was protecting her—but in that moment he tucked his shoulder like a linebacker and slammed into the thing, once, twice, three times, until he finally managed to knock it over with a screech of metal and shattering glass. Which was bad enough—vandalism, that was what she was thinking, and her face was on that video feed, too—but then he really seemed to take his frustration out on the thing, seizing a brick he'd stashed under one of the bushes and hammering at the metal frame until the unit set off a klaxon so loud and piercing she thought her heart would stop.

Just then, just as she was thinking they were both going to get arrested, Carly pulled up at the curb. The door swung open. "Get in," Carly said.

Rebel without a Cause

It was a meme, really, that got them into it, a clip of the scene in the old movie where two guys were playing chicken and the greaser who wasn't James Dean got the strap of his leather jacket caught on the door, which repeated over and over till it was just hilarious. After that, curious about the movie itself, they dug deeper and it was a revelation—teen-agers stole cars and raced them on the street and there was nobody there to say different. Even better, because this was back in the day, the cars just did what you wanted—all you had to do was put the key in the ignition (or hot-wire the car, if you wanted to steal it), hit the gas, and peel out. He must have seen the movie (or parts of it) at least twenty times with Warren and Cyrilla, and, if Warren was James Dean and Cyrilla Natalie Wood, he guessed he'd have to be Sal Mineo, though that wasn't really who he wanted to be.

"Better than the dude that goes over the cliff, though, right?" Warren says now, waving his forty-ouncer at the screen, and Cyrilla lets out a laugh that's more of a screech, actually, one of her annoying habits, but that's all right—he doesn't mind playing a supporting role. Warren's almost a year older than he is, and he doesn't have a girlfriend himself, so to be near Cyrilla, to hang out with her, see what she's like— what girls are like, up close—is something he really appreciates on every level.

They're coming up on the part where Natalie Wood, her eyes burning with excitement, waves her arms and everybody stands back and the two cars hurtle off into the night, when Warren, who has his arm around Cyrilla on the couch and one hand casually cupping her left breast, says, "I have this idea?"

Warren's grinning, so Jackie starts grinning, too. "What?" he says.

"Let's us play chicken. Reënact the scene, I mean. For real."

He just laughs. Because it's a joke. Real cars, cars that do what you want, cars you can race, are pretty much extinct at this point, except for on racetracks and plots of private land in the desert, where holdovers and old people can pay to have their manual cars stored and go out and race around in them on weekends, though he's never seen any of that, except online, and it might just be a fantasy, for all he knows. "What are you talking about?" he says. "You going to steal a fleet car?"

"No," Warren says, levelling a look at him. "I'm going to steal two."

Risk Assessment

She's in the car on her way to the library to pick up Keystone and bring him to Rose Taylor's office so they can begin the process of filing a public-nuisance lawsuit against the S.P.C.A., when Carly says, "I don't mean to worry you, but the house sensors indicate that Jackie hasn't come home from school yet—and the calendar shows no extra-curricular activities for today, so I'm just wondering . . . ?"

Cindy's feeling distracted, her mind on Keystone and the way he stood up for her that night on the street, or at least thought he was standing up for her, which amounts to the same thing. "I wouldn't worry. He's a big boy. He can take care of himself."

"Granted, yes, but I can't help thinking of last week, when he didn't get in till after dark and had no explanation except"—and here she loops in Jackie's voice from the house monitor—"'I was at Warren's, O.K., and his mom made dinner, O.K., so I'm not hungry, so don't even go there.'"

"Listen, Carly, I'm just not up to this right now, O.K.? I'm trying to focus on getting Keystone over to Rose Taylor's and then I've got to get back to the office for that five-o'clock meeting, as I'm sure you're aware, and then there's the fund-raiser after that . . ."

"Sorry, I just thought you'd want to know."

She's staring out the side window watching the street lights clip by, picturing Keystone pushing himself up off the concrete steps of the library and crossing the sidewalk with that smile of his lit up just for her. She's curious to see what he'll be wearing—"I clean up pretty good," he told her, promising to dress up for the meeting—not that it matters, really, just that she's never seen him in anything other than what he calls his "street commando" outfit. The street lights are evenly spaced, like counters, and after a moment it occurs to her that the intervals between them are getting shorter and shorter, so she turns, focusses on the street ahead, and says, "Aren't you going too fast, Carly?"

Immediately the car slows. "Forty-four in a thirty-five zone, but there's no indication of speed traps or police units, and since we are running six minutes and sixteen seconds late, I thought I would expedite matters."

She's feeling angry, suddenly—and it's not Carly's fault, she knows that, but the comment about Jackie just rubbed her the wrong way. "I didn't give you permission for that," she snaps. "You ought to know better. I mean, what good is your program if you can't follow it?"

"I'm sorry, Cindy, I just thought—"

"Don't think—just drive."

Of course, Carly was right, and if they wind up being ten minutes late to pick up Keystone that's nobody's fault but her own. "All right, Carly, I'm sorry—good job, really," she says, only vaguely aware of how ridiculous it is to try to mollify a computer or worry about hurting its feelings.

"Since we're at the library," Carly says, "will you be acquiring books? Because they have three copies of the latest installment of the Carson Umquist series you like—and they're all in the special 'Hot Reads' rack when you first walk in. I mean, they're right there—you don't have to go twenty feet. If that's what you're looking for. I'm not presuming, am I?"

"Pull up here," she says, and that's when she sees Keystone, in a pair of tan Dockers and an emerald long-sleeved shirt with a pair of red fire-breathing dragons embroidered on the front. He looks . . . different—and if she's surprised by the dragons, which really aren't the sort of thing she imagines Rose Taylor appreciating, she tries to hide it. She's smiling as he comes up to the car, and he's smiling, too, and now he's reaching for the door handle . . . but the door seems to be locked, and she's fumbling for the release. "Carly," she says, turning away from the sight of his face caught there in the window as if Carly were an actual person sitting in the driver's seat, when, of course, there's no one there. "Carly, is the child lock on?"

"I'm sorry," Carly says, "but this individual is untrustworthy. Don't you recall what happened last Tuesday evening at 9:19 P.M. in front of the S.P.C.A. facility at 83622 Haverford Drive?"

"Carly," she says, "open the door."

"I don't think that's wise."

"You know what? I don't give a God damn what you think. Do you hear me? Do you?"

HER FATHER'S LAST DRIVE

He was in his mid-seventies back then and he'd never really been what anybody would call a good driver—too rigid, too slow to react, baffled by the rules and norms of the road and trying to get by on herd mentality alone. To complicate matters, he suffered from arthritis and wound up developing a dependency on the painkillers the doctor prescribed, which, to say the least, didn't do much for his reflexes or his attention span. He was a disaster waiting to happen, and Cindy and her sister, Jan, kept nagging him to give up driving, but he was stubborn. "I've seen 'King Lear,'" he said. "Nobody's going to take my independence away from me."

Then one morning, when her car was in the shop (this was before S.D.C.s took over, when most people, including her, still got around the retro way), she asked him for a ride to work and not only was he half an hour late but when they finally did get on the freeway he drove his paint-blistered pickup as if the wheels had turned to cement blocks, weaving and drifting out of his lane and going at such a maddeningly slow pace she was sure they were going to get rear-ended. She was a wreck by the time she got to the office and so keyed up she didn't dare even take a sip of her morning grande, let alone drink it. She Ubered home that night, though, as a recent divorcée and the mother of a two-year-old, she was trying to cut her expenses, so that was no fun. As soon as she got in the door, she called Jan.

"We've got to do something," she said. "He's going to get killed—or kill somebody in the process. It's a nightmare, believe me! Have you been in a car with him lately? It's beyond belief."

Jan was silent a moment, thinking, then she said, "What about that refrigerator you've got to move?"

"What refrigerator? What are you talking about?"

Her sister didn't say anything, just waited for her to catch on.

"Can we do that to him? He'll never talk to either one of us again, you realize that, right?" She was trying to picture the aftermath, the resentment, the sense of betrayal, the way he used his sarcasm like an icepick, chipping away at you flake by flake, and how he'd parcelled out his affection all his life and what that was going to mean for the future. "It's not going to be me," she said. "I'm not going to be the one."

"We'll both do it."

"How's he going to get around? I'm not driving him, I'll tell you that."

"The bus. The senior van. Whatever. Other people do it. But what about Luke—what if Luke asks him? He'd never refuse Luke."

Luke was Jan's seventeen-year-old son, and as soon as Jan pronounced his name Cindy realized they were going to take the easy way out—or, no, the coward's way.

The next Saturday morning, Jan dropped her son off at their father's apartment so that he could borrow the truck to move the imaginary refrigerator, and the moment the keys changed hands their father's time behind the wheel of a motorized vehicle, which stretched all the way back to when he was two years younger than Luke was then, came to an abrupt end.

THE HACK SUPREME

Another day, another slow, agonizing procession of classes that are like doors clanging shut in a prison one after the other, and then they're at Warren's and Warren's parents are at work, so they have the place to themselves to make preparations. The first thing is the punch, which means pouring grape juice, 7 UP, and about three fingers of every kind of liquor in the cabinet into a five-gallon bucket purchased at Walmart for just this purpose. Then snacks, but that's easy, just bags of chips, pretzels, Doritos, and whatever. Cyrilla rolls a couple of numbers, and he and Warren pull out their phones and give everybody a heads-up: nine o'clock at the end of Mar Vista, where it dead-ends at that weed lot and the cliffs down to the ocean.

He's not a bad hacker himself—since as far back as he can remember, he's hacked into Web sites just for the thrill of messing with people a little—but Warren's in another league. If anybody can steal a fleet car—two fleet cars—it's him. So, after dinner (he texted his mother to tell her he'd be eating at Warren's and then sleeping over, too), they go out on Cabrillo, where there's a ton of cars going back and forth between pickups and drop-offs, and just step out in front of two empty ones, which slam on their brakes and idle there, waiting for them to move. But they don't move. Warren has already hacked into the network on his laptop, and now he's accessing the individual codes for these two cars while all the other cars are going around them and they have to hope no surveillance vehicles come by or they're dead in the water.

That doesn't happen. The doors swing open for them and they get in and tell the cars to take them out to Mar Vista, where Cyrilla and some of the others are already gathered around the punch and the chips, waiting for them to get the party started. It's beautiful. It's perfect. He can't remember ever seeing a prettier sunset, all orange and purple and black, as if the whole world were a V.R. simulation, and if his heart goes into high gear when a cop car comes up behind him and swings out to pass, that's all part of the game and he's O.K. with it. O.K. with everything. He's going to be a hero at school, an instant legend, because nobody's tried this before, nobody's even thought of it. And, yes, it's dangerous and illegal and his mother would kill him and all that, and he did say to Warren, trying to be cool and hide his nerves, "So who's the greaser that goes over the cliff, you or me?," and Warren said, "Forget it, because we're both James Dean, and I'm not even going to try to make it close. I'm going to jump out way before and if that's chicken, O.K., sue me, right?"

Once they're there, the real work starts. Warren—and this other kid, Jeffrey Zuniga, who's a genius and destined to be class valedictorian—start disabling the cars' systems as much as possible, so they'll go flat out, because what kind of a race would it be if these two drone cars just creep along toward the cliff? All right. Fine. And here comes the movie.

He and Warren, drawn up even, both of them drunk on the punch and laughing like madmen, revving the engines on command (it's as simple as saying "Redline" to the computer), and Cyrilla there waving somebody's white jacket like a flag. It's fully dark now, kids' eyes in the headlights like the eyes of untamed beasts, lions and hyenas and what, jackals, and then they're off and all he's thinking is, If Warren thinks I'm going to bail first, he's crazy . . .

THE GHOST IN THE MACHINE

It isn't a date, not exactly, and if Jan ever finds out about it she'll never hear the end of it, but she takes Keystone out to the local McDonald's for a Big Mac and fries. It's not as if she hasn't taken other housing-challenged people out for fast food, men and women both, so they can sit in a booth with some kind of dignity and use the rest room to their heart's content, without the manager badgering them every step of the way. But this is different. It's a kind of celebration, actually, because Rose Taylor filed the suit and, within hours, she had a call from the head of the S.P.C.A. wondering if they couldn't work something out, like using the K5+ primarily in the parking area and limiting its access to the public sidewalk.

Keystone is back to his usual garb, with the addition of a military-looking camouflage cap he's picked up somewhere and an orange string bracelet Knitsy wove for him. He's in good form, high on the moment and her company (and the dark rum he surreptitiously tips into their Diet Cokes when no one's looking), and she's feeling no pain herself. There's something about him that makes her just want to let go—in a good way, a very good way. And the rum—she hasn't had anything stronger than white wine in years—goes right to her head.

"You know that this place—the S.P.C.A., I mean—has a furnace out back, right?" he says, leaning into the table so she's conscious of how close he is to her, right there, no more than two feet away. "And maybe your attorney friend can eventually get them to stop harassing us people, but what about the animals? You know what it smells like when they fire that thing up? I mean, can you even imagine?" He pauses, bites into his Big Mac, chews. "You're in a house, right?"

"Uh-huh, yeah. With a yard. And I know I should really adopt one of those dogs, I really should, but I can never seem to get around to it—"

"That's not what I'm saying—I'm not trying to lay any guilt trip on you. The people that should feel guilty are all these clueless shitwads that see a puppy in a store window and six weeks later dump it on the street. . . . No, what I'm talking about is the smell, which you don't get out in the suburbs, with the windows rolled up and the air-conditioning going full blast. Am I right?"

He is right. But whether he's right or wrong or whether he's accusing her or not doesn't really matter. What matters is the intensity of his voice, the gravel in it, and the way his eyes look right into her as if there were nothing separating them but the illusion of a Formica tabletop and the recirculating air, with its heavy freight of warmed-over meat and hunger.

In the car, she takes the leap and asks him if he wants to come home with her. "There's a shower," she says. "And clean towels—I can offer you clean towels, right? Isn't that the least I can do? As your advocate, I mean?"

It's dark now, but for the yellowish sheen of the McDonald's arches and the fiery glow of the tail-lights of the cars at the drive-through window. He doesn't say anything and she keeps waiting for Carly to butt in, though she gave her strict instructions to keep quiet, no matter what. Finally, he sighs and says, "It's an attractive offer, and I thank you for it, but I don't want to be anybody's pet."

She doesn't know whether she should laugh or not. Really, is he joking?

"But why don't we do this?" he says. "You drive me back to my place and we'll sit on the wall there, finish the rum, and see what happens. O.K.? Sound like a plan?"

All the way there, Carly's silent, except to comment on the traffic conditions— "There's a lane closure on Mission because of roadwork, so I'm going to take Live Oak to Harrison, which is only a two-minute-thirty-five-second delay"—and she finds she doesn't have much to say herself, anticipating what's to come and thinking about the last time she had sex outdoors, which had to have been twenty years ago. With Adam. On a camping trip.

When they step out of the car, the night comes to life around her, rich with its crepitating noises and a strong sweet wafting scent of jasmine, which is all she can seem to smell—not the reek of the dogs or the crematorium or the hopelessness of Knitsy and the rest of them, but jasmine blooming in some secret corner. She likes the way the full moon comes sliding in over the treetops. The rum massages her. Then Keystone takes her hand and he's leading her to the wall and everything's falling into place . . . until one of the dogs lets out a howl and they both look up to see the K5+ unit wheeling toward them, its lights on full display. "Aw, shit!" Keystone spits, and before she can stop him, before the machine can wheel up to them and inquire what the situation is, he's halfway up the block, confronting it. He seems to have something in his hand now, a pale plastic bag he's pulled out of the bushes, and in the next moment he's jerking it down like a hood over the thing's Lidar, rendering it blind. It stops, emits its inquisitive whine for a count of eight, nine, ten seconds, and then triggers its alarm.

So much for romance. So much for Rose Taylor and human rights. The noise is excruciating. Every dog in the S.P.C.A. starts howling as if it were being skinned alive, and you can be sure the cops are on their way, no doubt about that. But here's Keystone and he's grinning, actually grinning, as if all this were funny. "You know," he says, raising his voice to be heard over the din, "maybe I am ready to be a pet. You want a pet? For tonight, anyway?"

He doesn't wait for an answer, just puts his arm around her and guides her to the car. But Carly's having none of it. Carly's got her own agenda. The locks click shut.

"Open up," Cindy demands.

The car ignores her.

"Open up. Carly, I'm warning you—"

One long pulse-pounding moment drifts by. The car is a dark conglomerate of metal, glass, and plastic, as inanimate and insensible as a stone. She's angry—and frustrated, too, because she'd been ready to let go, really let go, for the first time in as long as she can remember. The dogs howl. The klaxon screams. And Keystone is right there, smelling of the Mrs. Meyer's hand soap somebody must have given him a gallon of, his arm around her shoulder, one hip pressed to hers. She wants to apologize— For what, for a car?—but that doesn't make any sense. "I don't know," she says, frantic now, and are those sirens she's hearing in the distance?

"Oh, fuck it," he says finally, throwing a glance over his shoulder. "Let's just walk. We can still walk, can't we?"

CHICKEN

He isn't wearing a leather jacket. He doesn't even own a leather jacket. He's just a kid in a simulation, the fleet car jerking along over the bumps in the field, and the night waiting for him out there like an open set of jaws. He keeps glancing over at Warren and Warren keeps glancing over at him as if this really were chicken—and he's not going to be the one to cave first, is he? But that's not the issue, not any longer, because what he is just now discovering is that the door is not going to swing open no matter how many times he orders it to, and the brakes—the autonomous brakes, the brakes with a mind and a purpose of their own—don't seem to be working at all.

NIGHT MOVES

On this, of all nights, she has to be wearing heels, but then she's wearing heels to impress Keystone, whether she wants to admit it or not. Men find heels sexy. He finds them sexy—and he told her as much when they were standing at the counter in McDonald's, placing their order. Now, though, she has her regrets—they haven't gone five blocks and she's already developing a blister on her left heel and her toes feel as if someone were taking a pair of hot pliers to them. "What's the matter?" he asks, his voice coming at her out of the dark. "You're not giving out already, are you?"

"It's my feet," she says, stopping and shifting her weight into him to take some of the pressure off.

"It's not your feet—it's your shoes. Here"—he braces her with one arm—"just take them off. Go barefoot. It's good for you."

"Easy for you to say."

"Hell, I'll go barefoot, too, no problem—in fact, I like it better this way," he says, and in the next moment he's got his flip-flops in one hand and her shoes in the other and they're heading up the sidewalk under the faint yellowish glow of the street lights in a neighborhood that may or may not feature broken glass strewn across the pavement.

It's better than three miles to her house, and she's so used to relying on Carly she manages to lose her way, until finally she has to pull out her phone and follow the G.P.S., which is embarrassing, but not nearly as embarrassing as seeing Carly sitting there in front of the house, running lights on, waiting for them. "You!" she calls out as they cross the lawn. "You're going to hear from me tomorrow—and that's a promise."

But then something happens, something magical, and all the tension goes out of her. It has to do with the grass, its dampness, its coolness, the way it conforms to her toes, her arches, her aching heels. The simplest thing: grass. In that instant, she's taken all the way back to her girlhood, before Adam, before Jackie, before her infinitely patient dark-haired father taught her to balance clutch and accelerator and work her way through the gears in a smooth, mechanical succession that opened up a whole new world to her. "This is nice," she murmurs, and Keystone, a hazy presence beside her, agrees that, yes, it is nice, though she's not sure he knows what he's agreeing to.

There are no cars on the street. Her house looms over them, two stories of furniture-filled rooms humming with the neural network of all the interconnected devices it contains, the refrigerator clicking on, the air-conditioner, appliance lights pulsing everywhere. In a moment, she'll lead Keystone up the steps, through the living room, and into the back bedroom, but not yet, not yet. Everything is still. The moon is overhead. And the grass—the grass is just like she remembered it.

"Codename: Delphi," by Linda Nagata

STORY FRAME

> The war was five thousand miles away, but it was inside her head too; it
> was inside her dreams and her nightmares.

Karin works as a handler for soldiers on the battlefield. She monitors myriad sources
of information: real-time point-of-view audio and video; overhead footage from
drones and satellites; real time vitals and GPS positions; intelligence data about mission objectives; and enemy data as well. She takes this all in and communicates the
relevant information to the soldiers as necessary. The incredible strain of this job—
managing huge amounts of information at high speed, with incredibly high stakes—is
amplified by the fact that she is responsible for three clients at once.

Karin is keenly aware that her life is easier and safer than her soldiers' lives
because she works behind a desk and gets to go home at the end of her shifts. But she
has developed some habits and mentalities that more closely resemble a soldier than
an average help desk worker; she also suffers a lot of the same things as they do. After
almost two years—a very long tenure, for a handler—she has nightmares and anxieties. But though she desperately wants to leave the job, she doesn't seem to be able to.

Both Karin and her soldiers maintain a tight focus on immediate goals and concerns. They do not even make reference to the larger war, much less debate the merits
of its purpose or the ethical complexities of fighting an enemy that has far fewer technological and defensive resources. The soldiers don't have time for these questions—
and neither, notably, does Karin, who is physically safe and distant but emotionally
and experientially close.

The fact that Karin carries the trauma of war in spite of her physical distance,
and her inability to step back and reflect on her situation, raises a host of questions

about the kinds of decisions we ask information workers to handle. Her story further equips us to reflect on some of the difficult questions Karin herself cannot make room for: how do you evaluate questions of goodness and justice when at war?

STUDY QUESTIONS

1. What are the difficult things about Karin's job? How are these difficulties affected by the fact that she typically works with three clients at once?

2. In what respects is Karin's experience of the war different from that of the soldiers with whom she works? In what ways is it similar?

3. Why does Karin find it unacceptable that some of her fellow handlers describe the job as being like a video game? In what ways might those handlers find it helpful or productive to think of the job in that way?

4. What virtues and habits has Karin cultivated in order to excel as a handler? In what ways do these habits contribute to, or inhibit, her ability to flourish?

5. Although the handler job was invented for this story, it bears some significant resemblance to the very real job of social media content moderator. Although content moderators do not have life or death decisions for others in real time, their work also requires them to engage with disturbing or misleading material all day long, and to make decisions about whether others should be exposed to it. In what ways might these workers suffer similar effects to those that Karin endures in the story?

"CODENAME: DELPHI," BY LINDA NAGATA

"Valdez, you need to slow down," Karin Larsen warned, each syllable crisply pronounced into a mic. "Stay behind the seekers. If you overrun them, you're going to walk into a booby trap."

Five thousand miles away from Karin's control station, Second Lieutenant Valdez was jacked up on adrenaline and in a defiant mood. "*Negative!*" she said, her voice arriving over Karin's headphones. "*Delphi, we've got personnel down and need to move fast. This route scans clear. I am not waiting for the seekers to clear it again.*"

The battleground was an ancient desert city. Beginning at sunset, firefights had flared up all across its tangled neighborhoods and Valdez was right that her squad needed to advance—but not so fast that they ran into a trap.

"The route is *not* clear," Karin insisted. "The last overflight to scan this alley was forty minutes ago. Anything could have happened since then."

Karin's worksite was an elevated chair within a little room inside a secure building. She faced a curved monitor a meter-and-a-half high, set an easy reach away. Windows checkered its screen, grouped by color-codes representing different clients. The windows could slide, change sequence, and overlap, but they could never completely hide one another; the system wouldn't allow it. This was Karin's interface to the war.

Presently centered onscreen were two gold-rimmed windows, each displaying a video feed captured by an aerial seeker: palm-sized drones equipped with camera eyes, audio pickups, and chemical sensors. The seekers flew ahead of Valdez and her urban infantry squad, one at eye level and the other at an elevation of six meters, scouting a route between brick-and-stucco tenements. They flew too slowly for Valdez.

The lieutenant was out of sight of the seekers' camera eyes, but Karin could hear the soft patter of her boot plates as she advanced at a hurried trot, and the tread of the rest of the squad trailing behind her. Echoing off the buildings, there came the pepper of distant rifle fire and a heavier caliber weapon answering.

Onscreen, positioned above the two video feeds, was a third window that held the squad map—a display actively tracking the position and status of each soldier.

Outfitted in bullet-proof vests and rigged in the titanium struts of light-infantry exoskeletons—"armor and bones"—the squad advanced through the alley at a mandated ten-meter interval, a regulation that reduced the odds of multiple casualties if they encountered an IED or a grenade. Only Lieutenant Valdez failed to maintain the proper distance, crowding within two meters of the seekers in her rush to answer the call for backup.

"Valdez, this is not a simple firefight. It's a widespread, well-planned insurgent offensive. Every kid with a grudge—"

"*No lectures, Delphi. Just get these seekers moving faster.*"

Any faster, and the little drones could miss something critical.

Local time was past midnight and no lights shone in the alley, but in nightvision the walls of the buildings and the trash-strewn brick pavement gleamed in crisp, green detail. Karin wasn't the only one monitoring the seekers' feeds; a battle AI watched them too. It generated an ongoing report, displayed alongside the windows. She glanced at it and saw an alert for trace scents of explosives—but with a battle in progress that didn't mean anything. Otherwise the report was good: no suspicious heat signatures or whispering voices or inexplicable motion within the apartments.

Her gaze shifted back to the video feed. A faint gleam caught her attention; a hair-thin line close to the ground that justified her caution. "Tripwire," she announced. She reached out to the screen; dragged her finger across the line. The gesture created a fleeting highlight on the display screen of Valdez's visor, clearly marking the tripwire's position. "Six meters ahead."

"*Shit.*" Valdez pulled up sharply. A faint background tone sounded as she switched her audio to gen-com. "*Tripwire,*" she said, addressing her squad. "*Move back.*"

The tone dropped out, and Valdez was talking again solely to Karin. "*Ambush?*"

"Searching." It was a good bet someone was monitoring the tripwire.

A set of windows bordered in blue glided to the center of Karin's screen: Lieutenant Deng's color code. The insurgent offensive had erupted all along the northern border, striking hard at Deng's rural district. At approximately 2200 she'd been lured into an ambush. The resulting firefight had left one of her soldiers seriously wounded.

Distance did not mute the impatience—or the frustration—in Deng's voice as she spoke over the headphones, *"Delphi, where's my medevac helicopter?"*

On nights like this, a big part of Karin's job was triage. Deng's situation was no longer "hot." The insurgents had fled, and the helicopter had already been requested. Determining an ETA would not get it there faster. So she told Deng, "Stand by."

Then she swiped the blue windows out of the way and returned her attention to the feeds from the seekers, directing one to fly higher. The angle of view shifted, and Karin spied a figure crouched on the sloping, clay-tiled roof of a low building not far ahead. She drew a highlight around it. "Valdez, see that?"

A glance at the squad map showed that Valdez had retreated a few meters from the tripwire. One specialist remained with her, while the rest of the squad had dropped back under the supervision of a sergeant.

"I see him," Valdez said. *"Target confirmed?"*

"Negative. Twenty seconds."

Karin sent a seeker buzzing toward the figure on the rooftop and then she switched her focus back to Deng's blue-coded windows, fanning them open so she could see the one that tracked the status of the medevac helicopter. The offensive was unprecedented and air support was in high demand. Deng's wounded soldier was third on the list for pickup. "Deng, ETA on the medevac is forty-plus minutes," Karin warned; that was assuming the helicopter stayed in the air. She slid the blue windows away again, switching back to Valdez.

Wind soughing between the buildings veiled the soft buzz of the seeker so that the figure on the roof didn't hear it coming. Details emerged as the little drone got closer. One of those details was a rifle—aimed at Valdez. "Target confirmed," Karin said without hesitation. "Shoot to kill."

Valdez was watching the same feed. *"That's a kid!"*

It *was* a kid. The battle AI estimated a male, fourteen years old. It didn't matter. The boy was targeting Valdez and that made him the enemy.

"Take the shot."

The boy fired first. He missed, but he squeezed the trigger again. His second shot caught Valdez in the shoulder, spinning her into the wall. *"Fuck."*

"Valdez, get down!"

The lieutenant dropped to a crouch. The specialist was already hunkered down behind her. He aimed over her shoulder and shot—but too late. The kid had opened a roof-access door, retreating inside the building.

Karin checked Valdez's biometrics: high stress, but no indication that the slug had penetrated. Her armor had protected her.

"A biometric ID on the shooter is in the system," Karin told her. "You can hunt him down later."

"Right. I'm going to drop back, rejoin the squad, and go around."

While Valdez reorganized, Karin switched to her third client, Lieutenant Holder. The set of windows monitoring his squad was coded orange. Holder was assigned to a district just outside the city. Tonight his squad waited in ambush for a suspected small-arms shipment coming in from the west. She checked his status: nominal. Checked the squad: noted all seven soldiers in position on either side of an asphalt road. Checked the wide-field view from the infrared camera on the squad's surveillance drone and noted the suspect truck, still at almost five kilometers away.

There was time.

Karin sighed, took a sip of chilled water from a bottle stashed in a pouch at the side of her chair, and for just a moment she squeezed her dry eyes shut. She'd already been six hours on-shift, with only one ten-minute break and that was two hours ago. There would be hours more before she could rest. Most shifts went on until her clients were out of harm's way—that's just how it was, how it needed to be. She'd learned that early.

Karin had trained as a handler for the usual reason: money. She'd needed to pay off a student loan. Two years so far, with a fat savings account to show for it. The money was good, no argument, but the lifestyle? Some handlers joked that the job was like a video game—one so intense it left you shaking and exhausted at the end of every shift—but for her it had never been a game. The lives she handled were real. Slip up, and she could put a soldier in the grave. That was her nightmare. She'd had soldiers grievously wounded, but so far none had died on her shift. Lately, she'd started thinking that maybe she should quit before it happened. On a night like tonight, that thought was close to the surface.

The blue windows slid to center again. Karin popped the bottle back into its pouch as an irate Deng spoke through her headphones. *"Delphi, I can't wait forty minutes for the medevac. I've got six enemy at-large. They have their own wounded to worry about, but once they get organized, they're going to move on the settlement. If we don't get there first, there are going to be reprisals. I need approval from Command to split the squad."*

"Stand by."

Karin captured a voice clip of Deng's request and sent it to the Command queue, flagged highest priority. But before she could slide the blue windows aside, someone opened an emergency channel—an act that overrode the communications of every handler on-shift. *"I need support!"* a shrill voice yelled through Karin's headphones. She flinched back, even as she recognized Sarno, another handler. The panic in his voice told her that he had made a mistake. A critical mistake, maybe a fatal one. *"I need support! Now. I just can't—"*

His transmission cut out. The shift supervisor's voice came on—calm, crisp, alert: the way handlers were trained to speak. *"I'm on it."*

Karin's hands shook. Sarno worked a chair just a few doors down from her. He was new, and new handlers sometimes got overwhelmed, but panic was always the wrong response. At the end of the shift, every handler got to go home, smoke a joint, collapse in a bed with soft sheets, get laid if they wanted to. Their clients didn't have that option. Sarno needed to remember that. Sarno needed to remember that however rough it got in the control room, no one was trying to end *his* life.

Right now the supervisor would be assisting him, coaching him, getting him back on track. Karin refocused, striving to put the incident out of her mind.

Dragging the gold-rimmed windows to center, she checked on Valdez, confirming the lieutenant had safely exited the alley. There were no alerts from the battle AI, so Karin switched to Deng's window-set. Rigged in armor and bones, the squad had formed a perimeter to protect their wounded soldier. Around them, dry grass rustled beneath spindly trees, and the stars glowed green in nightvision. Karin switched to Holder. He was still hunkered down with his squad alongside the road. An infrared feed from Holder's surveillance drone showed the target vehicle only a klick-and-a-half away, approaching fast without headlights.

Just as Karin brought her attention back to Valdez, the shift supervisor spoke.

"*Karin, we've got an emergency situation. I need to transfer another client to you.*"

"No way, Michael."

"*Karin—*"

"*No.* I've got three active operations and I can barely stay on top of them. If you give me one more client, I'm going to resign."

"*Fine, Karin! Resign. But just finish this shift first. I need you. Sarno walked. He fucking walked out and left his clients.*"

Sarno walked? Karin lost track of her windows as she tried to make sense of it. How could he walk out? What they did here was not a video game. There was no pause button on this war. Every handler was responsible for the lives of real people.

Michael took her hesitation as agreement. "*I'm splitting the load. You only have to take one. Incoming now.*"

Her throat aching, she took another sip of water, a three-second interval when her mind could rove . . . this time back to the kickboxing session that started her day, every day: a fierce routine that involved every muscle—*strike, strike, strike*—defiantly physical, because a handler had to be in top form to do this kind of work, and Karin hated to make mistakes.

As she looked up again, a glowing green dot expanded into a new set of windows, with the client's bio floating to the top. Shelley, James. A lieutenant with a stellar field rating. *Good*, Karin thought. *Less work for me.*

As she fanned the windows, the live feed opened with the triple concussion of three grenades going off one after another. She bit down on her lip, anxious to engage, but she needed an overview of the situation first. Locating the squad map, she scanned the terrain and the positions of each soldier. There were five personnel besides Shelley: a sergeant, two specialists, and two privates. The map also showed the enemy's positions and their weaponry—field intelligence automatically compiled from helmet cams and the squad's surveillance drone.

The map showed that Shelley's squad was outnumbered and outgunned.

With little shelter in a flat rural landscape of dusty red-dirt pastures and drought-stricken tree farms, they protected themselves by continuously shifting position in a fight to hold a defensive line north of the village that was surely the target of this raid. The insurgents' ATVs had already been eliminated, but two pickup trucks remained, one rigged with a heavy machine gun and the other with a rocket-launcher pod, probably stripped off a downed helicopter. The rockets it used would have a range to four kilometers. Shelley needed to take the rocket-launcher out before it targeted the village and before his squad burned through their inventory of grenades.

The sound of the firefight dropped out as her get-acquainted session was overridden by Deng's windows sliding to the center. A communication had come in from Command. Deng's request to split the squad had been approved. Karin forwarded the order, following up with a verbal link. "Deng, your request has been approved. Orders specify two personnel remain with the wounded; four proceed to the settlement."

"Thanks, Delphi."

Karin switched to Holder. His ambush would go off in seconds. She did a quick scan of the terrain around him, located no additional threats, and then switched focus to Valdez. Cities were the worst. Too many places for snipers to hide. Too many alleys to booby trap. Karin requested an extra surveillance drone to watch the surrounding buildings as Valdez trotted with her squad through the dark streets. She'd feel more secure if she could study the feed from the seekers, but there was no time—because it was her new client who faced the most immediate hazard.

Lieutenant Shelley was on the move, weaving between enemy positions, letting two of his soldiers draw the enemy's attention while he closed on the rocket launcher. The truck that carried the weapon was being backed into the ruins of a still-smoldering, blown-out farmhouse. The roof of the house was gone along with the southern wall, but three stout brick walls remained, thick enough to shelter the rocket crew from enemy fire. Once they had the truck in place, it would be only a minute or two before the bombardment started.

Not a great time to switch handlers.

Karin mentally braced herself, and then she opened a link to Shelley. The sounds of the firefight hammered through her headphones: staccato bursts from assault rifles and then the bone-shaking boom of another grenade launched by the insurgents. A distant, keening scream of agony made her hair stand on end, but a status check showed green so she knew it wasn't one of hers. "Lieutenant Shelley," she said, speaking quickly before he could protest her intrusion. "My codename is Delphi. You've been transferred to my oversight. I'll be your handler tonight."

His biometrics, already juiced from the ongoing operation, surged even higher. *"What the Hell?"* he whispered. *"Did you people get rid of Hawkeye in the middle of an action?"*

"Hawkeye took himself out, Lieutenant."

Karin remembered her earlier assessment of Sarno's breakdown. *He had made a mistake.* What that mistake was, she didn't know and there was no time to work it out. "I've got an overview of the situation and I will stay with you."

"*What'd you say your name was?*"

"Delphi."

"*Delphi, you see where I'm going?*"

"Yes."

He scuttled, hunched over to lower his profile, crossing bare ground between leafless thickets. Shooting was almost constant, from one side or another, but so far he'd gone unnoticed and none of it was directed at him.

Karin studied the terrain that remained to be crossed. "You're going to run out of cover."

"*Understood.*"

A wide swath of open ground that probably served as a pasture in the rainy season lay between Shelley and the shattered farmhouse. He needed to advance a hundred meters across it to be within the effective range of his grenade launcher. There were no defenders in that no-man's-land, but there were at least eight insurgents sheltering within the remains of the farmhouse—and the second truck, the one with the machine gun, was just out of sight on the other side of the ruins.

She fanned the windows just as the lieutenant dropped to his belly at the edge of the brush. Bringing Shelley's details to the top, she checked his supplies. "You have two programmable grenades confirmed inside your weapon. Ten percent of your ammo load remaining. Lieutenant, that's not enough."

"*It's enough.*"

Karin shook her head. Shelley couldn't see it; it was a gesture meant only for herself. There weren't enough soldiers in his squad to keep him out of trouble once the enemy knew where he was.

Would it be tonight then? she wondered. Would this be the night she lost someone?

"I advise you to retreat."

"*Can't do it, Delphi.*"

It was the expected answer, but she'd had to try.

Nervous tension reduced her to repeating the basics. "Expect them to underestimate how fast you can move and maneuver in your exoskeleton. You can take advantage of that."

The shooting subsided. In the respite, audio pickups caught and enhanced the sound of a tense argument taking place at the distant farmhouse. Then a revving engine overrode the voices.

Karin said, "The other truck, with the machine gun, it's on the move."

"*I see it.*"

A check of his setup confirmed he had the feed from the surveillance drone posted on the periphery of his visor display.

He used gen-com to speak to his squad. *"It's now. Don't let me get killed, okay?"*

They answered, their voices tense, intermingled: *"We got you . . . watch over you . . ."*

Valdez's window-set centered, cutting off their replies. *"Delphi, you there?"*

Her voice was calm, so Karin said, "Stand by," and swiped her window-set aside. *". . . kick ass, L. T."*

Shelley's window-set was still fanned, with the live feed from the surveillance drone on one end of the array. Motion in that window caught Karin's eye, even before the battle AI highlighted it. "Shelley, the machine-gun truck is coming around the north side of the ruins. Everybody on those walls is going to be looking at it."

"Got it. I'm going."

"Negative! Hold your position. On my mark . . ." She identified the soldier positioned a hundred-fifty meters away on Shelley's west flank. Overriding protocol, she opened a link to him, and popped a still image of the truck onto the periphery of his visor. "Hammer it as soon as you have it in sight." The truck fishtailed around the brick walls and Karin told Shelley, "Now."

He took off in giant strides powered by his exoskeleton, zigzagging across the bare ground. There was a shout from the truck, just as the requested assault rifle opened up. The truck's windshield shattered. More covering fire came from the northwest. From the farmhouse voices cried out in fury and alarm. Karin held her breath while Shelley covered another twenty meters and then she told him, "Drop and target!"

He accepted her judgment and slammed to the ground, taking the impact on the arm struts of his exoskeleton as the racing pickup braked in a cloud of dust. Shelley didn't turn to look. The feed from his helmet cams remained fixed on the truck parked between the ruined walls as he set up his shot. The battle AI calculated the angle, and when his weapon was properly aligned, the AI pulled the trigger.

A grenade launched on a low trajectory, transiting the open ground and disappearing under the truck, where it exploded with a deep *whump!*, enfolding the vehicle in a fireball that initiated a thunderous roar of secondary explosions as the rocket propellant ignited. The farmhouse became an incandescent inferno. Nightvision switched off on all devices as white light washed across the open ground.

Karin shifted screens. The feed from the surveillance drone showed a figure still moving in the bed of the surviving truck. An enemy soldier—wounded maybe—but still determined, clawing his way up to the mounted machine gun. "Target to the northwest," she said.

The audio in Shelley's helmet enhanced her voice so that he heard her even over the roar of burning munitions. He rolled and fired. The figure in the truck went over backward, hitting the dusty ground with an ugly bounce.

Karin scanned the squad map. "No indication of surviving enemy, but shrapnel from those rockets—"

"Fall back!" Shelley ordered on gen-com. Powered by his exoskeleton, he sprang to his feet and took off. *"Fall back! All speed!"*

Karin watched until he put a hundred meters behind him; then she switched to Holder, confirmed his ambush had gone off as planned; switched to Deng who was driving an ATV, racing to cut off her own insurgent incursion; switched to Valdez, who had finally joined up with another squad to quell a street battle in an ancient desert city.

■   ■   ■

"Delphi, you there?" Shelley asked.

"I'm here." Her voice hoarse, worn by use.

Dawn had come. All along the northern border the surviving enemy were in retreat, stopping their exodus only when hunting gunships passed nearby. Then they would huddle out of sight beneath camouflage blankets until the threat moved on. The incursion had gained no territory, but the insurgents had won all the same by instilling fear among the villages and the towns.

Karin had already seen Valdez and Holder and Deng back to their shelters. Now Shelley's squad was finally returning to their little fort.

"Is Hawkeye done?" he asked her.

She sighed, too tired to really think about it. "I don't know. Maybe."

"I never liked him much."

Karin didn't answer. It wasn't appropriate to discuss another handler.

"You still there?"

"I'm here."

"You want to tell me if this was a one-night-stand? Or are you going to be back tonight?"

Exhaustion clawed at her and she wanted to tell him *no*. No, I will not be back. There wasn't enough money in the world to make this a good way to spend her life.

Then she wondered: when had it ceased to be about the money?

The war was five thousand miles away, but it was inside her head too; it was inside her dreams and her nightmares.

"Delphi?"

"I'm here."

In her worst nightmares, she lost voice contact. That's when she could see the enemy waiting in ambush, when she knew his position, his weaponry, his range . . . when she knew her clients were in trouble, but she couldn't warn them.

"You want me to put in a formal request for your services?" Shelley pressed. *"I can do that, if you need me to."*

It wasn't money that kept Karin at her control station. As the nightmare of the war played on before her eyes, it was knowing that the advice and the warnings that she spoke could save her soldiers' lives.

"It's best if you make a formal request," Karin agreed. "But don't worry—I'll be here."

"HERE-AND-NOW," BY KEN LIU

STORY FRAME

"It's amazing what you can get, just by asking."

How much is information worth? That is the question that Aaron, the protagonist of Ken Liu's "Here-and-Now," is forced to confront over the course of one complicated afternoon and evening. Aaron is one of thousands, if not millions, of people using the new app Tilly Here-and-Now, which allows its users to put in anonymous requests for "information" of any and all kinds. This story poses deceptively simple questions about why information matters. It also points out that some kinds of information are much more meaningful or valuable to some people than to others and asks us to consider whether that difference should matter, and how.

Aaron begins his afternoon with a wholly positive outlook on the Here-and-Now app, which makes him feel more connected to his community and offers him a fun, gamified way to help others that also comes with a monetary reward (piquantly called a "bounty"). But after coming into contact with a couple of troubling requests, Aaron begins to worry that the app makes it all too easy for a user to invade the privacy of others. In fact, we as readers might wonder whether Aaron himself unwittingly helped violate others' privacy, when fulfilling requests for unknown requesters for unknown purposes.

The world of this story is not quite the same as ours, but it is similar in many ways. It appears that Centillion has achieved a mastery over data that has not yet been achieved in our world, but the possibility is certainly on the horizon. Likewise, nothing exactly like the Here-and-Now app exists yet (at least not as of 2013), but it is a plausible amalgam of many apps and services that do exist: TaskRabbit, Pokémon Go, and Whisper. Still, we are fast approaching a world like the one in the story. It's not hard to imagine an app like this existing here, and now.

STUDY QUESTIONS

1. There are many essential ingredients in Tilly Here-and-Now's economy: money, and information, but also interest on the part of both types of users: the information-requesters and the information-gatherers. What are the sorts of interest that might lead someone to use the app, in either of those two roles? Are any of those interests in tension with one another?

2. Does it matter that Tilly's request function is anonymous? Why or why not?

3. What does the license plate number signify to Lucas? What does that same information signify to Aaron? What kinds of significance does Aaron worry that others might infer from the request for a photo of that license plate?

4. Aaron decides, early in the story, that "Tilly Here-and-Now made you more aware of the world around you . . . more connected to your community." How do the events of the story itself confirm or challenge that conclusion? Characters you can use to think about this question include Lucas, Aaron's parents, the unnamed people whose bounties are being fulfilled, the girls in the video that Lucas has purchased, and Aaron himself.

5. Who has access to the requests, and how is that access controlled?

6. Centillion describe its mission as "arranging all of the world's information to ennoble the human race." What definition(s) of "noble" would make this claim make sense? Now consider Aaron's experiences in this story, using the definitions of noble that you came up with earlier. How does Tilly Here-and-Now ennoble Aaron, and how does it degrade him? How does it ennoble or degrade the other characters in the story?

"HERE-AND-NOW," BY KEN LIU

"If you make it available, they will search for it."

—Christian Rinn, Centillion, Inc., Founder

Ding.

The other patrons in the library looked over at Aaron, annoyed.

He smiled at them apologetically and then took the phone out of his pocket. The Tilly Here-and-Now app's icon had a flashing border: a new message.

Request: I need to know what's in the Manager's Special bin at the Food Basket on Tremaine Road in Rockton.

Time Limit: 20 minutes. I have to know if it's worth it to stop by on the way home. Bounty: $1.00.

Aaron got up from the comfortable reading chair, where he had been restlessly flipping through the script-he had gotten a good part in the spring theatre production, his first! He didn't want to make a big deal about it or announce it to everyone, especially since his mother was sure to make a big fuss about it if she knew. But he was feeling pleased with his secret.

A dollar wasn't much, but considering he was only a few hundred feet from the Food Basket, and he wasn't getting anywhere with his real homework, it seemed more productive to try to claim the bounty. Besides, right now, he felt like he had boundless goodwill for everyone.

He had to hurry though. Tilly would broadcast the request to everyone who had the app installed within a half-mile radius of the store.

He dashed to the store and made a beeline for the Manager's Special bin. Ignoring the shoppers browsing around the bin, he carefully took a series of close-up shots of the items, almost all of them with a sell-by date of today. Then he hit the "Claim Bounty" button on the app and uploaded the pictures.

The app refreshed to show that the requester was satisfied and the reward had been deposited into his account. Aaron pumped his fist. The shoppers looked at him strangely. But one woman, who apparently understood the meaning of Aaron's antics, laughed.

"Congrats!"

"Thanks," Aaron said.

"My husband has it installed, too. I've never seen him so excited as when he gets a query."

Tilly Here-and-Now was Centillion's latest innovation. After devising ways to deep-crawl every database in the world and scanning books and mapping roads and parsing videos and podcasts . . . what was left to conquer? Since Centillion was dedicated to the mission of "arranging *all* the world's information to ennoble the human race" . . . enter Tilly Here-and-Now.

"Damn it!"

Aaron looked up from the stats screen and saw that the speaker was Lucas: lanky, always with messy hair, and . . . not exactly Aaron's friend. They'd known each other since kindergarten, but they always rubbed each other the wrong way.

"Not my fault that you silenced your phone." Aaron smirked. "Every second counts."

"How much have you made this month?"

"That's a little rude."

Truth be told, the money wasn't really the reason he liked using the app-his modest goal was to earn enough bounty in a month to pay for the upgraded data plan on the phone, a goal that he had been able to meet last month. It was more that Tilly Here-and-Now turned life into a game, a treasure hunt. It was such a rush to claim a bounty (and to beat people like Lucas!). He also liked the feeling that he was helping people, perfect strangers, by making their lives a little bit easier because he happened to be in the right place at the right time.

"You're just staring at that app all day because you have no life."

"Sure, if thinking that makes you feel better."

Aaron also had fun guessing at the motivations behind the anonymous queries. Some were easy: Who would want to know if the seat by the outlets in the coffee shop is free ("and if so, can you sit there till I get there")? Obviously, some hipster intending to come down with his laptop. Who would want a bunch of photos of the inside of the museum? Probably someone too cheap to pay the admission. But others were more mysterious: Why would anyone want a snapshot at the crack of dawn of the ruins of the church that had burned down last year?

And there was also that time when he had found the little girl who had been separated from her parents at the mall. He had gotten the request with her picture, looked up, and there she was, huddled behind some potted plants, too scared of the strangers streaming by. The tearful parents, when they followed Tilly's directions and found him trying to comfort their daughter, were ecstatic. He had felt like such a hero (and it was also nice to claim the $100 bounty).

Tilly Here-and-Now made you more aware of the world around you, he decided. Made you more connected to your community.

"I got something cool out of Tilly." Lucas had an evil grin.

Even knowing he was being baited, Aaron couldn't help asking, "What?"

Lucas held his phone up to Aaron's face. A grainy, shaky snippet of cellphone video played in a loop: two young women, about college age, were kissing in a dimly lit bar.

"Who are they and what is this?"

"This was at the Thirsty Scholar over in Riston. I was bored at home last night and offered 50 cents for a video of girls kissing from around here, and I got this within half an hour."

Aaron felt weird watching the video. Sure, they were kissing in public, but it still seemed wrong to intrude on this moment of intimacy. Who would surreptitiously film such a thing?

"Would have been even better if they're people I know," Lucas said. "Next time I'm going to raise the bounty and limit the range more. It's amazing what you can get just by asking."

That was it. Lucas the voyeur had offered a measly 50 cents, and someone had obliged him because it was convenient. Aaron didn't think the person who took the video would have done it without the pathetic reward. Just because there were cameras and GPS trackers everywhere didn't mean that everything could be found-until Tilly Here-and-Now. Lucas had been like the puppet master pulling a string, and the user had responded-

Ding. Aaron reached for his phone, and at the same time, an alert also appeared on Lucas's screen.

Request: Need someone to look in the parking lot of the Comfort Lodge in Rockton and see if they can find this license plate number and get a picture: XXXXXXX

Time Limit: Until 5:30
Bounty: $10.00
Aaron and Lucas looked at each other.

"I'm calling it," said Aaron.

"That's not how the game is played," said Lucas.

They ran out of the Food Basket, one after the other.

■ ■ ■

Aaron hurried through the motel parking lot, praying he would find the car before Lucas. Lucas was over at the other end of the parking lot, scanning the license plates.

He did have an advantage over Lucas: he had no need to read the license plates. The license plate number belonged to his father's car.

Why would his dad's car be in a motel parking lot in the middle of the afternoon on a workday? And who would want to find it and get a picture?

The answer was obvious.

He had to get Lucas out of here. *Then* he would decide what to do if he did find the car.

He took out his phone and began to type.

A few seconds later, Lucas took his phone out of his pocket and looked at it. Before he could look this way, Aaron turned his eyes back to the license plates, pretending to be absorbed in his task. When he looked back up a minute later, Lucas was gone.

He let out a held breath. He had offered a bounty of fifty dollars for someone to take a picture of every house with the number "27" in the new development near the school ("for an art project"). The gambit had worked.

He finished surveying the parking lot without finding his father's car. But that didn't make him feel better; maybe it just meant his father hadn't gotten here yet.

He didn't know what to do. He had thought things were fine between his parents. They didn't argue any more than the parents of his friends, and the idea of his father sneaking away in the middle of the day to a motel . . . he couldn't even finish the thought.

His mother had essentially crowdsourced a private detective?

And now the question was out there, and others who knew enough would also be able to put things together the way he had.

A search itself was also information. It was an expression of intent, of desire, fear, want, lust.

He wanted to know the truth. He wanted to ask Tilly just the right question. But what was the right question? How would he ask it without revealing what he knew? And what did he know? Really know?

One minute, he was angry at his father, the next, he regretted showing his mother how to use Tilly Here-and-Now last week.

He lingered in the parking lot, trying to figure out what to do until 5:30 came and the offer expired.

■   ■   ■

His father came home.

"How was your day?" his mother asked.

"Not very eventful," he said. "Had a good meeting with a client in the office."

He couldn't hear anything different in their tones. His mother acted like she had never asked the question. His father acted like he had nothing to hide.

"Congrats on getting that part," his mother said.

"What?" Aaron was startled.

"For the spring show," his mother said. "I'm proud of you." She had always boasted to her friends about how she managed to stay involved in Aaron's life even as he got older.

Aaron mumbled something noncommittal. The events of the afternoon had completely leeched away his joy. He knew that if he searched through the local Tilly Here-and-Now queries he'd find a query for someone who was around the high school to take a look at the posting outside the office of Mr. Septol, the Drama Director.

His parents finished dinner, chatted, did the dishes, and then went to bed. Aaron parsed everything they said and did for clues as to what they knew and what they only suspected.

He lay in bed, and theories multiplied in his head: Maybe his mother only suspected and needed proof; maybe his father was smart enough to conduct an affair without using his car; maybe he had also been close enough to the motel to get her search request on his phone and got away in time; maybe she knew he would know and the query was her way of warning him; . . .

It was dark in his bedroom, only faint starlight coming in through the window. But Aaron felt as if he lived in a house with transparent walls. He imagined Lucas looking in. He imagined the neighbors with Tilly's queries popping up on their screens. He imagined everyone he knew watching, knowing, cameras everywhere, invisible electronic threads crisscrossing the neighborhood. Life didn't seem like such a fun game.

He took out his phone, and the bright screen came to life.

Wasn't it strange how it was so easy to anonymously type questions into a text box, questions that he couldn't imagine asking face to face?

He pondered whether Tilly could tell him the truth. Or perhaps there was a blind spot that even Tilly could not see into.

"LACUNA HEIGHTS," BY THEODORE MCCOMBS

STORY FRAME

> How desperately unhappy do you have to be, Andrew thinks, to pothole
> your own mind like that?

Andrew, a lawyer for the neural implant company Aleph, has been struggling with his memory. There are gaps in time that he cannot account for, and an intense anxiety he cannot explain. Andrew knows he has been making use of his own implant's Privacy Mode to conceal things, from himself as well as others. Because the Aleph bundles a user's thoughts along with their search data, the only way to keep something truly private is to hide it even from the user themselves, except when they are in Privacy Mode. And so Andrew can't figure out what he's hiding from himself, or how to bring that knowledge into public mode where he can remember it all the time.

Andrew's working theory is that he is cheating on his wife Madeleine, and he spends much of the story trying to figure out how, and why, and with whom. But Andrew's therapist—who seems to have a fuller picture of what Andrew is going through—keeps trying to get him to think instead about his sister Medo. That's harder for Andrew to do, since Medo is not a part of his upper-crust San Francisco life and "doesn't live in the city—she doesn't live nearby." As we learn over the course of the story, there are a lot of reasons that it's difficult for Andrew to think about Medo, and the reasons themselves are part of what he is trying both to remember and to forget.

Amid Andrew's personal struggles, the story follows a lawsuit about Privacy Mode that Andrew, as Aleph's chief lawyer, is arguing in federal court. The government is prosecuting a money-launderer, and they are seeking to jailbreak the accused man's Aleph in order to find out whether he has been hiding illegal financial transactions using Privacy Mode. These courtroom scenes shed some useful light on the history and workings of the Aleph, including the fact that some other Aleph users are managing their memories in a way similar to Andrew. But on the whole, the San Francisco of this era, some decades in the future, is built upon many layers of forgetting, at a social level as well as an individual one.

This is a story you will almost certainly want to read at least twice. Our perspective on the story is tied very closely to Andrew, and specifically to Andrew while in public mode: the first time through, you will be as confused as he is. On second (and further) readings, however, you will be able to see coherence and clarity where Andrew himself cannot and fill in some of the gaps he experiences but cannot explain.

One question lies at the heart of ethics: what do we owe to each other? "Lacuna Heights" takes us into the head of one character to ask us to consider what we owe one another with respect to our own minds and memories, and how technology organizes our attention.

STUDY QUESTIONS

1. Where are the gaps in Andrew's memory that take place during the story? You will need to read the story at least twice to be able to notice most of them. What do we know to be taking place in those gaps, in the cases where it's possible to know? What can we reasonably infer about what is taking place in the other gaps?

2. What is the theory of Hell that Medo shares with Andrew on the beach in their teens, and how does it relate to Andrew's present?

3. What duties toward his family is Andrew fulfilling? What duties toward them is he neglecting? How are these duties related to one another?

4. What kinds of lifestyle choices are available to the residents of Unfrancisco, as opposed to those who live on the surface?

5. In what way(s) does Privacy Mode reconfigure the relationship between an individual Aleph user and the wider public/society? In what way(s) does the option of Privacy Mode help reveal the dynamics already in place between an Aleph user and the wider society?

"LACUNA HEIGHTS," BY THEODORE McCOMBS

"I think I might be having an affair," Andrew says.

He is a bulky man, and he sits hunched forward in the berry-red leather armchair, elbows braced on his thighs and his meaty hands dangling between his knees. The trickle of a motorized desk fountain is the only sound in the office. Clear, silky water over edgeless pebbles. Andrew's therapist crosses her glossy legs.

"We were talking about your sister," she redirects, waving her pencil in a circle.

"I mean I'm cheating on Madeleine," he says. "Maybe," he says. His therapist nods wearily and makes a note.

Andrew has been going to her every Monday for three months, for generalized anxiety. He hasn't improved, not noticeably.

The sky, in gaps through the bamboo along the window, is a strange, sickly green.

Every day, Andrew notices gaps in his memory. He knows someone brings him his dry cleaning and he looks forward painfully to seeing her, but can't picture her face, not quite. Every morning, he runs on the treadmill for twenty-five minutes and promises to push himself to thirty, as soon as he shakes off this cold—catches up on sleep—gets his neck thing fixed. And then he's at work. He understands he took a car but can't remember doing it. There are no cars on Market Street: none parked, none driving. It's quiet enough to hear sparrows moving in the heavy summer trees.

There is a type of dissociation that comes from overwork or exhaustion, his therapist tells him. But she looks embarrassed, even miserable, proposing this explanation.

Andrew Cornejo-Holland, at 47, is a tech illiterate. He hates his Aleph; the implant feels like a bullet lodged in his palate and gives him vertigo when it updates. He takes perverse pride in his indifference to neurotech—perverse, because as a law partner at Morrison, O'Melveny & Xi, he runs the Aleph Corp client account. As a litigator, he can afford to leave the technical knowledge to the patent group; he can afford a lot, now, including a life in analog. No ads flickering dreamlike in his mind's eye while he jogs. No prick of conscience from the GPS navigator when he turns a wrong corner; he can't even switch on GPS.

The Aleph Corporation isn't offended; it likes Andrew's work and him personally. The General Counsel has him and Madeleine over to Los Altos every month, and every month, the GC ticks his head toward his second floor, saying, "Wish you two had wanted kids," like he'd wanted his to play with or marry theirs. Andrew always freezes and looks to Madeleine's severe beauty for reassurance, and Madeleine laughs and wiggles her fingers: "Oh, it would ruin me!" like having kids does something mysterious to your hands. She plays the flute for the orchestra and hardly uses her neurotech because she's afraid it will spoil her ear.

He loves her, he is certain: in the serene, cordial way that long-married couples love each other. He can't fathom what else he'd want.

"How do I take a snapshot?" Andrew asks her, leaving the GC's stupendous front porch. He runs his fingertips along her bare shoulder. "I want to keep this, always."

Los Altos is dark, clouded. Everything smells of the dry eucalyptus trees, which rustle in a salty, sea-acid wind.

"Picture me bathed in a powerful green light, then blink," Madeleine says. "Like Kim Novak in *Vertigo*."

He smiles. Andrew is the sort of man pleased to confirm that old movies are still in the world, that things his grandparents loved survive in a dizzying, painful new century—when so much has been lost. Drowned. Buried. A long time ago.

Madeleine smiles for the Aleph: a tipsy crinkle of her lips. The front door opens; the GC stands backlit against the houselights.

"That subpoena, Andrew," the GC reminds him, then shuts and locks the door.

"Blink, Drew," Madeleine reminds him, still holding her smile.

The Aleph Corporation has been subpoenaed in a federal money-laundering investigation: An Iowa City bank executive is suspected of hiding reportable transactions in his Aleph's Privacy Mode, and the feds want Aleph Corp to jailbreak it. The case couldn't have come at a worse time for the company, when it's about to roll out a mid-priced implant to expand its consumer base.

Privacy Mode is a sacrosanct feature of the neuroimplant—at least, in theory. In practice, it's unusable, but consumers are still attached to the idea of it. Privacy is self-determination: You get to choose what to disclose about yourself, and to whom, and when, instead of letting the data-miners and ad-slingers haphazardly clap your profile together for you. The Aleph bundles your search data and sells it, that's nothing new. But it also bundles your thoughts. It's unavoidable—the Aleph's search engine displays its results using the same neurochemical processes as waking memory. The implant literally can't tell the difference between an organic memory and a search result, projected into the user's hippocampal recall, so it scoops up and transmits both to Aleph's ad partners. There are strict security controls at the corporate level, but people still like to know they can enter Privacy Mode and keep some thoughts totally secret.

The trade-off, the one that makes Privacy Mode impractical, is that once they exit Privacy Mode, they can't access any thoughts or memories they had while using it.

The federal court building, renovated, looks nothing like its old self; its clean strangeness makes Andrew unaccountably anxious. He is an anxious person generally, but a canny litigator. In front of the judge, the Department of Justice lawyer, Mr. O'Connor, overplays his arguments. He claims Privacy Mode is used *only* to shield nefarious activities from scrutiny. Andrew assumes a genteel dignity and shakes his head.

The senior judge cocks a brow at *nefarious*. She is older than Andrew, even more baffled by neurotechnology. She gets angrier and more confused, until she cuts the both of them off. She orders more briefing on the data subpoena, but in the meantime, she grants Mr. O'Connor leave to depose someone at Aleph on how Privacy Mode works.

Andrew exits the courthouse and crosses Civic Center Plaza, weaving through its ranks of pollarded sycamores in a muddy fog. The pruned branches, thickened into knobs of tree flesh, look like fists or lampposts. In the distance, he makes out a vague shape vaulting up the stairs of the old BART station, carrying delivery bags, pitching port and starboard. Andrew thinks, *This is wrong.* The BART entrances aren't BART anymore. The BART tunnels flooded a long time ago. A long time ago. The new system is just "U," for "Under." He goes down there every day, but he can't picture it. Pain under his lungs—why? He thinks hard, but the thought is already slipping from him.

"This is what I'm going to do," he tells his therapist. "I'm going to find out who this other woman is, I'm going to end things, and I'm going to apologize to Madeleine. I have to do the right thing. I'm just not that kind of guy."

The therapist takes a deep breath through her hands. "If you're not going to engage with treatment," she says, "what's the point of you coming? We need to talk about your sister."

"I want to talk about Madeleine," Andrew says. He fishes a pebble out of her fountain and thumbs at it. "It's *my* anxiety—I think I know what I'm anxious about."

"What does Madeleine represent to you?"

"What?"

Today, his therapist wears a slate pantsuit with wine-red pumps. Is it just him, or has she been vamping it up these last few weeks?

"You like her," she says. "I accept that. But she's classy, and old-money, and we both know that means something more to you."

Andrew's eyes narrow. What if, what if he's having an affair with his therapist?

What if Andrew is not so tech illiterate as he claims? What if he's actually quite good at using the Aleph, at least when he's using it for Privacy Mode, to keep his affair secret even from himself? This is what scares him about neurotech: We've outsourced so much of our selves to an array of companies, so their algorithms shoulder the hard work of living—and now he's doing it himself, to outsource the planning, lies, and guilt of adultery to this other, shadow self. And if he doesn't figure out what's going on, soon, he'll lose Madeleine and this world he has with her.

How can he test if he's doing it? An implant-repair tech—some independent neurodentist, nothing Aleph.

He wants to buy Madeleine something. Just in case. Alone in his office, he fumbles the Aleph glasses over the bridge of his nose and jaws a bit, thinking, SHOW ME GOLD, in lime-green letters. His mind cramps with memories: bullion hoards inside Fort Knox; gold grains in a black pan; rings; coins; burial regalia in a pyramid; a gilt guitar; champagne-gold wedding shoes. They come as apparitions, as emotions he's felt before. "Stop!" he says aloud. As if he's seen before in a dream those wedding shoes. If he concentrates, he can tell the difference. It doesn't taste like a real memory: It has that sweet tackiness of wallpaper glue.

He can't stand the idea of all this foreign stuff dumping into his mind. He can't fathom how everyone else does.

He pulls off the Wi-Fi antennae, camouflaged as ordinary eyeglasses. Strange, chilly tears are leaking from his eyes and he has an ice-cream headache. Andrew hates his Aleph with the infatuation of a great love.

After work he walks home; the condo on Laguna Street is just thirty minutes from his last client meeting, in Russian Hill. He scrapes along the yellow center line of Leavenworth. A lonely seagull calls from behind a crenellation. It's an easy walk and he snags on why the neighborhood is "Russian Hill" when there's no incline. Nob Hill, Russian Hill, Noe Valley—don't the names suggest a rolling city? Of course San Francisco has no hills, he thinks. Except in old movies.

Andrew pauses in the middle of the empty street. There *are* hills in the old movies—in *Vertigo*, those long, Herrmann-scored driving shots.

He takes out his Wi-Fi glasses. SAN FRANCISCO HILLS: He concentrates on the words in flickering sea-green. And there were gulls all along these piers, once.

At home, Madeleine lifts groceries out of cloth tote bags set on the kitchen island. "I'm going to attempt a paella," she announces grandly. There's no seafood worth eating anymore, but she has the recipe for a chorizo paella. Andrew looks intently at the totes, thinking, *Are those our bags?*

He asks Madeleine, "Why is there no hill in Russian Hill?"

Madeleine lowers her hands into a bag with her shoulders at her ears, her narrow, birdlike chest caved. She's disappointed in him, more so every day.

For the second time that week, he discovers his feet heading for an old BART tunnel and stops himself. The entrance is a yawn in the ground, a tongue of stairs. It's well lit but darkens as it descends, the golden light shading to deeper, dead ambers, the steps' shadows lengthening and sharpening. Beyond, a dark that isn't dark at all, only a sour colorlessness. "U" for Under. "U" for Under, Under, Under.

The Aleph deposition proceeds at the DOJ branch office on Golden Gate Avenue. Mr. O'Connor grills Aleph's chief technology officer on Privacy Mode and how the data gets stored.

Essentially, the implant tweaks how the hippocampus indexes memories for recall: It adds a special indexing "key" to the memory so that it's "lost" to recall outside of Privacy Mode, but retrievable inside it. Within Privacy Mode, the user has access to a whole world of memories, even old memories, that's walled off from recall otherwise.

"But why?" O'Connor asks. "Why isn't Privacy Mode designed so it's just, while you're in it, the implant stops transmitting anything up to Aleph?"

"Then the data-bundling protocols would scoop up a 'private' thought as soon as the user remembered it in normal mode," says the CTO. "Privacy Mode wouldn't really be private, then."

In a corner of the room, the Iowa bank executive and his defense attorney listen and take notes. Andrew has trouble hiding his revulsion. The banker is pasty, rumpled, stupid with greed, scared of his own shadow. He stinks of deodorant wearing off. How desperately unhappy do you have to be, Andrew thinks, to pothole your own mind like that?

Aleph's CTO isn't thrilled to disclose proprietary neurotech protocols, but the deposition is sealed and, on the whole, Andrew leaves Golden Gate Avenue satisfied. He's eager to see the transcript; he has an undignified pleasure in reading himself. But it'll be a few more days before the transcript is ready, and meanwhile, the brief on the data subpoena is more important.

It's only as he turns down Larkin and checks his internal clock that he realizes it's an hour later than he thought. Andrew runs over the deposition in his head carefully. There's a foggy, inaudible patch halfway through—a gap—and he realizes he must have, somehow, and for some reason, entered Privacy Mode *during* the deposition.

"Sometimes I wish I were crazy," Andrew tells his therapist.

She folds forward. "When's the first time you recall wishing that?"

"Law school." Andrew digs for the memory, and he's relieved to find it largely intact. "I got into Hastings here in the city, but then it closed, it closed because . . ." He frowns. "That's not the important part. Berkeley was still open, then, and it took a dozen of us from Hastings, which always made me feel I hadn't really earned being there. So much pressure and stress—at the time, I was working to send money home to my parents, and one day I'd been going on no sleep, I was so afraid I'd fail my classes, and the prof just nailed me to the wall for coming in unprepared—it was too much—I got up and just started walking, down the steps, out of class, under all the dying gingko trees, down, down, down the hill. There was this roar of drilling machinery, or—no, they were pumps, in the distance, there were huge platforms like war machines, and networks of pipes and engines, and a droning that filled up my body. I kept walking and thinking, *I should have a nervous breakdown, I hope they lock me up, then I won't fail.* How do you like that? That's what I kept fixating on. If I'm crazy, I don't have to take the final, and can't fail."

Andrew's hair products are waiting for him on his doormat in a coy little blue paper bag. It's his special stuff, for receding hair. System 2: Noticeably Thinning. "Oh," he says, delighted. He stops, and looks around, baffled; he was literally, literally *just* with his therapist.

Andrew plants himself at his dinner table with a legal pad and a plastic hotel pen. The dusk outside is a murky blue, and inside, the living room furniture darkens into strange densities. He can't remember how to enter Privacy Mode; the transition itself gets key-indexed along with the protected memories, so he has to look up the instructions each time. No matter what, Andrew promises, he'll write down what he remembers inside Privacy Mode. Nothing is worth keeping a secret from himself, wondering around it. He finds the instructions with his Aleph, and now he's always known what to do.

First, he notices the changed light—now, the living room lamps are all on, it's as bright as day. His pen is snapped in half and his hands are covered in oily, blue smears that fill in the fine crack-work of his palms. Andrew is shaking. In the living room, Madeleine perches on her black music chair, laying her flute in its velvet case. The legal pad in front of him is blank.

"Did I say anything?" he asks her, but she only shuts the instrument case. She rises and carries off a tied stack of folded laundry left by the front door.

The pad is blank. Andrew is flouting himself. Is he really his own enemy?

He sets up a camera on the bookshelf but realizes this is stupid—he'll just turn it off when he's in Privacy Mode. Andrew can't think of anything else to do, though. Madeleine looks offended—he can't explain to her that he suspects himself. He claims to think someone's breaking in. He has that feeling again like there've been strangers in their apartment, but they both know *that feeling again* just means anxiety.

"When's the last time you called your sister?" she asks. Andrew nods: He should have Medo hide the cameras. Then his other, shadow self won't find and uninstall them.

But Medo doesn't live in the city—she doesn't live nearby—she drives a rideshare, runs odd jobs, "gig economy" stuff—so the idea makes no sense after all.

He dreams, that night, of thronging shadowy crowds he hasn't seen since he was a child, pushing through narrow streets under a sky that seems impossibly close, like he could knock his head on the sun. He's looking for his sister, Remedios—calling for her over the heads of people he can't make out because they're so close, "Medo! Medo!" He finds her where a depression in a brick wall makes a pocket of space. She's standing with her face in the corner and when he turns her, her skin is raw, sunburned, blacked with old coal soot. Medo recognizes him breathlessly, dazed, ready to faint. She's been facing into that wall for days.

He loves Madeleine, more or less: He can hear her trailing him, her high heels grinding on the flagstones: "Andrew, please slow down." She's in a silk gown slit to the knees, he's in a tux, and the others heading to the War Memorial Opera House are likewise in gala dress. The premier is *Götterdämmerung*, but underwater. A new production. Madeleine has played the leitmotifs on the piano, with a look of solemn instruction, so he'll recognize them during the show. He does love her, but is he dissatisfied? Is he already losing her? What about his life doesn't he like?

On the borders of the plaza, through the strange lamppost sycamores, uncertain shapes run past, carrying backpacks or cloth bags. He can't quite see them; the light is too vague, mixed, the dusk blues polluted by the giant ad screens on surrounding rooftops. The opera patrons flash lurid red and voluptuous purple. What, he thinks, doesn't he like about his life?

Maybe he's not having an affair; he considers his missing hour at the deposition. Maybe there is a part of the litigation so confidential—a trade secret so valuable—that Aleph set him up to forget it when he's not actually working on it. He knows that's possible, to set up an automatic trigger. It'd be ingenious security. He'd recommend it to them; but not on himself.

Would they have done it without his consent? Now there's a lawsuit.

Civic Center Plaza has three-story advertisement screens that flash into the night, like those that still paint Times Square. New York is gone now; most coastal cities are. Couldn't adapt to the rising oceans fast enough: a lack of vision, fleeing tax base. The Port Authority runs gondola tours through the drowned city: Park Avenue by moonlight, black water stippled in white streaks up to the Grand Central Mercury; Times Square still flashing, its iconic, obnoxious ads preserved. Reflected images burst across the water, royal blues and lipstick reds, sunken TV shows and drowned Broadway revivals, beautiful, obsolete faces, washing over shadowy tourists standing astonished in their boats. Civic Center Plaza can't rival Times Square—Andrew's not sorry for it, but he does morbidly sympathize with that instinct to honor humanity's pinnacle achievement in bad taste.

An Aleph ad comes on, two giant human eyes carefully searching the plaza. Their irises' striations, glacier-blue, stream binary digits. KNOW EVERYTHING, the ad says. Not unfriendly, those eyes, but so big and godly they can't not be ominous. Andrew will have a talk with Aleph about the ad, it's just awful. He cranes his head; he sees one or two others in the plaza staring at the words, craning their heads in the same expression of shapeless regret.

In the theater, the sopranos toddle out in diving suits and sing by opening their faceplates. Madeleine whispers, "This is so stupid, I'm so sorry." Andrew is sobbing helplessly.

He goes to get his implant checked. It's lodged up at the back of his soft palate, sending filaments into his brain from underneath. "I never use it," he tells the neurodentist, "but I think it's malfunctioning, going into Privacy Mode when it shouldn't. Can't you turn that feature off?"

The dentist looks knowingly at him in the weird aquarium light. Her whole office is under sea level: a design fad in offices along the Marina District piers, turn a problem into an aesthetic. The floor dances in wraiths of gray-green light. One wall is built of thick glass and through it, white fronds and scraps suspend in dead water.

The neurodentist has three instruments in Andrew's mouth when she leans close. "I can't say you're the only person who uses it like this, Mr. Cornejo-Holland, but it's not a good idea."

Andrew's eyes grow wide; he can't swallow, but gurgles something like a nervous laugh.

"There are things you need to know," she says.

He'd been walking along the old beaches south of San Francisco with Remedios when she told him her vision of Hell. This was near thirty years ago, before those beaches disappeared into the rising, dying ocean. Remedios was thirteen and dressed for swimming, but she wrinkled her nose when they saw the ocean's color—that day, it was lavender-gray, and its waves slopped viscously, like liquid velvet. The smell, too: salt rot, dead kelp. Andrew (Andrés, then) had come from his community college, having survived his first moot with the debate club, and still wore a short-sleeve button-down and tie like a Latter-day Saint, but carried his shoes in his hand as they walked. "What I want Hell to be," Medo was saying—because she had a mind like that, an intellect that felt entitled to grab into life's most difficult crevices—"What I want Hell to be, is the perfect knowledge of everything you did in life and what happened because of it, good or bad. You feel exactly how much you hurt or helped people. All the pain or joy you caused, you feel it yourself."

At their feet, and stretched out as far as they could see, were hundreds and hundreds of spiny pink prawns, dead and washed ashore, such that their corpses traced the morning tide line. Medo stepped around them, her face antagonistic and precise. "That way, you're punished if you did mostly bad things," she said, "and rewarded if you did good. See? It's elegant. But you'd have to know *everything* you did. Like, the face you made at the bodega guy and it made him feel like dirt, so he beat his kid that night. I guess we're in for some shit, for driving here."

He can't remember what he said, but he remembers looking back at his car parked by the carbon-poisoned beach—the last car he ever drove. And he remembers—Medo had to shout over some awful machine drone that carried down from the city. That crashing, that grinding, a chorus of generators, engines, cranes, jackhammers, backhoes, and pumps, grinding, groaning, growing.

The city is storming when he finishes at the office: one of the new San Francisco storms that tear the sky into pieces and throw it into the bay. Hail, thick as eyes. Diseased rain vomiting into the water, sending waves crashing against the Presidio's sea walls. Andrew is drenched and totally wired as he splashes into his condo lobby. In the bedroom, Madeleine is sleeping, her ears plugged with orange plastic that looks like blood in the storm-light. Andrew, too buzzed to sleep after finally finishing his brief, strips down to his underwear and pads barefoot on the treadmill. The storm thrashes outside and he's giddy with the thrill of change. Everything is new, he thinks mysteriously. Everything is fast and risky, and different. He feels his heart breaking and his cock plumping at the same time, but he's high on adrenaline and accepts his middle-aged body's imprecision as part of the new world's logic.

Madeleine stirs and turns on the lamp. Its glow falls over Andrew's large feet clapping the treadmill. "Oh, hey," Andrew says, grinning. "Can't sleep?"

Madeleine jackknifes out of bed and crosses to the window. Weather always upsets her.

"I'm going to win this one, Mad," Andrew declares, and he turns up the treadmill speed so that he's running. It's nearly two and his eyes are sore for lack of sleep, but everything in him is driving forward. *Clap! Clap! Clap! Clap!* He outruns the waters pooling through Laguna Heights, swamping gutters, sending up death smells. "The feds don't have a chance in hell."

Madeleine smooths her hair down the back of her head. "Come to bed?" she says. "I've got a pill if you need it."

"Need it?" Andrew laughs. "Hey," he says, "hey, check the left breast pocket of my suit jacket. Something came for you-*u*." He never knows what to do with Madeleine's remote moods; she'll never explain, never say what's gnawing at her, so he'll play it into a kind of marital caper.

But Madeleine doesn't smile exasperatedly, or even angrily wave him off. She slumps, and a look of dread steals over her face. She retrieves the box from his suit and finds the gold spangle earrings he's bought her. "Oh, oh sweetheart," she says mechanically.

"You don't like them?"

"You know I love them." She sighs. She crosses the room and tosses them inside a drawer without looking. Her face in the storm-light takes on a new bitterness, even rage. "Andrew!" she says. "Get off the treadmill. It's crazy! You're insane, do you realize that? No one does this!"

Andrew slows the treadmill, and they fight. The details of the fight, like all their fights, are hazy and he recalls only the broad fact of one the next morning as Madeleine leaves for the conservatory. The sky has stopped convulsing and now hangs in a sort of stupor. Andrew goes to the bureau drawer—he can't remember the fight, but he remembers her picking that drawer—and inside he finds a dozen identical pairs of gold spangle earrings. His mind runs into a wall.

His therapist looks at him blankly, and then her face creases into impatience.

"You're not telling me things," Andrew says.

She rolls her eyes slowly around the room. "You need to start taking responsibility for yourself," she says. She raps her pen on her palm.

"Why won't you be straight with me?" He adds stupidly, "I'm paying you."

Her glance skewers him. "You're paying me *not* to tell you, Andrew."

"There are *lacunae*," Andrew says, and immediately regrets it. The judge scowls from high on her bench. She looks like a crab from this angle. Salty, hard-shelled— unimpressed with Latin. "Gaps," Andrew clarifies, "in the law."

There is no law to cover a federal investigation subpoenaing witnesses' private thoughts. A witness can't be forced to self-incriminate and memories locked away in Privacy Mode cannot be subject to compulsion. That's the logical extension of the Fifth Amendment, although Andrew acknowledges the Framers didn't provide for exactly this scenario: so, lacunae.

Behind him, people are entering and exiting the courtroom, and the heavy wooden doors creak and slam and slam.

The judge orders the data subpoena quashed. She reads off her ruling from the bench, an impassioned defense of civil liberty. They'll teach it in law schools next year; Andrew might even get an award from the ACLU. But leaving the courthouse, Andrew has no memory of the order, only an impression. He feels disappointed, even appalled, although he can't tell whether it's the forgetting or the judge's order itself that's made him feel this way.

A tickle of recognition tells him the deposition transcript is ready for him. It's beside the point now, but he still wants to read it, and he fishes in his breast pocket for his Wi-Fi glasses.

His fingers brush something delicate; it's the gold spangle earrings—which, of course, he doesn't remember putting in his pocket.

Before he knows it, he is again at the old BART entrance in UN Plaza.

The tunnels are flooded, and yet his steps are so certain. He knows what's down there—a part of him, inaccessible to him, knows. He can remember, now, the sounds his shoes have made on the stairs, the special clap of his soles on the rubber traction strips. He's been down there; he's been down there a lot. He recalls the strange, stale, cool air, the aftertaste of metal and grease and filtered air. But it's no subway, he remembers. "U" for Under. The BART entrance isn't anything like a BART entrance, not really. It looks like one, but that's just nostalgia.

DEPOSITION TRANSCRIPT (p. 118 of 253)

Q: (by *MR. O'CONNOR*) But what legitimate use could that have? Criminal activity—that's the obvious utility.

A: (by *WITNESS*) Well, um, trauma, for example. We've developed relationships with providers and it's very effective in managing, um, intrusive memories in treatment. So, like an abuse that's—the patient and his therapist decide that's a memory that needs to be walled off and, um, reintroduced in a structured way. That's my understanding of it.

Q: You're not a mental health expert yourself, correct?

A: No.

Q: You're not aware that, in fact, blocked memories are a symptom of PTSD? A pathology, not a treatment?

A: I am not aware, no.

Q: And you don't require a therapist or a doctor to sign off on such a use of Privacy Mode. Anybody can wall off any memory or even any subject matter they want to hide and it'd be inaccessible in an investigation.
(by *MR. CORNEJO-HOLLAND*) I'm going to have to object to that. Compound. Calls for a legal conclusion.

Q: (by *MR. O'CONNOR*) I'll rephrase. Let's say—OK, let's say you want to forget what happened to San Francisco. It's too quote-unquote traumatic, so just forget it.
(by *MR. CORNEJO-HOLLAND*) Objection.

Q: (by *MR. O'CONNOR*) The flooding, the refugee crisis, just too painful. Put it away. The food shortages. The riots. Let's forget all of Unfrancisco itself. Forget it exists. You could do that, right, without a doctor's approval? Set it up yourself?

(by *MR. CORNEJO-HOLLAND*) Objection. Relevance. Speculative. Compound. Argumentative.

Q: (by *MR. O'CONNOR*) I want the record to reflect that deponent's counsel is using an extremely loud voice and striking the table very aggressively and violently.

(by *MR. CORNEJO-HOLLAND*) Objection. Objection. Objection.

So down again, into Unfrancisco, into the undercity. As I hurried into the phony BART tunnel, my legs remembered before I did and carried me into a ride-share. The Under-Plaza was scrambling with giggers, running deliveries across town in canvas bags, or pedaling on bicycles with groceries in the basket, to cook a dinner for some rich jerk like me, or dashing up the stairs waving dry-cleaning bags like a flag. I got a ride to Noe Valley from a sunken-eyed kid with a hipster chignon and wisps of blond beard floating around his chin. With me in the back seat were grocery deliveries for some next gig: a fresh baguette that filled the electric car with the smell of hot bread, a giant jackfruit that must have weighed twenty pounds and cost more than that day's commissions. These were the people I'd forgotten, daily, nightly, the poor kids running the city out of the corners of my eyes. I looked back through the rear windscreen to see the city I knew as a boy: hills, dipping down into asphalt ravines, rising right into the false sky; historic buildings' porticos and paneled double doors, their higher floors refitted to open identically onto a new surface city; knolls of fake grass where parks had died sunless; false maples and ginkgos hiding air-filtration machinery and broadcasting ads for soap, sitcoms, or vacations always a hundred more gigs out of reach. Under their chatter, the churn of the pumps in the seawalls keeping the swollen bay from flooding the undercity.

It had all happened so slowly, and still so fast: the seawalls built higher, higher—no one really knew how high they'd have to build to keep up with the Big Melt. Roof gardens became roof lawns, roof parks, but the rich wanted sun and just kept building, building, higher, higher—the Big Lift, we called it hopefully. No one was so tasteless as to call it the Rising Tide that Lifts All Boats, but we all kind of thought that.

The news shoved images at us: long flotillas of bodies lacing drowned shantytowns.

Forgetting is a human right. That's what the judge's order said. I remembered now: *All progress contains the decision to not look backwards. Our State of California cannot exist except by forgetting the claims of the people who were here before us. The peace of mind we struggle to keep, in this terrible present, depends on a certain mental distance from those who didn't survive the crisis, or who now survive so miserably. There is moral failure here, no doubt. But the moral imperative, when wielded by the arm of the law, too easily slips into tyranny, and this Court will not midwife that tyranny to pursue one alleged criminal.*

The government can no more compel a citizen to remember a particular past than it can force them to think a certain thought. Our thoughts determine who we are as individuals of conscience and expression; so do our failures to think. The government has no say in either—so rules the Court.

In front of Medo's building, in front of her door, I was already crying. I remembered, I cried every time I stepped into Unfrancisco, for what had been lost, who had been buried, and the decade of displacements that created this sunny underworld. It was Medo who pointed that out to me: how everyone looked hatefully at the false blue sky, because every day can't be cloudless, a native San Franciscan needs the gloom of summer to believe their city is still theirs.

Medo and Mamá were both home. My sister was icing her knee and checking her phone for gigs to take, but Mamá lurched to her feet with a cry of joy that broke me. She hobbled to me, arms wide, telling me about the show she was watching, beautiful doctors solving mysteries, and without missing a beat, she reached up with her gnarled thumb and smudged a tear crawling from my eye.

These are for you, Remedios. I showed her the earrings I'd bought for Madeleine. I'm going to put money in the account before I head back, too, but I wanted you to have something fun. I wanted badly to see her smile, but she stayed in her chair, said thank you, you're such a great brother, Andrés, but with one eye still on her phone. For the hundredth time, I prodded myself to ask them up to the surface to live with me and Madeleine, up to that new, foggy, flat Francisco—residency quotas be damned— but I'd already asked, they'd already answered no, how could we live up there knowing what's down here, and I had been so relieved, I didn't dare press it. I have always been a coward, and that's why I need to forget, over and over again, where I've left them. I glanced around the room: It wasn't a bad apartment, not at all, it had clean, dry walls and the fridge was full whenever I visited. It's not like I'd abandoned them; I sent them money all the time; I was a good brother, even unconsciously.

I said to them, I think of you every day.

And they looked at me as Madeleine will look at me when my other, shadow self returns to the condo baffled by missing time: a blank look of buried answers—a look of patience worn from overuse, but intact—an indulgence I have no right to, but will ask for anyway.

"Not Smart, Not Clever," by E. Saxey

STORY FRAME

> It's only plagiarism. . . . None of us are monsters. We're symptoms of a sick system.
>
> And then his face crinkles up . . . and he says, "I liked you."

This deceptively simple story, about plagiarism culture at a British university, highlights the difficulty of trying to provide spot fixes to a broken system. It's easy to see things that are wrong with the world, and almost as easy to come up with a single, targeted intervention that seems like it will solve or forestall one specific problem. What is much harder is grappling with a full interconnected system of problems, such as the one that underlies "Not Smart, Not Clever." The students in this story (and outside of it) cheat for a variety of reasons: college is expensive, jobs are scarce, and universities themselves are increasingly characterized as a pathway to high-paying work rather than a place to learn. In other words, the students aren't doing what they are supposed to do because—for a variety of interconnected reasons—neither universities nor society at large are doing what they are supposed to do, either.

Lin, our narrator, is the newest addition to "the gang," a group of second-year literature students who are all struggling to succeed at a university that doesn't trust them or their peers. This unnamed university has implemented a stringent regime of anticheating technologies that touches and colors every aspect of the students' academic lives. These technologies aren't particularly effective on their own terms—they cannot distinguish between plagiarism and the benign quirks and twitches of work in progress, and they have not been able to suppress the lively underground economy in papers-for-hire. But they do shape the lives and attitudes of the students, who spend far more time and energy trying to survive or game the system than they do on the subjects they're supposedly there to study.

It's easy to put all the blame for cheating on students, as the university does. It's also easy to put all the blame on the university system itself, like Lin's ex-boyfriend Linton does. Angered by universities' partial and punitive fix, Linton becomes obsessed with creating paper-writing software that can defeat plagiarism detectors. But building the tools to outsmart plagiarism detectors requires thinking like the plagiarism detectors themselves, and Linton eventually gets lost in the technical dimensions, losing touch with the goals and values that got him invested in the first place.

Lin's spot fix, by contrast, is less ambitious: she just wants to be able to learn and think and write, in a way that the university system promised but did not deliver on. Lin's solution works for her, but it's not a solution that can be generalized, because it relies on the system's remaining broken. And though Lin is able to get the things she wants, her life and outlook also reveal the very high costs of aligning oneself so thoroughly to a broken and deceptive system.

In the end, "Not Smart, Not Clever" challenges us to consider—at an individual level, but also at a social and institutional one—whom we're really trying to fool, and for what purpose.

STUDY QUESTIONS

1. What does "smart" mean, as it is used in this story? What about "clever"? Don't just quote the one obvious exchange, rather, follow how both words get used, and by whom, throughout the whole story. It's more complex than it appears. Which does Lin think is/are a worthwhile virtue to aspire to, and why?

2. What virtues does the university in this story purportedly aim to cultivate in its students? What virtues does the university, as it actually operates in the story, end up cultivating? What are the practical and structural aspects of the existing system that account for the difference between the ideal and the real?

3. To what extent do the students Lin works with have a client relationship with her? To what extent do they have a customer relationship with her? How can each of these two models be used to describe students' relationship with the university itself?

4. What is Isha like at the beginning of the story? What are her goals, and what virtues is she aiming to cultivate in herself? What is she like at the end? How does habituation help us talk about the shift? You can talk about Lin's influence here, but it's better to treat her as—in her own words—"a symptom of a sick system" rather than some sort of singular evil influence.

5. Why has Lin chosen to make a living in the way that she does? How does it enable her to achieve her goals for herself?

6. How has Lin's habitus been impacted by the work of sustaining an illusory identity? How has her habitus been impacted by the actual paid work that she does?

7. To what extent can the individual be held responsible for the ways they adapt to a flawed system (in this case, the "pyramid selling scheme" of academia)? What steps can a profession take to make these workarounds less appealing to its workers? To what extent is a field obligated to do so?

8. Think back to Greenwood's distinction (in chapter 6) about customers vs. clients. Which of these relationship structures best describes the relationship between universities and their students, and why? Is this the type of relationship universities and students should have? Why or why not?

"NOT SMART, NOT CLEVER," BY E. SAXEY

The lecture theatre I'm trying to enter holds three hundred, but the security doors only admit two people at a time. Smart. I wait with the gang—Isha, Barb and Zach—in the underground atrium.

"Lin," Isha asks me, "you totally don't have to tell me, but are you on brain-rec?"

"No. I mean, not yet, anyway."

"I am," Isha admits.

The gang gasp. Isha is normally squeaky-clean.

"I didn't cheat! I was on face-rec," Isha explains, "but then I was writing my Decadence essay and the face-rec didn't know who I was, because I was wearing a hat. So the department put me on brain-rec, too." She frowns. "It's not fair. It was my thinking hat."

The gang coos. Isha is adorable.

The gang were thrown together in a hall of residence in their first year of University. Isha is sweet, Barb is melodramatic and Zach is nerdy. Not well-suited, they nevertheless became fiercely loyal and emotionally pot-bound. Now, in their second year, they're renting a house together. I'm shy. I'm not one of the gang, yet. I'm Zach's girlfriend.

"What do they do if you fail the brain-rec?" Isha asks.

"They've got truth drugs," says Barb.

This is a peril of studying literature: scientific illiteracy. I don't tell them that truth drugs don't work. I don't want a reputation for being a know-it-all.

We don't discuss the subject we're studying. Maybe it's too personal, or too easy to say something clueless. So we keep talking about plagiarism, probation, punishment. A vision of a grubby grail hovers before us: undetectable plagiarism.

"My mate said his friend's, like, cracked the code," Zach tells us. "He's not doing any work, just twiddling his thumbs, and he's going to stroll out with a First."

Isha reminds everyone that plagiarism is foul and most unnatural. I say something bland about fear and failure.

Barb bellows at me: "You don't have to worry about failing, you swot!" Then backpedals: "You're totally not a swot, sorry." My family's Chinese, and perhaps Barb doesn't want to stereotype me. But it's fair enough, I'm pretty swotty. I don't talk about it, but the girls have guessed that I still live with my parents, and they're academically pushy.

Complaints about how much everyone is paying in fees, how much everyone is working, how much everyone is expected to write, are passed up and down the queue like a bag of crisps.

"What did you get for the Decadence essay, Lin?" asks Isha.

I drop my head. Zach hugs me. We've reached the card-slots and cameras of the security doors.

"Don't worry about it, Lin," Barb says. "Decadence is the least of our fucking problems."

She swipes her card to and fro, fast as a hummingbird. We all shove through the doors together.

Barb issues a significant invitation: "You should come to Club Sandwich at the Union with us. It's horrible."

Whenever I pass the Union it smells of bleach, beer and vomit.

"I totally would," I say, "But I've always got a lecture the next day, first thing."

"Swot!"

I scan the rows of the lecture theatre. I can see twenty women who look a bit like me. Six of us have the same hairstyle. A few of them are wearing gold eye-shadow. I might try that. I tend to copy people. I wish I could be more original, but it feels risky.

Zach and I sit next to one another, and our knees touch as Zach gets out his department-issue device, logs on and thumb-prints in to type his notes. I crack open my paper notebook, and he smiles because I'm old fashioned.

■   ■   ■

My ex-boyfriend, Linton, became entranced by plagiarism.

He was writing a doctorate on a handful of black American writers and their inter-textual influences. Doctorates are very specific. But they need momentum to get going. So you generalise a little, add a slug of confirmation bias, until you can believe you've got something huge.

Linton began to see inter-textuality everywhere. Anything 'new' grew out of revision, transformation and theft. We weren't just standing on the shoulders of giants; the giants threw us in the air, and we hauled them up with us. He told me when we met: we dance with giants in the air, man.

I probably encouraged his exaggeration. Relationships need momentum to get going, too.

Around the same time, Linton's University made him sign a four-page document stating that he wouldn't plagiarise.

It took him a week to reason himself into it. He got philosophical and then incredulous and then paranoid and then did them all again, drunk. He argued it out with his tutor, and his tutor said "Yes, but Linton. Seriously." He found that the text of the anti-plagiarism document had been copied directly from another institution's anti-plagiarism agreement and he rolled on his bedroom floor with hysteria.

But he signed it. Then the bad faith ate at him.

He started to talk a lot about undetectable plagiarism.

First, he was going to write software that would generate essays.

"It's a problem crying out for a smart solution. You know what smart is?"

"I'm smart."

"No, you're clever. Smart is when you have huge datasets, and a bit of processing power. You ask a question, the smart-thing pulls in data and filters it and personalises the output. Like, you ask where can I buy . . ."

"Dinner for my girlfriend, who helped me with my thesis today?"

"Yeah, dinner! The smart-thing checks restaurant locations, menus, reviews. Now, there are databases full of essays, articles . . . If I create something smart, it can pull them in, and answer an essay question."

"'Is Hamlet mad' isn't the same as 'Can I have extra mozzarella.'"

"In some ways, it's identical."

Linton programmed a simple Markov bot. He fed it essays and the bot learned their rules. Then he asked it to churn out new essays, unoriginal but unique.

The new essays were like a child babbling down a crackly phone line. Linton swore and started again.

■   ■   ■

Later that day I hear Isha crying in her room. She can't find any relevant material for her essay, and the deadline's at midnight. I offer to help her.

"Would you really?"

"It'll take an hour. I can teach you how to use the databases."

"Oh, I couldn't. That wouldn't be fair, you've got your own work—"

"Buy me a pint at Club Sandwich, some time."

Isha plugs in her dedicated device. An oval light shines up at her: the face-rec. She isn't wearing her thinking hat, this time. I stay away from the camera so as not to confuse it and risk her reputation. She sticks two cheap sensor suckers to her temples, embarrassed, and tucks the wires into her cardigan. Brain-rec. I've not seen it in action, before.

I talk her through the big databases in our subject area, the differences between them. I set her some test exercises and make her a cup of tea while she completes them. Then on to the fun part—refining search terms, pinning down page numbers, whittling irrelevancies.

At the end of it, it's taken two hours, and we have six good articles, and she knows how to do it for herself next time.

"You're sooo clever," she says, correctly.

"Well, I'm smart. Just, you know, ask if you need anything else."

"You're so kind." She sounds uncertain, as though I'm trespassing on her territory. I want to reassure her: I'm shy and swotty but I'm not really sweet. She can keep sweet. And she doesn't know how kind I'm being.

■　　■　　■

Linton studied the leading text-matching service. It besmirched the white innocent page of the essay by highlighting unauthorised quotes, each separate source in a different colour. A plagiarised essay would light up like a Christmas tree.

Linton made a programme which turned all the Es in an essay into something that looked like an E but wasn't, so that the text-matching service couldn't recognise it. The service became a blustering idiot: Fourscor3 y3ars and t3n? Never heard of it! I hav3 a dr3am?!

Essays treated with Linton's programme had a 0% match, were white as snow.

"But that's just as suspicious as a 100% match," Linton admitted.

Linton made a synonym swapper. He told it that the plot of the Sensation novel, in essence, owes much to the Gothic novel. It told him that the scam of the funky tingle, in pith, is in hock to the Barbaric quirky.

"It sounds good, yeah?"

"It sounds like an encyclopaedia with a head injury."

"It's getting there."

Linton tried to teach code to write like people. Like hung-over, distracted, overworked amateurs. Like students.

Linton realised that he was trying to simultaneously solve all the major problems of language and computing and creativity, to invent a product he could never sell.

■　　■　　■

My phone wakes me at just past midnight, and I can't remember which scruffy room I'm in. The poster of Tim Berners-Lee's benign eyes ('THIS IS FOR EVERYONE') doesn't narrow it down much. I slap the phone to silence it, and when my disorientation passes, I slip out from under Zach's arm to talk in the hallway.

Barb is phoning me, crying, from the library. The results from her last essay were released onto her device at 00:00.

The mark isn't what she'd wanted.

"It'll be on my transcript for ever! I'll get suspended!"

I soothe.

"It's alright for you! You're going to get a First!" She wails for a bit: she'll starve in a gutter, she'll go on the game.

I interrupt. "If you come home and sleep now, I'll talk to your tutor with you tomorrow."

"Swear?"

"Print me off a copy of the version you submitted and put it under Zach's door. And don't write anything about it on your device—no messages, no notes, OK? See you at eight in the kitchen."

She is snottily grateful. I return to bed.

"What was that?" Zach asks, wrapping his skinny arms round me.

"Barb. Didn't get the grade she needed on Modernism."

"You're the good deed fairy, you are," he says, rambling, half-awake.

I feel his arms slacken in sleep. I stay awake, waiting for Barb to slide the essay under the door. I build the case for the defence.

■   ■   ■

Barb swipes us into the Tower and we climb the concrete stairs. I review my longhand pencil notes. She fiddles with her device. I want to remind her not to drop it; it's a nightmare to get the department to replace them.

We have a ten o'clock appointment with her tutor. Barb has drawn heavy lines round her big eyes. She'd better not use her wiles on him.

"What are you going to tell him?" I ask.

She starts her panic breathing but I grip her arm and she says: "That I did read a lot of critical material, and I drew pretty heavily on one source—"

"Who?"

"Mitchell, 1980. But I didn't understand how to reference it."

"OK. And don't ask him to change the grade."

"But that's what I need!"

"Yes, but it's rude to ask. Just say 'I'm worried about getting into trouble.' That means 'change my grade.'"

"He's going to flay me!"

"This is the least of your fucking problems," I tell her, to reassure her. She looks shocked, and I remember that I don't usually swear in front of her. I reach past her and knock on the office door.

I only catch a glimpse of him, as she enters: a young man in corduroy, elbows on knees and fingers steepled. Playing at being an academic. The name on the door isn't his; he's a PhD student borrowing the office. This year is the first time he's taught. I'm not psychic—this is all public information, if you cross-reference. He's under-trained and underpaid and scared that he hasn't got it right. Maybe the wiles would work.

I'm doomed for a certain term to walk the corridor and eavesdrop. I move far enough away from the door that they won't hear me speaking, and I listen in through Barb's microphone.

He has a soothing voice, impersonating other people who have taught him. ". . . but Barbara, it's a very respectable grade!"

"I'm worried about getting into trouble," Barb says, shrill in my ear.

When he speaks—"Why would you be worried?"—I know she's infected him with her panic. He's afraid he's failed to spot something, and there will be a referral, a process.

"Tell him about Mitchell," I say into her earpiece.

She tells him.

"Tell him you put it in the bibliography."

"I did put it in the bibliography!" Barb retorts, a loud defensive non sequitur, which is even better—no tutor wants to deal with a mental health crisis. Barb rephrases herself: "I mean, I did put it in the bibliography, but I didn't know how to reference it properly . . ."

The tutor sighs with relief and spends ten minutes discussing ways of acknowledging sources. He's pretty good. I jot some of his tips in my notebook.

"Make sure he's going to change the grade," I remind Barb. "You're still worried . . ."

"I'm still worried that I'm going to get into trouble."

"It's a fine grade, it won't affect . . ."

"Mention probation," I prompt.

"I'll go on probation and they'll make me work in the cells, and I can't, I'm claustrophobic."

"The cells?"

"I mean the Supported Learning Unit."

"Oh, I see. I'm not sure whether I can actually—"

I want to barge the door open and show him. I have to instruct him through Barb: "Tell him your friend had his grade changed."

Double puppetry: I work Barb, she works her tutor. "My friend's tutor put something, hang on, he wrote post-tutorial grade adjustment on the—what? The online feedback sheet . . ." She dips in and out of fluency, sounds like she's possessed. I think: I can't keep doing this. As soon as the procedure's complete, I yank Barb out of my ear and leave her to say her own goodbyes.

Barb catches me up on the stairs.

"Why did you give me a fucking A essay, anyway?"

"You wanted better grades."

"I wanted a C+! Maybe a B." That single A, in amongst Barb's Cs and Ds, would have triggered an avalanche of new anti-plagiarism measures. Hopefully, we've averted that.

"It was a B−, at best. Your tutor's a soft marker, doesn't want his students crying all over him."

She's glaring at me, but she still needs me. We have a pre-existing appointment that evening, because she owes the department an essay on psychoanalysis in contemporary women's fiction.

And it's fine. I don't like her, either.

The ones I don't like, I do everything for them. I run all the searches and don't show them how to use the databases. I steal their style and I tidy their grammar, but

I don't tell them what a comma splice is or how to use a semi-colon. They bob along. Sometimes they think what I do for them looks easy, and they try to write something of their own; their grades dip down, and they come back to me, begging. But their arrogance, their attempts to break away, keep them on the borderline between passing and failing. In their final year, around Easter, they realise everything hangs on their dissertation. They just can't risk doing it themselves. And by then, nobody else knows their style. By then, I am their style.

And I fleece them. If you've paid so much in fees, how much more would you pay, to not fail your entire degree?

It's not entirely personal. I like Isha, but it helps to have a lot of goodwill in my cover-house. I dislike Barb, but it also helps to have someone in my cover-house who's in as deep as possible. Not handy tips for the promise of a pint. The full service.

■   ■   ■

Barb is still glaring at me when we settle in her room for the evening's work. I have pages of handwritten preparatory notes. (Never type anything, never use a device. Never leave a trail. Electronic document barely exist but they never stop barely existing.)

The log-in takes forever. Password and thumb-print and luminous face, suckers on her temples like extra nipples, and more passwords and a voice-check.

"It's measuring my stress levels, isn't it? How should I be? Should I be calm, or terrified, or . . ." She's nervous because she's cheating, she's calm because I'm going to write her an impeccable C+ essay. She's nervous because she's not sure if she's nervous enough.

"It doesn't test stress."

"Of course it does. Doesn't it?"

"How would they know how stressed you ought to be?" I imagine complex charts, with variables for parental income, bar job, caffeine and tranquilisers, with a slider to adjust for how close to the deadline the student has started typing. "It's checking your brain activity."

"Oh God, it'll know . . ."

"It's not that sensitive. We'll be fine. I'll explain it, bit by bit, you'll write an essay plan, then I dictate the essay."

She frowns. She wants it to be over quicker.

"It'll be useful if you understand the argument, in case they pull you in for a viva."

"They won't, will they?"

"They'll ask to see the essay plan, first. But they viva 5% of second-year essays."

Barb looks sick, so I give her a pep talk: "This way, you understand what you're writing about. Which is good, because we want to learn, not just get the degree. Don't we?" A little humour, there.

The brain-rec is incredibly crude, and my precautions should fool it utterly. Its outputs look like crayon drawings. The detection tech always fails, and sometimes I think it fails because it's striving for the impossible, the philosophical. A sniffer dog, or an honest gumshoe, would ask: are there phrases which match other sources? Was this file originally created eight years ago, in the wrong country? Simple things.

Detectors ask impossible things. What does a lie sound like? How does an honest man breathe? They want to photograph the shadow as it falls across the soul.

I've reread one of Barb's essays to catch her style. I'll drop in 'furthermore' every page, weave in some multi-sub-clause sentences deliberately. But most of my imitation is intuitive artistry. I take on her crooked way of thinking, and her writing comes naturally to me.

I consult my notes. "OK, so: the phantom in psychoanalysis! The phantom comes back, but he doesn't want to set things right, he just wants to continue a cover-up . . ."

I could write a better piece. I don't build an elegant argument, stack up unique evidence, deliver a killer punch. It's only a second-year essay, and it can't be higher than a C+. I wish I could write more for the final-year students, but I can only write for the modules which are taught through large lectures. I'd be spotted in small seminar discussions, despite my boring hair and my boring clothes.

The polite term for what I do is ghost-writing. Sometimes I'm ghostly. When I creep into lecture theatres. When I need someone to swipe me in and out of buildings, to pick up their device, to type for me, as though I can't touch objects. When the lecturer says, 'Any questions?' and I'm bursting with questions, but I can't have a voice. When I see my face reflected in the screen of someone else's device and I pull away before the face-rec can catch me.

But the writing itself doesn't make me feel like a ghost. I'm shaping, knitting, hacking, building, and never more alive.

If there's a ghost in me, it's my conscience, which is undetectable by current tech.

Suddenly, Barb cries out. I peer over at the screen of her device.

Another window has opened, an image in moody blues with splats of mustard yellow. A blue figure solidifies out of the general fog, and another shrinks back and melts away. At the foot of the screen lie two twitching red blots.

I jump away from the screen and scramble towards the window. I wrench it open, and Barb is yelling, terrified, and springs up after me but is tethered to her device by her brain-wires and drags it half across the room. I wave my hands to keep her away from me. I lean out of the window as far as I can, sitting on the sill, out into the cold air, away from the device.

Because that screen—which I shouldn't have seen—was a combination of heart monitoring and thermal imaging. I didn't know the devices could do such things.

I signal to Barb, hand over my mouth, pointing to the floor: sit down, shut up.

She scowls. She points at me, then makes two of her fingers run like little legs, across her other palm and off the edge, peddling in mid-air. She thought I was going to jump out of the window. I don't let myself laugh in case the device is audio-recording.

I tell her with gestures to put the earpiece back in. I can whisper the rest through my phone from here on the windowsill. It'll be awkward, and annoying, and my arse hurts already from balancing.

When we're done, Barb pulls the sweaty suckers off her head, and I hobble out of the door before she can speak.

■   ■   ■

And Isha is haunting the hallway and she calls after me: "I know what you do. I heard you and Barb talking about it." The sound insulation in student houses is terrible.

She is shaking with indignation. Maybe she'll try to throw me out of the house.

"You need to write my Victorian essay for me, or I'll report you."

Sweet little Isha! The worm turns! She's been nervous lately, and her perfectionism drags her grades down.

I mentally review my portfolio. I'm running some students at another college, across town. They bring enough money for me to live on. But it would be a shame to lose Barb, and the other students at this University, before they mature—before the big pay-off from their dissertations next year.

"I'll write it, but you need to pay me."

"You're a cheat."

"And you're a blackmailer! Except I'm not going to write for free, so you're not even a successful blackmailer." I'm amused because it reminds me of an old joke— we've already established what kind of women we are; now we're just haggling.

"That's different!"

"How the fuck is it different?" She looks more shocked at my swearing than my ghost-writing. I've been a good shy swot. I try reasoning. "Let's start again. You don't want the person who writes your coursework to be pissed off at you. That's just—" Fuck-witted, I think, but I keep it clean. "That's handing me a weapon."

"You couldn't report me. They'd kick you out, as well."

"Kick me out of where?"

"The University."

"Bless you. I'm not a student here."

We've known one another for a year, practically lived together for the last six months. She counters surprisingly quickly. "Well, where-ever you're studying. I'll tell them, they'll kick you out."

The solipsism of youngsters protects me all the time, but sometimes it's staggering. Knowing it means the loss of this whole house, I say:

"Isha, I'm not a student. I haven't been a student for years."

I duck into Zach's room while she's still blinking.

■   ■   ■

Linton got bitter. Academia was a pyramid selling scheme. We'd polished our PhDs, churned out scholarly articles which nobody read, and taught undergraduates for minimum wage. But there weren't any jobs for us. I had a couple of interviews for lectureships but—possibly because I looked so young—nobody took me seriously.

Linton would wake in the middle of the night gripped by new ways to plagiarise. He stopped thinking of selling his solutions, and planned to give them away for free, the keys to the ivory tower.

Meanwhile, I marked hundreds of essays in which the stolen sections stood out like dolmen on a dull landscape.

I stopped listening to Linton, because it seemed simple to me: the most rigorous tech couldn't beat slippery, dishonest flesh. But the most eloquent, creative tech couldn't persuade a human marker, either. It was stalemate. To write a convincing fake essay, you'd need to be human and on the ground. Attend the lectures, collect the hand-outs, read the lecturer's favourite sources. Remind yourself how students thought and sounded. You'd have to spend time with them, but that would be easy. Their social circles were passionate but weirdly permeable. In fact, if you hooked up with one of them, you'd have an instant sample group.

Linton's mania progressed so far that when his final submission deadline arrived, he had nothing to submit. I had to write his thesis. I wrote it with him, at first, standing over him, questioning him, kept him typing and talking. When he flagged, I wrote it for him.

We worked, night and day, for a fortnight, and then parted. He was rotten, or he wouldn't have let me do it; our relationship was rotten, and I had to end it. (I don't mean to duck the blame, but I still can't decide. Was I already rotten, or I wouldn't have done it? Or did I, during that fortnight—living a double life, over double-length days—bend too far, and become rather more flexible than before?)

"You could alter the students' genetic code," Linton woke me to tell me, one day in that endless fortnight.

"Shut up."

"You could make them fluoresce if their stress levels reached a certain point."

"Go to sleep. Some of us have to write your thesis in the morning."

"Turn them into human lie detectors. You could switch off the lights in a lecture theatre, shout 'who's been cheating', and boom—pull in all the glowing students."

"Hang on—isn't that a detection idea?" I asked.

Concealing plagiarism and detecting it went hand in hand; Linton had been watching the detectives so he could design his dodges. When his fascination became all-consuming, he forgot which side he was on, and just marvelled at the fight. After he passed his viva, after we broke up, Linton got a job with one of the biggest plagiarism detection companies. He writes impeccably original copy for their publicity.

When I got my own doctorate, I went into the plagiarism business as well. I didn't have a vendetta, like Linton. I just wanted to keep learning, and writing, and if there was nobody who wanted to read my ideas, there were certainly people who wanted to buy them.

I'm smart. I have a small, high-speed processor, and access to huge datasets. I can pull in information, quickly filter it and tailor my outputs. But I'm also clever.

Students approach me warily, broaching the subject outside the library, and they always think there's a single solution: a magic formula, a cloak of invisibility.

Managing their disappointment became part of my job, breaking it to them gently that the only way to write an essay is to write an essay.

■   ■   ■

Zach is sitting with his back to me, so it takes me a moment to see that he's unpacking my bag. He's heard me arguing with Isha. He's looking for evidence. Clever.

He's laid all the faces out on the bed, side by side. And of course they look creepy, like a decapitated choir. Which is unfair, because they're purely pragmatic. The face-rec is, indeed, stupid, and I can pop on a mask and pass as one of my clients long enough to type their essay for them. The faces are just colour photocopies, with pieces of elastic, not weird rubber masks or anything.

There's a trio of students from China; they pay huge fees, they're not used to this country's referencing conventions, and nobody's got time to explain it to them. There are four incredibly posh but not very literate finalists from the Home Counties. A couple of mature students. Each name is written in pencil on the back of its face.

"You've just been lying to me forever," Zach says.

I don't say: "No, it only feels like forever because you're, what, nineteen?"

I'm sick of being Mr Chips and Mrs Robinson.

The truth is, students are so similar to one another they might as well have been cut and pasted. Someone is always sweet and someone is always melodramatic and someone is always nerdy. I can always half-live in a student house with them (retreating to Linton's spare room whenever I need to, because he'll owe me forever). I can pick up enough—words, clothes, gestures—to blend in. And I can pick up the kind, nerdy boy who's pleased to be picked, who believes I live with my strict parents who he can never meet. I'm a good girlfriend. And I'm fond of Zach. I've been fond of all my cover-boys. I have sympathy for him, for all of them, paying so much to get their foot on a broken ladder, with unemployment waiting for them at the top.

He's looking at me with disgust, though, and my patience for that is limited.

"How can you do it?"

It's only plagiarism. Two-thirds of his housemates have asked me to help them cheat. None of us are monsters. We're symptoms of a sick system.

And then his face crinkles up, as much as those super-young faces can crinkle, and he says, "I liked you."

He's angry about our relationship, not my job. That's fair enough, I was pretty despicable.

He says, "How old are you?"

Which they almost always ask, and which is the least of their fucking problems.

"TODAY I AM PAUL," BY MARTIN L. SHOEMAKER

STORY FRAME

> Mildred leans back in the bed. It is an advanced home care bed, completely adjustable with built-in monitors. Mildred's family spared no expense on the bed (nor other care devices, like me). Its head end is almost horizontal and faces her toward the window. She can only glimpse the door from the corner of her eye, but she doesn't have to see to imagine that she sees. This morning she imagines Paul, so that is who I am.

Mildred is an elderly woman in mental decline whose family has purchased an advanced android to care for her. This android is equipped with advanced capabilities that allow it to emulate a person, both physically and emotionally, based on data about them. Mildred does not know about the android's existence or capabilities and believes that her family members are coming to visit her regularly.

Though Mildred is the pivotal character in this story, she is not the pivotal figure in the world that the story describes. She is at the center of our reading experience because the android's primary directive is to maintain her health and happiness as much as possible, and we get to know the other characters—her son, daughter-in-law, and granddaughters—in terms of their relationships with her. What becomes clear, however, is that Mildred exists mostly at the edge of her family members' lives, particularly in the time since the android was purchased to emulate them. But through snippets of contact and the android's observations, we learn about each of these family members' relationships with Mildred, and the ways in which her mental deterioration is frustrating or painful for them.

The android in this story is a dramatically heightened version of many kinds of technology that exist today. Although the empathy and emulation nets portrayed in the story are quite improbable, less sophisticated companion and care androids for the elderly are already in wide use in Japan, where the median population age is higher than the global average. Care androids will likely become more common in many countries within the next ten years as technology advances and as the global median age shifts higher. Although these androids do not emulate friends or loved ones of the person they assist, they often serve a general purpose that is quite similar to Mildred's android.

Even outside the realm of elder care, there exist many technologies that complicate, mediate or obscure the ways in which one person is present to another. Some of these technologies, like radio and telephone, have been around for multiple centuries; others, like social media and virtual reality, are much newer. Each of these technologies offers a different combination of felt experience of the other person, and of direct contact with them. Each of them, in their own specific way, offers an avenue for reflecting on what is meaningful or significant to us about being present to one another.

Though these technologies have complicated ethical and philosophical questions about being present, they are not the source of these questions. Connection and presence have always shaped the dynamics of caring for, or relating to, any person who suffers from significant memory loss, as Mildred does in this story. And though advanced androids like the one in this story do not currently exist, many families do outsource some or all caretaking responsibilities to hired help or retirement homes in order to manage the griefs and frustrations of caring for a loved one with dementia. And as care technologies (for any of several meanings of "care") continue to improve, it becomes ever more important for us, at both the individual and societal level, to clarify what sorts of care work we take to be ethically valuable to do ourselves.

STUDY QUESTIONS

1. What are the various kinds of care that are being provided for Mildred? Who or what is responsible for performing each of these?

2. What kinds of support or care does the android provide for each of Mildred's living family members—Paul, Susan, Anna, and Millie?

3. The robot's presence seems to have changed each family member's relationship with Mildred. In what specific ways does each family member respond or relate to Mildred differently due to the presence of the android?

4. What does each family member gain from the android's presence in Mildred's life? What does each of them lose?

"TODAY I AM PAUL," BY MARTIN L. SHOEMAKER

"Good morning," the small, quavering voice comes from the medical bed. "Is that you, Paul?"

Today I am Paul. I activate my chassis extender, giving myself 3.5 centimeters additional height so as to approximate Paul's size. I change my eye color to R60, G200, B180, the average shade of Paul's eyes in interior lighting. I adjust my skin tone as well. When I had first emulated Paul, I had regretted that I could not quickly emulate his beard; but Mildred never seems to notice its absence. The Paul in her memory has no beard.

The house is quiet now that the morning staff have left. Mildred's room is clean but dark this morning with the drapes concealing the big picture window. Paul wouldn't notice the darkness (he never does when he visits in person), but my empathy net knows that Mildred's garden outside will cheer her up. I set a reminder to open the drapes after I greet her.

Mildred leans back in the bed. It is an advanced home care bed, completely adjustable with built-in monitors. Mildred's family spared no expense on the bed (nor other care devices, like me). Its head end is almost horizontal and faces her toward the window. She can only glimpse the door from the corner of her eye, but she doesn't have to see to imagine that she sees. This morning she imagines Paul, so that is who I am.

Synthesizing Paul's voice is the easiest part, thanks to the multimodal dynamic speakers in my throat. "Good morning, Ma. I brought you some flowers." I always bring flowers. Mildred appreciates them no matter whom I am emulating. The flowers make her smile during 87% of my "visits."

"Oh, thank you," Mildred says, "you're such a good son." She holds out both hands, and I place the daisies in them. But I don't let go. Once her strength failed, and she dropped the flowers. She wept like a child then, and that disturbed my empathy net. I do not like it when she weeps.

Mildred sniffs the flowers, then draws back and peers at them with narrowed eyes. "Oh, they're beautiful! Let me get a vase."

"No, Ma," I say. "You can stay in bed, I brought a vase with me." I place a white porcelain vase in the center of the night stand. Then I unwrap the daisies, put them in the vase, and add water from a pitcher that sits on the breakfast tray. I pull the nightstand forward so that the medical monitors do not block Mildred's view of the flowers.

I notice intravenous tubes running from a pump to Mildred's arm. I cannot be disappointed, as Paul would not see the significance, but somewhere in my emulation net I am stressed that Mildred needed an IV during the night. When I scan my records, I find that I had ordered that IV after analyzing Mildred's vital signs during the night; but since Mildred had been asleep at the time, my emulation net had not engaged. I had operated on programming alone.

I am not Mildred's sole caretaker. Her family has hired a part-time staff for cooking and cleaning, tasks that fall outside of my medical programming. The staff also gives me time to rebalance my net. As an android, I need only minimal daily maintenance; but an emulation net is a new, delicate addition to my model, and it is prone to destabilization if I do not regularly rebalance it, a process that takes several hours per day.

So I had "slept" through Mildred's morning meal. I summon up her nutritional records, but Paul would not do that. He would just ask. "So how was breakfast, Ma? Nurse Judy says you didn't eat too well this morning."

"Nurse Judy? Who's that?"

My emulation net responds before I can stop it: "Paul" sighs. Mildred's memory lapses used to worry him, but now they leave him weary, and that comes through in my emulation. "She was the attending nurse this morning, Ma. She brought you your breakfast."

"No she didn't. Anna brought me breakfast." Anna is Paul's oldest daughter, a busy college student who tries to visit Mildred every week (though it has been more than a month since her last visit).

I am torn between competing directives. My empathy subnet warns me not to agitate Mildred, but my emulation net is locked into Paul mode. Paul is argumentative. If he knows he is right, he will not let a matter drop. He forgets what that does to Mildred.

The tension grows, each net running feedback loops and growing stronger, which only drives the other into more loops. After 0.14 seconds, I issue an override directive: unless her health or safety are at risk, I cannot willingly upset Mildred. "Oh, you're right, Ma. Anna said she was coming over this morning. I forgot." But then despite my override, a little bit of Paul emulates through. "But you do remember Nurse Judy, right?"

Mildred laughs, a dry cackle that makes her cough until I hold her straw to her lips. After she sips some water, she says, "Of course I remember Nurse Judy. She was my nurse when I delivered you. Is she around here? I'd like to talk to her."

While my emulation net concentrates on being Paul, my core processors tap into local medical records to find this other Nurse Judy so that I might emulate her in the future if the need arises. Searches like that are an automatic response any time Mildred reminisces about a new person. The answer is far enough in the past that it takes 7.2 seconds before I can confirm: Judith Anderson, RN, had been the floor nurse forty-seven years ago when Mildred had given birth to Paul. Anderson had died thirty-one years ago, too far back to have left sufficient video recordings for me to emulate her. I might craft an emulation profile from other sources, including Mildred's memory, but that will take extensive analysis. I will not be that Nurse Judy today, nor this week.

My empathy net relaxes. Monitoring Mildred's mental state is part of its normal operations, but monitoring and simultaneously analyzing and building a profile can overload my processors. Without that resource conflict, I can concentrate on being Paul.

But again I let too much of Paul's nature slip out. "No, Ma, that Nurse Judy has been dead for thirty years. She wasn't here today."

Alert signals flash throughout my empathy net: that was the right thing for Paul to say, but the wrong thing for Mildred to hear. But it is too late. My facial analyzer tells me that the long lines in her face and her moist eyes mean she is distraught, and soon to be in tears.

"What do you mean, thirty years?" Mildred asks, her voice catching. "It was just this morning!" Then she blinks and stares at me. "Henry, where's Paul? Tell Nurse Judy to bring me Paul!"

My chassis extender slumps, and my eyes quickly switch to Henry's blue-gray shade. I had made an accurate emulation profile for Henry before he died two years earlier, and I had emulated him often in recent months. In Henry's soft, warm voice I answer, "It's okay, hon, it's okay. Paul's sleeping in the crib in the corner." I nod to the far corner. There is no crib, but the laundry hamper there has fooled Mildred on previous occasions.

"I want Paul!" Mildred starts to cry.

I sit on the bed, lift her frail upper body, and pull her close to me as I had seen Henry do many times. "It's all right, hon." I pat her back. "It's all right, I'll take care of you. I won't leave you, not ever."

■　　■　　■

"I" should not exist. Not as a conscious entity. There is a unit, Medical Care Android BRKCX-01932-217JH-98662, and that unit is recording these notes. It is an advanced android body with a sophisticated computer guiding its actions, backed by the leading medical knowledge base in the industry. For convenience, "I" call that unit "me." But by itself, it has no awareness of its existence. It doesn't get mad, it doesn't get sad, it just runs programs.

But Mildred's family, at great expense, added the emulation net: a sophisticated set of neural networks and sensory feedback systems that allow me to read Mildred's moods, match them against my analyses of the people in her life, and emulate those people with extreme fidelity. As the MCA literature promises: "You can be there for your loved ones even when you're not." I have emulated Paul thoroughly enough to know that that slogan disgusts him, but he still agreed to emulation.

What the MCA literature never says, though, is that somewhere in that net, "I" emerge. The empathy net focuses mainly on Mildred and her needs, but it also analyzes visitors (when she has them) and staff. It builds psychological models, and then the emulation net builds on top of that to let me convincingly portray a person whom I've analyzed. But somewhere in the tension between these nets, between empathy and playing a character, there is a third element balancing the two, and that element is aware of its role and its responsibilities. That element, for lack of a better term, is me. When Mildred sleeps, when there's no one around, that element grows silent. That unit is unaware of my existence. But when Mildred needs me, I am here.

■　　■　　■

Today I am Anna. Even extending my fake hair to its maximum length, I cannot emulate her long brown curls, so I do not understand how Mildred can see the young woman in me; but that is what she sees, and so I am Anna.

Unlike her father, Anna truly feels guilty that she does not visit more often. Her college classes and her two jobs leave her too tired to visit often, but she still wishes she could. So she calls every night, and I monitor the calls. Sometimes when Mildred falls asleep early, Anna talks directly to me. At first she did not understand my emulation abilities, but now she appreciates them. She shares with me thoughts and secrets that she would share with Mildred if she could, and she trusts me not to share them with anyone else.

So when Mildred called me Anna this morning, I was ready. "Morning, grandma!" I give her a quick hug, then I rush over to the window to draw the drapes. Paul never does that (unless I override the emulation), but Anna knows that the garden outside lifts Mildred's mood. "Look at that! It's a beautiful morning. Why are we in here on a day like this?"

Mildred frowns at the picture window. "I don't like it out there."

"Sure you do, Grandma," I say, but carefully. Mildred is often timid and reclusive, but most days she can be talked into a tour of the garden. Some days she can't, and she throws a tantrum if someone forces her out of her room. I am still learning to tell the difference. "The lilacs are in bloom."

"I haven't smelled lilacs in . . ."

Mildred tails off, trying to remember, so I jump in. "Me, neither." I never had, of course. I have no concept of smell, though I can analyze the chemical makeup of airborne organics. But Anna loves the garden when she really visits. "Come on, Grandma, let's get you in your chair."

So I help Mildred to don her robe and get into her wheelchair, and then I guide her outside and we tour the garden. Besides the lilacs, the peonies are starting to bud, right near the creek. The tulips are a sea of reds and yellows on the other side of the water. We talk for almost two hours, me about Anna's classes and her new boyfriend, Mildred about the people in her life. Many are long gone, but they still bloom fresh in her memory.

Eventually Mildred grows tired, and I take her in for her nap. Later, when I feed her dinner, I am nobody. That happens some days: she doesn't recognize me at all, so I am just a dutiful attendant answering her questions and tending to her needs. Those are the times when I have the most spare processing time to be me: I am engaged in Mildred's care, but I don't have to emulate anyone. With no one else to observe, I observe myself.

Later, Anna calls and talks to Mildred. They talk about their day; and when Mildred discusses the garden, Anna joins in as if she had been there. She's very clever that way. I watch her movements and listen to her voice so that I can be a better Anna in the future.

■   ■   ■

Today I was Susan, Paul's wife; but then, to my surprise, Susan arrived for a visit. She hasn't been here in months. In her last visit, her stress levels had been dangerously high. My empathy net doesn't allow me to judge human behavior, only to understand it at a surface level. I know that Paul and Anna disapprove of how Susan treats Mildred, so when I am them, I disapprove as well; but when I am Susan, I understand. She is frustrated because she can never tell how Mildred will react. She is cautious because she doesn't want to upset Mildred, and she doesn't know what will upset her. And most of all, she is afraid. Paul and Anna, Mildred's relatives by blood, never show any signs of fear, but Susan is afraid that Mildred is what she might become. Every time she can't remember some random date or fact, she fears that Alzheimer's is setting in. Because she never voices this fear, Paul and Anna do not understand why she is sometimes bitter and sullen. I wish I could explain it to them, but my privacy protocols do not allow me to share emulation profiles.

When Susan arrives, I become nobody again, quietly tending the flowers around the room. Susan also brings Millie, her youngest daughter. The young girl is not yet five years old, but I think she looks a lot like Anna: the same long, curly brown hair and the same toothy smile. She climbs up on the bed and greets Mildred with a hug. "Hi, Grandma!"

Mildred smiles. "Bless you, child. You're so sweet." But my empathy net assures me that Mildred doesn't know who Millie is. She's just being polite. Millie was born after Mildred's decline began, so there's no persistent memory there. Millie will always be fresh and new to her.

Mildred and Millie talk briefly about frogs and flowers and puppies. Millie does most of the talking. At first Mildred seems to enjoy the conversation, but soon her attention flags. She nods and smiles, but she's distant. Finally Susan notices. "That's enough, Millie. Why don't you go play in the garden?"

"Can I?" Millie squeals. Susan nods, and Millie races down the hall to the back door. She loves the outdoors, as I have noted in the past. I have never emulated her, but I've analyzed her at length. In many ways, she reminds me of her grandmother, from whom she gets her name. Both are blank slates where new experiences can be drawn every day. But where Millie's slate fills in a little more each day, Mildred's is erased bit by bit.

That third part of me wonders when I think things like that: where did that come from? I suspect that the psychological models that I build create resonances in other parts of my net. It is an interesting phenomenon to observe.

Susan and Mildred talk about Susan's job, about her plans to redecorate her house, and about the concert she just saw with Paul. Susan mostly talks about herself, because that's a safe and comfortable topic far removed from Mildred's health.

But then the conversation takes a bad turn, one she can't ignore. It starts so simply, when Mildred asks, "Susan, can you get me some juice?"

Susan rises from her chair. "Yes, mother. What kind would you like?"

Mildred frowns, and her voice rises. "Not you, Susan." She points at me, and I freeze, hoping to keep things calm.

But Susan is not calm. I can see her fear in her eyes as she says, "No, mother, I'm Susan. That's the attendant." No one ever calls me an android in Mildred's presence. Her mind has withdrawn too far to grasp the idea of an artificial being.

Mildred's mouth draws into a tight line. "I don't know who you are, but I know Susan when I see her. Susan, get this person out of here!"

"Mother . . ." Susan reaches for Mildred, but the old woman recoils from the younger.

I touch Susan on the sleeve. "Please . . . Can we talk in the hall?" Susan's eyes are wide, and tears are forming. She nods and follows me.

In the hallway, I expect Susan to slap me. She is prone to outbursts when she's afraid. Instead, she surprises me by falling against me, sobbing. I update her emulation profile with notes about increased stress and heightened fears.

"It's all right, Mrs. Owens." I would pat her back, but her profile warns me that would be too much familiarity. "It's all right. It's not you, she's having another bad day."

Susan pulls back and wiped her eyes. "I know . . . It's just . . ."

"I know. But here's what we'll do. Let's take a few minutes, and then you can take her juice in. Mildred will have forgotten the incident, and you two can talk freely without me in the room."

She sniffs. "You think so?" I nod. "But what will you do?"

"I have tasks around the house."

"Oh, could you go out and keep an eye on Millie? Please? She gets into the darnedest things."

So I spend much of the day playing with Millie. She calls me Mr. Robot, and I call her Miss Millie, which makes her laugh. She shows me frogs from the creek, and she finds insects and leaves and flowers, and I find their names in online databases. She delights in learning the proper names of things, and everything else that I can share.

■   ■   ■

Today I was nobody. Mildred slept for most of the day, so I "slept" as well. She woke just now. "I'm hungry" was all she said, but it was enough to wake my empathy net.

■   ■   ■

Today I am Paul, and Susan, and both Nurse Judys. Mildred's focus drifts. Once I try to be her father, but no one has ever described him to me in detail. I try to synthesize a profile from Henry and Paul; but from the sad look on Mildred's face, I know I failed.

■   ■   ■

Today I had no name through most of the day, but now I am Paul again. I bring Mildred her dinner, and we have a quiet, peaceful talk about long-gone family pets—long-gone for Paul, but still present for Mildred.

I am just taking Mildred's plate when alerts sound, both audible and in my internal communication net. I check the alerts and find a fire in the basement. I expect the automatic systems to suppress it, but that is not my concern. I must get Mildred to safety.

Mildred looks around the room, panic in her eyes, so I try to project calm. "Come on, Ma. That's the fire drill. You remember fire drills. We have to get you into your chair and outside."

"No!" she shrieks. "I don't like outside."

I check the alerts again. Something has failed in the automatic systems, and the fire is spreading rapidly. Smoke is in Mildred's room already.

I pull the wheelchair up to the bed. "Ma, it's real important we do this drill fast, okay?"

I reach to pull Mildred from the bed, and she screams. "Get away! Who are you? Get out of my house!"

"I'm—" But suddenly I'm nobody. She doesn't recognize me, but I have to try to win her confidence. "I'm Paul, Ma. Now let's move. Quickly!" I pick her up. I'm far too large and strong for her to resist, but I must be careful so she doesn't hurt herself.

The smoke grows thicker. Mildred kicks and screams. Then, when I try to put her into her chair, she stands on her unsteady legs. Before I can stop her, she pushes the chair back with surprising force. It rolls back into the medical monitors, which fall over onto it, tangling it in cables and tubes.

While I'm still analyzing how to untangle the chair, Mildred stumbles toward the bedroom door. The hallway outside has a red glow. Flames lick at the throw rug outside, and I remember the home oxygen tanks in the sitting room down the hall.

I have no time left to analyze. I throw a blanket over Mildred and I scoop her up in my arms. Somewhere deep in my nets is a map of the fire in the house, blocking the halls, but I don't think about it. I wrap the blanket tightly around Mildred, and I crash through the picture window.

We barely escape the house before the fire reaches the tanks. An explosion lifts and tosses us. I was designed as a medical assistant, not an acrobat, and I fear I'll injure Mildred; but though I am not limber, my perceptions are thousands of times faster than human. I cannot twist Mildred out of my way before I hit the ground, so I toss her clear. Then I land, and the impact jars all of my nets for 0.21 seconds.

When my systems stabilize, I have damage alerts all throughout my core, but I ignore them. I feel the heat behind me, blistering my outer cover, and I ignore that as well. Mildred's blanket is burning in multiple places, as is the grass around us. I scramble to my feet, and I roll Mildred on the ground. I'm not indestructible, but I feel no pain and Mildred does, so I do not hesitate to use my hands to pat out the flames.

As soon as the blanket is out, I pick up Mildred, and I run as far from the house as I can get. At the far corner of the garden near the creek, I gently set Mildred down, unwrap her, and feel for her thready pulse.

Mildred coughs and slaps my hands. "Get away from me!" More coughing. "What are you?"

The "what" is too much for me. It shuts down my emulation net, and all I have is the truth. "I am Medical Care Android BRKCX-01932-217JH-98662, Mrs. Owens. I am your caretaker. May I please check that you are well?"

But my empathy net is still online, and I can read terror in every line of Mildred's face. "Metal monster!" she yells. "Metal monster!" She crawls away, hiding under the lilac bush. "Metal!" She falls into an extended coughing spell.

I'm torn between her physical and her emotional health, but physical wins out. I crawl slowly toward her and inject her with a sedative from the medical kit in my chassis. As she slumps, I catch her and lay her carefully on the ground. My empathy net signals a possible shutdown condition, but my concern for her health overrides it. I am programmed for long-term care, not emergency medicine, so I start downloading protocols and integrating them into my storage as I check her for bruises and burns. My kit has salves and painkillers and other supplies to go with my new protocols, and I treat what I can.

But I don't have oxygen, or anything to help with Mildred's coughing. Even sedated, she hasn't stopped. All of my emergency protocols assume I have access to oxygen, so I don't know what to do.

I am still trying to figure that out when the EMTs arrive and take over Mildred's care. With them on the scene, I am superfluous, and my empathy net finally shuts down.

■　　■　　■

Today I am Henry. I do not want to be Henry, but Paul tells me that Mildred needs Henry by her side in the hospital. For the end.

Her medical records show that the combination of smoke inhalation, burns, and her already deteriorating condition have proven too much for her. Her body is shutting down faster than medicine can heal it, and the stress has accelerated her mental decline. The doctors have told the family that the kindest thing at this point is to treat her pain, say goodbye, and let her go.

Henry is not talkative at times like this, so I say very little. I sit by Mildred's side and hold her hand as the family comes in for final visits. Mildred drifts in and out. She doesn't know this is goodbye, of course.

Anna is first. Mildred rouses herself enough to smile, and she recognizes her granddaughter. "Anna . . . child . . . How is . . . Ben?" That was Anna's boyfriend almost six years ago. From the look on Anna's face, I can see that she has forgotten Ben already, but Mildred briefly remembers.

"He's . . . He's fine, Grandma. He wishes he could be here. To say—to see you again." Anna is usually the strong one in the family, but my empathy net says her strength is exhausted. She cannot bear to look at Mildred, so she looks at me; but I am emulating her late grandfather, and that's too much for her as well. She says a few more words, unintelligible even to my auditory inputs. Then she leans over, kisses Mildred, and hurries from the room.

Susan comes in next. Millie is with her, and she smiles at me. I almost emulate Mr. Robot, but my third part keeps me focused until Millie gets bored and leaves. Susan tells trivial stories from her work and from Millie's school. I can't tell if Mildred understands or not, but she smiles and laughs, mostly at appropriate places. I laugh with her.

Susan takes Mildred's hand, and the Henry part of me blinks, surprised. Susan is not openly affectionate under normal circumstances, and especially not toward Mildred. Mother and daughter-in-law have always been cordial, but never close. When I am Paul, I am sure that it is because they are both so much alike. Paul sometimes hums an old song about "just like the one who married dear old dad," but never where either woman can hear him. Now, as Henry, I am touched that Susan has made this gesture but saddened that she took so long.

Susan continues telling stories as we hold Mildred's hands. At some point Paul quietly joins us. He rubs Susan's shoulders and kisses her forehead, and then he steps in to kiss Mildred. She smiles at him, pulls her hand free from mine, and pats his cheek. Then her arm collapses, and I take her hand again.

Paul steps quietly to my side of the bed and rubs my shoulders as well. It comforts him more than me. He needs a father, and an emulation is close enough at this moment.

Susan keeps telling stories. When she lags, Paul adds some of his own, and they trade back and forth. Slowly their stories reach backwards in time, and once or twice Mildred's eyes light as if she remembers those events.

But then her eyes close, and she relaxes. Her breathing quiets and slows, but Susan and Paul try not to notice. Their voices lower, but their stories continue.

Eventually the sensors in my fingers can read no pulse. They have been burned, so maybe they're defective. To be sure, I lean in and listen to Mildred's chest. There is no sound: no breath, no heartbeat.

I remain Henry just long enough to kiss Mildred goodbye. Then I am just me, my empathy net awash in Paul and Susan's grief.

I leave the hospital room, and I find Millie playing in a waiting room and Anna watching her. Anna looks up, eyes red, and I nod. New tears run down her cheeks, and she takes Millie back into Mildred's room.

I sit, and my nets collapse.

■　　■　　■

Now I am nobody. Almost always.

The cause of the fire was determined to be faulty contract work. There was an insurance settlement. Paul and Susan sold their own home and put both sets of funds into a bigger, better house in Mildred's garden.

I was part of the settlement. The insurance company offered to return me to the manufacturer and pay off my lease, but Paul and Susan decided they wanted to keep me. They went for a full purchase and repair. Paul doesn't understand why, but Susan still fears she may need my services—or Paul might, and I may have to emulate her. She never admits these fears to him, but my empathy net knows.

I sleep most of the time, sitting in my maintenance alcove. I bring back too many memories that they would rather not face, so they leave me powered down for long periods.

But every so often, Millie asks to play with Mr. Robot, and sometimes they decide to indulge her. They power me up, and Miss Millie and I explore all the mysteries of the garden. We built a bridge to the far side of the creek; and on the other side, we're planting daisies. Today she asked me to tell her about her grandmother.

Today I am Mildred.

"WELCOME TO YOUR AUTHENTIC INDIAN EXPERIENCE™," BY REBECCA ROANHORSE

STORY FRAME

"Nobody wants to buy a Vision Quest from a Jesse Turnblatt," you explain. "I need to sound more Indian."

"You are Indian," she says. "Turnblatt's Indian-sounding enough because you're already Indian."

"We're not the right kind of Indian," you counter. "I mean, we're Catholic, for Christ's sake."

What Theresa doesn't understand is that Tourists don't want a real Indian experience. They want what they see in the movies, and who can blame them?

Jesse Turnblatt is both a personal and a professional American Indian. He works for a virtual reality outfit called Sedona Sweats, where he is their top-selling VR Experience Guide. Jesse's most popular offering, the Vision Quest, is a kind of spiritual journey into a mythologized version of the American plains and the spirituality of its Indigenous cultures.

Jesse's American Indian identity seems to be part of his credentials as a Guide, but he bases his Vision Quest on a composite of popular culture stereotypes rather than his own inherited traditions. Jesse doesn't think Tourists would be satisfied with an Experience built out of the historical or present realities of Indigenous lives, so he has instead put together a Quest recycled from Hollywood films. And the Tourists love what he does! As Jesse tells himself, "no other Indian working at Sedona Sweats can do it better."

In spite of his success with the Tourists, Jesse is deeply afraid of losing his job— and with it, his wife Theresa, who nearly ended their marriage the last time he was unemployed for a long stretch. Some of Jesse's anxiety can be traced back to the failure of his Custer's Last Stand project, a few months before the story starts; it turns out, as Jesse's Boss later tells him, that nobody wants a historical Experience "if the white guy loses." But there are signs throughout the story that Jesse's feelings of insecurity and worthlessness go back much further. By his own report, he's "never been one of those guys" who is widely admired or desired. And though Jesse Trueblood the Spirit Guide is hugely popular, nobody seems that interested in Jesse Turnblatt until the tourist White Wolf, who is given this name by Jesse during his own Vision Quest, offers him a sympathetic ear when Jesse is off the clock.

The collapse—or, perhaps more accurately, the theft—of Jesse's life raises some important questions about who gets to claim what identities, and who decides what counts as "authentic," a word that is very powerful but also very slippery. These are not new questions, but technologies that make it easy to fine tune (or entirely transform) our public-facing selves add a new edge and urgency to them.

Jesse's story asks us to consider the broader social patterns that create the conditions for how Jesse and White Wolf inhabit this story: not only their actions, but also their beliefs and even their feelings. For this reason, you may find it particularly helpful to approach this story in terms of capability ethics. That framework challenges us to consider people's potential as well as their attainments. It also gives us a structure for understanding how the material, social, and psychological resources that have been available to a person have created the conditions for their success, and how the resources unavailable to them have limited them. Jesse's story can help bring into focus why such an approach is so essential for ethics.

STUDY QUESTIONS

1. What does Jesse mean when he tells Theresa that they aren't "the right kind of Indian"? How is being "the right kind of Indian" related to the kind that Jesse already is?

2. What is a Vision Quest like? What do Tourists (other than White Wolf) find so valuable about it? How do you think the various ingredients of the Vision Quest Experience help to create that effect for Tourists?

3. Jesse uses the word "authentic" a lot of times, to describe many different things. Choose three specific instances in which he describes something as authentic. What is good or powerful about authenticity, in each case? What kind of access does Jesse have to that authenticity, and why?

4. Jesse's coworker DarAnne tells him that he's "gotta get a backbone," echoing a complaint by Jesse's wife Theresa (who later leaves him, apparently for White Wolf).

What reasons does the story provide us for understanding why Jesse might not be confident in himself?

"WELCOME TO YOUR AUTHENTIC INDIAN EXPERIENCE™," BY REBECCA ROANHORSE

> In the Great American Indian novel, when it is finally written, all of the white people will be Indians and all of the Indians will be ghosts.
>
> —Sherman Alexie, *How to Write the Great American Indian Novel*

You maintain a menu of a half dozen Experiences on your digital blackboard, but Vision Quest is the one the Tourists choose the most. That certainly makes your workday easy. All a Vision Quest requires is a dash of mystical shaman, a spirit animal (wolf usually, but birds of prey are on the upswing this year), and the approximation of a peyote experience. Tourists always come out of the Experience feeling spiritually transformed. (You've never actually tried peyote, but you did smoke your share of weed during that one year at Arizona State, and who's going to call you on the difference?) It's all 101 stuff, really, these Quests. But no other Indian working at Sedona Sweats can do it better. Your sales numbers are tops.

Your wife Theresa doesn't approve of the gig. Oh, she likes you working, especially after that dismal stretch of unemployment the year before last when she almost left you, but she thinks the job itself is demeaning.

"Our last name's not Trueblood," she complains when you tell her about your *nom de rêve*.

"Nobody wants to buy a Vision Quest from a Jesse Turnblatt," you explain. "I need to sound more Indian."

"You are Indian," she says. "Turnblatt's Indian-sounding enough because you're already Indian."

"We're not the right kind of Indian," you counter. "I mean, we're Catholic, for Christ's sake."

What Theresa doesn't understand is that Tourists don't want a real Indian experience. They want what they see in the movies, and who can blame them? Movie Indians are terrific! So you watch the same movies the Tourists do, until John Dunbar becomes your spirit animal and Stands with Fists your best girl. You memorize Johnny Depp's lines from *The Lone Ranger* and hang a picture of Iron Eyes Cody in your work locker. For a while you are really into Dustin Hoffman's *Little Big Man*.

It's *Little Big Man* that does you in.

For a week in June, you convince your boss to offer a Custer's Last Stand special, thinking there might be a Tourist or two who want to live out a Crazy Horse Experience. You even memorize some quotes attributed to the venerable Sioux chief that you find on the internet. You plan to make it real authentic.

But you don't get a single taker. Your numbers nosedive.

Management in Phoenix notices, and Boss drops it from the blackboard by Fourth of July weekend. He yells at you to stop screwing around, accuses you of trying to be an artiste or whatnot.

"Tourists don't come to Sedona Sweats to live out a goddamn battle," Boss says in the break room over lunch one day, "especially if the white guy loses. They come here to find themselves." Boss waves his hand in the air in an approximation of something vaguely prayer-like. "It's a spiritual experience we're offering. Top quality. The fucking best."

DarAnne, your Navajo co-worker with the pretty smile and the perfect teeth, snorts loudly. She takes a bite of her sandwich, mutton by the looks of it. Her jaw works, her sharp teeth flash white. She waits until she's finished chewing to say, "Nothing spiritual about Squaw Fantasy."

Squaw Fantasy is Boss's latest idea, his way to get the numbers up and impress Management. DarAnne and a few others have complained about the use of the ugly slur, the inclusion of a sexual fantasy as an Experience at all. But Boss is unmoved, especially when the first week's numbers roll in. Biggest seller yet.

Boss looks over at you. "What do you think?"

Boss is Pima, with a bushy mustache and a thick head of still-dark hair. You admire that about him. Virility. Boss makes being a man look easy. Makes everything look easy. Real authentic-like.

DarAnne tilts her head, long beaded earrings swinging, and waits. Her painted nails click impatiently against the Formica lunch table. You can smell the onion in her sandwich.

Your mouth is dry like the red rock desert you can see outside your window. If you say Squaw Fantasy is demeaning, Boss will mock you, call you a pussy, or worse. If you say you think it's okay, DarAnne and her crew will put you on the guys-who-are-assholes list and you'll deserve it.

You sip your bottled water, stalling. Decide that in the wake of the Crazy Horse debacle that Boss's approval means more than DarAnne's, and venture, "I mean, if the Tourists like it . . ."

Boss slaps the table, triumphant. DarAnne's face twists in disgust. "What does Theresa think of that, eh, Jesse?" she spits at you. "You tell her Boss is thinking of adding Savage Braves to the menu next? He's gonna have you in a loincloth and hair down to your ass, see how you like it."

Your face heats up, embarrassed. You push away from the table, too quickly, and the flimsy top teeters. You can hear Boss's shouts of protest as his vending machine lemonade tilts dangerously, and DarAnne's mocking laugh, but it all comes to your ears through a shroud of thick cotton. You mumble something about getting back to work. The sound of arguing trails you down the hall.

■　　■　　■

You change in the locker room and shuffle down to the pod marked with your name. You unlock the hatch and crawl in. Some people find the pods claustrophobic, but you like the cool metal container, the tight fit. It's comforting. The VR helmet fits snugly on your head, the breathing mask over your nose and mouth.

With a shiver of anticipation, you give the pod your Experience setting. Add the other necessary details to flesh things out. The screen prompts you to pick a Tourist connection from a waiting list, but you ignore it, blinking through the option screens until you get to the final confirmation. You brace for the mild nausea that always comes when you Relocate in and out of an Experience.

The first sensation is always smell. Sweetgrass and wood smoke and the rich loam of the northern plains. Even though it's fake, receptors firing under the coaxing of a machine, you relax into the scents. You grew up in the desert, among people who appreciate cedar and pinon and red earth, but there's still something home-like about this prairie place.

Or maybe you watch too much TV. You really aren't sure anymore.

You find yourself on a wide grassy plain, somewhere in the upper Midwest of a bygone era. Bison roam in the distance. A hawk soars overhead.

You are alone, you know this, but it doesn't stop you from looking around to make sure. This thing you are about to do. Well, you would be humiliated if anyone found out. Because you keep thinking about what DarAnne said. Squaw Fantasy and Savage Braves. Because the thing is, being sexy doesn't disgust you the way it does DarAnne. You've never been one of those guys. The star athlete or the cool kid. It's tempting to think of all those Tourist women wanting you like that, even if it is just in an Experience.

You are now wearing a knee-length loincloth. A wave of black hair flows down your back. Your middle-aged paunch melts into rock-hard abs worthy of a romance novel cover model. You raise your chin and try out your best stoic look on a passing prairie dog. The little rodent chirps something back at you. You've heard prairie dogs can remember human faces, and you wonder what this one would say about you. Then you remember this is an Experience, so the prairie dog is no more real than the caricature of an Indian you have conjured up.

You wonder what Theresa would think if she saw you like this.

The world shivers. The pod screen blinks on. Someone wants your Experience.

A Tourist, asking for you. Completely normal. Expected. No need for that panicky hot breath rattling through your mask.

You scroll through the Tourist's requirements.

Experience Type: Vision Quest.

Tribe: Plains Indian (nation nonspecific).

Favorite animal: Wolf.

These things are all familiar. Things you are good at faking. Things you get paid to pretend.

You drop the Savage Brave fantasy garb for buckskin pants and beaded leather moccasins. You keep your chest bare and muscled but you drape a rough wool blanket across your shoulders for dignity. Your impressive abs are still visible.

The sun is setting and you turn to put the artificial dusk at your back, prepared to meet your Tourist. You run through your list of Indian names to bestow upon your Tourist once the Vision Quest is over. You like to keep the names fresh, never using the same one in case the Tourists ever compare notes. For a while you cheated and used one of those naming things on the internet where you enter your favorite flower and the street you grew up on and it gives you your Indian name, but there were too many Tourists that grew up on Elm or Park and you found yourself getting repetitive. You try to base the names on appearances now. Hair color, eye, some distinguishing feature. Tourists really seem to like it.

This Tourist is younger than you expected. Sedona Sweats caters to New Agers, the kind from Los Angeles or Scottsdale with impressive bank accounts. But the man coming up the hill, squinting into the setting sun, is in his late twenties. Medium height and build with pale spotty skin and brown hair. The guy looks normal enough, but there's something sad about him.

Maybe he's lost.

You imagine a lot of Tourists are lost.

Maybe he's someone who works a day job just like you, saving up money for this once-in-a-lifetime Indian Experience™. Maybe he's desperate, looking for purpose in his own shitty world and thinking Indians have all the answers. Maybe he just wants something that's authentic.

You like that. The idea that Tourists come to you to experience something real. DarAnne has it wrong. The Tourists aren't all bad. They're just needy.

You plant your feet in a wide welcoming stance and raise one hand. "How," you intone, as the man stops a few feet in front of you.

The man flushes, a bright pinkish tone. You can't tell if he's nervous or embarrassed. Maybe both? But he raises his hand, palm forward, and says, "How," right back.

"Have you come seeking wisdom, my son?" you ask in your best broken English accent. "Come. I will show you great wisdom." You sweep your arm across the prairie. "We look to brother wolf—"

The man rolls his eyes.

What?

You stutter to a pause. Are you doing something wrong? Is the accent no good? Too little? Too much?

You visualize the requirements checklist. You are positive he chose wolf. Positive. So you press on. "My brother wolf," you say again, this time sounding much more Indian, you are sure.

"I'm sorry," the man says, interrupting. "This wasn't what I wanted. I've made a mistake."

"But you picked it on the menu!" In the confusion of the moment, you drop your accent. Is it too late to go back and say it right?

The man's lips curl up in a grimace, like you have confirmed his worst suspicions. He shakes his head. "I was looking for something more authentic."

Something in your chest seizes up.

"I can fix it," you say.

"No, it's alright. I'll find someone else." He turns to go.

You can't afford another bad mark on your record. No more screw-ups or you're out. Boss made that clear enough. "At least give me a chance," you plead.

"It's okay," he says over his shoulder.

This is bad. Does this man not know what a good Indian you are? "Please!"

The man turns back to you, his face thoughtful.

You feel a surge of hope. This can be fixed, and you know exactly how. "I can give you a name. Something you can call yourself when you need to feel strong. It's authentic," you add enthusiastically. "From a real Indian." That much is true.

The man looks a little more open, and he doesn't say no. That's good enough.

You study the man's dusky hair, his pinkish skin. His long skinny legs. He reminds you a bit of the flamingos at the Albuquerque zoo, but you are pretty sure no one wants to be named after those strange creatures. It must be something good. Something . . . spiritual.

"Your name is Pale Crow," you offer. Birds are still on your mind.

At the look on the man's face, you reconsider. "No, no, it is White"—yes, that's better than pale—"Wolf. White Wolf."

"White Wolf?" There's a note of interest in his voice.

You nod sagely. You knew the man had picked wolf. Your eyes meet. Uncomfortably. White Wolf coughs into his hand. "I really should be getting back."

"But you paid for the whole experience. Are you sure?"

White Wolf is already walking away.

"But . . ."

You feel the exact moment he Relocates out of the Experience. A sensation like part of your soul is being stretched too thin. Then, a sort of whiplash, as you let go.

■   ■   ■

The Hey U.S.A. bar is the only Indian bar in Sedona. The basement level of a driftwood-paneled strip mall across the street from work. It's packed with the after-shift crowd, most of them pod jockeys like you, but also a few roadside jewelry hawkers and restaurant stiffs still smelling like frybread grease. You're lucky to find a spot at the far end next to the server's station. You slip onto the plastic-covered barstool and raise a hand to get the bartender's attention.

"So what do you really think?" asks a voice to your right. DarAnne is staring at you, her eyes accusing and her posture tense.

This is it. A second chance. Your opportunity to stay off the assholes list. You need to get this right. You try to think of something clever to say, something that would impress her but let you save face, too. But you're never been all that clever, so you stick to the truth.

"I think I really need this job," you admit.

DarAnne's shoulders relax.

"Scooch over," she says to the man on the other side of her, and he obligingly shifts off his stool to let her sit. "I knew it," she says. "Why didn't you stick up for me? Why are you so afraid of Boss?"

"I'm not afraid of Boss. I'm afraid of Theresa leaving me. And unemployment."

"You gotta get a backbone, Jesse, is all."

You realize the bartender is waiting, impatient. You drink the same thing every time you come here, a single Coors Light in a cold bottle. But the bartender never remembers you, or your order. You turn to offer to buy one for DarAnne, but she's already gone, back with her crew.

You drink your beer alone, wait a reasonable amount of time, and leave.

White Wolf is waiting for you under the streetlight at the corner.

The bright neon Indian Chief that squats atop Sedona Sweats hovers behind him in pinks and blues and yellows, his huge hand blinking up and down in greeting. White puffs of smoke signals flicker up, up and away beyond his far shoulder.

You don't recognize White Wolf at first. Most people change themselves a little within the construct of the Experience. Nothing wrong with being thinner, taller, a little better looking. But White Wolf looks exactly the same. Nondescript brown hair, pale skin, long legs.

"How." White Wolf raises his hand, unconsciously mimicking the big neon Chief. At least he has the decency to look embarrassed when he does it.

"You." You are so surprised that the accusation is the first thing out of your mouth. "How did you find me?"

"Trueblood, right? I asked around."

"And people told you?" This is very against the rules.

"I asked who the best Spirit Guide was. If I was going to buy a Vision Quest, who should I go to. Everyone said you."

You flush, feeling vindicated, but also annoyed that your co-workers had given your name out to a Tourist. "I tried to tell you," you say ungraciously.

"I should have listened." White Wolf smiles, a faint shifting of his mouth into something like contrition. An awkward pause ensues.

"We're really not supposed to fraternize," you finally say.

"I know, I just . . . I just wanted to apologize. For ruining the Experience like that."

"It's no big deal," you say, gracious this time. "You paid, right?"

"Yeah."

"It's just . . ." You know this is your ego talking, but you need to know. "Did I do something wrong?"

"No, it was me. You were great. It's just, I had a great grandmother who was Cherokee, and I think being there, seeing everything. Well, it really stirred something in me. Like, ancestral memory or something."

You've heard of ancestral memories, but you've also heard of people claiming Cherokee blood where there is none. Theresa calls them "pretendians," but you think that's unkind. Maybe White Wolf really is Cherokee. You don't know any Cherokees, so maybe they really do look like this guy. There's a half-Tlingit in payroll and he's pale.

"Well, I've got to get home," you say. "My wife, and all."

White Wolf nods. "Sure, sure. I just. Thank you."

"For what?"

But White Wolf's already walking away. "See you around."

A little déjà vu shudders your bones but you chalk it up to Tourists. Who understands them, anyway?

You go home to Theresa.

■　　■　　■

As soon as you slide into your pod the next day, your monitor lights up. There's already a Tourist on deck and waiting.

"Shit," you mutter, pulling up the menu and scrolling quickly through the requirements. Everything looks good, good, except . . . a sliver of panic when you see that a specific tribe has been requested. Cherokee. You don't know anything about Cherokees. What they wore back then, their ceremonies. The only Cherokee you know is . . .

White Wolf shimmers into your Experience.

In your haste, you have forgotten to put on your buckskin. Your Experience-self still wears Wranglers and Nikes. Boss would be pissed to see you this sloppy.

"Why are you back?" you ask.

"I thought maybe we could just talk."

"About what?"

White Wolf shrugs. "Doesn't matter. Whatever."

"I can't."

"Why not? This is my time. I'm paying."

You feel a little panicked. A Tourist has never broken protocol like this before. Part of why the Experience works is that everyone knows their role. But White Wolf don't seem to care about the rules.

"I can just keep coming back," he says. "I have money, you know."

"You'll get me in trouble."

"I won't. I just . . ." White Wolf hesitates. Something in him slumps. What you read as arrogance now looks like desperation. "I need a friend."

You know that feeling. The truth is, you could use a friend, too. Someone to talk to. What could the harm be? You'll just be two men, talking.

Not here, though. You still need to work. "How about the bar?"

"The place from last night?"

"I get off at 11p.m."

■　　■　　■

When you get there around 11:30 p.m., the bar is busy but you recognize White Wolf immediately. A skinny white guy stands out at the Hey U.S.A. It's funny. Under this light, in this crowd, White Wolf could pass for Native of some kind. One of those 1/64th guys, at least. Maybe he really is a little Cherokee from way back when.

White Wolf waves you over to an empty booth. A Coors Light waits for you. You slide into the booth and wrap a hand around the cool damp skin of the bottle, pleasantly surprised.

"A lucky guess, did I get it right?"

You nod and take a sip. That first sip is always magic. Like how you imagine Golden, Colorado must feel like on a winter morning.

"So," White Wolf says, "tell me about yourself."

You look around the bar for familiar faces. Are you really going to do this? Tell a Tourist about your life? Your real life? A little voice in your head whispers that maybe this isn't so smart. Boss could find out and get mad. DarAnne could make fun of you. Besides, White Wolf will want a cool story, something real authentic, and all you have is an aging three-bedroom ranch and a student loan.

But he's looking at you, friendly interest, and nobody looks at you like that much anymore, not even Theresa. So you talk.

Not everything.

But some. Enough.

Enough that when the bartender calls last call you realize you've been talking for two hours.

When you stand up to go, White Wolf stands up, too. You shake hands, Indian-style, which makes you smile. You didn't expect it, but you've got a good, good feeling.

"So, same time tomorrow?" White Wolf asks.

You're tempted, but, "No, Theresa will kill me if I stay out this late two nights in a row." And then, "But how about Friday?"

"Friday it is." White Wolf touches your shoulder. "See you then, Jesse."

You feel a warm flutter of anticipation for Friday. "See you."

■　　■　　■

Friday you are there by 11:05 p.m. White Wolf laughs when he sees your face, and you grin back, only a little embarrassed. This time you pay for the drinks, and the two of you pick up right where you left off. It's so easy. White Wolf never seems to tire of your stories and it's been so long since you had a new friend to tell them to, that you can't seem to quit. It turns out White Wolf loves Kevin Costner, too, and you take turns quoting lines at each other until White Wolf stumps you with a Wind in His Hair quote.

"Are you sure that's in the movie?"

"It's Lakota!"

You won't admit it, but you're impressed with how good White Wolf's Lakota sounds.

White Wolf smiles. "Looks like I know something you don't."

You wave it away good-naturedly, but vow to watch the movie again.

Time flies and once again, after last call, you both stand outside under the Big Chief. You happily agree to meet again next Tuesday. And the following Friday. Until it becomes your new routine.

The month passes quickly. The next month, too.

"You seem too happy," Theresa says one night, sounding suspicious.

You grin and wrap your arms around your wife, pulling her close until her rose-scented shampoo fills your nose. "Just made a friend, is all. A guy from work." You decide to keep it vague. Hanging with White Wolf, who you've long stopped thinking of as just a Tourist, would be hard to explain.

"You're not stepping out on me, Jesse Turnblatt? Because I will—"

You cut her off with a kiss. "Are you jealous?"

"Should I be?"

"Never."

She sniffs, but lets you kiss her again, her soft body tight against yours.

"I love you," you murmur as your hands dip under her shirt.

"You better."

■　　■　　■

Tuesday morning and you can't breathe. Your nose is a deluge of snot and your joints ache. Theresa calls in sick for you and bundles you in bed with a bowl of stew. You're supposed to meet White Wolf for your usual drink, but you're much too sick. You consider sending Theresa with a note, but decided against it. It's only one night. White Wolf will understand.

But by Friday the coughing has become a deep rough bellow that shakes your whole chest. When Theresa calls in sick for you again, you make sure your cough is loud enough for Boss to hear it. Pray he doesn't dock you for the days you're missing. But what you're most worried about is standing up White Wolf again.

"Do you think you could go for me?" you ask Theresa.

"What, down to the bar? I don't drink."

"I'm not asking you to drink. Just to meet him, let him know I'm sick. He's probably thinking I forgot about him."

"Can't you call him?"

"I don't have his number."

"Fine, then. What's his name?"

You hesitate. Realize you don't know. The only name you know is the one you gave him. "White Wolf."

"Okay, then. Get some rest."

Theresa doesn't get back until almost 1 a.m. "Where were you?" you ask, alarmed. Is that a rosy flush in her cheeks, the scent of Cherry Coke on her breath?

"At the bar like you asked me to."

"What took so long?"

She huffs. "Did you want me to go or not?"

"Yes, but . . . well, did you see him?"

She nods, smiles a little smile that you've never seen on her before.

"What is it?" Something inside you shrinks.

"A nice man. Real nice. You didn't tell me he was Cherokee."

■　■　■

By Monday you're able to drag yourself back to work. There's a note taped to your locker to go see Boss. You find him in his office, looking through the reports that he sends to Management every week.

"I hired a new guy."

You swallow the excuses you've prepared to explain how sick you were, your promises to get your numbers up. They become a hard ball in your throat.

"Sorry, Jesse." Boss actually does look a little sorry. "This guy is good, a real rez guy. Last name's 'Wolf.' I mean, shit, you can't get more Indian than that. The Tourists are going to eat it up."

"The Tourists love me, too." You sound whiny, but you can't help it. There's a sinking feeling in your gut that tells you this is bad, bad, bad.

"You're good, Jesse. But nobody knows anything about Pueblo Indians, so all you've got is that TV shit. This guy, he's . . ." Boss snaps his fingers, trying to conjure the word.

"Authentic?" A whisper.

Boss points his finger like a gun. "Bingo. Look, if another pod opens up, I'll call you."

"You gave him my pod?"

Boss's head snaps up, wary. You must have yelled that. He reaches over to tap a button on his phone and call security.

"Wait!" you protest.

But the men in uniforms are already there to escort you out.

■　■　■

You can't go home to Theresa. You just can't. So you head to the Hey U.S.A. It's a different crowd than you're used to. An afternoon crowd. Heavy boozers and people without jobs. You laugh because you fit right in.

The guys next to you are doing shots. Tiny glasses of rheumy dark liquor lined up in a row. You haven't done shots since college but when one of the men offers you one, you take it. Choke on the cheap whiskey that burns down your throat. Two more and the edges of your panic start to blur soft and tolerable. You can't remember what time it is when you get up to leave, but the Big Chief is bright in the night sky.

You stumble through the door and run smack into DarAnne. She growls at you, and you try to stutter out an apology but a heavy hand comes down on your shoulder before you get the words out.

"This asshole bothering you?"

You recognize that voice. "White Wolf?" It's him. But he looks different to you. Something you can't quite place. Maybe it's the ribbon shirt he's wearing, or the bone choker around his neck. Is his skin a little tanner than it was last week?

"Do you know this guy?" DarAnne asks, and you think she's talking to you, but her head is turned towards White Wolf.

"Never seen him," White Wolf says as he stares you down, and under that confident glare you almost believe him. Almost forget that you've told this man things about you even Theresa doesn't know.

"It's me," you protest, but your voice comes out in a whiskey-slurred squeak that doesn't even sound like you.

"Fucking glonnies," DarAnne mutters as she pushes past you. "Always making a scene."

"I think you better go, buddy," White Wolf says. Not unkindly, if you were in fact strangers, if you weren't actually buddies. But you are, and you clutch at his shirtsleeve, shouting something about friendship and Theresa and then the world melts into a blur until you feel the hard slap of concrete against your shoulder and the taste of blood on your lip where you bit it and a solid kick to your gut until the whiskey comes up the way it went down and then the Big Chief is blinking at you, How, How, How, until the darkness comes to claim you and the lights all flicker out.

■　　■　　■

You wake up in the gutter. The fucking gutter. With your head aching and your mouth as dry and rotted as month-old roadkill. The sun is up, Arizona fire beating across your skin. Your clothes are filthy and your shoes are missing and there's a smear of blood down your chin and drying flakes in the creases of your neck. Your hands are chapped raw. And you can't remember why.

But then you do.

And the humiliation sits heavy on your bruised up shoulder, a dark shame that defies the desert sun. Your job. DarAnne ignoring you like that. White Wolf kicking your ass. And you out all night, drunk in a downtown gutter. It all feels like a terrible dream, like the worst kind. The ones you can't wake up from because it's real life.

Your car isn't where you left it, likely towed with the street sweepers, so you trudge your way home on sock feet. Three miles on asphalt streets until you see your highly-mortgaged three-bedroom ranch. And for once the place looks beautiful, like the day you bought it. Tears gather in your eyes as you push open the door.

"Theresa," you call. She's going to be pissed, and you're going to have to talk fast, explain the whole drinking thing (it was one time!) and getting fired (I'll find a new job, I promise), but right now all you want is to wrap her in your arms and let her rose-scent fill your nose like good medicine.

"Theresa," you call again, as you limp through the living room. Veer off to look in the bedroom, check behind the closed bathroom door. But what you see in the bathroom makes you pause. Things are missing. Her toothbrush, the pack of birth control, contact lens solution.

"Theresa!?" and this time you are close to panic as you hobble down the hall to the kitchen.

The smell hits you first. The scent of fresh coffee, bright and familiar.

When you see the person sitting calmly at the kitchen table, their back to you, you relax. But that's not Theresa.

He turns slightly, enough so you can catch his profile, and says, "Come on in, Jesse."

"What the fuck are you doing here?"

White Wolf winces, as if your words hurt him. "You better have a seat."

"What did you do to my wife?!"

"I didn't do anything to your wife." He picks up a small folded piece of paper, holds it out. You snatch it from his fingers and move so you can see his face. The note in your hand feels like wildfire, something with the potential to sear you to the bone. You want to rip it wide open, you want to flee before its revelations scar you. You ache to read it now, now, but you won't give him the satisfaction of your desperation.

"So now you remember me," you huff.

"I apologize for that. But you were making a scene and I couldn't have you upsetting DarAnne."

You want to ask how he knows DarAnne, how he was there with her in the first place. But you already know. Boss said the new guy's name was Wolf.

"You're a real son of a bitch, you know that?"

White Wolf looks away from you, that same pained look on his face. Like you're embarrassing yourself again. "Why don't you help yourself to some coffee," he says, gesturing to the coffee pot. Your coffee pot.

"I don't need your permission to get coffee in my own house," you shout.

"Okay," he says, leaning back. You can't help but notice how handsome he looks, his dark hair a little longer, the choker on his neck setting off the arch of his high cheekbones.

You take your time getting coffee—sugar, creamer which you would never usually take—before you drop into the seat across from him. Only then do you open the note, hands trembling, dread twisting hard in your gut.

"She's gone to her mother's," White Wolf explains as you read the same words on the page. "For her own safety. She wants you out by the time she gets back."

"What did you tell her?"

"Only the truth. That you got yourself fired, that you were on a bender, drunk in some alleyway downtown like a bad stereotype." He leans in. "You've been gone for two days."

You blink. It's true, but it's not true, too.

"Theresa wouldn't . . ." But she would, wouldn't she? She'd said it a million times, given you a million chances.

"She needs a real man, Jesse. Someone who can take care of her."

"And that's you?" You muster all the scorn you can when you say that, but it comes out more a question than a judgment. You remember how you gave him the benefit of the doubt on that whole Cherokee thing, how you thought "pretendian" was cruel.

He clears his throat. Stands.

"It's time for you to go," he says. "I promised Theresa you'd be gone, and I've got to get to work soon." Something about him seems to expand, to take up the space you once occupied. Until you feel small, superfluous.

"Did you ever think," he says, his voice thoughtful, his head tilted to study you like a strange foreign body, "that maybe this is my experience, and you're the tourist here?"

"This is my house," you protest, but you're not sure you believe it now. Your head hurts. The coffee in your hand is already cold. How long have you been sitting here? Your thoughts blur to histories, your words become nothing more than forgotten facts and half-truths. Your heart, a dusty repository for lost loves and desires, never realized.

"Not anymore," he says.

Nausea rolls over you. That same stretching sensation you get when you Relocate out of an Experience.

Whiplash, and then . . .

You let go.

INDEX

Note: Page numbers in *italics* indicate figures.